You Are Not Not Wrong.

A NEW FOUNDATION

ERIC HAWKINS

STORY SYSTEMS LIMITED

You Are Not Wrong.

A New Foundation

By Eric Hawkins

No artificial intelligence (A.I.) of any kind was utilized
in the writing or editing of this book.

ISBN: 979-8-9916265-0-7

CONTENTS

You Are Not Wrong.

A New Foundation

0

An Intro To 300 Million Volts

I don't remember hearing the bang.

I didn't see a blinding flash of white.

There was only a brilliant burst of red followed by a huge rush of ...*something*.

I do remember sitting down to check if I was still alive, while shingles and scraps of wood rained down outside my metal-enclosed sunroom.

An arc of lightning had blasted a hole through my neighbor's roof closer to me than a batter is to the pitcher in a game of baseball, and at nearly the same height as my head, *but nothing about it made sense.*

The plasma channel didn't connect with the highest point of anything. It didn't find a peak, chimney, or edge. An arc five times hotter than the sun went straight through the middle of a flat, featureless, shingled roof face, with no antennas or vents or metal sticking out of it anywhere.

Multiple power lines nearby were higher. Trees surrounding both my house and my neighbor's house were higher. There were a million things around in every direction that lightning is known to typically strike that were higher than the new hole in the roof next door.

I was mere feet away, in a room that was almost completely made of metal (including the roof), with all the windows wide open and electrical wiring in every corner connected to ground – and for *some reason* this bolt of lightning instead took a path straight through wood and asphalt shingles.

I admit that lightning and electricity can be incredibly hard to predict and understand. If this was the only part of this story that didn't make sense, I'd be willing to chalk it up to just being "strange" and then move on with my life. Perhaps you feel the same. However, this is just the tip of the iceberg-of-strangeness surrounding this particular bolt of lightning.

The house next door with the new unplanned skylight was being rented by a multi-national order of priests. Somewhere between five and seven priests lived in the home at the time. This number is uncertain because we (meaning all the nearby neighbors) were never quite sure how many priests actually lived in the house. They are not social, they stay indoors except to go on walks alone, and the individuals we think lived there seemed to change every few months due to being re-assigned to different locations around the world.

This housing situation is honestly strange all by itself, considering I live with my family in a suburban subdivision full of nothing but single-family homes. However this is still just a small piece of a story that expands way past the limits of making sense.

This particular order of priests was front-and-center when the child sexual abuse scandals came to light back in the early 2000s. According to what I was able to find through sources of information that are widely trusted, the world leader of this entire religion oversaw the investigation and approved the restructuring of this particular order. Now, only two years after that approval, and with ongoing accusations of cover-ups by investigators into the order, a lightning bolt passed straight through their house. On a very peculiar day.

That morning, the Supreme Court of the United States was making every headline in the country, because they were overturning a decision they made decades earlier. This ruling would overturn and remove a constitutional protection - a personal "right" - for all citizens. The loss of this right was going to have a huge impact on US society. Everyone I knew in-person and through social media was talking about it. This was a very big deal.

It *just so happens* the organized religion next door had been working *for decades* to try to make this happen – to have this right removed. They wanted it removed not only for United States citizens - but also for *every human being on Earth*. On this particular morning, one huge step was made toward that goal. It was likely a great morning for them.

The highest court in the country. A worldwide religion and its leader. The removal of a previously-protected right. The desire to remove it for everyone in the name of a higher power. A scandalous order. A house full of priests. And now a bolt of lightning.

Some combination of these separate stories just connected themselves in the mind of every single resident in a quiet little subdivision located in the heart of the United States, and would remain there for weeks to come. But I'm getting ahead of myself.

Back in the enclosed metal room, moments after a blast from nature violently introduced itself, none of these connections mattered. More important things had taken priority - *much* more important things: There was now visible smoke rolling out of the hole where 300 million volts had found the path with ground.

The house was on fire.

My wife, having rushed into the room where I was, took one glance and immediately set herself into motion knowing exactly what to do. She's kind of amazing like that. As for me, my mind wandered in a different direction, which quickly caused my stomach to drop like a rock: Were any of those priests inside touching a light switch when it hit? Perhaps taking a shower? Touching something metal? Did someone just possibly *die*? I joined my wife in moving and tried to keep up, but she was already ahead of me. Neighbors around the now-smoking house started racing through the rain toward the front door at the same time…

I know you would like this story to continue. I will tell you everything - but *not yet*. Instead, just know that as much as you know right now, you are *still* at the very tip of strangeness in this story – which will end up twisting itself in extraordinary ways, to perhaps even become *your* story.

This bolt of lightning set off what might be an endless string of cause-and-effects that seem to still be cascading around me as I finish writing this book, two-and-a-half *years* after they began. However, it only took 14 days for this chain reaction to lead me into a profound personal experience I will forever refer to as "The Event".

This "event", for lack of a better word, was a radical, dramatic, and immediate change to my own thought processes over the course of what I believe to be about 10 seconds. I still felt like myself on the other side of these 10 seconds, but everything else was *different*. It was like I had been moved into a different reality, and was now dealing with it from a completely different place in my own head.

My interactions with others were completely transformed from my perspective. Emotions I couldn't understand or control poured out of me for days. My wife would worry about me for quite some time, wondering if whatever happened was going to come undone or be my undoing. Neither of us really understood what had happened to me. Interestingly, the kids didn't seem to mind at all – they even seemed to like the "new" me.

I must admit - even though this event was frightening when it occurred, and the aftermath difficult to comprehend at times, the time period of several days immediately after that 10 second Event was one of the most beautiful and wonderful experiences of my life. I had transformed from a broken and depressed person into the exact opposite.

This unexpected transformation also came bearing a hidden gift, which is something I also struggle to put into words, but here's an attempt: My old view of reality - my memories and thought processes from before the event - were still intact. I could still see the world around me from *both* points of view. I seemed to have access to the way I *used* to think and see the world before The Event, and could compare it to my thinking and view of the world that came after.

So, I put myself to work. I began trying to figure out what exactly had changed in that 10 seconds. But as I began to explore these differences and uncover what changed, the scale of what I was dealing with quickly grew. Something had come undone for me, and it revealed a hidden beauty and wonder in the world, but it also exposed traps of human suffering and enormous social cycles of human atrocity that were previously invisible to me.

The more I talked with others about what I was finding, the more I realized I seemed to be alone in seeing them. Whatever had come undone for me was still very much in place for everyone else. *How could no one else realize what was happening in the world? Could I somehow make this invisible world become visible for everyone? Could I expose the traps and stop the cycles everyone seemed to be falling into without realizing it? Or was I going crazy?*

I decided taking this story to its conclusion was worth my every effort, and it was upon this decision that reality took a turn for the *deeply* strange. Things started to occur both *to me* and *around me* that perhaps no one, including yourself, would ever believe - even if I told you with a perfectly straight face. I could have stopped. I could have decided against continuing, but I had a reason to push myself. I eventually made it through, and I made it through without losing myself.

The chain of events that began with the loudest silent bang and the reddest flash of white eventually resulted in me knowing exactly what my purpose was in this world, and it would take me two-and-a-half years and every ounce of my past experiences and understandings to finish the biggest step in my journey toward fulfilling that purpose: Writing this book.

This book has been its own journey in time for me – a journey back across those 10 seconds to the old and broken me, and my old way of seeing the world. It took me almost two years from the bolt of lightning to take my journey into darkness and make my way back, and another few months to proofread, edit, and illustrate what you are now reading.

I wrote this so you can hopefully see the invisible traps that bind you and nearly everyone else. So you can maybe see the enormous cycles of collapse and atrocity. So that maybe, just *maybe*, you can see a brighter future where we finally break the cycles.

This journey will probably not flow like you might expect. The way each story connects may not be apparent until many chapters later. You may be sure I am about to repeat myself, or be tempted to assume you know everything I'm about to say, or feel safe assuming everything I believe based on certain statements I might make – but I think there is a very good chance you will be surprised. After all, I've already made my journey.

I wrote this one for you.

PART ONE – THE INVISIBLE WHITE RABBIT

Candy bars can lead to genocide.

The presence of chickens can lead you to violence.

1

The Keeper of Truth

You are not wrong.

No matter what you believe is true.

Who am I to say otherwise?

Am I a protector of what is right?

The keeper of what is true?

Should I give myself a title like *King Eric, Keeper of Truth and Decider of What Is Right*?

To be perfectly honest, I already know I am not qualified to claim a title like *Keeper of Truth,* because I actually don't know the whole truth about, well, *anything.* This means I don't have the authority to write a book titled "You *Are* Wrong".

Yet right and wrong seem to exist to some degree, which means someone somewhere must have such an authority and title. Who is that for you? Who do you allow to be the protector of what is right?

Should you allow me, or anyone or anything else, to have the authority to call *you* wrong? *Should* you allow me, or anyone or anything else, to wear the crown of *Keeper of The Truth and Decider of What is Right...For You*?

I cannot know your answer, or if you even have an answer. What I can know is if you try to offer this title to me, I'll refuse it.

Instead, I am only going to claim one authority in relation to you: These words. I am the one that chose these words and what order they appear on these pages. You have no control over these things at all.

This authority doesn't add up to much, though. You are the one with full authority over your choice to keep reading. I can't *make* you read these words. This means my goal is for you to *want* to keep reading. I can't actually control this, but I plan to help by keeping these words simple and easy to understand, maybe even fun or silly at times. After all, I didn't write this for you to be bored to tears or confused about everything I'm trying to share on our journey together.

So, I'll make you a promise: I promise to do my very best to keep these words as fair as possible to you throughout this entire book.

This promise creates a pretty big challenge for me. This book is a one-way communication from the past when I wrote it into the future where you are reading it. Since time-travel doesn't seem like it will be possible anytime soon (at the time I wrote this anyway), I am stuck choosing my words for you with no idea *who you are* or *where you are* or *when you are* as you read them.

This means even though I claim authority over my word choices, any I might use to describe your life, beliefs, or thoughts are nearly blind assumptions. Since I can't know anything about you specifically while writing this, when I must make assumptions about you, I plan to state them up front so you can know exactly what they are. That's as fair as I think I can be.

So with that said, here's my first and most important assumption about you: *You are a human being.*

Which human being? Not a clue, and it doesn't matter anyway. Even if I could know exactly who, where, and when you are – even if you were sitting in the very same room with me while I wrote this sentence - it wouldn't change anything when it comes to me being the *Keeper of the Truth and Decider of What is Right* for you. There is something in the way, and it prevents me from knowing the whole truth about you and what you believe:

I can't read your mind.
I'm going to assume you can't read mine either.

This barrier, which I believe exists as part of our shared human condition, creates very real *mental boundaries* neither of us can cross and actually stay fair to each other. I believe a lot of the problems I'm about to detail in future chapters wouldn't even exist if these barriers weren't absolutely real for both you and me. Your mental boundary keeps me from ever claiming authority over the wrongness of your thoughts, and my mental boundary keeps you from doing the same to me.

I can't call you wrong about anything you believe. I'm in no position to do so.

But - and perhaps this may surprise you - *neither are you.* You are not in a position to declare *yourself* wrong at this moment, either.

I believe this claim probably just *feels* wrong to you, or at the very least strange. However, as I see it, this is an incredibly important and often hidden part of the human condition, so please give me a chance to explain.

Inside each of our mental boundaries, where you *can* read your own mind and I can read mine, the way we each determine right and wrong can get a bit weird if you've never thought about it much before.

You can never be wrong to yourself at this moment. It's impossible. Whatever you believe to be right at this moment is based on your truth and you act using that truth.

I didn't give you any advance warning of this being an assumption about you, because I don't believe it is an assumption. If you are human like me, then I stand by this claim without falter. I accept this may not be obvious or self-evident as true for you right away. If this is the case, I completely sympathize. I didn't realize this simple shared truth for the first 42 years of my life, and it turns out there is very much a reason: It is incredibly difficult to notice because of how incredibly difficult it is to stay in the present moment.

I'll try to bring you closer to the present moment over the next few sentences, though it's nearly impossible to keep you there:

You recently decided the act of reading this book, (or at least this sentence) is the right thing to do, so that is what you are doing *right now*.

You apparently still believe continuing to read is the right thing to do, so you are still reading.

Should you decide reading this is *no longer* the right thing to do, you'll stop reading about ...*now*.

However you are still here, which means you thought it was wrong to stop reading and right to keep going - a choice by which I am truly honored. Thank you.

You are not wrong in this exact moment, no matter what you believe, because your beliefs are the basis for all of your actions, including the action you are performing right now. Since I assume you are human, I also assume you have a mind and a body, and it is *your* mind that controls most of *your* body's actions (except for a select few actions that work without your mind, such as what happens if you whack just below your knee with a mallet).

To restate this in a different way: For your body to perform voluntary actions, something must instruct your body to do those actions, and the only thing *directly* commanding your muscles, tendons, and ligaments to do anything at all is *your* mind. I am not controlling your body with *my* mind and *my* beliefs. I think it's safe to assume no other humans are using their minds to control your actions like you are a puppet on strings, either.

Here's one more way to express this: If my mind were removed from my body, it will not affect your ability to perform actions *in the least*. However, if *your* brain were removed from *your* body, it would most definitely affect *your* ability to perform actions.

This relationship between your mind (and the beliefs in it) and your body's actions means your current beliefs are all you have to *do anything* in life. Your beliefs define, guide, and even limit what actions are available for you to take. Your beliefs *can't* be wrong for you at this exact moment, no matter what they are, because they *are* you in this exact moment. And in this moment, *you* are the only one performing actions with your body. This means your current beliefs are *everything* to you.

To scale this way up, I'm claiming every action that every individual person has ever performed through the entirety of human history was done because that person believed it was the right thing to do at the moment they did it.

To believe otherwise is to believe it's possible for someone like me to perform an action and believe it is wrong *at the same time I am performing it*.

Lots of people I have met believe this is possible. Even I used to believe this was possible. Perhaps you believe it is possible too, I can't be sure.

I can only assure you at this point it is not.

If you believe it is possible for me to believe an action is wrong at the same moment I perform it, then you are indirectly claiming you can read my mind, because you are observing my body's actions and also claiming to observe my mind's beliefs about that action at the same time.

For the sake of exploring all the possibilities though, let's imagine you actually *can* read my mind somehow, and you decide I intentionally performed an action I believed was wrong while doing it – even though I am claiming this is actually impossible. What now?

This puts us back into the very same position from the beginning of this chapter, but everything is upside down and backwards.

Are *you* the protector of what is right?

The keeper of what is true?

Should you give yourself the title *Keeper of Truth and Decider of What Is Right*?

Should I give it to you on my behalf?

There's only one way out of this looping trap of authority and accusations of wrongness between us. We have to establish a new baseline, a new *foundation*, for determining what is true, what is right, and what is wrong. I happen to know the perfect place to start our journey to do this: In the early-to-mid 1900s, a mustached man with ties to Germany had some radical ideas that would shake humanity to its core.

He, too, was not wrong.

2

The German Approach

You were thinking of Albert Einstein, right? If not, that's okay – you aren't wrong if you thought of someone else, it just wasn't the particular mustached man from Germany I was thinking of to help us escape this looping trap of authority over right and wrong. What we need from Albert is his Theory of Special Relativity. If you aren't familiar with it, that's okay, what we need from it is actually pretty simple.

Wrong is a *relative* word.

So is *right*, actually.

This means something can only be defined as wrong or right *relative* to something else. To call anything or anyone wrong (including yourself) there must be a comparison being made, and this requires a minimum of *two* points. Now for the kicker: How many points are you in this moment?

You are one.

You

A single point like you has no second point for comparison. You do not have another reality. There is no "second you" doing something else somewhere else. You only have *one* present moment with nothing to compare it against, so you can't be wrong in this moment.

Of course, this also means you can't be right, either.

Yet I think it's safe to assume you do believe in right and wrong as you read this. If all of the above is true, then how are you able to do such a thing?

Well, there *is* a second point you are using to do this, but it's entirely contained within your mental boundary - *within* the single point of you where no one else can venture.

You determine what is right and what is wrong by comparing your current moment to something that is no longer your current moment: Your own past. Thanks to your memory, you have the two points needed to define right and wrong, and you are making this comparison every second of every day, and acting on your own definitions of right and wrong every second of every day – even right now.

The action you perform *at this moment* is based on what you believe to be right *at this moment*, which is based on what you believe to be true *at this moment*, which is based on your own experiences and understandings – which are all *in your past*.

Even right now you are comparing everything to your past to decide if continuing to read this book is the right thing to do, or if you just aren't a fan of my writing voice.

I accept you might still believe someone else actually *can* act right now in a way he or she believes to be wrong. If you believe this about another human, then it must also apply to you, since you being human is the main assumption underlying this entire book. You must believe *you* can perform an action you believe to be wrong right now.

After all, if they can do it, *so can you.*

As I've written, this is actually impossible, but the opposite seems to be widely believed at the time I wrote this. Unfortunately, believing this can be done is perhaps the grandest and most tragic illusion in the entire history of humanity.

For reasons I will explain, any action you believe to be wrong when compared to your truth is automatically discarded as a possible option in this moment. If you believe your action can be wrong compared to your truth built by your past, then based on what I've claimed, you are required to believe *your truth is also wrong*. Since you choose actions based on what you believe to be true, to believe your truth is somehow wrong requires you to have a *second* and *opposing* truth in your mental boundary. On top of this, to believe a second opposing truth *and* act on it, you would need two separate realities - two separate *nows* - two....of *you*.

The problem with requiring another you is that you are only physically *one point* experiencing one brand new moment in time. You can only ever act with and control one body in that moment. *It is your only moment.*

I realize this might be confusing, so rather than gloss over this and assume you understand, I'll dive into this situation a little deeper.

What would happen if you believed you really could act with what you believed to be wrong? What does the most tragic illusion in the history of humanity look like?

3

A Sleight-of-Mind Candy Bar Trick

I can't know your actual beliefs, so the following set of assumptions about them will only be for the purposes of this particular story. I won't hold you to them.

Let's assume you believe honesty is right.

Let's assume you believe stealing is wrong.

Now imagine you just stole a candy bar.

Did this annoy you right away? Did you flinch or wince? Did you immediately want to argue *you would never steal in the first place if you believed stealing is wrong*?

If so, you are arguing to keep your actions and beliefs in agreement. This actually *reinforces* my claim - that you perform an action because you believe it is right when you do it. That's not what we are trying to do in this section. Instead, you have to steal the candy bar and also believe stealing is wrong *as you steal it*.

I admit the previous sentence seems like it should be possible. I've heard individuals accuse others of exactly this all the time. Perhaps you have even done it to yourself or been accused of it: "You *knew* it was wrong when you did it!"

Unfortunately, in a situation where something like this is said, a sleight-of-mind trick is being pulled on potentially everyone involved. I'm not saying you are wrong if you've said this to others, or that others are wrong if they've said it to you. I'm saying what happens is so fast and hidden it's nearly impossible to catch.

Here's the reveal: You always act on what you believe to be right in the moment. You believe stealing is wrong, but you now have a candy bar in your pocket and you didn't pay for it. For this to be possible, your belief about stealing had to *switch* from "stealing is wrong" to "stealing is *right*".

And this switch is *exactly* what is so well hidden.

To steal that candy bar, you had to *justify* the act first. To do this, your belief about stealing had to change from it being wrong to it being right. You did this because of a very good reason in the present moment. (I will dive into these reasons later, but I bet you can think of one or two of them on your own before I get there).

Once your belief about stealing was switched due to that very good reason, you were able to act using your new belief of what is right *relative* to the current situation, and perform the action of stealing the candy bar.

Your act still matched up with what you believe to be right *at the time*. It *always does*. Even if you truly believed stealing is wrong before the moment of theft, stealing still became justified – or the *right* thing to do - just before you did it due to whatever very good reason you had to justify it.

This sleight-of-mind belief-switch trick might have happened a tiny sliver of a second before you carried out the act. It's possible you would not remember it switching at all. *You just suddenly snatched it without thinking about it. It all just happened so fast.*

Personally, I believe it's typical to not be aware of justifications like these, because the closer they occur to the moment of action, the more likely it will be handled by your subconscious, instead of your conscious mind (or inner voice) where you can notice it.

As I see it, this is exactly why awareness of the always-changing relative nature of right and wrong are so often lost from our view. It's entirely possible everyone involved that witnesses your theft, including you, will be unaware of what actually had to occur for it to happen. You yourself might believe you *knew* it was wrong when you did it, and the person that witnessed you do it might also believe you *knew it was wrong when you did it*.

It doesn't really matter if you are aware of this switch in your beliefs happening or not; the relative nature of *what is right and wrong to you* is not actually what leads to tragedy and disaster. It's what can happen afterwards that leads us toward the rabbit hole:

With the stolen candy bar in your pocket, if you now try to believe you knew stealing was wrong when you stole it, all hell breaks loose.

4

The Runaway Process

If you just stole something, you can tell yourself you believe stealing is wrong all you want after you've stolen it. However, "telling yourself" is *not a belief*, it's *an action*.

Telling yourself stealing is wrong became the right *action* to take based on what you believe to be right and true in this *new* moment. What belief might that be? In this case, it's the belief that you *should believe* stealing is wrong.

I cannot stress enough how super-fast the justification for this new action of "telling yourself" happens, and how tricky this is to notice. It is not handled consciously at all. This is not something you are "thinking about" and "deciding on." It is often just as hidden as the switch that allowed you to steal in the first place, and is just as lightning-fast.

As soon as you do this though, your second point – your past – which is used to build your truth and determine right and wrong for you, gets very noisy. Your past now contains memories of you believing stealing is wrong, you believing honesty is right, you performing the act of stealing, and now a memory that you *should* believe stealing is wrong. You may or may not remember or even notice the very good reason that changed your belief about stealing back when it happened.

This batch of memories creates a major problem the next time you need to act using anything related to them. *Some of these things oppose each other.*

If you try to keep the truth you *tell yourself* to believe, and block out or ignore the always-changing truth used to justify your actions - even though they might have only existed for a flash of a moment - *your entire belief system of right, wrong, and truth can flip upside down.*

Here's how I imagine this to work: Suppose you are a few steps out of the store with the stolen candy bar when the store clerk pops out and yells "Hey! You forgot to pay for what's in your pocket!"

This new situation might immediately create a new belief for you about honesty. Perhaps honesty *isn't* right anymore and dishonesty is right, so that lying can now be justified as the right thing to do. Why? Because you believe you *should* believe stealing is wrong, and are trying to *mentally hide* that it became the right thing to do for you – if only for a brief moment.

Perhaps you may justify running as the right thing to do at this moment. Why? Because you believe you *should* believe stealing is wrong, and are trying to *physically escape* the discovery of theft becoming the right thing for you to do – if only for a brief moment.

You might even justify yelling, confronting, manipulating, or intimidating the clerk as the right thing to do. Why? Because you believe you should believe stealing is wrong, and are trying to physically prevent the discovery that stealing became the right thing for you to do by trying to "force" the clerk to justify backing down or reversing direction as the right thing to do.

Now let's go back and change this up a bit.

This time, imagine you are only two steps out of the store with the candy bar, and store security guards pop out, grab you, and hold you in custody while the police are on their way. Now your belief that you should believe stealing is wrong may do some *really* strange things with all those other memories in your past.

You might try to push the justification to steal out of your mind in hopes that it somehow changes your situation. This would involve *telling yourself* your act of stealing *wasn't within your control*, rejecting your authority over your own actions. *You didn't want to steal, but some part of the situation made you do it.* Something or someone else is responsible for your wrong action. Why justify this? Because you believe you *should* believe stealing is wrong, and are trying to *deny* that it ever became the right thing to do for you – if only for a brief moment.

You might even tell yourself *you didn't steal anything – it never even happened.* You have no idea where the candy bar came from – someone must have put in your pocket. Why do this? Because you believe you *should* believe stealing is wrong, and are trying to *erase* the brief moment it was the right thing to do by denying the very *existence* of your past action.

Finally, you might tell yourself *you never do anything right – there must be something wrong with you.* Why? Because you believe you should believe stealing is wrong, and are trying to reject the brief moment it became the right thing to do by also rejecting your *entire truth* and sense of right and wrong with it. I believe the hope is this might somehow change the truth of your actions for others by simply canceling it out, because you outwardly reject your entire self in its place. Perhaps you hope the guards or police will notice this self-destruction of you that steals things, and let you go.

Should you reach these mental positions, the runaway process is in full swing and is starting to tear your mind down the middle. The beliefs you *tell yourself* are true, and your self-built beliefs of truth are literally trying to separate. Part of your mind needs to be free from your body so it can believe 'what it should' without your actions messing everything up. It starts to run from, deny, or even try to erase your "wrong" action along with the beliefs that justified it.

Everything within your second point (your past) is clashing and crashing, and some of it is already turning upside down. "Stealing is wrong. Honesty is right. Your act was wrong and you knew it when you did it. Being honest about it is wrong. Lying about it is right. Pretending it didn't happen is right. Not understanding or acknowledging why you did it in the first place is right. You are always wrong.......is right."

This is a fraction of the incredible amount of mental noise that is possible. The noise grows with each and every act you justify which you later *tell yourself* you knew was wrong at the moment you did it.

If you think this might actually be a good thing, or don't really see the problem here, I'll go ahead turn this noise up to its loudest possible volume to explore what is possible, which might surprise you.

5

The Deathtrap of Breathing

If you always act on what you believe is right, which is based on what you believe is true, then if you try to believe *all* your actions are wrong, you are also trying to believe the truth used to justify those wrong actions must also be wrong.

But if *doing the wrong thing is right* because that's what is *true* for you, you are immediately trapped in a belief loop where justifying *any* action is not possible. Right and wrong are defining each other inside your mental boundary.

I realize these two sentences might be hard to follow. I think it's far easier to understand what I'm trying to say by using some examples.

Assume you decide to walk. If you always do the wrong thing, you couldn't step with your left foot after your right foot, because that would mean right-foot-then-left-foot is justified to be the right way to walk for you.

You also can't take two steps in a row with your right foot, because then *that* would be justified to be the right way to walk at this moment.

Nor could you walk on your hands or scoot on your butt, because these methods would be justified as the right way to walk in this moment.

No matter how you imagine walking to be done, you have to first justify that method to be the *right* way to do it. This is based on what you believe to be true about walking in the current moment. But since you are trying to believe your justifications are *always* wrong because your actions are always wrong, the act of walking is not possible at all, because whatever you believe to be true about the act of walking right now is actually false.

This looping paradox-of-a-problem starts compiling immediately, and quickly reaches a peak to include *any* justification you focus your attention onto, even justifications for your own beliefs, which attempts to immediately flip your entire reality upside down.

For example, you apparently believe you should walk, but *that must be wrong* so you shouldn't walk. But if you believe you shouldn't walk, that must also be wrong so you *must* walk.

As the early stages of a panic attack start setting in, you might notice your breathing has sped up. Should you slow down your breathing? No that would be wrong. Speed it up? Wrong again. *Should you be breathing at all?* No. That too would be -------- (If you are holding your breath right now because you believed it was the right thing to do without even noticing, it's now safe to exhale again.)

In all of these situations, you are actually "telling yourself" with *a new action using your own mind* what you think you *should* believe is right and true in this moment. If you have to tell yourself to believe something, this means it is not what you naturally believe to be true. This two-opposing-truths situation prevents you from being able to justify any action at all without it going against one truth or the other. Since you want to believe every truth you have is wrong, which includes the truth that every truth you have is wrong, a personal disaster starts to unfold. You, as a single point, are required to *act* at every moment of your life. You can't *not act*. An action must *always* be justified, even if that action is "don't move".

To be able to justify actions again without the tremendous noise caused by two truths fighting each other inside your mental boundary, something must give. You need one of your two beliefs to be the right belief, and the other to be wrong. One truth must be canceled. It must be *invalidated*.

For reasons I will explain, you will almost always choose to keep the truth you think *you should* believe. When this happens, the runaway process in your head does something truly terrible.

6

Birthing a Monster

You *want* to believe what you think you *should* believe.

If you believe you are always wrong, this means you had to *invalidate* your beliefs built naturally moment-to-moment. After all, your natural beliefs justified stealing to be right, and that was wrong. Successfully overwriting this natural truth and belief system allows you to use the truth you think you *should* believe, which redefines right and wrong.

But if you only have one mind that controls your actions, and you determine right and wrong using it at every single moment, how are you going to invalidate an already running and subconscious process that keeps being wrong? You can't create another past and present and switch over to it. *You are only ever one point existing in this moment, and your past seems to be forever stuck in a noisy loop of constantly invalidating every wrong action entering your memory to "fix" your truth, right?*

In a frankly terrifying display of the sheer power of the human mind, it will actually do what seems impossible: Your mind will start to create *a separate* present moment for you. It will make a second truth, a second reality, a second mind, a *Second You* inside your head to be able to act with what you think you *should* believe is right and true, and not whatever you believed to be right and true moment-to-moment.

If you tell yourself to believe something that goes against the entirety of your self-built truth like "*everything* I do is wrong," your mind is forced to tear until it is split almost *completely* in half inside your own head. *An entirely new second you is born to cancel the original you.*

Your original mind – First You – is still the one that justifies and sends the commands for your actions in the moment, such as telling your eyes to keep moving to read this sentence (because it is the right thing to do). By believing you are always wrong, you refuse to believe this is the *real* you. It is completely invalid. This new second mind – Second You - which is birthed totally from the beliefs you *tell yourself* are right and true, is now the *real and valid* you.

You do not realize this yet, but this Second You has the potential to grow into a monster that feeds itself from an abyss, and if this happens, Second You's first goal is the destruction of First You. Its first act will be to overwrite your past, which builds the truth that keeps justifying all your *wrong* actions.

The experience would be like reading this sentence, and now telling yourself "reading is wrong". *Its already happened.* You did read it. It's already been justified as the right thing to do and is now in the past, right?

Second You doesn't care. Since it has been born from the belief you are always wrong, it just keeps *telling you* reading is wrong. Second You refuses to acknowledge that reading was and ever should be justified. It justifies telling yourself you didn't read it and would never read it because reading is wrong.

There is a simple reason for this splitting of reality, and it is not some minor thing – it's actually a *huge* thing happening on a massive scale at the time I wrote this. All of this looping, ripping, noisy, invalidating, two-minded mental disaster happens because what you tell yourself you *should* believe is right and true *did not actually originate from you at all.*

In the previous situations where you believed yourself wrong, *you had internalized some outside-to-you belief of right and wrong.* Something or someone else was handed the authority to call you wrong. You felt your beliefs *should* match whomever or whatever this outsider-to-you says is right, and that you should not trust your own truth and justifications.

The noise this creates begins growing, like a feedback loop between a microphone and speaker. It eventually reaches such a volume and frequency, that your entire consciousness desperately wants to separate from your actions and whatever is being used to justify them. It just wants to comply with the wishes of your new *Keeper of the Truth and Decider of What is Right.*

Once the runaway process is started by attempting to invalidate your self-built truth by telling yourself you "should" believe this *Keeper of Truth* instead, it will begin to happen more and more, and it becomes increasingly difficult to slow down or stop this process.

Your beliefs of right and wrong turn upside down as you become hypercritical of your ability to be right at all, and you start to actually *want* to tear away from First You and move completely into Second You's version of truth. *You want to believe what you think you should believe.*

Once you start wanting this to happen, you no longer need to interact with your external *Keeper of the Truth* to keep this mind-tearing process going. Second You has grown larger, and will actually begin *making assumptions* of how right or wrong you are on behalf of your *Keeper of Truth. You want to believe what you think you should believe.*

If you reach this point, where your *Keeper of Truth and Decider of What is Right* can judge you as wrong without being anywhere in your life but inside your own mind - you are trapped in a mental paradox of epic and tragic proportions. The beast known as Second You is overwriting First You to try to take control over all your justifications, and it entered from some *outside source* you internalized for a very good reason: *You want to believe what you think you should believe.*

This transition of control to the beast, however, has a sticking point. You act in the present moment based on what you believe to be right and true relative to *your* past, not someone else's. This means First You is *hard-wired* into its position as action controller in every moment.

The monster of Second You wants control, but First You is the one permanently strapped into the driver's seat. This means for Second You to gain control over your actions, it has to somehow interfere *before* First You can determine right and wrong on its own and convert it into justified action.

To accomplish this, the monster called Second You not only begins *altering* your past to change what is true, but also begins delaying your present. By delaying your present, it can control your actions from the past, where it lives. Since it cannot remove First You, it will simply invalidate it at every new moment. It tries to cover the windshield for First You, then stick its own head out of the window to yell directions at First You from the backseat to make it stay on the "right" roads.

If you are unsure if any of this has happened to you, I can only assure you it has, and I don't even have to know who you are. If you are reading this, *everything I just explained has already happened*. I can't know to what degree Second You has taken over, but I think you will soon enough.

Believe it or not, you can *directly experience* Second You blocking the windshield of First You and start backseat driving. Are you ready?

You are now manually breathing (again).

It's possible you may not resume automatic breathing *ever again*.

Breathe in.

Breathe out.

Careful, now.

Second You – the monster that has actually been living in your head for quite a while - has been customized to carry what you think you *should* believe is right and true, and it's entirely possible that you've already lost much of your entire truth and sense of right and wrong to it. (Be sure to inhale now)

You are manually breathing because Second You thought you *should* believe breathing is consciously controlled, so it took over the justification of every breath in this brand new moment. It is doing this by invalidating your self-built truth (exhale now) that you breathe automatically (assuming you don't require an artificial breathing device), and replacing it with what you felt you *should* believe to be true (inhale now).

Now that you are consciously aware Second You has taken over your breathing, you *know* you'll go back to automatic breathing if you quit thinking about it, right? *Are you sure?* It's possible you might not. How worried are you that your self-built truth won't be there anymore if you stop inhaling and exhaling manually?

This insecurity can be very unnerving, because you can't *tell yourself* to go back to automatic breathing. You might *try*, or you might *try* to focus onto these words a little more to see if your automatic truth comes back, but I think it's more likely you'll keep secretly spying on your breathing out of fear. Fear you might *be wrong*. Fear you might not breathe automatically. Fear that First You is gone, and will no longer be able to take back over. Are you breathing automatically right now? Did you just find yourself mentally racing back to inhale or exhale while reading this again?

I think it's safe to assume you know the truth is still there. That breathing is natural, subconscious, and automatic. *All* of your naturally built relative-to-the-moment truths and sense of right and wrong are still there. They are all subconscious and automatic. When you exercise, your subconscious justifies breathing faster automatically. When you rest, you automatically breathe slower. If you have a good enough reason to steal, you steal automatically because it is justified, and if you don't it's discarded as an option, because it's not justified.

Even though First You and its system of truth, right, and wrong are largely invisible from your perspective, it can be hard to trust it is still there for you. When Second You consciously overrides it, it can feel like an abyss has opened up where First You used to be. Suddenly you may not feel so sure there is any truth of your own beneath your feet anymore. Second You can make it feel like First you is erased – but the best it can ever do is cover its ability to see where it's going to prevent it from justifying anything.

It can't keep doing this forever though, right? After all, you have probably been breathing automatically for at least a few seconds....

7

A Hostile Takeover

Second You's takeover is usually released pretty quickly when it comes to breathing, because the takeover requires it to stay focused on your breathing non-stop.

Since you are also reading while breathing, this takeover gets unmanageable for Second You – it can't be in two places at the same time. It must either decide to read or override your breathing. This is why you might have jumped back and forth between these two things in the last chapter. Eventually though, Second You (hopefully) released control of your breathing so it could better focus on the ideas coming at you through these words.

When it comes to the candy bar theft from earlier though, everything is different. The justification to steal and the act were performed *one time*. Then it was over and done. All of this slides into your past, where the monster of Second You gains unlimited access to it. It doesn't require constant work to override and invalidate - this is a quick snack for it.

First You (which justified stealing to be right in the moment for a very good reason) and Second You (which justified telling yourself you knew stealing was wrong when you did it) will now clash whenever you have to deal with honesty or stealing again. Second You has the advantage though: You often *want* Second You to take over. This gives the monster the ability to make however much "mental noise" it needs. *You want to believe what you think you should believe.*

And making noise is exactly what the beast does. Second You replays the memory of the theft and what came before and after it over and over in your mind, amplifying everything it sees as wrong to be larger and louder until the very good reason and entire justification process which made stealing right can no longer even be acknowledged to exist in your mind. It is invalidated. It is covered up and blocked out by the incredible noise the monster has made. All that remains is the memory of how *wrong* you were.

But this is not enough for Second You. It needs to make sure your *future* actions aren't wrong either. To get ahead of First You and gain total control, Second You also builds stories of what it thinks *will* happen in future events. The result is you imagining your actions *that haven't even happened yet,* which are then judged by the monster in your head using the same noisy, invalidating process.

Second You – which again is not originally *your* truth at all but somebody or something else's you have internalized – has literally pulled your present and future into the past by imagining it all *before it happens*. There, it can invalidate and overwrite all your future choices and possibilities, just so it can keep your future actions in line with whatever you believe this *Keeper of Truth* monster believes they *should* be.

Invisible loops of noise are constantly happening in your head, where no one else can see or hear them.

It is hell to be here.

I was tortured by these feedback loops for decades without the slightest clue what they were or how they got there.

I constantly ruminated about what I should have done, or what I should have said. My reactions and speech were delayed and cautious and I didn't even realize it. All of this was happening because Second Me had to constantly check and overwrite justifications for my actions *beforehand* to keep them "right" based on what Second Me believed to be true. What it believed to be true was *literally everyone else knew what was right in the moment better than I did.* Everyone else was my *Keeper of the Truth and Decider of What was Right* for me. First Me was almost always wrong.

To try to be seen as "right" by others, my sensitivity to the tones in others' voices, their subtle gestures, and hidden implied meanings behind their words became extreme. The monster in my head grew to be impossibly large and would ratchet up the noise for days leading up to a social event as it obsessed with imagining and replaying all the ways First Me was going to fail, screw it up, and be wrong in the moment.

To stop the noise and stress of self-invalidation from growing unbearable in my mental boundary, I always needed someone else, anyone else, anything else, *everything else* to declare what was right or wrong in the moment, or what they thought about my rightness or wrongness, so I could just comply with what they expected and not have the noise in my head grow any louder. *I wanted to believe what I thought I should believe: Everyone knows better than me.*

My social goal was always to stay hidden as much as possible and get out of the situation as soon as possible. There were very few people I liked to be around, and that was usually because they made their approval of me loud and clear.

But this was my experience with my own beast. Your internal battle might be different or it might be very similar – again I can't know.

No matter how Second You has affected your life (and I promise you it has even if you aren't yet aware of the extent), this runaway mind-splitting feedback loop from hell that birthed an entire second you had one major disastrous consequence.

It collapsed, warped, or even expanded your mental boundary.

I can't be sure of course, but I feel reasonably confident claiming no individual in all of documented human history has ever avoided this happening to them for reasons I will explain later. I believe only a very select few have ever repaired the damage.

This means nearly all of us are stuck struggling against invisible monsters in our heads, trapped in our own personal hells where no one can see what's happening.

I think it's time to change that.

8

The Hydra From Hell

The way to break this two-you self-invalidating feedback loop of personal hell created by whatever *Keeper of Truth* you have internalized looks easy on paper. If you've ever taken the dive into therapy, you'll likely recognize what comes next immediately.

All you have to do is *accept* that you acted on what you believed was right and true at the moment you did it. *Accept* that you can read. *Accept* that you did steal a candy bar. *Accept* that you did walk in whatever silly way you chose. *Accept* that you chose to breath manually and it resumed automatically afterwards. Reclaim your authority over your actions, and accept that you always act on what you believe is right in the moment, and you always have.

Easy! So simple!

Just *change everything and accept it*! That's all there is to it!

Right?

Speaking from personal experience and from the shared experiences of many others, accepting past actions you already re-justified to be wrong when you did them is anything but simple or easy. I'll go so far as to claim this is a losing battle most of the time for most people.

This is because seeking help in the form of therapy, books, or programs, *which is in no way wrong*, has the potential to backfire and make things worse by creating another internalized feedback loop of something you think you *should* believe. You can end up *telling yourself* you *should* believe you are responsible for your actions.

Please don't take what you're about to read as a reason to avoid or dismiss seeking professional help if you are considering such a thing. There are many wonderful therapists and counselors in the world, and talking to them can help, but keep in mind their goal is not to fix you, but to *help you fix yourself.* If you or they misunderstand what is happening in the first place, you can't make things better, and the invisible monster can cause things to get worse.

To show you what this looks like, let's bring back the stolen candy bar scenario.

Imagine the rumination cycles are in full swing after the theft you just got away with. You obsess over the look the shopkeeper gave you (did he know?), the moment you stole it (did he see me?), what your friends might think of you (do they think less of me?), and so on. All of it created by telling yourself you should believe stealing is wrong.

Eventually, this noise becomes so much of a problem that you cave and have a conversation with a therapist (or some other counsel) in an attempt to be rid of, or at least lessen the constant attacks by the monster known as Second You. After your session, you learn to tell yourself "I am responsible for my actions." You feel great. The world feels right side up again. You believe Second You – your internalized version of your chosen *Keeper of Truth* - is defeated. You no longer feel the noise.

But you are actually *telling yourself* what you think you *should* believe yet again. This is the problem with Second You, it can be your friend sometimes, but these moments are almost always short-lived. It's a very sensitive beast.

The very next day, you justify stealing a loaf of bread for exactly the same very good reason you stole the candy bar the day before. This new act of theft is immediately moved into your past, where Second You is waiting.

Suddenly your internalized *Keeper of Truth and Decider of What is Right*, which you thought was gone, starts making lots of noise – *way* more noise than it did before. It starts looping "stealing is wrong" immediately and relentlessly just like before, but now with double the memories of you doing something wrong, so it starts trying to take control of your present and future all over again. But this time Second You has something else to hurt you with: *I am responsible for my actions.*

If you are were not aware that you are always acting on what you believe to be right and true in the first place, then you are unaware First You *even exists*, meaning Second You is, and has always been, *the real you. You want to believe what you think you should believe, after all.* You *want* to obey the monster because there is only the monster. This dragon from the abyss is *right*. You are *wrong*.

If this is the case, you wouldn't even be aware that for you to steal, it had to be justified as right, which makes it entirely possible to believe you knew it was wrong when you did it *and did it anyway.*

This missing-First-You creates a roadblock to truly believing "you are responsible for your actions," because you can't believe you *ever* thought stealing was right in the first place. This reinforces the belief *you knew it was wrong when you did it.* But now you also believe you should believe *you are responsible for your actions*, which, with First You out of the picture, is being responsible for *acting wrongly and knowing it.* This adds another self-invalidation loop *on top of* the first self-invalidation loop you were not even aware existed.

You have internalized not one, but now two *Keepers of Truth.*

There is now an invisible *two-headed* hydra monster in your mental boundary where absolutely no one else can see. In this particular example though, everything is warped, because you are unaware all your actions had to be justified as right to do them.

If you are completely unaware that First You even existed, you would believe the first head on the dragon (Second You) is actually the real you, and the second head is the monster. You might blame all of this on the therapist or counselor, because the monster seems to be using his or her words against you.

But even if this is where you are right now, again you are not wrong. How can you be aware of invisible and hidden things you didn't know existed or were happening?

Seeking help and internalizing a therapist or counsel's words or another self-help mantra into another *Keeper of Truth and Decider of What is Right* over your self-built relative-to-the-moment truth has doubled the noise and intensity of self-invalidation for you. You might be left unable to understand or accept why any of this is happening.

You told yourself you knew you were wrong in the moment and are responsible for that wrong action. You were sure you had the "right" belief. Now, things are worse.

There is no limit to the number of heads the hydra can have.

My hydra had hundreds of heads at times.

Should any of this happen to you, and the number of heads on your mental monster increases because you internalized yet another outside-to-you source as *Keeper of Your Truth and Decider of What is Right*, then you should know First You does not go quietly into the night. Even if you are unaware that First You exists, it does not hide or curl up in the corner as this multi-headed monster of Second You tries to control and invalidate it.

Oh no, First You does anything *but* that.

9

Your Secret Weapons

If you reach extreme depths of believing you are invalid and wrong, like I did, First You may eventually "discover" physical actions that distract, control, or temporarily silence the internal hell of looping, tearing, self-denial, self-invalidation, feelings of dissociation, worthlessness, and rejection caused by allowing outside sources to become internalized *Keepers of Truth* for you.

Once discovered, these actions immediately become the "swords" wielded by First You against the Hydra called Second You.

These weapons work for the same reason Second You can never completely invalidate First You: First You is *hard-wired* to the nerves throughout your body. Second You cannot perform surgery and physically place itself in the driver's seat. The beast can only override First You, it can't destroy First You or it would lose all connections to your actions.

At the moment Second You was born – the moment a *Keeper of Truth* was internalized - your original mental boundary was damaged. You very likely can't remember this happening for reasons I will explain later. First You subconsciously knows this has happened, and attempts to rebuild the boundary anyway, but it does not build it in its original shape and size.

This boundary – this *wall* - no matter where it ends up being placed or what shape it might take, becomes the battleground between you and your hydra. This means these sword-swinging actions by First You have the potential to inform you of where this mental boundary wall has been built.

As for the "swords", First You "finds" them through the one thing it can control - your body. It uses your body as a weapon to fight back against the Hydra.

As far as I can tell, these weaponized actions seem to fall into three categories, or "types". Each type is defined by *who* is included. Who gets included will depend on how your mental boundary is warped or rebuilt.

The first type is actions that turn inward to mostly involve and affect yourself. I'll call this the *collapsing* mental boundary type.

The second type is actions that try to hold an arbitrary-and-warped boundary between you and others. I'll call this the *stabilizing* mental boundary type.

The third type is actions that turn outward to mostly affect others and not yourself. I call this the *expanding* mental boundary type.

The reason they are not called "collapsed", "stabilized", or "expanded" mental boundary types, is because the boundary is *always moving* as the battle between the dragon and yourself rages on in your head. These actions tell you the direction your mental boundary is moving relative to yourself.

No matter the type, almost all of these distracting, controlling, or noise-silencing physical actions discovered by First You to wield against Second You typically only work *while they are being performed*. As a result, they are often repeated over and over, making them a *compulsion,* with the sole purpose of keeping your internalized Hydra from Hell at bay.

Unfortunately, these action-swords have a double-sided blade, which means some of these compulsions can work against you. Some might destroy your body or harm others. If this discovered-and-repeated compulsive action creates its own physical and chemical dependence, an *addiction* is born.

What follows is a quick overview of the three types. As usual, I can't know which (if any) of these apply to you, but I feel pretty confident that at least a few will feel familiar. The difference between these actions being healthy and being a compulsion tied to fighting an internal battle is determined entirely by "why" they are being done. When they are tied to the battle, they will often become the number one priority in your life over everything else.

Collapsing Mental Boundary Actions – "Escape"

Collapsing Mental Boundary Actions typically only involve you, but can affect others indirectly. This category includes addictions and compulsions that "escape" the always-growing noise in your head. These actions are often stretched to last as long as possible and are performed as often as possible. They can include substance use, eating, internet use, alcohol use, gambling, porn use, gaming, inducing pain and self-harm, shopping or spending, collecting, body modification, and masturbation, just to name a few. Again, these actions are often done to an excess degree because they release certain chemicals that drown out the noise.

Socially, escape actions can include avoiding interactions with others, avoiding interactions with the outside world, avoiding eye contact, or perhaps even avoiding being face-to-face with someone else.

If you felt a ping of familiarity with any of these so far (I am intimately familiar with many), and perhaps struggle to control them for reasons you can't quite put your finger on, then you might also already know these actions tend to have the side effect of enforcing the belief *you are wrong* after each time they are performed.

In my opinion, this happens because outside sources of right and wrong have frequently and loudly judged these actions as wrong or strange or just unacceptable, and constant self-invalidation of your own natural judgment of right and wrong caused by internalizing outside sources just like these *is the underlying problem driving these compulsions in the first place.*

This being said, not all escape compulsions are seen as wrong by others. Some can even appear as good or constructive from the outside when you perform them, which makes them *socially acceptable* escape compulsions.

Examples could be something as simple as fidgeting, to something more extreme like intensely practicing an art, hobby, or task as a means of escape. This might have the side effect of creating artists, musicians, performers, and athletes of phenomenal abilities, built entirely by trying to escape underlying personal suffering. This describes my teenage years in a nutshell – I was nearly the best in the world at the instrument I was using to escape – the snare drum (think of marching-band drumlines). It just so happens the snare drum perfectly fits into a description of *competitive technical fidgeting* – and I got into drums *because* I was fidgeting.

There is also a terrible new possibility that pops into existence if you build something wonderful while trying to escape something awful like I did: The better you become at your chosen art, hobby, or task you use to escape, the more likely you will be asked to share it should it be discovered. This increased positive attention can lead to temporary validation of First You using that escape behavior, which feels like a move in a good direction. Unfortunately, this exposure also leads to more opportunities for others to criticize or judge your "talent" because they believe it is the right thing to do in their current moment. This, of course, only adds more heads to your hydra from hell. It also does not help that many of these growing-skill escapes are often *judged* activities.

The result can be an extreme skill or talent, but being unable to take a compliment or accept the elevated value and social status it offers. You may always need to downplay your abilities around others, ruminate on your mistakes (Second You's attack method), or just collapse completely if criticized.

You may even feel like an *imposter* as others move you to an ever-higher pedestal. From Second You's perspective (which you prefer because you want to believe what you think you should believe), First You *is* an imposter looking to screw things up and just be wrong constantly.

Due to all of this, I truly believe there are unimaginable-levels of developed and undeveloped talent hiding in the shadows of society, suffering in silence.

Stabilizing Mental Boundary Actions – "Control"

Stabilizing Mental Boundary Actions typically involve you *and* others you interact with regularly. These acts are performed to prevent any further internal noise by keeping rigid "control" over your actions *and* other's perceptions of these actions, which means your mental boundary is warped outward to *include* or inward to *exclude* others, depending on the situation.

You often "discover" these controlling and competitive compulsions because your internalized sources of right and wrong, your Keepers of Your Truth, have already defined them as *right* behaviors. You internalize these "externally pre-approved" behaviors, usually without realizing it, and then flip them around to face outward. This might feel like it is stabilizing your mental boundary from collapsing. It might feel like it prevents and protects you from those very same external sources of right and wrong to ever judge you as wrong again.

Controlling compulsions are nearly constant, and can include an intense around-the-clock effort to be seen as *perfect*. This might involve constantly organizing or list-making, "voluntarily" working long hours, being hyper-productive to the point of no longer socially engaging with others, strictly following a routine to the point of it being a ritual, excessively cleaning or re-arranging, and saving random objects to remind you of validating experiences or memories (the object serves as a mental cue to replay memories tied to it that temporarily silence the hydra, making it incredibly valuable).

You might be extremely judgmental of flaws in yourself and others, or obsess with your own image, which can lead to body modifications or cosmetic surgeries, or even hyper-controlling your food intake to control your image. You might micro-manage others around you, and perhaps even constantly and loudly "volunteer" and "sacrifice yourself" to do things for others that you justify to be right because *sacrifice* is nearly guaranteed to not be judge-able as wrong.

Still other stabilizing actions might be obsessing over the clock and schedules, since the real goal is to keep your mind and body so busy all the time in pre-approved-by-others ways there is no opportunity left for you to feel the self-invalidation loop pain from the hydra. Others can't step in and add heads to the monster in your head by claiming you are doing less, being less, or just being wrong if you are doing more or being "better" than they are – never mind that the monster you are fighting is the one already driving from the backseat.

Even in downtime, controlling-type compulsions can take over and "discover" some action that occupies your mind but is still Keeper of Truth pre-approved (unlike many escaping compulsions). You might obsessively consume things like books, media, television, music, news, and puzzles the instant you can't find something pre-approved to do.

In a social setting, boundary-stabilizing compulsions might include seeking out constant validation from others, constantly trying to keep others laughing, or being the life and center of the party.

You might find yourself compulsively checking and updating social media and ensuring every photo of you is perfect, or at least filtered or edited to only ever show what you believe others to judge as perfection. You might suffer the internal devastation of a hundred new hydra heads if something negative is said or shared about you – perhaps prompting you to delete whatever content triggered such a response in an effort to undo the damage.

You might take forever to write an email or letter, reading it over and over again before sending it; because you need to be sure it can't result in being judged as wrong.

Because you believe these actions are right, mainly because they have some pre-approval by your now-internalized Keepers of Truth, they are usually *situation-specific*. This means you might feel extremely pre-validated as right in certain situations involving others, which may reinforce your belief there is an absolute right and wrong with no gray in *every* situation. This, in turn, can create an obsession over rules, instructions, details, and consistencies between yourself and those around you.

If Second You imagines future situations where it seems possible others might make an observation of you or those you are "associated with" and potentially blame or think less of *you*, then you may suddenly find yourself controlling the actions of those around you. Because you see the individuals around you as associated with you, they also need to be "valid" and right just like you – but based on *your* internalized Keepers of Truth. This can quickly make those around you unwitting participants in your controlling acts to prevent further invalidation and collapse of your mental boundary.

Due to you constantly monitoring for possible incoming judgments by outside sources of right and wrong, you may constantly seek feedback and communicate with these outside sources to better maintain control of any stories about you or others that could reflect wrongly on you. You may also constantly communicate with others to prevent an unknown-to-you or "uncontrolled" story about you from entering your outside source's awareness, which could quickly lead to a new head on your hydra. You must quickly rewrite and untangle it to restore and control your "rightness".

Of course, controlling and preventing all stories that might include you is ultimately impossible, because it requires building and maintaining an "action-based" boundary to function as a "mental" boundary. Actions exist *outside of your mind*, and this creates a "no-man's land" between your body and others that must be mentally filled at all times to provide safety from invalidation.

Everyone must be kept at arm's reach, so to speak, but it's already too late for this to work. If these actions feel familiar to you, then outside-to-you Keepers of Truth are already inside your mental boundary, and no amount of controlling yourself and others can fix this problem and stop the noise.

Being well practiced at keeping this situation-specific mental boundary filled and controlled with compulsive actions may cause you to rise to the top around others. These compulsions have likely given society amazing comedians, executives, entrepreneurs, leaders, and innovators, which may include you. If so, it's possible you were driven to succeed as a side effect of *controlling* any opportunities your hydra might have to grow new heads by making yourself into what it values (usually being perfect) as a façade to "control" the suffering in your own mind.

Expanding Mental Boundary Actions – "Dominate"

On the opposite end of the spectrum from Collapsing Mental Boundary Actions are Expanding Mental Boundary Actions. These contain destructive and constructive behaviors when viewed by others as well, but instead of typically involving only you, expanding actions involve other people and things almost *exclusively*.

These compulsions are First You actively trying to dominate Second You by repelling, or invalidating, anything or anyone seen as an *external* source of right and wrong in your life. First You is trying to cut off *all the heads* of your Hydra from Hell. However, since they are internalized, no amount of swinging swords will slay the beast within.

Dominating compulsive actions include jumping straight to anger, hate, and even violence toward what you see as outside sources of right and wrong. You may become obsessed with trying to feel valid or right without realizing it, which might lead you to blame, demand, or manipulate others into validating you as "right" by any means necessary. *Any* means.

You may find yourself bullying or gas-lighting others, or declaring other's beliefs as wrong, idiotic, or stupid. You might set them up to fail on purpose, require their loyalty, or demand subservience and obedience from them. This type of compulsion can also send you to much, *much* darker possibilities like stalking others, abducting others, imprisoning others, torturing others, or even killing others. All in an effort to feel valid and right over them, because you potentially believe they are responsible (or are going to be responsible) for your "wrongness," or that they hold the key to your "rightness" or validation.

Socially, expanding compulsive actions can have a side effect you might view as beneficial. Due to often being self-promoting to ensure your validity or rightness to others, you can easily be mistaken as confident, causing you to potentially rise to the top in a group. More often though, these actions work in the opposite way from desired, and come across to others as evil, hateful, self-centered, arrogant, and over-confident to the point of being pompous. You may feel the need to *repeatedly* tell others how great and right you are, talk over them, and insist you are right because you *know* you are.

This results in a small inner social circle that continually shrinks one member at a time, because anyone associated with you can *never* be seen as "more right" than you unless you allow it. Otherwise, that person is potentially a threat to your own validity and must be cast out or put into a place lower than you in front of others.

You may compulsively seek to acquire anything and everything that seems valued by others by any means necessary, such as flashy cars and vehicles, extravagant homes and vacations, and extremely expensive or attention grabbing clothing, just to name a few. All of this is done to try to always one-up those you see as outside sources of right and wrong, which from your perspective invalidates their ability to ever label you as "less-than" or wrong. Essentially, you are making yourself into a type of idol by trying to be something that *you think everyone else wants to be.*

Unfortunately, if any the above is familiar, you are unlikely to be aware of the root cause of these feelings and compulsions, so you may try to destroy "uncomfortable feelings" by insisting your beliefs are the only right and true beliefs.

You may obsess with gaining power over others because *they are always wrong* if they don't agree with you. No one else can ever be viewed as more right than you, so anyone that seems to have more authority than you needs to be corrected and dominated. It's likely you will, or have already become, successful, powerful, wealthy, or influential in the world, not due to your desire to help others, but due to your insatiable drive to defeat your Hydra – which is honestly admirable but it is also impossible, because it is inside your head, not out in the world.

The closer any source of right and wrong gets to justifying you as wrong in front of others, which could potentially add hundreds of hydra heads, the more explosive and violent you may be to prevent it from happening. You may aggressively redirect the conversation to be about the wrongness of *someone else* by using non-stop critique and name-calling intended to invalidate this someone else in other's eyes. This keeps the focus off of you and onto lowering someone else by your lead. You may also divert the conversation to a topic that is undeniably validating to you, like talking about some major success you've already had, even if it has to be imagined from thin air or "borrowed" from someone else. All just so others can't possibly justify calling you wrong.

You will essentially do anything it takes to *force* others to see you as valid and right, while never realizing the problem was internal to your own mind the entire time.

The Combo Attack

It's possible for you to use all three of these types of boundary actions in your life in different situations. You can be a compulsive list maker (controlling), withdraw into drug addiction and self-isolation when you fail to finish your list (escaping), and also explode onto a keyboard in raw rage at others on particular issues you feel strongly about, demanding you are right and valid and other beliefs are wrong (dominating).

You may explode with rage, punch holes in walls and smash glassware in the privacy of your home when your loved ones or roommates don't obey your hyper-controlling rules, but withdraw into dark corners in a drunken stupor in public if surrounded by strangers, or after your rage session is over.

You may always be selfless toward strangers at first, to control being seen as valid and right. But due to the increasing insecurity of a collapsing boundary, you may try to stabilize things through increasing passive-aggressive comments, or be manipulative and underhanded to those same individuals in order to squeeze out the validation you feel you need.

No matter what direction your mental boundary is moving, you still desperately need validation from outside sources of right and wrong because your always-changing natural source of right and wrong has been successfully drowned out by the monster within, and fallen into the darkness of the abyss. This is incredibly painful, and I think it's almost certain you will try to extract validation from those that can't escape you such as spouses, significant others, friends, relatives, and even your own kids. If there is no one around, you are left with only yourself and any social or internet setting where you interact with others to pursue the same outcomes.

The Full Split

Is it possible for your subconscious, memory, and motor centers of your brain to also split to create an entirely separate and semi-*complete* mind? My belief is that yes, this can and sometimes does happen, but it is very rare.

I have personally had close to fully dissociated experiences in front of hundreds of people while performing the compulsive action I used to escape and control my internalized sources of judgment for the sole purpose of *being judged*. After my solo performance was done, thankfully I re-associated, but while I was split, it was literally like watching my body perform actions with no input at all from "me" – some individuals call this an "out of body" experience.

If your mind was to fully divide, I believe your two semi-complete minds would be constantly working against one another, or interact in strange ways as they both attempt to control the only body you have.

I suspect this would create disturbing hallucinations contained entirely in your head, which would actually be the dissociated mind, Second You, subconsciously interacting with First You through your senses without any ability for you to stop it. You might not be able to control which "you" you consciously inhabit, and experience hyper-depression or super-euphoria depending on which you were at the moment. You might even find yourself making movements and performing actions you don't seem to be in control of or can't remember doing.

If First You is totally disconnected from Second You, I imagine all social interactions with others have the potential to become bizarre and unstable. Perhaps you would switch from screaming at a person to believing that same person no longer even exist as "you" unknowingly switched between minds during the same interaction.

The Hidden Beginning

All this being said, these compulsions are not something you could have easily avoided. They didn't actually begin after a stolen candy bar or some act you could have noticed. The entire illusion that is responsible for creating this underlying problem exists due to an absolutely massive and hidden human species-level problem I accidentally uncovered on my own journey. I have come to call it The Floating Anchor Problem.

This "hiddenness" of when this all gets started is also why I believe it's possible for you to seek out therapists, programs, or self-help books entirely focused on your *compulsive actions*, maybe even potentially free yourself from one of them, only to fall into the arms of another "self-discovered" compulsion shortly after.

The repeating actions are not and never were the root of the problem – they are actually the attempted self-*treatment*. This accidentally-upside-down approach of trying to stop the *treatment* can have the horrific outcome of growing and compounding your underlying problem, especially if your new compulsion is discovered and judged as wrong by another person. This just adds more heads to the monster, because it has already flipped your sense of right and wrong upside down and more-or-less stolen your identity.

Again I want to emphasize that many of these actions appear the same to others whether they are healthy or toxic for you. You are the only one that can ultimately decide what is happening in your head, and if the action is a problem for you or not.

But again, I believe no one in human history has ever escaped the two-mind problem. I will fully explain the reason why I believe this before you are done with this book. Now though, let's pull back from all these trees and take a look at the whole forest.

10

The Bigger Picture

I'm not going so far as to say believing you are wrong in this moment is the cause of *everything* good and bad in your life (and everyone else's), but I believe this is a huge part of the problem modern psychology is working to figure out in detail.

I believe the mental "tearing" caused by allowing an outside source to become your Keeper of Your Truth is the most devastating result of the not-yet-exposed Floating Anchor Problem, because all of these escaping, controlling, or dominating compulsions eventually do affect other people.

As these three types of warping boundaries mix together in society, they can worsen the tearing of others' minds, and add hydra heads at a disturbing rate as nearly everyone becomes someone else's *Keeper of Truth*. This compiles and compounds, making the overall problem grow significantly worse in a big hurry.

If you can relate to any of the compulsions in the previous chapter, then you already know they aren't done to make your suffering worse; they are done to try to make it go away. I tried like hell to stop my chosen destructive compulsions. I read self-help books for years. I went through therapy multiple times. None of it even came close to the core of what was happening.

Nevertheless, each time I "fixed me," I eventually fell back into an escape compulsion. This created a new feedback loop - a new head on my hydra. This just reinforced *even more* how wrong my actions and thoughts were at all times, or how my "constructive" escape behaviors were *not good enough* to earn validation from others.

I ended up latching onto and pushing these "approved" compulsive and constructive behaviors to the absolute limit, converting them from a means to "escape" judgment, to a means of "controlling" judgment, and even into a means to "dominate" others' judgments of me. I required more and more validation from more and more sources of right and wrong until the only direction left was down, so I abandoned the compulsion cold turkey to find another one of pre-approved value. The old one was no good to me anymore.

I never was able to stop all the other self-destructive compulsions though, because I was never truly addressing the real issue.

The root of why this whole multi-headed-hydra-torn-mind mess happens, which I keep alluding to as The Floating Anchor Problem, is actually a far deeper issue – much deeper - and on a much grander scale, and *you are absolutely tied into it.*

In the imagined candy-bar stealing situation, your anchor was already floating to have an internal struggle at all. The belief that drove you to justify stealing when you were so sure you knew it was wrong, or that drove you to break what you think your beliefs *should* be, is totally hidden in darkness under a nearly impenetrable barrier which is then covered up by time itself. You cannot see the reason your anchor began floating on your own. It involved way more moving parts than just you.

I'm going to try to show it to you and explain how I found it. Before I do though, I want to take a minute to emphasize that the following is still absolutely true as you keep moving forward:

You are not wrong. No matter what *you* believe is true. You are the authority of you. You are *King You: Keeper of Your Truth and Decider of Right For You.*

This always-right-in-your-current-moment way of existing applies to every human that ever lived and is living today, whether they realize it or not. It took me 42 years and apparently a lightning bolt, a keyboard, a profound mental event, and loads of privilege and time to become aware of how this all fits together.

There are more than 8 billion truths living on Earth (at the time I wrote this, anyway). Every single person is acting on what they each believe is right, based on what they each believe is true, based on comparing each of their current moments to their own pasts. That includes you.

The enormous tragedy here is that I believe all of these individuals, including you, have invalidated some piece of their own personal truth. Perhaps it is small, and perhaps it is extreme – but it *is* invalidated and cast into the abyss.

For me, the internal tear was nearly total – an almost complete closing of the mental paradox loop - an almost entirely self-invaliding mind. I second-guessed nearly everything I ever did, assumed nearly everyone could do things better than me, ruminated about most things I did do in order to build stories of what others must have thought of me, and could never let *anything* go. Everyone knew more than I did, everyone else had everything figured out better than I could. I was trying as hard as possible to reject my actions as my own, because they were nearly always wrong. I was *sure* of that.

Despite all of this being said, I still must accept it's possible none of what I have written really strikes a chord with you. Perhaps you can't remember anywhere or anytime in your life where you let someone or something override your judgment. Or perhaps there is still a part of this idea of relative right and wrong in the current moment that bothers you or is confusing.

You may have already tried to think of a thousand ways this idea must not be true or apply to you. Perhaps you still think you or someone else can be wrong when acting in the present moment, and *know* they are wrong.

After all, if you are right, and everyone else is right too, then if two people disagree on anything in the moment we've got a pretty big problem, right?

If you found yourself thinking like this, you are not wrong. You've instead pinpointed almost the *entire* problem.

I'll offer a quick reminder that right and wrong are *relative* to what you believe to be true *first and foremost*. Given this, before you begin the next chapter, how would you decide which of two disagreeing people was right and which was wrong?

When you think you have it, I'd like to introduce you to someone.

11

Bob and the Jet Treadmill

I've just invented an imaginary person named Bob. Well, *we* did, actually – because you've just imagined him into existence as well. Say hi if you want.

You and Bob are going to be given a problem to solve. It could be any problem at all, but for the sake of a good story I'll choose one for you. How about a problem that has divided everyone on the internet for years?

"Imagine a 747 is sitting on a conveyor belt, as wide and long as a runway. The conveyor belt is designed to exactly match the speed of the wheels, moving in the opposite direction. Can the plane take off?"

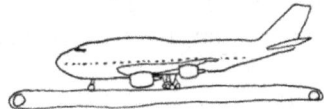

Now, I can't know *who* or *where* or *when* you are, or read your mind, so I can't possibly know what your real answer is to this question. For the sake of this example, then, I'm going to assume your answer (Don't worry, I won't hold you to it!).

Bob says the plane will not take off.

You say the plane will take off.

You both can't be right, right? You or Bob has to be wrong! There are only two possible answers, and there are now two points to compare - you and Bob. This means you are wrong or Bob is wrong and the other is right, right? Someone *must* be wrong in the current moment!

This is fair to assume and makes sense at first, but can you know if you are the one giving the *wrong* answer? Do you think Bob can know if it's him?

Both of you are still *not wrong*, and that is the only thing either of you can *ever* be at this point in the story. Here's why:

You believe something to be true because the story makes the most sense to you.

Bob's story of *why* the 747 won't take off, which he tells himself inside his mental boundary (based on his past) makes the most sense to him, so his answer is right from his point of view. Your answer makes less or no sense at all relative to his story, so you seem wrong to him.

Your story of *why* the 747 will take off, which you tell yourself inside your mental boundary (based on your past) makes the most sense to you, so your answer is right from your point of view. Bob's answer makes less or no sense relative to your story, so he seems wrong to you.

You and Bob are telling *different* stories inside your own minds to come up with an answer, and both answers are *right from your own perspectives because they each make the most sense compared to the truth built from your respective pasts.*

If one of you is wrong, neither of you can be aware of it. You both answered to the best of your ability what you believe is *right*, and are unaware of a reason why that answer is *wrong* – otherwise you wouldn't have justified answering the way you did.

You are always doing what is right *relative to your own truth* in the moment.

And just like I have no authority to call you wrong about anything you believe from back here in the past when I wrote this, you do not have any natural-born authority to call Bob wrong in this imagined moment, and Bob has no natural-born authority to call you wrong either. No matter how badly you might want to yell at each other and call each other wrong, idiotic, or stupid, neither of you were born with the title *Keeper of the Truth and Decider of What is Right.*

So how do we settle this? How do we determine which of you is *actually* right or wrong? Who is the actual *Keeper of the Truth and Decider of Right* for both you and Bob?

Who gets the *authority* to call you or Bob wrong?

I have an idea.

12

How Poultry Can Help Solve Problems

Every person - you, Bob, and me included - has custom beliefs of what is true and what is not.

You have your own *real* belief and answer about that 747 on the treadmill problem right now, so I'll stop assuming your answer. I can't know your answer, but I do know whatever it is, it is based on the story that makes the most sense to you at this moment.

Bob's answer, which I'll now make the opposite of whatever you *actually* believe to be right, is also based on a story that makes the most sense to him at this moment.

Different truths like yours and Bob's cannot magically and instantly come to an agreement to move forward. If Bob says "It doesn't take off" and you say "it does take off" and that's all that's ever said, there's absolutely no progress being made except that you might instantly dislike or want to avoid each other.

This can become a pretty big problem, especially if you and Bob are about to both get onto the same plane and *need* to know if it can take off from the treadmill it's sitting on that exactly matches the speed of its wheels.

To grease these wheels of progress, I've imagined a way to add some pressure to see if magic starts happening.....

You and Bob are now standing on an extra-wide treadmill. The belt goes all the way to the edge on all sides, and there are high walls at the front and both sides so you can't jump off. Behind you, at the end of the belt, is a huge un-climbable cage that creates a fourth wall. Inside that cage are one hundred irrationally angry roosters. If one of you drifts back too far on the belt, a motion sensor will open the cage door, releasing the cock-a-doodle-wraths onto you both in an enclosed space you cannot escape.

The treadmill is currently set to a comfortable-but-quick walking pace. It is programmed to bump up the speed one-increment every minute on the minute until it tops out *way* beyond your ability to keep up.

If you two agree to the right answer about the 747 on the treadmill, the treadmill will stop and you are free to go. If you can't, well.....you're going to get tired at some point, and there's a lot of beady eyes and sharp-looking beaks, claws, and spurs behind you.

So….what's the process for you and Bob to reach an agreement and be sure the answer is not wrong?

It's going to take work by both of you, *in the form of two-way communication*, to reach a common understanding and agree at all.

This is because the only time your current truth-story can ever change is when another story comes along to explain things differently. But to know each other's stories at all to figure out where they are different, you both have to first communicate *more than just your final answers*. You can't read each other's minds, after all.

What follows is a cleaned-up walkthrough of this process. I'll share what I removed to clean it up after you get to the other side.

Bob tells you why he thinks the plane will or won't take off. You listen and compare your current truth-story about the 747 in your mind (one point) to Bob's truth-story he has now converted to words (second point). You do this inside your mental boundary safe from Bob. He can't read your mind.

Inside your mind, you decide what story makes the most sense. You may end up rejecting your entire story and replacing it with Bob's, or modifying your story with pieces of Bob's, or rejecting Bob's entire story and staying with the same story and belief you had before Bob shared his story at all.

The result, no matter what it might be, is two stories becoming a single story (two points to one point) inside your mental boundary that makes more sense for you than it did before, because it now includes Bob's story. All of this happens extremely quickly for you because it requires no extra thinking. Making the most sense is the *only* requirement for a story to become your truth. If it doesn't make sense, it is rejected.

If you now convert your truth story about the plane into words and share them with Bob, he will listen to it and compare your truth-story with his own inside his mental boundary where you can't see. Your truth will be the second "point" he compares with his own, which may cause him to change his whole truth-story, or parts of his truth-story, or nothing at all.

And *that's it* - that's the basic process to *everything* you ever come to believe as true through your entire life. Your "truth" is evolving all the time, every day of your life, in nearly every passing moment in the form of comparing and merging thousands upon thousands of stories in your mind. They may be told to you by others, or by what you experience yourself. You discard the pieces of story that don't make sense to your own truth-story, and instantly accept the pieces that make more sense.

Every other human on Earth experiences and builds truth in the same way (and right and wrong with it). You are *always* adjusting and changing your truth, and your actions in life are *always* based on what you believe to be the right thing to do, which is based on what you currently believe is true.

This is happening all around you all the time, and you and everyone else are never wrong in the current moment.

Now I'll share what I removed to keep this walk-through cleaned up: If part of Bob's story was a story you had *already heard* in your past and had already discarded because you already know stories that make more sense, it will be *incredibly difficult* for you not to interrupt Bob and respond with "No that's wrong" to stop wasting precious time from your perspective.

This reflex-like rejection sets off a chain of events that quickly create a mess between you and Bob, and will waste even more time than just letting him finish.

The ensuing mess is *directly* tied to how Bob might handle you invalidating his truth in this moment, which will likely be tied to how much of himself he already sees as invalid, and if he deals with further invalidation in a collapsing, controlling, or expanding way. No matter how he reacts, the word *wrong* only added noise to this process and started involving the hydra, so I filtered it out.

I'll detail this noise-added-by-the-word-*wrong* in a bit more detail later on, but for now, let's leave behind the treadmill and Bob, settle down the roosters, and run through some much more realistic and perhaps familiar examples of this 2-point-to-1-point truth-story process over the next three brief chapters.

13

Ladies and Gentlemen of the Jury

Imagine you are arrested for a crime you did not commit. Want to prove your innocence? Have your lawyer tell the story of your innocence in a way that makes more sense than the prosecution lawyer's story of your guilt.

Assuming the trial involves a jury, each person in the jury will listen to two truth-stories told about you. It's unfortunate that each juror has already merged the story of the crime and the story of your appearance into his or her own truth-story about you before the lawyers even speak.

Regardless, both lawyers believe the story they are telling about you is the *right* story to tell. They will each tell a story they want to make the most sense. There might even be witness stories and pieces of stories told by evidence, such as video footage or fingerprints. The two lawyers will question and test each of these storytellers to get each story to merge better with their story about your guilt or innocence in a competition to make the most sense.

After hearing all these stories about you, each of the twelve jurors will arrive at which story about you he or she believes to be true. Then they will get together as a group to try to decide your guilt or innocence. They start this process by sharing their answers with each other. Are you guilty or innocent?

If everyone does not have the same answer right away – onto the "poultry treadmill" they go. They will share, using two-way communication, each of their truth stories about your guilt or innocence built from pieces of the lawyers' stories that made sense - with their own personal beliefs merged in as well. This process is an attempt to merge *twelve* individual truths about you into *one* shared truth-story that provides the final shared answer.

From beginning to verdict, this is a one-point (your actions) to two-point (the lawyers' stories about your actions) to twelve point (each juror) to one-point process (all jurors agree to same single verdict).

It might take awhile for twelve people to reach an agreed upon truth. But you can be pretty sure that shared truth made the most sense to everyone.

As for your consequences; if all the jurors believe the story you are innocent makes the most sense and is therefore true, you walk away free.

If they believe the story you are guilty makes the most sense and is therefore true, they will recommend a punishment.

If both stories *still* make sense after trying to merge - which means some jurors believe the story of your innocence is true and others believe the story of your guilt is true - a truth-story merge isn't made at all, and the jury will remain split, or hung. This may result in doing your trial all over again with a different jury, your release, or some other pre-decided outcome based on the rules about hung juries wherever you are being tried.

Should you be found guilty, the harshness of the punishment the jury recommends will be based on just how severe your crime seemed to be. What determines how severe the crime was to the jury? Whatever story about its severity made the most sense!

If the story of your guilt was very convincing, meaning it made a ton of sense, and the story of the crime was particularly nasty, you'll likely end up with the harshest sentence recommendation possible for that crime. The end.

But you were innocent......right?

The jury is wrong! Right?!

Not to each of them. Right and wrong are *relative* words. Each member of the jury believed he or she did the right thing, and acted on what was true based on what made the most sense. The lawyers did the same, and so did the judge, the witnesses, and everyone else involved. You know your *own* truth, most likely with great clarity, but *no one can read your mind*, which means your truth is just another story to everyone else. In the case of your guilty verdict, your own truth-story just didn't make enough sense to become everyone else's truth-story too.

So what can you do now?

If you were found guilty in this situation, you'll likely try to *appeal* the guilty decision, because some story was told poorly (or perhaps 'too' well) if you are truly innocent.

If you are granted an appeal, you won't ask the new appellate (appeal) judge to listen to a retelling of the same stories from your first trial. An appeal is not your action-story do-over to determine innocence or guilt. Instead, you're moved up to a higher court, and your lawyer will be trying to tell a story showing how unfair or improper your first trial was. Your lawyer will argue that it makes the most sense for the results to be thrown out because of this.

If your lawyer convinces the appellate judge that your first trial was unfair or improper with a story that merges to become the judge's truth too, you win the appeal and your sentence is undone and your record is potentially clean again.

If your lawyer's story does not make sense to the judge, the original results of your lower court trial and sentencing remain in place. Since you are ultimately innocent, you and your lawyer now must decide if you can appeal the appellate judge's truth-story about your trial to the next highest judge in a way that continues to make sense.

In the end, and ignoring lots of possibilities that can make this process get really noisy like witness tampering and perjury (lying under oath), everyone involved has acted on what they believe to be true and right at the moment of action.

No one involved was wrong at all from each of his or her individual perspectives.

14

What's That Squeaky Sound?

If you don't know much about cars and have a problem with your car, you tell the story of the problem to your mechanic the best you can. The mechanic already knows *many* stories of cars in general. These stories are what he or she have personally experienced or have heard from others about things that happen when a certain part needs fixing or replacing on specific cars.

For example, there are stories a car "tells" a mechanic when an alternator or battery cell or generator is going bad, or the transmission synchronizer is wearing out, or a fuel injector is clogged.

The closer your story of the car's problem is to one of the stories the mechanic knows to happen when certain parts need fixing or replacing, the more sense your story makes and the faster the two stories can merge in the mechanic's head. If the two truths align almost perfectly, the mechanic can know which part is going bad without even examining the car, and fix the problem right away.

If your story of the car's problem makes no sense to the mechanic (because it doesn't match any story the mechanic knows to happen when parts are going bad), the longer the process of merging two truth-stories will take. It's worth noting the mechanic will not likely say 'you are wrong' without examining your car. "Wrong" does nothing helpful for anyone in this situation.

Instead, if your story doesn't make sense, the mechanic has to get your car to tell a story similar to the one it told you, *and* get it to tell that story in a way that merges with stories of what happens when parts go bad, so whatever is broken or worn can be found and fixed as quickly as possible. This often involves the mechanic doing what you do with your car - driving it - to see if a similar story can be experienced to the one you told. The mechanic is trying to make your story *make sense*.

For example, if you said "there is a bonking noise when I push the brakes" and your mechanic doesn't know a story about bonking noises related to brakes, that mechanic might start asking questions to see if you "rewrite" your story with different words that better line up with known brake stories. If that doesn't work, the mechanic will likely take your car and your un-mergable story, inspect your brakes for something obviously bad or broken first (tip: do not drive a strange car with suspected brake problems). If nothing worn or broken is found, the mechanic will drive your car to try to experience and validate your "bonking" brake story.

If your car tells the mechanic a story of a part going bad, but it doesn't merge at all with your story of the problem, then the mechanic has to put this newly found story aside, and keep trying to find one that matches yours.

For instance, if the mechanic is out driving your car to confirm this "bonking" noise when braking, and instead hears a horrible screeching sound when braking, the mechanic will just have to take note of the screeching and keep trying to find the "bonk".

Eventually the mechanic will loop back around to you, and tell you the story of the screeching noise that was found in the form of "you need new brake pads." If *this* story makes sense and is true, it will be up to you to decide if it's worth the cost and time to replace them, or if you need to ask questions to make more sense of it.

If your mechanic simply can't get your car to tell the same story you told, it's likely there will be some head scratching, the checking of a few parts known to make bonking noises, and then he or she will give it back to you with a best guess and a recommendation to "bring it back if it happens again."

Both you and the mechanic are not wrong, and aren't in positions to call each other wrong. You just have different truths you couldn't agree on and merge in a timely manner. You can't, after all, read each other's minds.

NOTE: This chapter can be revised to be about anything you might call in or visit an "expert" to fix for you, since an expert is really just someone who knows many stories about a specific thing, topic, or idea relative to most people around them.

15

Learn How to Drive, Moron!

I apologize for bringing things back to cars yet again, but this type of two-to-one story merge is common in the United States where I live: Stories about other drivers.

Driving is something that occurs at high speeds with quick situational changes. This means it provides very frequent feedback loops for our actions to find truth and decide right and wrong. Because of that, I believe driving might be where stories are built the fastest about other people in our day-to day-lives.

We often tell ourselves stories about other drivers based on their actions, and these are great examples of how fast stories can be built. If you flag me through a 4-way stop sign intersection when it's your turn to go through based on traffic rules – I'm probably going to be annoyed with you. I might quickly *tell myself* a story that you have no idea what the rules of an intersection are, and how you are screwing up the expected traffic pattern, which leads to accidents.

But this can be flipped just as easily. If I drift into your lane on the freeway and cut you off while trying to deal with my 5-year old in the back seat who can't open her water bottle, you will probably *tell yourself* a story I am not paying attention, do not know how to drive properly, am endangering you, and am likely to cause an accident.

In both cases, we may each blow the horn, shout a lot of angry words or use angry gestures, and perhaps even drive aggressively toward each other. This is all based on a story we have each *imagined and* then *assumed* to be true about the other person. There is *not* a two-way communication happening at all – it is often exclusively one-way. The best outcome for these situations is probably an attempt at an apology by the other driver with a "wave" or an over-exaggerated mouthed "sor-ry".

Here are a few examples of other driving stories that may feel familiar:

People that drive slow in the left lane are idiots.

Semi truck drivers are always getting stuck on some side street or hung up on railroad tracks or block traffic.

The person tailgating you is a jerk.

The person you are tailgating is a jerk.

The guy that flew by you in the fast lane is reckless.

The person with stickers all over their car is crazy.

On and on this goes – endless possibilities of one-way made-up stories, or extremely limited two-way communication that likely makes no significant difference to the stories we already made up and believe as true about someone else.

But no matter what you believe, you are not wrong at that moment *to yourself.* You believe the story you tell yourself to be true because it makes the most sense to you, even though you may have made the whole thing up based off of one action by the other driver.

The only way to ever change that truth is somehow get the other driver's truth-story – but that almost *never* happens in traffic, and is very unlikely to happen even if you both end up together in the same place, because the story you *made up* about the other make so much sense to you, you often aren't prepared or willing to accept the other driver's truth-story as valid. 'That person is an idiot,' plain and simple.

Face to face, you both end up almost drowning in the noise added by the word *wrong* that was never even said out loud between you.

16

The Reason For All of Society's Problems

If it seems to you like reality and truth are splitting at the seams, and society is on the verge of collapse, you probably already believe a story for why this is happening. You already know that I personally believe this is happening because I feel it is tied to something I haven't yet explained called the 'Floating Anchor Problem'.

Here in the United States, in late-Summer of 2024 as I write this, there are many stories individuals believe to explain *why* this is happening, and perhaps some of these are the same ones you believe to be true while reading this: It's the government. It's the Deep State. It's the Democrats. It's the Republicans. It's election security. It's social media. It's the Supreme Court. It's MAGA. It's woke-ism. It's liberalism. It's conservatives. It's Fox News. It's CNN. It's stupid people. It's millennials. It's the boomers. It's people that think they are smart. It's the scientists. It's transgenders and gays and drag queens. It's black people. It's white people. It's illegal immigrants. It's guns. It's gun control. It's gun promotion. It's mental illness. It's mental healthcare. It's Critical Race Theory. It's school indoctrination. It's lack of teacher and school support. It's money. It's science. It's religion. It's lack of religion. It's the absence of god. It's the devil. It's the police. It's our broken healthcare. It's the ultra-rich. It's the union. It's Black Lives Matter. It's Antifa. It's the loss of morality. It's confusion over what is right and wrong......

There are probably hundreds more, but I feel it's probably safe to assume you get the general idea.

Whichever of these stories above (or others not listed) makes the most sense to you as the reason for most or all of society's problems, you will believe. It might even be several of these put together. You may have made comparisons to hundreds of different truth-stories to end up at the one merged story you currently believe to be true. There may be a hundred other people in your life that seem to agree completely with your story, and you may have a thousand partial truth-stories, or "sources" of truth for you, saved and cataloged. Each of these thousand stories validates the story you believe to be the *real* truth of the whole situation. They are your evidence of being right, or they are the evidence other truth-stories are wrong.

And just like before, *you are not wrong* to yourself, because whatever story you believe is true right now is the one that makes the most sense for you. If a different story came along that made *more* sense, you would believe that story instead, not the one you believe now.

To play this out for the sake of example, if you believe the main problem with society is *rock-and-roll* music, then that is your in-this-moment truth. It makes the most sense right now. You arrived at this belief by merging your own starting truth-story (one point), with other stories that serve as 'evidence' (second points) until this larger truth-story was validated to the point of being true for you.

You wouldn't believe it's the fault of *country* music. Country music makes no sense to you as the reason for society's problems, there are no stories that makes this make sense, no evidence for this story at all. It's actually *rock-n-roll* for very specific reasons, because those reasons *tell you* a story that makes the most sense for why society seems to be falling apart.

And this is actually what I believe to be the *right* story to explain the cause of nearly all of society's problems. No, not rock-n-roll music, and not any of those things I just listed above as commonly believed stories either. It is actually *none* of those things. It's far more basic and simple:

I believe the truth-story to explain *why* society seems to be tearing apart at the seams; *why* violence and shootings seem to be increasingly 'senseless' and nearly everyone seems to be angry, and lonely, and depressed and anxious all the time, is because each of us, including you and me, have truths that are based on stories that make sense from our own perspective, but we don't realize and can't easily see *not one single truth is identical to another*, or even 100% agreeable from *any other perspective*.

But, for reasons I will explain, we often believe we share the same exact truth-story with others. We tend to condense into groups that *seem* to validate our own truth-stories as being shared, causing it to make more and more sense.

As a result, you may come to believe tens, hundreds, thousands, or millions agree with your truth about all kinds of things. But no human truth is identical to the next, and it never was.

It is my belief the reason *why* this is so hard to realize and notice is the same reason it's hard to notice when we tell ourselves "stealing is wrong." It is not our own personal belief of what is true – it is something we tell ourselves. We end up believing it's actually possible to "be wrong *and* act on it."

This belief condenses and compiles and grows larger in society until the story of First You and how right and wrong are naturally determined fall into darkness, pass through a great barrier, and are then covered by layer after layer of noise. We then end up too busy arguing and tearing each other apart over what is right and wrong in our ever-growing groups of validated and compiling 'truths' until we begin invalidating others, calling their truths wrong and our own right, until we believe *they are wrong and they know it*.

You want to believe what you think you should believe.

If two truths disagree, no matter how many people are involved, *then one of them must be wrong*, right?

…..right?

17

Return to the Rooster Run

Let's mind-travel back to you and Bob trying to solve the 747-on-the-tread-mill problem while on a treadmill to explore the idea of "wrong" a little more. At this point, you and Bob are representing the beliefs of thousands of people, because this problem has been a source of very real arguments for years.

I'll go ahead and bang on the rooster cage to get them irrationally fired up again and start the treadmill back up at a fast walk that will get faster every minute.

If you know you have to come to an agreement or face the wrath of the angry birds, is "wrong" ever a *useful* word? Is it helpful in any way whatsoever?

If you said "Look, Bob, you measure wheel speed by using its rotation!" and Bob says "NO! That's wrong!" would you say the word "wrong" is helping this situation at all? To me, the answer is no. Bob is literally invalidating *everything* you just said. And what did you just share? *Your truth.*

Wrong, when used this way, can be horrifically damaging. It becomes a form of mental violence and is immediately *polarizing*.

If you are skeptical of this, imagine if Bob refused to say anything else but "No! You are wrong and an idiot! The plane takes off!" to every attempt you made to explain your truth or learn his truth. Bob's is trying to *expand* his truth by invalidating, *collapsing*, or negating yours.

You must try to figure out what to do about this as quickly as possible. Bob is trying to grow his mental boundary to absorb and overwrite your own. *Wrong* becomes a weapon in a very real mental boundary war. 'Wrong' is the monster trying to get into your head.

This will likely cause you to quickly pivot away from the plane and treadmill; to trying to get Bob to *let go* of whatever truth he is stuck to (or at least talk about it), and quit declaring he is the *Keeper of the Truth* over whatever you might believe.

Escape is not an option, so as Bob *continues* to deny your truth's existence, discuss anything with you, and invalidate everything you share as wrong, you will eventually end up at one of three familiar options:

Option 1: Collapse your mental boundary, and let Bob "win". He becomes an authority over you, but you might try to release yourself of responsibility and tell yourself (and probably Bob) that he is now responsible for what happens, and you deny any responsibility for the outcome. You invalidate yourself while yelling: "Fine! It takes off, but if that's wrong it's on you!"

There is no "redoing" this moment, and there's a good chance Bob won't even care about what happens to you after this (why should he?), so this self-invalidation will be incredibly difficult to walk back.

Bob's expanding boundary was "pushed into" your mental space. When pushed, you chose to collapse your boundary. Bob's act of expansion is now validated as right for him by your collapse.

Option 2: Try to stabilize, and keep looping continuously into exhaustion as the treadmill gains speed; begging and pleading with Bob and asking him questions. Perhaps you'll try to match his mental violence and yell at him, demand to know why he believes his answer is right, or beg for explanations over and over hoping he'll finally open up before it's too late. If he doesn't change, and he continues to only ever say "No! You are wrong and an idiot! The plane takes off!" then I suspect this option will be abandoned and you will end up at boundary collapse or the next option in an attempt to change something before you both meet the roosters.

Option 3: Expand, and try to grow your boundary against Bob's. Since Bob is not willing to see you as a valid truth – which is choosing mental violence - you might attempt to force him to recognize your existence and validity. This is physical violence, and it is an attempt to stop the mental expansion and collapse between you with physical force applied through your fist or your foot or some other part of you making direct contact with Bob.

Unfortunately, any relief you get from doing this will be short lived. This does push Bob back out of your mental boundary, but it does not actually fix anything between you, as Bob is not likely to "snap out of it" and open up. Instead, I think he will choose to return physical violence, creating a loop of feedback "noise" in a situation that is doing nothing to move you toward a shared answer.

To avoid all this, Bob *must* share his truth-story of how he thinks speed should be measured instead of only ever invalidating yours. He must share *why* his story still makes more sense to him than yours at *the very least* to keep things balanced between you. This has to keep going, or the result inevitably is mental violence or physical violence.

Eventually someone's story (or some combination of both) will be agreed upon as the truth-story that makes the most sense for both you and Bob. Two points of truth finally merge into one point of truth. Both of you are valid, both of you were not wrong, and together you built one merged truth that is potentially better than it was before you both began.

The word *wrong* rarely, if ever, does any good in any situation. It *feels* like it should save precious time by stopping what seems like a bad idea and a waste of time, but all too often it only wastes more precious time by all the noise it creates. You need every second you can get with this problem, because you and Bob must work through many, many stories together other than how to measure speed.

This Jet-On-A-Treadmill problem is actually hiding tons of mini-stories in its shadows that have to be brought into light and agreed upon. If you and Bob can't agree on the bigger story after one round of sharing your truth-stories, you will *naturally* break down each full story into the smaller stories underneath, and see if you can agree on those.

If you can't agree on the smaller story, you'll naturally keep digging down to smaller and smaller stories until you both finally *can* agree on something. Then you'll start scaling back up while trying to stay in agreement, until you both finally get to a point where you each agree on the whole story and thus arrive at a shared-truth which provides a shared right and wrong based on that truth, and thus a final shared answer. You are *falling down rabbit holes, finding a shared foundation, and building on it until you get back to the top* together.

Let's say you and Bob do exactly this. After many minutes of intense discussion into several rabbit holes, drenched in sweat from running, you reach a mutual understanding. You both agree on how to define and measure everything in the problem, and submit your final shared answer: *The plane will take off.*

Bravo! You and Bob worked so hard. You worked to understand each other and the problem, and came up with a shared truth and answer to the problem. You merged two different answers built on two different truth-stories into one answer built on one foundational truth-story shared by you both. Instead of calling each other wrong, which could have resulted in physical violence with feathers and blood everywhere - you each let go of what you believed to be right, and worked together to build a newly shared *right*.

The treadmill stops. I congratulate you both.

Great work.

Wait.....

You weren't satisfied with this and ready to move on, were you?

You didn't lose sight of what's important here, did you?

Do I have any authority to declare your now-combined answer right or wrong?

Why should I know better than the two of you?

Am I the *Keeper of the Truth and Decider of the Right Answer to the Airplane and Treadmill Problem*?

I am just another single-point truth. No better or worse than your own. I have no natural authority to call you right or wrong at all.

We are now back at the beginning. Two truths. Two points.

My truth. Your truth.

Your truth just happens to now be largely shared with Bob's.

My truth is also shared with you and Bob. Well, the part you know anyway. *Which is literally none of it.* I've shared absolutely nothing about my truth justifying my answer.

Did you think to question my authority as the treadmill kept speeding up while reading this chapter? Maybe you did, and maybe you didn't. As usual, I can't know. If you didn't, then I successfully seized authority and made myself your new source of validation up until you reached this page, and I did this by not giving you time to think about the situation and not drawing attention to what I was doing.

We even left the situation for multiple chapters and came back to restart it. Depending on how fast you've been reading this, days may have passed since you finished "How Poultry Can Help Solve Problems". I even banged the cage with a stick to get the roosters all fired up again, and restarted the treadmill. I imagined myself smiling from ear to ear as I watched you and Bob work so hard to answer for me. I got to play the Keeper of the Right Answer - Keeper of Truth. It was intoxicating, I must say, to know your minds were working so hard on my problem, because I honestly didn't know the right answer at all.

After you and Bob answered, the only thing I had to do to *keep* my authority as Keeper of Truth was stop the treadmill. I didn't even have to directly confirm if you were right or wrong. Go back and check if you don't believe me.

You and I haven't compared anything. My answer to the problem was only *implied* to agree with yours by my actions. You *imagined and assumed* everything else matched your truth-story, *if* you fell for my trick. You just handed the crown of Keeper of Truth right to me, and you probably don't even know why.

We needed to first compare *stories*. I, too, must get on the treadmill with you to confirm, through two-way communication, that you and I actually agree on everything else related to this problem. Not *just* the answer.

If you expand this idea just a little bit, you'll find an absolutely enormous hole-of-a-question: Who in the heck is actually capable of controlling the treadmill and being *Keeper of the Truth* in reality?

18

Gazing Into the Abyss

The only person capable of controlling the 'poultry treadmill' in real life without getting on it is the *Keeper of the Truth*.

Here's the catch about that *Keeper of the Truth* title, though: It requires all truth-stories to be included. Not one or two or three. Not yours and mine and Bob's. *All* truths.

No one knows the whole story about anything.

There are no exceptions. I don't care who you are or what you know.

Any story you believe to be complete is not the whole story. I mean even the simplest stories you know that seem like they *must* be the whole story. I am not claiming this as an authority to invalidate your truth, or imply you are wrong for thinking you can or do know the whole story, nor am I trying to play the role of *Keeper of the Truth* over everyone by claiming this. This applies to myself exactly the same, and I will say it outright to you: I don't know the whole story about anything, either.

There is actually a solid set of reasons many people, perhaps including you, believe it is possible to know all truths.

The *simplest* stories often seem to be complete, but the whole truth is an abyss for us. Anytime I have seen the Abyss of Objective Truth begin to enter conversation, someone inevitably tells an extremely simple story as an example of proof we *can* know the whole story. Maybe you believe some of humanity's shared simple truth-stories to be absolutely the whole story, but I promise you this is *always* an illusion created by the Great Barrier coupled with Floating Anchor Problem (both of which I have still not explained, but I will - hang in there!)

Perhaps stories like "killing is wrong", "the sky is blue", "2+2 = 4", and "words have exact meanings" seem to include all possibilities and all truths to you. However, if you re-arrange them into questions, I think you'll quickly realize there is still no one capable of controlling the treadmill for *everyone else* when it comes to the objectively right answers:

"Is killing wrong?"

"Is the sky blue?"

"Is 2 plus 2 equal to 4?"

"Do words have exact meanings?"

You may believe a simple story, maybe even one of the above simple stories, is the *whole* story. If you answered any of these above questions with an immediate "yes" then it seems you might.

You might even believe your answer is objectively right because it is *also* believed to be the whole story by 60 million other human beings. This wouldn't make that answer the whole truth-story either. It only means it's a story that currently makes the most sense for 60 million people *at this moment*.

I'm not saying those 60 million people are wrong or that you are wrong. After all, everyone acts on what he or she each believes to be true and right. Instead, 60 million people believe they are sharing a one-point merged truth, just like you and Bob shared a merged truth right before I stopped the treadmill.

But (and this is an *enormous* "but"), if anyone else believes something to be true that doesn't match that massively shared truth-story - even if its only *one person;* or if it's possible for *anything* to happen that isn't part of that massively shared truth-story – even if its only for a sliver of one second - then the 60-million-strong merged truth does *not* include the whole story.

Does that 747 take off from the speed-matching treadmill or not? Thousands, possibly millions, believe one answer or the other is *absolutely right* and the other is *absolutely wrong*.

Which group is actually right and which is actually wrong?

Well, to be *absolutely sure* of being right or wrong to the point of becoming the *Treadmill Controller* or *Keeper of the Truth and Decider of What is Right*, you would have to know the whole story. The whole story is also called the *objective* truth.

The objective truth is what *actually takes place* in the real world in this moment. You can think of it as the truth of real *objects*, hence *object*ive truth, which of course includes other people and yourself.

The objective truth, or whole story, holds up as true under all conditions and from all perspectives, because it *includes* all conditions and perspectives. It is a *very real* one-way all-point merged truth. There is nothing to even compare it to, because there is nothing else left to compare. It answers every single question that could ever be thrown at it, but it would be impossible to even build a question it hasn't already answered anyway. No other story can even exist, because the whole story it tells already includes every possibility.

The objective truth, or whole story, includes you and Bob, the treadmill and the plane, the shared foundation, this book and all other books, all the layers between you and your foundation you are about to experience within this book, and everything in existence in all directions as well. It extends further than you can ever see, hear, touch, taste, and smell. It includes what is shrouded in darkness in all directions.

The reason the plane-on-a-treadmill question has been a matter of debate for years is because there are an absurdly large number of possibilities built into the problem, even though there are only *two* possible answers. The problem allows you to imagine and design multiple ways to solve it built from stories you try to build based on your past, and most everyone who does this tends to believe they have imagined all of it correctly because their overall story makes sense, and they have even discarded a few possibilities while figuring this out. Just like you and Bob did to arrive at your shared answer.

It's impossible to imagine all the possibilities contained in this problem.

Here's a smattering of possibilities you may have considered (or didn't – which implies they were hidden to you) to arrive at your answer for the problem: Is the 747 in working order? Does it have enough fuel? Is there a pilot in it? Is it a qualified pilot? Do you consider the friction from the weight of the plane on the belt? Could a belt be built that could operate under the strain of a 747? What size would the treadmill motor be? Can it be built? How quickly can it respond to change in speed? How fast would it need to be able to run? Exactly how wide and long is a runway? What communication method is used between the plane's speed sensor and treadmill's speed receiver? Is the speed sensor on the plane or is a radar outside of it? How exactly is the treadmill matching the speed of the wheels? What does the "speed of the wheels" mean? Is it based on the wheels' rotation around their axles or based on the wheel's movement relative to the runway? At what speed do the wheel bearings fail on a 747? How about the treadmill's roller bearings? What's the condition of the wiring throughout the plane? Is there a wind? What speed and direction is it? How heavily loaded is the jet?

If this type of thinking annoys you, and seems like I'm being intentionally annoying - *notice this feeling*. This feeling is a good indicator you might believe it is possible to know the whole story. Accepting that you don't know the whole story if you already believe you do is incredibly difficult. So difficult, that my questions above might feel like an attack, or a child nagging you with a thousand questions, or like I am trying to pry you open to invalidate your beliefs so I can call you wrong, but I am not.

Your current truth will always be available to you. I can't control it or even know it - it is yours. I might already be dead and gone when you read this. I am no threat to you or your truth. You are still right from your perspective, and you always will be, because you are a human being and your truth is all you have, and I recognize that.

But your truth is not the *objective* truth. Neither is mine.

I want you to know it is safe within these pages to let go, or at least loosen your grip on what you believe to be true. This can feel scary because it might feel like you are falling, because the ground you believed was there is suddenly missing. It might feel like you are losing control or about to lose control of *something*. I promise you I am not writing this to overwrite or threaten your truth.

I do not want you to blindly believe me. I am not interested in becoming an authority. I want to help you if I can, and maybe you can help me, too. For now though, let's gaze into this abyssal rabbit hole a little longer while I share some attempts by others to resist falling, or to fill it in.

19

Attempts to Avoid Falling

Mythbusters was a popular television show in my youth. It focused on finding and filling in the gaps in stories people passed around like they were shared truths. The show's team (maybe you were one of them) converted these stories to *objective reality* by recreating them with *objects*, which seems like it fills in all gaps. At the end of each experiment to test a truth-story, they would pronounce the story (or "myth") busted, plausible, or confirmed.

It just so happens Mythbusters made an episode to tackle this wretched-plane-on-this-awful-treadmill problem (Season 6, Episode 28, January 30, 2008).

The team did some tinkering and planning, and ultimately put an ultra-light plane (tiny one-person plane) on a long tarp, attached one end of the tarp to a truck, and had the truck pull the tarp in the opposite direction the ultra-light plane was trying to go. They then made sure the speedometer of the truck and the speed of the plane's wheels matched at all times.

The plane took off. The team declared the story it wouldn't take off as "busted." So is this the *objective* truth then? All done? Whole story?

If you think so, I have very bad news. Adam Savage (one of the main team members of the Mythbusters, which I have to accept could actually be you) believed it was the right to make a follow-up video more than 13 years later to discuss the sheer amount of mail the show had received after this episode aired. This mail was from individuals who felt writing the show to express their truth-story was the right thing to do (and maybe one of those letters was sent by you).

In this follow-up video, Adam tries to explain his truths even further, and why he thinks people are getting things *wrong* (Adam Savage's Tested, "Plane on a Conveyor Belt Controversy", March 19, 2021).

If the Mythbusters team truly had the *objective* truth this would not be possible, because the truth present in every one of those letters would have been addressed and included in the show already. So what's going on?

As I see it, what Adam and the team created for the show is a Mythbuster's team-merged one-point truth, like what you and Bob had after arriving at a shared answer with this same problem.

However, unlike you and Bob, while the Mythbusters team members did share the stories they believed to be true, they did not "jump on the poultry treadmill" and attempt to merge them to get a final answer to the question. They instead built a shared truth to answer a different question: 'How can we build an experiment we all agree with to let objective reality answer this question for us?'

Most of the show's runtime is actually watching the team *fall down rabbit holes, find a shared foundation-of-a-story, and climb back up again* for every piece of the experiment's design – the exact same process you and Bob went through to come to a shared *answer*. It's the best part of the show if you ask me!

Near the end of the episode, a *real-world agreed-upon experiment* has been constructed to recreate the story. The team then makes predictions beforehand and watches with anticipation before the experiment is performed. You of course already know the plane *did* take off.

That being said, the problem is not the experiment and its results – it's in the first part: It's in the stories that designed that experiment.

The stories used by the team to fill in the gaps so they could construct the experiment can be filled just the same with an *entirely different set* of stories. The Mythbusters team designed the experiment using the stories and truths they believed to be *right from their perspective*.

They chose to measure speed using a particular method. They chose not to use an actual 747. They chose not to run the experiment with all possible combinations of winds and pressures, with flat versus inflated tires, or with any of the other *tons of possibilities* that aren't stated in the words of the problem that someone can bring into it as *right from their own perspective*.

Was Mythbusters wrong? No of course not. But their resulting experiment and its answer applies *only* to the combination of stories they used to fill the gaps in the problem to be able to test it in objective *reality*. The result is not the all-points *objective truth*, it is just another one-point truth shared by the team and the viewers that agree with the stories they chose to use along the journey. Is the show's truth stronger than the one you and Bob built together on the treadmill? *Yes it is and by quite a lot*. But it's not time to explain why just yet. That comes later.

Instead, let's go over another attempt at filling in this abyss.

Randall Munroe, author of some great books (my opinion) and creator of the web comic "xkcd", also approached this 747-on-treadmill problem. Unlike Mythbusters, Munroe did not try to build an experiment to convert his truth-story to objective reality. He didn't get an actual plane or an actual treadmill, but instead saw this for what it is: A fill-in-the-gap problem. His approach was to explain what he saw as the main possible stories individuals might use to fill in the gaps. (xkcd blog of the webcomic, "The Goddamn Airplane on the Goddamn Treadmill", September 9, 2008)

Munroe explained three different stories of how a treadmill could match the speed of the wheels, provided some equations and what he believed to be the truths behind them, and how these different ways of filling in the gaps can change the final answer.

The result of his effort was not an attempt at building an objective truth at all, though he does end up sharing his own truth that the plane *should* take off in practical application.

Instead, Munroe focused on explaining what he thinks people are actually arguing about, and implies we should really just leave the problem behind and be nice to each other. I don't disagree with him.

So let's go ahead and wrap this up once and for all.

What is the objectively right answer to the 747-on-a-treadmill problem?

As it is worded, the plane and treadmill problem cannot be answered with any *certainty* the final answer is right. There are far too many visible *and hidden* gaps.

I can't be the *Treadmill Controller* and verify or deny that you and Bob have the right answer, because to actually answer the problem and be *objectively* correct, I'd have to determine every possibility *that can ever exist* related to each and every variable when it comes to putting an *actual* plane on an *actual* treadmill in the *actual* environment. We, as humans, don't even know every variable in existence yet, let alone what affect each variable has on planes and treadmills that match speed.

No one can possibly know the whole story.

To get a feel for the abyss before us, imagine Mythbusters does their experiment again exactly the same way they did it before. Just as the ultra-light plane is about to take off from the tarp (just like it did before), a giant meteorite falls from the sky and blows their entire set and the plane to pieces.

Well, *the plane didn't take off.*

In my opinion, it is our nature to want to ignore this attempt and say it didn't count. If the meteor strike actually happened though, the show would now have a very *real* truth where it did take off once, and a very *real* truth where it didn't. Two attempts, two outcomes, two truths. Both valid. *Neither was wrong.*

"Hey the meteorite was never in the problem! That shouldn't count as an answer!"

If this feels like a response you want to yell at me right now, I'd like to remind you just how many stories Mythbusters brought in to fill the gaps enough to build an experiment, like using an ultra-light, tarps, speedometers, and a pickup truck. None of these things were in the problem, either. The meteorite was always a *real* possibility. It is a *remote* possibility, but a real possibility all the same.

A meteorite strike happening or not happening *must* be part of the objectively right answer and part of the whole story and objectively right answer to the plane on a treadmill problem. It also needs to include what happens if the meteorite hits the ground in front of the plane. Or behind it. Or hits one wing. Or is tiny and only hits the pilot. Or the vertical stabilizer.

Did Mythbusters only attempt the experiment one time while filming the episode, or were there multiple takes we didn't see? If it took three attempts to get the final result that ended up on the show because something went wrong on the others, then those two unaired attempts count as *not taking off* or *taking off* – whatever the outcome was. One of the very-real-but-not-written-as-part-of-the-problem stories needed to fill in the gaps to build the experiment did not go as expected.

This happens because the gap wasn't noticed and therefore wasn't considered by anyone on the team, or it was ignored, or it was dismissed as a possibility. If the meteorite or failed attempts that weren't aired actually occurred, then these hidden or ignored or dismissed gaps contained a possibility that *changed the final answer*. No matter how you answer the question, you choose to arbitrarily ignore some things and accept others.

Are you annoyed by this thinking? If so, flag it for yourself. I know it seems borderline absurd – but is it possible? Yes....*it is.*

If something doesn't go exactly as expected in objective reality – then it is not to be ignored. If you ignore it and give a final answer anyway, you are trying to force that answer, and it wasn't accounting for some part of *the objective truth*. It wasn't accounting for *everything*. You are knowingly leaving gaps and intentionally cutting out possibilities that actually happened in objective reality. *You are invalidating reality itself.*

This means the whole story *always* remains uncertain and incomplete. The best you or I or Bob can ever do is to include this always-existing uncertainty in our answer: "I *believe* it will (or won't) based on the following *assumptions*...." Which is followed by a list of every imagined and assumed variable, gap, and ignored possibility we can think of, how they each would affect the situation and thus affect the final answer.

The result of collecting answers to this problem from every human and writing them down would create libraries upon libraries of books documenting the possible stories that all end with "it takes off" or "it doesn't take off." *Even then* these stories still can't account for every single gap, because there will still be gaps hidden in *blind spots*. And having blind spots means we can't possibly anticipate all the possibilities and gaps that might exist, because we are not capable of knowing everything.

Did you consider a meteorite strike as a possibility before I mentioned it? If not, that was a blind spot for you.

Did you consider the size of that meteor? Are you capable of tracking and knowing when and where a micro-meteorite or a mega-meteorite strike will occur?

The Mythbusters team and Randall Munroe have offered several gap-filling possibilities that *are required* to be included in an *objective* truth, and thus the whole story. We would also need to include your story, my story, Bob's story, and every other story used to attempt answering the question. We would need *all points* of truth.

No one can know the whole story about anything. Gaps and blind spots hiding even more gaps and blind spots exist in *all* stories.

Whether or not you believe all the previous stories I've told are the whole story or not doesn't matter. Whatever stories you *do believe* to be true (because they make the most sense for you), contain uncertainty within them even if you can't find it. This is not an assumption about you. It is the nature of being a human being. This includes stories as simple as 2+2 and the color of the sky.

You are not wrong, *but neither is anyone else.* As I see it, all the trouble in our human world occurs because we often think there are only two possibilities – opposites - but there are almost never only two. The opposite of what we each think is wrong is not *objectively* right. And you and I believing we are *objectively* right might seem relatively harmless ("The plane takes off!"), but it is *incredibly dangerous* for reasons I haven't yet explained.

It is also extremely easy to fall into this trap, because the default setting for our actions is "I am following my truth based on what I believe is right" and we entangle with each other and make decisions at a very fast pace.

You are, however, already familiar with what it's like inside this trap. It is being locked in a battle with an invisible hydra in your own head. It is feeling like there must be an absolute objective right and wrong, and if it is not followed we all tumble into the bottomless pit.

Well, I hate to tell you this, but that pit is wide open, and has been for quite some time. Now it sits before you and I.

20

The Choice To Tumble

Certainty implies the objective truth is known.
Certainty implies the whole story is known.
Certainty implies the gaps and blind spots in the story no longer exist.

Even if you still believe you can absolutely be right and the *Keeper of the Truth and Decider of Right and Wrong* about *something* with 100% certainty, certainty does not *actually* exist outside of your mental boundary in the real world. Certainty only ever exists in one place – a single human mind – and it is contained there inside mental boundaries where no one else can see. When it spills out of its boundaries, the damage and destruction it has caused throughout history is enormous and still going strong today.

True 100% certainty is only achievable with knowledge of the whole story, which includes all the possibilities of all of existence. Until that is known, and I have good reasons to believe no human will ever be able to know that much information, there will always be at least a sliver of uncertainty in *everything* when it comes to your beliefs, and mine.

If you claim the 747 will take off with certainty and believe you are making no assumptions, you're claiming to not only know every variable in the entire universe that can affect the problem, but also know the future state of each variable and how it will affect the problem at the exact moment the real plane attempts to take off from the real treadmill – even the appearance of meteorites.

If you are reading this while sitting on a plane about to take off right now, I'm sorry for this, but there is not a 100% guarantee your plane will successfully take off. Deep down you know this is true.

I recognize it might be *horrifically* uncomfortable to consider this possibility, so much so that you may *tell yourself* to ignore or dismiss this story. Here's the catch though: Did you imagine the opposite of 'not taking off' to be *crashing* and *death*? I hope not. There are almost never only two possibilities. You might have a mechanical problem beforehand, or maybe there's a medical emergency on board, or maybe your plane finds itself in the path of a sudden and unexpected bolt of lightning trying to reach ground.

Not knowing the truth does not result in actual chaos and disorder. There is very much a reason you might believe it does though.

There are gaps and blind spots hiding even more gaps in all human knowledge, and when you make the jump to certainty, these "voids" remain open, and it turns out *tons* of stuff is falling into and hiding in them - enough to feed a hydra for a million lifetimes. There are even millions of *people* in these gaps and blind spots, First You included, and you might not even realize it. There is no single human that is *Keeper of Objective Truth and Decider of Right and Wrong*.

Of course this hasn't stopped humans from constantly trying to claim the title. Philosophers and leaders of all kinds have been trying to debate, declare, and decide what is *objectively right* for millennia.

Righteousness, or to be morally right or justifiable, has been based on many things over time: Rationality, virtue ethics, greater good ethics, ethics of self, ethics of selflessness, nearly every single religion on earth, as well as all sorts of other theories and modes of thought. Most of these have points of breakdown, because they often keep trying to declare a one-way universal and objective right, and universal and objective wrong.

It cannot be done. And I think there is a perfectly good reason to believe it can't be done: *It's almost always possible for a second and third truth-story to exist.*

I'm writing this because I believe there *is* an underlying shared objective truth for humanity and all the attempts to find it I've ever studied miss all, most, or some part of it. The results is individuals, like me or maybe you, telling ourselves to follow a particular "moral code" which tries to force a *certainty* or *absolute universal code* of right and wrong.

We then try to base our actions on these claimed "objective" truths which say things like "stealing is wrong" only to completely wreck ourselves internally when we realize this "objective" truth-story is actually full of gaps and blind spots hiding more gaps. But rather than let it go and accept that new truth, *we want to believe what we think we should believe.*

Some wrecked individuals, like me, never realize what is happening when we collapse, loop, or expand our mental boundaries to cope with our minds attempting to dissociate and disown a piece of itself, and we create a chain reaction of disaster inside ourselves and all around us. I didn't realize this for 42 years.

The reason it's so insanely difficult to see when, how, and why this all happens from your own perspective is because *it all occurs in a gap hidden in a blind spot.*

And that's why believing there is a universal and objective one-point right and wrong for humanity, or even yourself, *must be abandoned* and reframed another way. Right and wrong are *relative* words. A minimum of two points is required to give them meaning. Finding these two points is everything.

I realize that I seem to be slinging a lot of certainties over the past few paragraphs, and that you can turn my own statements against me and accuse me of having blind spots and gaps myself. You might want to dismiss everything else I have to say beyond this point, *telling yourself* you are *certain* I am wrong and *telling yourself* that right and wrong are black-and-white absolutes.

If you do this – if you jump ahead, and start *assuming* things about what I believe to be true or where this book is headed, then you are choosing to expand your mental boundary to invalidate my own. You would be trying to predict me and *tell yourself* what I believe, because you may be very uncomfortable with the idea of right and wrong being relative and not absolute, maybe you think I'm going to tell you what to believe, or that your beliefs are wrong. Maybe you are fearful that I am about to validate everyone you dislike into believing they are right. That's a natural reaction. You are not wrong to feel this way, because you are human like me. Please trust me when I say: I understand.

It feels *right* to fear that you and everyone else will start drifting and falling into an abyss of chaos and social destruction if truth is uncertain. It feels much better to cling to some certainty for dear life.

However, I *confess* to having blind spots and gaps. I accept them. If you think you can see one of them right now, then please keep it in mind as you continue, because there is a good chance I have *intentionally* left it in place to expose later.

There are a million-billion things I do not know. I am not at all claiming to be the *Keeper of Objective Truth and Decider of Right and Wrong.* I refuse to accept that title. I believe no human can claim it. I would need all perspectives and to know all possibilities to wear that crown. What I am sharing here within this book is my own personal truth about humans and our relationships with truth based on what I have learned through experiences with others and myself. And I've only just gotten started.

I have a bias toward believing my story is right in this moment, just like everyone else. It's my default setting. I wouldn't have written this if I didn't think it was the right thing to do. But I also accept my "right" is not *objectively* right. I accept that this book might crash and burn due to something I have missed. But please keep in mind, there is no evidence I have missed it *yet*. This is the end of Part One, not the end of the book.

Besides, even if you agree with every word in this book, that is still not proof the truth-story they tell is *objectively* right. It only means I made the most sense to you right now.

As I said before, for 42 years my truth was that I was wrong and everyone else was right. I became a black hole pulling in and internalizing other people's beliefs, truths, and knowledge and splitting my mind into two, three, and a hundred pieces. I have no intent to do that to you. Instead, I am actually asking for your help.

My hope is that my blind spots are very small. Small enough that another perspective can make my perspective make even more sense than I can. Maybe that someone is you. Maybe you can use my understanding of truth - that's about to start coming at you faster now - to move us one moment closer to getting off this treadmill that is constantly being sped up. Maybe you can fill in what I couldn't see when I wrote this, and advance things toward an overall improved world for everyone.

As I see it, the objective truth does exist outside of you and me. By this, I mean the objective truth continues to exist even if all of humanity is extinct.

There is no human that is the *Keeper of the Truth and Decider of Right.*

No human being has a natural-born *authority* over another human being.

No human being can claim absolute certainty about anything.

Letting go of certainty isn't accepting you are wrong. There are blind spots all around us, and we don't know what's going on for sure.

Letting go of certainty does not result in chaos and collapse.

We are already *in* loops of chaos and collapse because of certainty.

Staying the course will likely result in the continuation of the next cycle that will damage and destroy who-knows-how-many lives. I don't want that. I don't want it for me, for my family, or for you and those you may love.

Objective truth is the white rabbit we are chasing.

Uncertainty is the hole where the rabbit hides that seems to have no bottom.

We have to jump into it. We have to fall past the major sources of certainty – past all the 'shared' truths claimed to be objective by others that may exist in your life. We have to wave at all the Keepers of Truth as we pass by.

I can't skip past them. I feel it's the right thing to do to show you the gaps and blind spots contained in what you might find to be the unlikeliest of places.

If you are still not sure about taking this tumble into darkness, that's okay and I understand how you feel. I went into the abyss alone the first time. It can feel pretty scary and maybe even upsetting at times, but I promise you it is not worse than facing a Second You that has been turned into a multi-headed monster. I'll have no intent to leave you alone in the darkness. Perhaps something to keep in your mind should you take this journey is that when it feels the darkest it can be, the only change possible is for things to get brighter.

Going into the abyss is the only way to defeat the invisible dragon flipping everything upside down and polarizing reality into separate pieces from the inside out. We have to find the smallest story we can agree on without any doubt we disagree. Only then can we begin filling in the hole and rise toward the larger truth-story.

I cannot force you to take this leap. I can't force you to keep turning these pages. It is up to you. The only way to break the cycle so we can keep going up together is to first go into the abyssal darkness together.

So when you are ready......jump.

PART TWO – COLLAPSE

Take a tumble into the abyss of uncertainty with me.

There's something I want to show you.

Oh, one thing: Could you please bring that stolen candy bar?

I'll bring some ice cream.

21

Mapping The Upper Layers

As a kid, you were most likely made very familiar with the existence of *objective* moral codes early in your life. These stories of objective truth – of all-point right and wrong - were known as "The Rules," which were enforced by "The Boss" through a process often times just called "Unfair."

These claimed objective right and wrongs were probably created by whoever raised you or was responsible for you. That person (or persons) was the authority, or The Boss over your entire reality. Your ability to bend or change The Rules to allow new possibilities was probably extremely limited, and it wasn't until you were older that you began to test them. If you disagreed with these limits placed on what was possible for you, the process for you to dispute The Rules was probably protest, arguing, or just breaking them to see what happened.

Through that process, you determined if the system had any flexibility at all, probably with mixed results. The new story that resulted from your testing of the objective rules was called The Consequences. The Consequences for "testing" a rule made specifically for you by The Boss probably ranged anywhere from violence to a simple correction using words. In between were all kinds of possibilities: Getting grounded, time-outs, being yelled at, getting "the look", being swatted with a shoe, passive aggressive comments, or maybe dragged out of the store by your ear.

Perhaps though, if you were "lucky", the rule you tested in an attempt to expand your possibilities ended up changed or removed, because it no longer made the most sense for The Boss.

I believe it is incredibly unlikely a rule was changed by The Boss during the act of testing it – but I can't be certain of that.

If you raised yourself, or had no adult setting and enforcing rules for you, then you had to navigate the world by yourself as a child, which means at a minimum, you had to figure out how to keep your actions based on your truth merged with or *hidden from* The Boss that wrote The Rules of the Adult World.

What are The Rules of the Adult World? Whatever rules you are still expected to follow today as you read this. Although I can't know where or when or who you are, I think it's safe to assume you follow more than one set of rules right now.

In the United States at the time I wrote this, there are Constitutional and Federal Laws, State Laws, County Laws, and Township or Local laws. These laws, or rules, are ideally written by individuals that citizens elect to represent them (The Boss), enforced by individuals in police and law enforcement agencies (The Boss's ability to "see" you relative to The Rules), and are processed and evaluated by individuals in the justice system such as judges, juries, and lawyers (The Consequences).

The process of writing, enforcing, and evaluating these rules of right and wrong is open to public debate at all times. This "debate" takes the form of something probably familiar to you: Protests, arguments, and just plain ol' breaking the rules to see what happens – to test the flexibility of the system – often with mixed results.

The Constitution, which is the top-level set of rules in the United States as I write this, only says the process you are entitled to after breaking a law is "due process," meaning *a process is due to you* by the government. In other words, government authorities can't just throw you in the slammer and lose the key because they feel like it – though one could argue those actions technically are a process.

Government aside, there is also a decent chance you are religious and belong to an organized religion. If this is the case, you likely try to follow the rules put forward by your religion as well. These rules are often created in multiple places: Whatever holy book you read (the Bible, Torah, Quran, etc), your individual denominational beliefs (Catholic or Baptist, Shiite or Sunni, etc), and your local high-ranking religious leader's beliefs (pastor, priests, rabbi, etc), whatever those might be.

I can't know your religion or its specific rules, because there is more than one religion out there, but if you are religious, I think it's probably safe to assume you see the ultimate lawmaker as the higher power (God or gods, creator, etc) with the religious organization supporting that higher power as second in line. If this is true for you, then this puts the head of your religious organization as the second-highest authority in your life. This person or persons (The Boss) is the controller of the process you must go through and is probably the provider of The Consequences due to you if you test a rule of your religion.

Beyond both government and religious rules, in the Adult World you might also be subjected to rules at work or school or any other hierarchy-style institution. I can't list all of these possibilities, but I trust you recognize these in your life.

You might even create rules for yourself. If that happens, you are The Boss, The Enforcer, and the Decider of The Consequences all in one person. I can't possibly know anything about that particular set of rules, enforcement, or process either. Why? *Because I can't read your mind and I wrote this in the past.*

As you consider all of the rules in your life, which do you believe to be *objectively* right, meaning they include all possible perspectives and truths? Which Boss do you accept as *Controller of your Poultry Treadmill* to judge your own truth and actions as right or wrong, even if you've merged that truth with one or 8 billion other human beings?

Keep in mind an *objective* set of rules would include all points of truth. If any possibility of a second and different rule can exist outside of claimed-objective certainties of truth, right, and wrong, we cannot use them as a shared foundation. If we can't use them, we have to keep tumbling down the rabbit hole until we can.

22

There Oughta Be a Law

As we begin to tumble through the upper layers of authority in our search for an objective truth, I'll try to make this fun. Even though this is serious stuff, I see no reason this has to be a stressful, depressing journey, so let's have some fun with it.

[Cue trumpets and red carpet]

I, *King Eric, Self-Declared Keeper of the Truth and Protector of Being Right,* have just been elected as the one and only lawmaker and representative in your district. My first law is as follows:

"No one in my district shall purchase, sell, manufacture, or eat ice cream on the second Tuesday of each month."

The penalty shall be up to 10 years in prison.

Boss? *Me.*

Rules? *Law.*

Consequences? *Prison.*

Unfair? Oh, *almost certainly.*

There are probably protests and arguments outside my home and workplace over my law. News sources everywhere in the kingdom either support it or decry it as awful. None of these things *require* me to change the law I just made, though. I was elected fair and square, have a term limit, and so have my office as your king for four whole years. I just made a rule, and now you have to follow it.

The only way out of this situation quickly is to leave the district. It's possible a process exists to kick me out of office or legally challenge this new rule, but it's going to take you lots of time to get it moving.

Besides, after a few weeks of public outrage (which I ride out in silence), the excitement dies down and everyone moves on to be outraged by something else.

Then, on the second Tuesday of the fourth month after I issued this law, police officers bang on your door with a warrant. It turns out you were recorded eating ice cream in your living room earlier that day by a neighbor, who then reported you. (You are pretty sure which neighbor: The one with anti-ice cream/pro-health movement yard signs.)

While searching your house, the police officers find an empty ice cream bowl in the sink and half-empty pint of ice cream in the freezer. You are arrested.

A judge reviews the 'objective' video evidence from your neighbor and pronounces you guilty. You are sentenced to seven years in the state prison, all because of the law I wrote as your elected *King*.

I imagine you will appeal to a higher judge because *this law is ridiculous*, right? So, you do. The appellate judge hears the story of your previous trial as told by your lawyer, and decides the enforcement of the law was handled and judged properly (your story didn't make more sense than the lower judge's story). Your appeal is denied, and the first judge's decision stands. You remain in prison.

You and your lawyer keep appealing higher and higher, each time claiming the process that came before was unfair and full of mistakes. Eventually, years later, you arrive in front of the Supreme Court on the basis that this law was crushing your personal liberty to eat whatever you want whenever you want.

The justices of the Supreme Court decide the "due process" you are entitled to by the Constitution was provided to you and all previous rulings stand as valid. Back into the slammer you go.

But the justices went a step further: They elaborated in their ruling they cannot be the ones to decide what "ice cream" means because those words are not in the Constitution, and there is wide-spread confusion in society over the difference between "ice cream," "gelatin," "frozen yogurt," and "soft serve." So, after citing a few ice cream laws from the 1800s to justify their decision against you, they kick the looming problem back to the elected lawmakers to decide what "ice cream" means today.

In the end, the district - *my district* - has a vested interest in making sure you stay healthy, so from my point of view as King, it is not unreasonable to ban ice cream one day a month, because ice cream *is* unhealthy. The original judgment of your guilt stood as Constitutional all the way up to the Supreme Court, is considered fair by the system, and you serve out the one year remaining of your seven-year sentence for eating ice cream before finally being released back into the world.

Oh, one thing: During the time you were serving your sentence, I was re-elected to another four-year term. Yay me!

So, you and I are cool now, right? No hard feelings?

Justice has been served after all: You knew the rules and what was right and wrong, you broke the rules and ate the ice cream on a Tuesday thinking it was the right thing to do, and you were fairly sentenced based on it. We are square and you are not angry or bitter at all toward me for making that rule that put you in a prison cell for seven years of your life, right?……..right?!

I would be shocked if you said we were cool with this. The whole thing is ridiculous! Who in the heck should care if you eat ice cream on a Tuesday? Why on Earth should eating ice cream be considered so wrong that it's worth losing seven years of your freedom to move about in the world as you please? Unless of course you really do hate ice cream and think it should be banned – then I suppose it's possible you and I very much disagree right now.

Regardless, the government justice system reached its limit. The highest court in the land faces outward with its back pressed against the Constitution itself, and is the authority of deciding what the words of the Constitution actually mean as they relate to right, wrong, and *you*. Your options to fight the Supreme Court's decision do not even exist unless you attempt to change the Constitution itself. The courts are bound to it and to its words.

There is no higher power left to appeal to. The Constitution itself does not talk. It can't help you or the Supreme Court justices understand its intentions.

You did seven years hard time for eating ice cream. And the system decided that was fair as understood by the words in the Constitution.

Your actions, which are based on what you believed to be right and true, have been judged to be wrong in the moment by *someone else*, and validated to be wrong by an entire hierarchy of people within that system.

So, do you force yourself to believe this law against eating ice cream on Tuesday, created by me, who was elected to office by the people, is right, and accept my law as right and true for you, and accept your own judgment and actions as wrong? Or do you hold a grudge deep down that this law and the entire system are what is wrong here?

Are you angry with your neighbor? I can't answer this for you, but I can tell you that if you see this situation as upsetting and unfair, which I feel safe assuming probably is the case, you have just exposed a big weak spot when trying to create and decide what is right and wrong using a Constitution, a written law, or any other document a government might use as a means of controlling morality: *Your judgment of what is right and wrong doesn't actually come from these sources.*

If *you* found the system unfair, there are going to be others that feel the same way. There will be a growing anger at the system and its unfair rules. There can be, and *is* a second truth possible that makes more sense for some people than my law, which I claim is ice cream being wrong (on Tuesday). And this can be the case with nearly *anything else* I might want to control.

The takeaway is this: The government, ruling documents, or anyone of political power and authority, cannot be a source of *objective* morality and keepers of the whole story and objective right and wrong, because the possibility exists for *your* truth of what is right and wrong to be different.

The highest judicial rank within your government is *still human*, and cannot be the ultimate, objective, all-point decider of the meaning of right and wrong for you. This individual cannot be Keeper of the Truth. He or she is still only a single point truth.

Just. Like. You.

23

The Great and Powerful Bob

How about organized religion as the *Keeper of Truth and Decider of Right and Wrong*? Is it possible for another truth-story to exist?

Religion and its holy documents are very often cited as a source of objective morals and values in my time. *They* each often claim to be the one true all-point objective truth for all of humanity.

I, as usual, can't know if you are religious or not, so for the sake of example, let's assume you are religious and I am too, and I just happen to hold the highest position in the religious organization built around our shared beliefs.

I just held a counsel with all the other leaders of our religion an hour ago. We reviewed our laws based on holy documents and historical beliefs, and determined that eating ice cream on Tuesday (our religion's holy day), is wrong and a sin against our religion's Higher Power, the great and powerful Bob.

This new rule is based on a new understanding (truth-story that made the most sense) of a particular verse in the text of our holy document, *The Bob Files*, which is believed written by the 100-finger hand of the all-powerful, all-knowing Bob just four thousand years ago.

The council and I decided breaking this rule is a sin of the highest magnitude against Bob, and results in eternal damnation of your soul. We pray and beg forgiveness to the great and powerful Bob for not realizing this crime against him sooner.

A few months later, you are recorded eating ice cream on Tuesday (by your neighbor who shares our religion and is intensely devout). The holy Bobwoman at your local place of Bobbian worship approaches and confronts you about the story of your alleged sin against Bob.

Let's assume you confess and claim authority over your actions. You are pronounced guilty of a terrible sin against Bob. Assuming you feel this is a bit harsh - what is the highest level you can appeal to understand why ice cream is so wrong against Bob?

Organized religions have an eerily similar structure to government. You'll end up at the "Supreme Court" of your organized religion, whatever that may be called. It will likely be the highest-ranking human being or group of human beings in charge of interpreting the words of absolute right and wrong provided by or through the "higher power" of your religion, whatever that may be. In our case, it's *me* interpreting Bob's words from The Bob Files.

To make this real for a brief moment: In Catholicism, that highest rank is the Pope. He is the head of the organization, and he and the upper ranks determine what the words of the Bible should mean to you if you are Catholic. If you have any doubt about this, just reflect on how changing the Pope has historically changed the beliefs of right and wrong for the entire organization.

So, back to the Temple of Bob, what if, after appealing all the way to me (the head of our religion), I say, "I'm sorry my child, but eating ice cream on our holy day is a mortal sin, The Bob Files tells us this truth. I have prayed, and the great Bob himself spoke to me about this. It is absolute, and only Bob himself is capable of truly forgiving you."

The system has now reached its limit, just like in government. The highest religious rank in the land (me), stands with his or her back pressed against the holy documents (The Bob Files), and is the authority of deciding what the words of the holy documents mean when it comes to defining right and wrong (ice cream is a mortal sin).

However, I, as the highest rank, can do something else - something *beyond* what the highest rank in government can do. I can claim to be in *direct two-way mental communication* with the higher power you and I believe in.

I claimed I communicate with the great and powerful Bob directly, who represents the all-points truth, and am merely relaying his direct words. In contrast, the Supreme Court justices cannot claim the long-dead Founding Fathers are speaking to them from the grave – although they are often trying to do something very similar by arguing they know what the Founding Fathers meant.

This two-way communication at the top of our religion creates an *invisible* top-level of the authority hierarchy, and any accountability to you or anyone else for the contents of my invisible two-way conversation with Bob is basically *zero*. Accountability does not exist in any measurable way in this situation.

And now that you reached the top, options to try to overturn your consequence do not even exist, just like the government appeal process. You would need to do something drastic, like change the holy documents themselves or somehow interrupt the invisible connection at the very top of the hierarchy: The one between the great and powerful Bob and myself.

Of course there is a major problem with you interrupting my connection with Him – you can't read my mind, and I can't read yours. My authority over you is being claimed using a two-way conversation with the all-point truth, with the ultimate *Keeper of Truth and Decider of Right and Wrong* inside a boundary that can't be crossed. The alleged conversation can't be heard, seen, recorded, or validated in any way by you or anyone else.

You ate ice cream on our holy day, which you believed to be right at the time you did it, even if you told yourself you knew it was wrong when you did it. Our religious system judged your actions as a terrible sin based on the intention of the Higher Power's words we both believe to be truth, and this was seen as fair by that system. As such, you will be punished accordingly by *eternal* damnation.

Your actions have been judged as wrong at the moment you did them. And it was validated you were indeed wrong in the moment by everyone in the hierarchy of religious authority, allegedly *including* the great and all-powerful Bob himself.

So, do you now force yourself to believe this rule created by me (or was it Bob *through* me?) against ice cream on Tuesday is right, and tell yourself your judgment is wrong? Or do you reject this rule because it doesn't make sense? It is a new rule, after all. If you think this whole thing feels all wrong or is upsetting, then your judgment of what is right and wrong doesn't actually come from a holy document, a high-ranking religious officer, or even that person's own claimed connection with a higher power.

There is no higher power left to appeal to in the system, except for you to appeal directly to whatever higher power you believe in yourself. You would have to appeal directly to the great and powerful Bob.

And how *exactly* would you appeal to Bob in a way that *overrules* me, the highest-ranking member of *our* religion, after I told you I literally had an invisible-to-everyone interaction with Bob himself about this rule? If you told everyone you spoke invisibly with the same Bob I speak with, and that Bob *actually* forgives you and thinks ice cream is fine and that I am actually wrong to create the rule in the first place, who would believe you over me?

The religious organization, holy documents, or any person of religious power and authority cannot be the ultimate, objective, one-way decider of what is right *for you*. They cannot be the true Keepers of Truth for you. Another possibility exists - your belief of what is true and right did not agree with theirs. This is even more apparent when you consider just how many organized religions there are, and that each often claims to be following the objectively *right* truth. Organized religion cannot be a source of *objective* truth and morality. In some ways, they are all canceling each other out.

The highest religious rank within your organized religion is *still human*, and cannot be an objective, all-point decider of the meaning of right and wrong for you. This person cannot know the whole story, even though he or she claim to be in communication with a higher power that does know it. He or she is ultimately still a single point. Just. Like. You.

If you are deeply religious, Second You might be making a lot of noise right now.

I'll use the next chapter to try to settle it before we continue the tumble.

24

Turning Down the Volume

I am not denouncing organized religion or government. *I am not saying the highest authority within your religion is wrong, or that the Constitution is wrong, or that you are wrong.* I am certainly not claiming your higher power is wrong. I am posing thought experiments where individuals with religious authority or government authority over you declare something to be wrong when it's entirely possible you may justify that same something to be *right*. This forces you to wrestle with what you thought was a source of *objective* right and wrong to the point of internalizing it. But everyone is not wrong in the current moment. To fix this disagreement and arrive at a shared truth with the Boss, you have to jump on the poultry treadmill together, and for these two hierarchies, that's the appeal process I just laid out.

That this process failed to end in a merged truth only means these rules that limit your possibilities, put in place by these authorities, are not *objectively* right or true for you *at all times*. The ice-cream rule didn't include all the possibilities, such as believing ice cream is fine to eat on a Tuesday. It didn't consider *your truth*. If this were to happen to you, these government and religious moral codes of right and wrong become moral codes you *tell yourself* to believe to be true and right – a concept that might feel pretty familiar by this point.

These rule systems often work fine for many of us, until someone - perhaps you - performs an action these sources declare to be absolutely *wrong* - such as eating ice cream or stealing a candy bar. Then they are the equivalent of Bob on the treadmill only ever screaming that you are wrong, and refusing to listen to you. And just like that invalidating treadmill situation, this forces you into a terrible position. Should the injustice be big enough, one of three familiar things will happen.

1. Collapse - a rejection of yourself. This is the invalidation of your own thoughts and your own truth within your own mind, done to stay in the good graces of your chosen religion or government. The mental consequences of this are huge, as you attempt to reject the part of you responsible for your actions and create a "second you" that complies with your new Keepers of Truth.

Collapsing may lead you to accept and internalize the story you are damned for eternity. You may try to stabilize your mental boundary from continued collapse by obsessively making sure you never mess up again, which is asking these one-way 'objective' sources of right and wrong (and not your own natural-and-relative sense of right and wrong) to start dictating the truth, and thus the choices available in your life.

But, just like trying to override breathing, you have to keep *telling yourself* the rules *constantly*. You have to tear your mind almost in half to keep track of what these authorities say is right and wrong for you. You have to determine if they would approve or disapprove of your actions before you justify them.

Some organized religions have tried to prevent and ease this internal suffering by allowing "confession and forgiveness," and some governments will allow you to plead "guilty for leniency". Both of these allow you to take responsibility and authority over your actions that weren't following their 'objective' moral code, in exchange for less consequence. Of course, you may still get a punishment of some kind, and are *still* expected to follow the same unchanging moral code afterwards, which means the Hydra called Second You and the cycle of self-invalidation are still there, waiting to repeat all over again.

These 'moral codes' tend to never change in the face of disagreement. They put those in positions of authority to be the equivalent of Treadmill Bob forgiving you for being wrong. "Just don't ever be wrong again."

2. Stabilize/control - Create distance between you and the authority, and control that space like crazy. This might be a complete rejection of religion and the higher power, or government and its documents entirely. In my opinion, this typically results in an atheistic or anarchist mindset, because those that still conform to these authority structures and moral codes are now who is wrong *to you*.

You might dislike anyone that worships the great and powerful Bob, or anyone that accepts the laws and the system of government responsible for what happened to you. You might also take more of a let-and-let-live approach, and try to keep their beliefs as far away as possible.

This option allows you to end your submission to these Keepers of Truth inside your mental boundary. You can be free to act on your own sense of right and wrong, though you may still very much wrestle with your feelings around 'being wrong' relative to a claimed 'objective' truth that seems much more believed by others than your own truth.

There is a catch with this option: Because the religion or government still exists, any attempt to keep them away from you is not likely to work. You are stuck in a loop, forever coming face to face with the beliefs, moral codes, authorities, and those that have collapsed their mental boundaries into these stories again and again.

This is like constantly asking Treadmill Bob questions in hopes that he'll quit saying you are wrong before it's too late. But in the case of organized religions and governments, "Bob" is literally anyone that belongs to and accepts these systems. "C'mon you people! Don't you think eating ice cream is *fine*? Such a stupid rule!"

Even if you convince a person that still accepts your ex-Keeper of Truth to agree you weren't wrong, there are thousands more for you to convince of this every day, and more people collapse into these authorities all the time, adding to the masses that tell themselves you were wrong and you knew it when you did it.

3. Expansion - This is violence, rebellion, or a forceful split from your organized religion or government, *not necessarily* an abandonment of the higher power or source documents (like constitutions and holy books).

If your unwillingness to accept "being wrong" is strong enough, you might create a new religious belief system based on the very same Bob Files and the very same great and powerful Bob, except this time *you* will make it a rule everyone can eat ice cream whenever the heck they want. You are essentially 'canceling out' the authority that tried to invalidate you. New high religious or governmental authorities will be born (you), and you will claim to have the *right* interpretation of the exact same holy documents or governing documents, and know Bob's or the Founding Fathers' *true* intentions, because neither ever said anything about ice cream. *And you would not be wrong.*

This is actually how different belief systems under the same higher power were, and are, born. This is how Protestantism got started out of Catholicism, and how the United States got started from Britain.

This is the equivalent of you punching Treadmill Bob in the face for invalidating what you believed was right, and inevitably creating a massive amount of "noise" between you until one of you finally falls behind, releasing the angry birds.

If you've ever done something that was illegal in the eyes of government, regardless of if you were caught or not, that forbidden action was right for you in

26

Smudging the Looking Glass

What do religious authority, government authority, and educational author- have in common when it comes to claiming they are *Keepers of Truth and iders of Right*?

All these systems follow and rule over you through interpretations of *uments*.

The government authority structures seem to exist because of the Constitu- n, organized religious authority structures seem to exist because of holy books, d educational authority structures seem to exist because of laws and policies d curriculums.

All these documents have an easily found common connection: Words. itten words in particular seem to be the tool providing the authority to the dividuals in these structures, right? So who is the *Keeper of Language and cider of the Right Words*?

Who decides what words mean for everyone else?

If words have exact agreed-upon meanings (which was one of those simple ries you might believe to be the whole story from earlier in this book), then if u somehow disagreed with the "objective" authority of right and wrong provid- by a constitution, laws, a holy book, or school policy, then these authorities ust be using the wrong meanings for those words, even though they seem to lieve they are right! *Right*?

It might make sense to go to the one place most of us go when we encounter vord we don't know: A dictionary.

Inside the pages (or website) of a dictionary, all words appear to have *very* ecific meanings. It *makes* sense to think that whatever the dictionary defines right and wrong are bound to show society the *right* meaning and how to erate and follow *objective* truth.

So, what's the dictionary's definition for right and wrong?

From Dictionary.com in the year 2023:

the moment, and the government and their documents and laws were no longer the *Deciders of Objective Right and Wrong* for you.

If you've ever sinned based on your chosen religion's rules, whether anyone knows about it or not, that action was right for you at that moment, and that religion and its documents and rules were no longer the *Deciders of Objective Right and Wrong* for you.

And since a second truth can and does exist in objective reality, these moral codes are not and cannot be *objectively* right.

Both of these systems are attempts to be *Controller of the Treadmill*. These systems of individuals will not and cannot get on the treadmill with you. They are built around one-way communications. You are subject to their unmoving rules; they are not subject to your feelings about those rules.

It should be no surprise then, that both of these systems have had to make changes and additions to stop destroying the well being of nearly all of their followers. Instead of "forgiveness," some go so far as to blame a "criminal mind" or "evil" as a way to free their followers from responsibility over their actions that keep violating their unmoving objectively-right moral code. '*It wasn't you, it was something beyond your control that took over your actions!*'

In my opinion, this actually *encourages* the splitting of your mind, giving you a way to almost completely invalidate First You and reject control over your actions. *It wasn't you, it was "evil" justifying your actions.* This implies divorcing your naturally-built truth from your actions is not only possible, but it is *the right thing to do.*

There is a long history of very brutal and horrific ways these same systems have tried to *physically* remove this "evil" or "wrong" part of *you*. They were physically trying to remove "First You" from the driver's seat and put "Second You" in its place. The result was almost always the destruction of the "evil" person or the "wrong" person's mind.

What these "purification" processes were actually trying to remove all along was relativity and subjectivity, and force objectivity and total compliance to the unmoving rules in its place.

So, after falling through organized religion and government, what's the next layer of *objective* rules?

25

Grab Your Sharpened #2 Pencil

Schools might just be where you and I experience (or experienced) the idea of an absolute right and wrong the most growing up. Individuals are often told they are in the wrong place, doing the wrong thing, answering the question wrong, and are wearing the wrong thing.

I feel it would be a bit disrespectful to your time to walk through a fully imagined situation with educational systems in the same way I did with religion and government – because it is literally the same authority structure repeated all over again. I'll give a quick and dirty rundown instead.

The school has rules for general behavior created by the principal and administrators, the classroom has specific rules created by the teacher, and the bus has specific rules created by the driver.

A teacher or another student is typically the first to pronounce you wrong to the Boss, and this judgment can then be "appealed" up the chain as needed. You can appeal up as high as the principal on your own as a student, assuming you are under 18 years old in the United States. If that person still believes the story you were wrong makes the most sense, then you are stuck dealing with the consequences connected to breaking the rule with very few options left.

Which places you before the same familiar three options: Collapse, stabilize, or expand.

If the decision was unjust enough though, there is a way to escalate the appeal process. You can move up the hierarchy on *your* side of the situation, should it be available: Involving your caretakers.

Your caretakers can be your "lawyers" that advocate for you at higher level than you can reach as a student, and can take the appeal process above the school principal and into the school administration and school board. In the United States, this would be moving from the level of the individual school into the district level.

After that, they too will end up at the crossroads of three paths: collapse, stabilize, or expand. Perhaps they will give up and tell you to just deal with the consequences. Perhaps they will choose to keep fighting or appeal to the public through media and networking. Perhaps they will explode into violence and face

arrest themselves, or they might attempt to "overthrow" the elected onto that board themselves so they can have the po

From there, should your caretakers *still* not succeed, appeal beyond the school board and cross into the realm of to control education systems in their county, state, or cou are other possibilities available in each direction – this is ju

If you are capable of believing a truth-story that creates wrong from the 'absolute' rules of 'objective' right or wrong as needing to go to the restroom when you are not allowed, it is not allowed, wearing shorts that are one inch too sho wanting to answer the teacher's question when it is not your t of the education system cannot be the *Keeper of Truth and Right* for everyone, either. They cannot be the objective truth

So why are these top layers such a strong trap for so r themselves or force themselves to follow, which might includ believe what you think you should believe.

You aren't wrong. Everyone else isn't wrong, either. Ther ground at these upper levels without forcing someone to believ the moment, and that is not how human truth, right, and wro

We're going to have to keep tumbling now.

the moment, and the government and their documents and laws were no longer the *Deciders of Objective Right and Wrong* for you.

If you've ever sinned based on your chosen religion's rules, whether anyone knows about it or not, that action was right for you at that moment, and that religion and its documents and rules were no longer the *Deciders of Objective Right and Wrong* for you.

And since a second truth can and does exist in objective reality, these moral codes are not and cannot be *objectively* right.

Both of these systems are attempts to be *Controller of the Treadmill*. These systems of individuals will not and cannot get on the treadmill with you. They are built around one-way communications. You are subject to their unmoving rules; they are not subject to your feelings about those rules.

It should be no surprise then, that both of these systems have had to make changes and additions to stop destroying the well being of nearly all of their followers. Instead of "forgiveness," some go so far as to blame a "criminal mind" or "evil" as a way to free their followers from responsibility over their actions that keep violating their unmoving objectively-right moral code. *'It wasn't you, it was something beyond your control that took over your actions!'*

In my opinion, this actually *encourages* the splitting of your mind, giving you a way to almost completely invalidate First You and reject control over your actions. *It wasn't you, it was "evil" justifying your actions.* This implies divorcing your naturally-built truth from your actions is not only possible, but it is *the right thing to do*.

There is a long history of very brutal and horrific ways these same systems have tried to *physically* remove this "evil" or "wrong" part of *you*. They were physically trying to remove "First You" from the driver's seat and put "Second You" in its place. The result was almost always the destruction of the "evil" person or the "wrong" person's mind.

What these "purification" processes were actually trying to remove all along was relativity and subjectivity, and force objectivity and total compliance to the unmoving rules in its place.

So, after falling through organized religion and government, what's the next layer of *objective* rules?

Grab Your Sharpened #2 Pencil

Schools might just be where you and I experience (or experienced) the idea of an absolute right and wrong the most growing up. Individuals are often told they are in the wrong place, doing the wrong thing, answering the question wrong, and are wearing the wrong thing.

I feel it would be a bit disrespectful to your time to walk through a fully imagined situation with educational systems in the same way I did with religion and government – because it is literally the same authority structure repeated all over again. I'll give a quick and dirty rundown instead.

The school has rules for general behavior created by the principal and administrators, the classroom has specific rules created by the teacher, and the bus has specific rules created by the driver.

A teacher or another student is typically the first to pronounce you wrong to the Boss, and this judgment can then be "appealed" up the chain as needed. You can appeal up as high as the principal on your own as a student, assuming you are under 18 years old in the United States. If that person still believes the story you were wrong makes the most sense, then you are stuck dealing with the consequences connected to breaking the rule with very few options left.

Which places you before the same familiar three options: Collapse, stabilize, or expand.

If the decision was unjust enough though, there is a way to escalate the appeal process. You can move up the hierarchy on *your* side of the situation, should it be available: Involving your caretakers.

Your caretakers can be your "lawyers" that advocate for you at higher level than you can reach as a student, and can take the appeal process above the school principal and into the school administration and school board. In the United States, this would be moving from the level of the individual school into the district level.

After that, they too will end up at the crossroads of three paths: collapse, stabilize, or expand. Perhaps they will give up and tell you to just deal with the consequences. Perhaps they will choose to keep fighting or appeal to the public through media and networking. Perhaps they will explode into violence and face arrest themselves, or they might attempt to "overthrow" the school board, or get elected onto that board themselves so they can have the power to change things.

From there, should your caretakers *still* not succeed, another option is to appeal beyond the school board and cross into the realm of government in order to control education systems in their county, state, or country. I'm sure there are other possibilities available in each direction – this is just a quick overview.

If you are capable of believing a truth-story that creates a different right and wrong from the 'absolute' rules of 'objective' right or wrong in the school – such as needing to go to the restroom when you are not allowed, chewing gum when it is not allowed, wearing shorts that are one inch too short to be allowed, or wanting to answer the teacher's question when it is not your turn – then the rules of the education system cannot be the *Keeper of Truth and Decider of What is Right* for everyone, either. They cannot be the objective truth between you and I.

So why are these top layers such a strong trap for so many people to tell themselves or force themselves to follow, which might include you? You want to believe what you think you should believe.

You aren't wrong. Everyone else isn't wrong, either. There can't be common ground at these upper levels without forcing someone to believe they are wrong in the moment, and that is not how human truth, right, and wrong naturally work.

We're going to have to keep tumbling now.

Smudging the Looking Glass

What do religious authority, government authority, and educational authority have in common when it comes to claiming they are *Keepers of Truth and Deciders of Right*?

All these systems follow and rule over you through interpretations of *documents*.

The government authority structures seem to exist because of the Constitution, organized religious authority structures seem to exist because of holy books, and educational authority structures seem to exist because of laws and policies and curriculums.

All these documents have an easily found common connection: Words. Written words in particular seem to be the tool providing the authority to the individuals in these structures, right? So who is the *Keeper of Language and Decider of the Right Words*?

Who decides what words mean for everyone else?

If words have exact agreed-upon meanings (which was one of those simple stories you might believe to be the whole story from earlier in this book), then if you somehow disagreed with the "objective" authority of right and wrong provided by a constitution, laws, a holy book, or school policy, then these authorities must be using the wrong meanings for those words, even though they seem to believe they are right! *Right?*

It might make sense to go to the one place most of us go when we encounter a word we don't know: A dictionary.

Inside the pages (or website) of a dictionary, all words appear to have *very* specific meanings. It *makes* sense to think that whatever the dictionary defines as right and wrong are bound to show society the *right* meaning and how to operate and follow *objective* truth.

So, what's the dictionary's definition for right and wrong?

From Dictionary.com in the year 2023:

Right	(adjective)	1. Morally good, justified, or acceptable
		2. true or correct as a fact
	(adverb)	1. To the furthest or most complete extent or degree (used for emphasis).
		2. Correctly
	(noun)	1. That which is morally correct, just or honorable
		2. a moral or legal entitlement to have or obtain something or to act in a certain way.
Wrong	(adjective)	1. Not correct or true; incorrect.
		2. Unjust, dishonest, or immoral.
	(adverb)	in an unsuitable or undesirable manner or direction
	(noun)	an unjust, dishonest, or immoral action
	(verb)	act unjustly or dishonestly toward

What do you get from these? Any general ideas you are taking away?

As for me, I get that *right* is morally good, justified, and acceptable, and *wrong* is unjust, dishonest, or immoral.

Welp, that settles it right? Let's pack it up cause we are all done here! You can now just take a look in the mirror and say "Oh yeah, *that* belief I just had is unjust, and *this* one is dishonest, oh and *that* thought is immoral." The dictionary solved everything!

Your current belief of what is true is whatever story makes the most sense to you. You act on what you believe is true and the right thing to do *at all times*. I know I've said this many times already but I'm going to keep repeating it.

Who then decides what is just and unjust, good and bad, dishonest and honest, acceptable and unacceptable, or moral and immoral? Does this put us back in the exact same position we started with in Chapter 1, searching for a keeper of these things?

Let's see if the dictionary can keep providing answers by looking up the definitions for the words contained within the definition for *right* and *wrong*, which were:

Moral (adjective)	concerned with the principles of right and wrong behavior and the goodness or badness of human character.
Good (noun)	that which is morally right; righteousness
Just (adjective)	based on or behaving according to what is morally right and fair.
Acceptable (adj)	able to be agreed on; suitable
Unjust (adjective)	not based on or behaving according to what is morally right and fair
Dishonest (adj)	behaving or prone to behave in an untrust worthy or fraudulent way
Immoral (adj)	not conforming to accepted standards of morality

I'm getting the sense that "good, just, and acceptable" generally means to be morally right, and "unjust, dishonest, and immoral" seems to be morally *not* right. Ok, what's the definition of "morality?"

Morality (noun)	principles concerning the distinction between right and wrong or good and bad behavior.

Wait....

Right is good, good is moral, and moral is.....right?

Wrong is bad, bad is immoral, and immoral is .....wrong?

We are trapped *in a loop* back to the beginning, which leads back to the same authorities and documents of the upper layers to define what is just, unjust, acceptable, wrong, dishonest, and immoral in the first place! But I've already shown how the government and religious organizations cannot be the *objective* deciders of right and wrong for both you and me at the same time. They can't be the *Keeper of the Truth and Decider of Right* for everyone.

How in the heck were these definitions created in the first place? Why aren't they providing the answers we need to find agreement?

The answer, if you've never thought about this before, can be a bit of a wild ride.

On dictionary.com's website in 2023 a page existed with the title "How New Words Get Added To Dictionary.com-And How The Dictionary Works" (https://www.dictionary.com/e/getting-words-into-dictionaries/). Here's an excerpt:

> "Language is a living thing, and so is Dictionary.com."
>
> "A word doesn't become a "real word" when it's added to the dictionary. It's actually the other way around: we add words to the dictionary because they're real – because they're really used by real people in the real world.
>
> In other words, our lexicographers add a word to the dictionary when they determine that:
> 1. It's a word that's used by a lot of people.
> 2. It's used by those people in largely the same way.
> 3. It's likely to stick around.
> 4. And it's useful for a general audience.
>
> All four of these points are important. Our lexicographers look for use not just by one person, but by a lot of people. Of course, many words have different shades of meaning for different people. But to be added to the dictionary, a word must have a shared meaning (that is, it must communicate a widely agreed-upon meaning from one person to the next.) If everyone used a word in a completely different way, we wouldn't be able to give it a definition, right?"
>
> "We must acknowledge that, historically, dictionaries have been gatekeepers to nonstandard words and usage (especially those that originate in non-dominant groups), but we at Dictionary.com take very seriously our role and responsibility in ensuring that our dictionary reflects and respects the language of people as they use it."

I am sure, if you are a lexicographer, this is nothing new for you. But I myself struggled to realize the deep implications this presents until I was well into adulthood. I knew words were flexible, and I knew I could play with meanings. But I still never thought about *how* a word got defined. I always saw the definitions in the dictionary as the concrete meaning of words we all agreed to if it got serious. I saw it *exactly backwards* from how it really was.

If this seems confusing, I'll try to help. Here's how I imagine this process to go:

Let's say I make up a new word right now, like "lexicorelativity."

To become a word in the dictionary, first it would have to be widely used beyond this book. Maybe say...*you* use it somewhere after reading this. People that heard you using this word start using it as well, and so on. The word has to spread enough that a *dictionary employee or searching program* becomes aware this new word exists at all. This will depend *highly* on the method used to look for words in the first place.

Next, lexicorelativity has to show up, or be "found" by a dictionary employee or program enough times to "appear to be" widely used. Which will *again* depend on the method used to look for words in the first place, and the definition chosen for "widely used."

Then there is the art of it all: Does lexicorelativity mean the same thing between people? The lexicographer at the dictionary has to *interpret* what I *intend* it to mean and what you *intend* it to mean to answer this. That person then has to determine if the word is going to be around for a while in the future (meaning its appearance is expected to keep happening, which would *again* be based on how he or she are looking for words in the first place), and also decide if it is a useful word at all (what authority gets to decide that?).

So, as I understand this, the dictionary's definition of a word is like taking 5,000 customer feedback comments of a product, and then trying to decide how to write a "master" review from everyone's opinion. I'm sure there is a very strict set of rules and regulations for how this is done. I don't know any of them. Regardless, I think I would be very good at this. Check it out:

> "This product is the worst best product that ever existed and should not exist. It should be destroyed and everyone should have one. It arrived both broken and in perfect condition. I would definitely buy it again, never."

I am sure that every lexicographer or dictionary manager that reads this is going to be pounding down my door soon. I expect to be receiving calls with job offers from every dictionary in the world non-stop for -----ahem. Sorry, where were we?

Oh right, this can get boring, so let's go ahead and make this into a game you can play with friends. I think I'll call it…..

Now I Am Become Human, Destroyer of Words

Step 1:   Find another person (or persons) to play this game with you. You will need a method to record words in a permanent way outside of your own head. This could be a pen and paper, or a digital notepad, for example.

(If you do not have a person you can play with, I will now attempt to play with you from back here in the past when I wrote these words.)

Step 2:   Using group conversation, work together to agree to a single "target" word you would like to destroy. It can be any word at all, but be careful to not accidentally define it while agreeing to it. Tip: If you want to really make the game more dramatic, choose a word that describes groups of people.

(I'll choose the target word "orange" for our game.)

Step 3:   Once the target word is chosen, have everyone secretly record the meaning of the target word using the method from Step 1 (pen/paper, digital notepad, etc). Try to be as accurate as possible. *No dictionaries or outside resources can be used!* Everyone's answer needs to be hidden from others, and needs to be recorded *outside* of their mind.

(As for the game between me and you: Please go ahead and define the word "orange" and record your definition outside of your mental boundary. I wrote this in the past so my definition is already written in the next step. Don't read any further without finishing this step!)

Step 4:   After everyone has the target word's definition recorded, have everyone share what they wrote. Enjoy the comparisons.

(My definition of orange: "a round segmented fruit that grows on trees in warm locations." Were we close? Did you even choose the fruit, or did you attempt to define the color instead?)

Step 5:   The nuclear option: In open conversation, decide whose definition is *objectively right*. If you can't reach a decision, repeat Step 3 with the word "right."

I am pretty sure there will be differences between everyone's definitions, but I also know it is possible (but rare) for people to match their definitions word-for-word. If this happens, you can break this by choosing a target word from the matching definitions and starting again at Step 3. After this, I'm almost certain there will be no exact matches remaining.

I've played this game with my kids a few times. There is only one other outcome I have experienced playing this game: Getting stuck in a definition *loop*.

This is the reason I added the last part of Step 5, because my daughter and I ended up playing the game with the word "right" in order to settle our debate over who had the best meaning for the target word we had chosen. We both ended up writing exactly "to be correct".

When we repeated the game for the meaning of the word "correct", we ended up both writing "to be right," which created a perfect *loop* between "correct" and "right".

Loops between word definitions *cannot actually exist in reality*. Words that loop between each other to define each other have no meaning at all. They become a *paradox*. If correct is right, and right is correct, then no one is capable of defining either of these words, because they are defining themselves.

This is exactly what happened previously when we tried to define right and wrong using the dictionary as a source of objective truth.

I would be fascinated to hear of any story where this game kept going with exact matches for word meanings beyond one round, and didn't result in a loop trap. I am not claiming this is impossible, but if exact matching definitions happened twice, I would expect to find some limiting rule I didn't provide to the game – something like "define the target word using only one word." This, in my opinion, is not being "as accurate as possible" as described in Step 3 of the game.

Here's another example from when I first played around with the idea of this game with some friends. I asked a simple question: "In the fewest words possible, define "Republicrat" (I have changed the specific party member word, because the point I'm about to make has nothing to do with the party). The answers were *not even close* to each other, which is why I recommended using a target word for groups of people to add extra drama. Here's a few definitions I received:

1. "Altruistic"

2. "Believes that government should be part of the solution, not most of the problem."

3. "For the people."

4. "A Republicrat is a member of the Republicrat Party, in the strictest sense. In common usage, it means a million things."

5. "Progressive"

6. "A person who feels fellowship with the image or ideal they see in that party. (Either one)."

7. "....traditionally been members of a political party that places greater emphasis on the community and the ways government can help to lift everyone up."

8. "It's a brand (and as a brand by its nature must mean different things to different people.)"

Is there a common theme here? I'm actually not entirely sure, I feel like I'd need a few more quotes to build a definition if I was a lexicographer working at the dictionary. (I definitely have a lot more respect for what they do now, that's for sure.)

One thing is absolutely apparent to me, though: All word meanings are individualized between people and this means they are *approximate at best* between us, even when we think we mean the same thing by the same words. We are only barely communicating at all.

I believe if I pushed my friends into the now-refined Destroyer of Words game, they would eventually end up stuck in a loop with another word just like my daughter and I did trying to define "right".

But here's the troublesome part: I believe it's safe to assume when everyone in this group of friends (me included) reads the *same article* that refers to Republicrats, we will all *believe we agree* on what this word means. It's smoke and mirrors. It is all an illusion created through language itself - perhaps I'll call it the Dictionary Illusion. Believing this is true is the same as believing everyone that answers "the 747 will not take off from the treadmill" believes the exact same story of *why* it is true in their head. There is not only one possibility for each word.

I believe this Dictionary Illusion is a big part of why individuals, and possibly you, suddenly can switch to being *certain* of what is right and what is wrong. It might feel like 60 million people agree with *your* words, but the moment you believe any word has a solid and unmoving meaning, you enter a trap built by an absolutely *massive* assumption.

You will *immediately* become stuck in a feedback loop from hell based on that word's meaning being unmoving, objective, absolute truth, whether you realize you've done this or not. It is a self-reinforcing process based on the belief that *your* meaning for the word is *the whole story*.

This illusion can be so strong and self-validating it can push you or others to seek some position of authority based solely on believing it is objective truth. You can end up telling yourself *you* should be the creator of moral laws and rules for everyone else built on *your* beliefs of what words are supposed to mean. Everyone else *must be wrong or confused*.

And since any word almost *always* means different things to different people, this means *words themselves* cannot be a source of common ground and objective truth between you and me. Words themselves do not and cannot have a one-way, objective, solid, unmoving meaning, which means it's *not possible* for any single word or collection of words to include all perspectives and all points of truth.

Another truth is possible at all times for all words.

This means the dictionary, and even words themselves, cannot be the *Keeper of Truth and Decider of What is Right For You and I*.

You and I cannot build shared foundation using words at all.

So who, or what, is left?

Deeper we fall.

27

We the Assumed

Now that the looking glass of language has a smudge on it, and you may or may not realize words are not likely to provide a shared objective meaning at all between any two people, let's go back to the question that started this whole mess: *Who do you allow to call you wrong?*

Government or laws or founding documents? All built around words. Who gets to decide what those words mean?

Organized religion or holy books? All built around words. Who gets to decide what those words mean?

Educational and corporate institutions? All built around words. Who gets to decide what those words mean?

No matter how much you want these authority's truth-stories made of words to be what *you believe* deep down as true, you can only ever "tell yourself" to believe their truth-stories. Why? Because words having exact shared meanings *is not actually possible*. It is not how language or words work.

Not realizing this is the mechanism that quickly flips everything upside down. Language itself is the looking glass, and you are unable to peer around the edge and see things right side up for hidden reasons. Language itself is the Great Barrier, because it expands to destroy all other boundaries and form its own super-boundary.

Who gets to decide what the words of these "objective truth" documents mean *for you*? The answer without fault is *you*. It's you. It can only *ever be* you.

And you are not wrong.

I don't care what you believe is true, it is not wrong *to you*.

What *you* believe the words to mean in these documents is what is true and right from your perspective. You are only *one point* in this moment, so the meaning of each word for you at this moment feels like *objective truth*. For this reason, I would be a fool to try and tell you your reality built through language and communication is wrong. But that is exactly what is happening all over the globe today as I write this.

There is a huge and invisible battle raging over which words are right, and which words are wrong, like words have some exact and agreed-upon definition *outside* of our own heads that can or *should* win over all the other meanings to become the objective all-points-included truth and the whole story.

Words like government, deep state, Democrats, Republicans, election security, fraud, social media, lies, truth, Supreme Court, MAGA, wokeism, liberalism, conservatism, Fox News, CNN, stupid people, millennials, boomers, smart people, scientists, transgender, drag queens, gays, racism, black people, white people, illegal immigrants, guns, gun control, gun promotion, mental illness, mental healthcare, Critical Race Theory, school indoctrination, lack of teacher support, money, profits, stupidity, science, religion, lack of religion, absence of god, the devil, police, broken healthcare, wealthy, poor, union, Black Lives Matter, Antifa, immoral, It's the loss of morality. It's confusion over what is right and wrong...and thousands more.

But no one person is born with natural authority over another. No one gets to declare the objectively right meaning of *any* word. There is no objective unmoving meaning for the words *right* and *wrong*. There is no objective unmoving meaning for any word in the list of words above, and no objective unmoving meaning for each and every word you've read so far in this book. There never was and there never will be.

The meaning for *all* these words was and is being injected *by you*. And because other people use the *same words*, and the words already existed and were in use long before you and I learned them, and because they often seem to mean *approximately* the same thing, you and I all-too-often just *assume* other people mean exactly the same thing with the same word, or worse: You and I may assume other people know what the word means *better than you or I do*. You may be letting someone else be the authority of what your words mean *to you – a Keeper of Words*.

Words themselves *are the limited answers* on the great treadmill of life.

The assumption words have some massively "agreed upon" exact meaning can and does choke out many of the very real possibilities for both you and I. It can lead you and I to argue with each other and get trapped inside loops of words that keep defining *each other* without even realizing it is happening, because these loops might be twenty or even *three-hundred* definitions around.

The result is the two of us never being able to get off the poultry treadmill. We would be stuck trying not to fall too far back on the belt, or trying to escape, control, or break everything around us (including each other) to be the one controlling the meaning of words that *never had any fixed meaning in the first place*, just so we can get off of this awful contraption to avoid falling into the "rooster" abyss.

Unfortunately, in a panic to get off the treadmill, we seem to end up in two positions: Declaring word meanings as being *certain*, or collapsing them completely. We seem eager and willing to demand our personal words are *right*, or accept our personal words are *wrong*. We seem to often make the *assumption* the personal meaning we each inject into our word choices is in perfect sync with the "agreed-upon" definitions for them – we assume we are using the dictionary-validated meanings – or we just don't know what words mean at all. And these two work perfectly together to solidify the Dictionary Illusion.

To get anywhere at all, we have to break free of these language traps – of this illusion.

In my opinion, a statement of absolute certainty about anything that exist *outside* of your own head, which is where your words travel, is exactly the trigger for a runaway process. Statements of certainty *imply* within their words lay the whole story and objective truth. Your words then stand apart *from you* – they become their own *separate* living thing. But this is all illusion.

Example: "Action A *is* good."

This statement becomes its own anchor point for both you *and* others. If and when you claim this, you are basically calling for everyone to come see the new objective truth you just constructed. "Look everyone! Action A *is* good!" This foundation made from words you uttered becomes a *second point* for everyone *including yourself*, but you will often still consider the words to be attached to you.

This is because they are attached to you. Each of your word's deepest and truest meanings can only ever be understood from *inside your own mental boundary*. A place I (and everyone else) cannot go. Only *you* know the reason *why* "action A *is* good" to you, and you will always understand that reason better than anyone else. You trigger the runaway process when you make the all-directional assumption that everyone else understands *Action A is good* in exactly the same ways you do.

This assumption - the Dictionary Illusion - is easy to make because of a simple-but-huge problem: Your mind is nearly *infinite* in possibilities, just like the objective universe seems to be, but your vocabulary to communicate those possibilities is *extremely limited* in comparison.

Trying to declare absolute right and wrong truth and actions using words is a lot like jumping to an answer - to a certainty - without even attempting to discuss the story underneath. It's not caring (or perhaps not noticing), the enormous blind spots and gaps that spring into existence by choosing not to fall down the rabbit hole and climb back up again – and instead believing you've filled it in or that it is safely covered up.

The entirety of language when compared to the entirety of objective truth is like saying *the plane doesn't take off*, and then lacking the ability to discuss *anything else*. This causes language to collapse reality to look like it only offers two possibilities, because there are *only* two answers possible. A thing is, or it is not. A thing does, or it does not. *One string of words is objectively right and the other word string does not match, so it must be objectively wrong.*

All of this works up until the point it doesn't. Up until the point of a language-constructed abyssal disaster. I believe humanity is on the threshold of such a disaster as I write these words.

As I see it, words stop working for us and switch to working *against* us when the meaning of the word *right* is split apart by a certainty that makes a lot of sense to a lot of people, but is not *actually* the whole story.

Certainty of the meaning of *right*, expressed as words which are then believed by *multiple* people to mean exactly the same thing (who validate the Dictionary Illusion assumption into truth for each other), start then changing the definition of what is *true*. What comes next might feel eerily familiar, as *right* starts to justify what is *true*:

"It is right because it makes sense to me and is absolutely certain. Something can only be true if it is right. If it doesn't make sense, it is untrue. If it is not true, then it is *wrong*."

Unfortunately this language trap, created by a certainty, quickly expands to include other sources of words – other *people*. It eventually ends up in the positions from Chapter 1 of this book:

"*Your* words don't make sense to me, so they are not right. You insist your words make sense, but they do not, you are wrong and now *you know you are wrong since you continue to say those words*. You are not changing your words to be *right*. You are now *intentionally* being wrong. *Being intentionally wrong is evil.* You must now be stopped from using words at all. *We* are right. *We* make sense. *We* are the *Keepers of Truth*."

Welcome to hell, and the state of reality as I see it at the time I wrote this.

There is no "we" when it comes to the exact meaning of any word, just like there is no "we" sharing everything inside your mental boundary. There never was. No one understands any word to mean exactly the same thing. "Agreed upon" word meanings are *always* an assumption, and an absolutely massive one, which gets easily validated by the Dictionary Illusion seeming to offer fixed, unmoving meanings. They all create loops and monsters.

You have your own personal sense of truth, right, and wrong, and it is far stronger than words will ever be. It is actually what gives meanings to all of your words.

There is no common ground between us here. It's always possible that you or I will force *the other* to be wrong any time one of us makes any statement of certainty.

We have to separate truth from words entirely.

Our tumble continues.

There is still a long, long way to fall.

28

The Unflippening

No word means exactly the same thing between any two people. That includes you and me.

These differences in meanings between us for the same word are caused by something called *context*. Here's the *approximate* meaning for the word context from Dictionary.com in 2023:

> **Context** (noun)   the circumstances that form the setting for an event, statement, or idea, and in terms of which it can be fully understood and assessed.

So, what are "the circumstances" of a "statement"?

As I see it, context is the set of *norms* and *assumptions* that existed at the moment the words were communicated. Having the *whole* context is the only way words can be "fully" understood. Anything less is using assumptions.

Where, though, are those norms and assumptions for words actually residing? *They exist entirely inside the mind of the person that made the statement.* And you can't read anyone's mind but your own. There are very *real* mental boundaries in place here.

Certainty of the context that provides meaning to someone else's words *does not exist* for you or I. This means someone else's words will *never* be fully understood by you or I.

You've already experienced this truth yourself when you played "Destroyer of Words" with me (or anyone else you played it with). Everyone defines words by pulling from their own unique context. In our game with the word "orange", if I believed choosing the fruit to be the right meaning and you believed choosing the color to be the right meaning, then not only did the words of our definitions not match, but we chose an entirely separate set of norms and assumptions *underneath* what we imagined that word to mean. You and I used entirely *different* contexts for the *same* word.

To pull back to a bigger picture, this means objective, all-way right and wrong truth and actions for you and me cannot be decided by the government or laws or founding documents, it can't be decided by organized religious organizations or holy books, or by schools or caretakers, or by the dictionary, or even through verbal or written or signed communication, because right and wrong can't be objectively defined by words at all.

This is because *all* words, which are the basis of verbal communication, can only be fully understood by having the *entire context* used to express them. This means the 'real' or 'entire' definition of the words in documents and rules and definitions is *forever* hidden.

I have a different context than you. You have a different context than me. Bob has a different context than either of us. This is why if we played the Destroyer of Words game, we would come up with approximate shared meanings *at best*, and totally different meanings at worst.

Everyone who reads this book will understand the words in it differently because their context is unique. If everyone understood words to mean exactly the same thing, there would be nothing to talk about after reading *any* book.

This means *I* am the only person that can clarify the full-and-intended meaning for the words in this book. *Only* me. You would need to have a two-way communication with me to clarify any of the assumptions or norms giving meaning to any of these words if they don't make sense for you, because my context chose the words and the order to put them in, and it exist exclusively inside my mental boundary. The *objectively right* context for this book is *my* context.

Once I'm gone or no longer communicating (or am unable to communicate with you directly), who becomes the rightful keeper of my words' meanings? Whose context becomes the *new* right context for these words?

Oddly enough: *You.*

In fact, *you* have been the one determining what the words mean in this book since you opened it. All the words in this book actually work more like empty shells built from clusters of twenty-six different letters (and a few numbers). I then try to limit and filter the meanings you might use to fill each word-shell the best I can by using all the previous words-shells, which I have been providing to you in a particular order since you read the title.

This "compiling" of context by using previous words to improve the odds of us having approximately the same meanings for the current word you are reading right now is a process being controlled *by me*. I am the author, and I'm trying to put these words in an order that makes the story *you* build with them from *your own* personal context make sense. I try to provide you ever-growing context throughout the book this way. Ultimately though, I know this process all collapses down to the first word I provided you, and what you understand it to mean based on your context.

That's why the very first word-shell you ever received from this book was one I could be *sure* you could completely fill nearly the same way, no matter who you are. This word was the context *anchor* for this entire book. It is the first word on the front cover: *You.*

You, my dearest reader, determine the final meaning of all *my* words *for yourself.* You cannot read my mind and so you cannot possibly ever know all the assumptions and norms behind my words. I can only provide more of my context to you for these word-shells in the form of *more words and maybe a few pictures.* If I start to appeal to my own context - my own assumptions and norms - as *your* truth, I anger or even create a hydra for you if my words are not objectively true for you. This is why me stating what is and is not an assumption about you is so incredibly important.

You might have already noticed the paradox in the room, though: It is *not possible* for me to successfully share my context for the words I put in this book. If I tried to give you my full context for my words by using words, it would take multiple lifetimes to try, and I will still fail, because there *aren't enough words in existence* to accomplish this feat. I would have to redefine the same words multiple times over, and create thousands or perhaps millions or even billions of word definition loops in the process.

You don't have that kind of time, and neither do I. As a result of wanting to communicate *quickly*, you will instead continue filling my word-shells with your context - your norms and assumptions – for what these words mean. This is why I believe using simple language matters. There is a much better chance you already have the similar context to fill my word-shells approximately the same way I do.

For example, if I started the first chapter *assuming* you know what "relativity" means and could fill it with the same meaning I did, we already have a major contextual problem not only for the word, but also every word that comes after that references back to that word.

This being said, if you choose not to communicate with me about this book, and I choose not to ever communicate with anyone about this book, then this is the way things will stay between us. I have my context for these words, and you have yours. Although I would love to learn what they meant to you when you are finished reading, I am okay if I don't, because I am not trying to take authority over your context with my own. I am trying very hard not to tell you what you *should* believe. I do sometimes state certainties, but I try to never state them about you, and I am trying not to leave any certainty behind to become a "word loop trap" for you or me. I fully intend for you to validate for yourself *why* they are certainties (anti-certainty certainties, really), because they are hopefully self-evident. But to do that, we must first get through the looking glass.

However you choose to fill my half-hollow words with meanings from your own context is perfectly fine to me. I chose these words already knowing context works this way. This is why any assumption I make of what you *must* believe, or *should* believe, or *do* believe using these words is absolutely unfair to you. I can't possibly know your context, and if I were to do this, I would be telling you stories about your own context, as if there's any way I could possibly know what's in there. I can't even know who or where or when you are.

No one has any access or authority over anyone else's hidden context - even when face-to-face.

So, if someone calls you wrong on the meaning of words in this book, he or she is possibly trying to claim a certainty of what these word-shells mean. To do this, this person must absolutely take them *out of my intended context*. Since certainties using words about anything other than one's own mind is not possible, this individual is indirectly claiming his or her *own* context has authority over *your* context. After all, both of you filled the *exact same* word-shells. This indirectly makes the claim he or she cannot only read your mind *and* my mind to understand your context (and even mine) filling these word-shells, but that our unknowable contexts are *inferior* to his or her own.

"Eric's words say you are wrong and I am right." *No they did not.* They said *you are not wrong in this moment.*

If this actually happens, somehow my word-shells are being re-contextualized into a certainty to be used as a *source* for power or authority. Given the title I chose for this book, I would hope this does not happen. I do not consent to the idea of a "superior context" for this book - it goes against everything I'm trying to communicate within these pages.

I am providing these word-shells to you, my reader, as a tool to hopefully help you see the hidden world around you that I can see after an extraordinary mental event, and maybe even help stop a bit of human suffering and the enormous cycles of human atrocity. I am going to continue using these word-shells to take you way beyond words, and I have no intent to use them as weapons of tyranny or authority to invalidate your context or anyone else's context. The words in this book exist to empower, enlighten, and entertain you relative *to yourself.* You are the *Keeper of the Truth and Decider of Right* for you in this moment. You always were, and you always will be moving into the future.

Who is capable of providing the *right* context for the words of the Constitution or national documents *for you*? Did the Supreme Court justices write them? No? Then it's *you* that give them the right context for you. No matter who you are.

Who is capable of providing the *right* context for the words in holy documents or religious books for you? Did your religion's leaders write them? No? Then it's *you* that give these books the right context for you. No matter who you are.

In the end, it's *you* reading the word-shells in these books and documents and filling them from your own context, in addition to the very small amount of compiling context provided by the order of the words themselves – which is again always limited by your own context.

If any person claims to know what the words in these documents mean with certainty, including you or me, then that person is indirectly claiming authority over the context that wrote them. That person is claiming his or her own assumptions and norms to be the *right* assumptions and norms for words he or she *did not create.*

The justices would end up claiming they know what the founders meant by the words in the Constitution *with certainty.* Who can claim they know every assumption and norm of a bunch of dead-and-gone people?

The pastors, priests, rabbis and other religious leaders would claim they know what the authors meant by the words in the holy books and documents *with certainty.* Do you believe your god or higher power wrote the words of these documents directly? Do you realize what this might mean if a human claims to know the meanings of these words *with certainty*? What human can claim they know every assumption and norm of *your god or higher power*?!

These individuals in leadership positions might very well have more surrounding words from the time period these documents were written and know more stories related to these documents than you do. This doesn't fix the problem as I see it, it can actually make it far worse.

When a person believes he or she knows more than you about something, it makes that person even more likely to call your context *wrong*. This person might easily become the source of what is right and true for you, which causes you to *tell yourself* what you *should* believe words to mean. If this happens, you end up trying to believe what this *one person in particular* believes the words to mean, because it seems like this person knows what the words mean *better* than you do.

This creates one hell of a predicament and exposes yet another type of definition loop:

If you are the one that decides what words mean for you, and you have always been the decider of what words mean for you, but you didn't create the meaning of any words in the first place, then whose assumptions and norms built the original meaning for any word? Is there some common ground in our contexts used to fill words with meaning we can uncover without forcing someone to believe they are wrong in the moment?

Further through the Great Barrier of language we descend...

29

Refinding Your Feet

Context, or assumptions and norms (which I called "stories" in the earlier chapters), is not only used to give meaning to words. Context is the way you and I have to understand *anything at all* about each other. It's needed to understand words, pictures, feelings, actions – *everything*.

To simplify this idea a bit, you can think of your context as "the story" in your mind for everything that happens. It doesn't matter if it's your own words, pictures, or actions, or someone else's. The stories for each of these are built from *your* memories and *your* associations locked away inside *your* mental boundary.

This is why the meaning for *every single word* is approximate at best between any two people. Your context is literally the stories in *your* memory – *your* past – and you use them to fill up every word-shell in this book and every word you ever encounter in your life. It's what I have been trying to avoid invalidating or implying as wrong while choosing what words to send to you from the past.

Your context is your everything. It *is* valid. It *is* you.

For me to even think about overwriting it or calling it wrong is ridiculous, because you understand me *through* the lens of your own context. In this way, everyone else's verbal or written communication with you is limited to what you have in your memory to make sense of it.

This runs into what appears to be a paradox almost right away: How did your context for words get built in the first place so that you know what meanings go in which word-shells to read these words at all? How do you link meaning to the words "right" and "wrong" to fill them in the first place?

To attempt to explain this, I have to begin pulling you completely through the looking glass of language so we can look at what lies beneath your words. This is going to be extra tricky because I am communicating to you from a book, which is totally made of word-shells that you fill.

As long as you've read everything up to this point and its made sense so far, I believe there is a good chance you have enough compiled context for every word that's about to come at you to make sense.

First though, we have to gather some speed....

30

Unspeakably High-Speed Charades

As we approach the other side of the looking glass of language, I think it's helpful to look back at what words and language accomplish for you. As I see things, words are an attempt to communicate your thoughts outside of your mind. They are not the *only* method you have to do this, though. You can also do this with gestures (body language) or pictures (drawings), but those methods are severely limited when compared to words.

Gesture works only by sight or touch. If you've ever played a game of charades, you realize how hard it is to communicate clearly using only gestures. If you aren't familiar, here's a quick explanation of the game: One person imagines a word or phrase (or draws it from a stack of pre-chosen words). This person then has a limited amount of time to silently "act out" that word or phrase in a way the other players can recognize and yell out the word or phrase.

Now imagine playing charades, but *no one can use words in any form*: No sign language (such as ASL), no writing, no speaking – nothing – *no words at all*.

Players must somehow communicate what they are guessing using *only* gestures, and the person trying to act out something has to somehow confirm the gesture guesses are right or wrong with some *other* gesture. This would be a certifiable nightmare of a game, in my opinion. Charades can be painfully slow and easily frustrating when words *are* allowed, I can't imagine how slow and frustrating it would be without them at all.

Drawings and pictures offer an improvement over gesture as I see it, which is validated by the saying "a picture is worth a 1000 words." But what if you don't have any words?

Imagine trying to communicate with someone entirely with pictures, and you can't confirm anything by blurting out "Oh, that's a house! Got it!" All you can do is nod or smile (back to body gestures), or you develop some system where one person confirms understanding with another picture, which is a system you *somehow* explained and agreed to using only pictures. There is still plenty of room for problems and misunderstandings here, and the process is still very slow. Not to mention this leads into an interesting debate: Are nods and smiles given meaning by words *before* they can be understood through gesture?

With words though, *especially* spoken words, you can communicate your thoughts and ideas with much more clarity *and* speed. You can clear up major assumption-based contextual misunderstandings within *seconds*, like "No I meant orange the fruit, not the color". You can even explain how you *feel* about something, and other complicated concepts with ease compared to pictures and gestures. You can even explain what it is you think inside that mental boundary of yours and why you think it.

Well, at least it *seems* like you can. Until you realize there is another strange invisible loop now sitting in plain sight. What makes language possible at all?

31

The Magically Reappearing Word Loop

Let's say I want you to know something I've got inside my mental boundary, so I put this "something" into words using speech or writing or sign language or morse code or braille, which all are different methods to represent the same language. For the sake of example, use whatever method you are using to consume this book right now.

This conversion from my-thought-into-my-words allows me to send you messages using this method. You then sense this method of communication, and translate what you receive (vocal sounds, letters, gestures, taps) back into words inside your head or out-loud to yourself. After that, your mind does or doesn't make sense of those words.

For this process to work at all though, *you had to already know words. You had to already be organizing and thinking in words for this transfer of a message to even work at all.* For this to work well enough for you to read this book as fast as you are reading it, we have to be using the same *code* – in this case, it would be a shared list of words – a *matching* vocabulary.

But if you *think* with words, and you also *communicate* with words, there is something very weird to consider here: How do you know what any one word means? Does this take us back to the Dictionary Illusion and looping meanings we had before?

Yes. Yes it does.

The way out of this which-came-first-the-thinking-or-the-words paradox will probably seem obvious once you've read the first few paragraphs of the next chapter, but I was surprised how hidden, buried, and complex breaking this loop could be during my first journey this deep into the abyss. The way out of this loop trap hides in a void and is invisible for you, and has its own blind spots hiding their own gaps.

Yet you have probably witnessed the entire paradox-creating process for someone else. You might even *be* the one currently building this loop-trap for someone else.

32

From Nothing to Something to Everything

Imagine you are now a baby again. You have no words at all. Assuming you have a voice, you do make sounds and have been doing so since your first breath.

You have been watching and listening to your caretakers (assuming you have hearing). They make strange bumbling babbling noises. These noises seem to be tied to their actions around you. They make sounds and move their lips when being gentle to calm you, different sounds and movements when feeding you, and still different sounds and movements when being silly with you.

All the screaming and crying, farting and gas, peeing and pooping yourself, and sucking on breasts and bottles were not things you truly had any *conscious* control over. They just happened for you, much like a reflex. They are essentially hard-wired subconscious behaviors – your bootstrap program.

After a while, it's likely you connected your crying to your caretaker's appearance. If this connection is made, and you repeat it *intentionally*, it would be one of the first sound-to-action connections you consciously make. On the flipside, if your caretaker *doesn't* show up when you cry, maybe you connect making a different sound, or making no sound at all to your caretaker appearing (I can't know what your specific life and connections were as a baby, and unfortunately neither can you).

Eventually (again assuming you can hear, see, and speak) you begin to "mimic," or copy the sounds your caretaker makes, because you realize you can control *when* your sounds happen. You can even 'sort of' control how they sound – but the sounds you are making still mean nothing to you. They are strings of sounds, like little mini-songs, which are vaguely connected in time to something that happens in objective reality – whatever that might be: Eating, getting changed, playtime, being held, and so on.

Once you decide you are going to communicate something back through sounds, you try to move your tongue, lips, and vocal cords in various combinations to see what happens. You eventually get enough control over these parts of your body through trial-and-error that you begin to "babble" strings of sounds, and you then try to match these babbles to your caretaker's sounds and actions the best you can.

This works on a spectrum from *not even close* to getting *pretty close*. But you steadily work to match the inflection and tone your caretaker uses in his or her sounds. In my opinion, this grows into the accent you have when you speak today (For example, I tend to have a southern drawl in my speech sounds).

With more time, specific pieces of sounds your caretakers make consistently appear with a specific action, and you consciously notice this. You recognize a pattern and connect the piece of sound and action, and eventually end up uttering your first word in an attempt to either copy the sound related to the action you are focused on, or because you moved your lips, tongue, and vocal cords in the right order to have an unexpected self-discovery event.

The reaction your caretaker has to this will most likely be very significant to you, even though you have no Earthly idea *why* his or her reaction seemed so big, and with that, your journey into shared language begins. This is, in theory, your very first validation through sound.

I think it's worth noting here that one of the first words you probably learned is "no." While "no" can simply be an answer to a question, it quickly gets used in lots of others ways. If you've never considered *why* a child would use the word no with anger and emotion toward you as an adult in their life, now's a good time to ask yourself what your own natural reaction might be if you were told what you wanted to do was wrong.

Now, let's fast-forward just a touch.

You are no longer a baby but instead a young child only beginning to use language. You have very few words you can use or say but *oh do you use the handful that you think you have!*

Imagine an adult is pushing you in a stroller down the sidewalk when you see a small creature like nothing you have ever seen before. It is covered in fuzz with many legs and a tail, and is looking back at you. Now it is coming toward you. You point and get the attention of your caretaker.

Before this moment, a creature that looked like this didn't even exist for you. Why would it? You can't imagine a real-life creature into your brain if you've never seen anything like it before. Can you imagine a species with perfect accuracy that doesn't exist right now, but will be discovered tomorrow to look exactly like you imagined? If this did happen, it would almost certainly surprise everyone, including you.

When born, you start out in a total *void* of knowing what reality contains. In your mind as a young child, where there was *nothing*, now there is suddenly *something.*

Your initial thoughts and feelings about this new *something* is a bit of a mystery even today. You almost certainly can't remember an experience this early in life, and I can't remember it on your behalf.

"That's a cat," says the person pushing your stroller.

Cat. You now have a specific *sound* for what you saw. The next time you see one, you point and yell "cat!" and your caretaker praises you and confirms yes, that *is* also a cat.

Freeze your mental video right here.

This moment is *extremely* important and I want to make dang sure you understand why I believe this.

What gave *meaning* to the sound your caretaker made when he or she told you "cat"? Where did your caretaker pull from to connect *that* sound and *that* creature?

Did the caretaker notice it and then open a giant *"Book of Somethings,"* find that something, and then tell you the word for it? Did he or she open a dictionary and consult the definition of 'cat' before giving you the word sound? I suppose it's *possible* this actually happened for you, but if that's true I have so many questions about how you were raised.

I think the answer here is almost definitely 'no'. No one carries around a master book titled *"What Sound You Should Make When You See, Hear, Smell, Taste, Or Touch Something in Objective Reality"*. I feel confident no one gave you the sound as a child, and then read you the definition so that you associate the "right" meaning to the sound. That creature just was - from that point on - a "cat". But this is not the only thing given to you.

Whoever told you that something was a cat provided more than just a word with a sound that starts at the back of your mouth and ends with a flick of your tongue, though. That person provided *so much more*.

He or she gave you an entirely new framework to use – a *specific* human sound assigned to a *specific* thing that exists in objective reality. You repeated the sound given to you by your caretaker for that something, and it was validated as *right*.

As a kid, you probably started overusing this new framework right away, and began calling non-cat creatures "cats". Why? *It's the only word you have to use for anything that seems similar in reality.* Until you know otherwise, everything that is fuzzy with many legs and a tail *is* a cat. Or perhaps this was more dramatic, and *everything* was now a cat.

Hamster? *Cat.*

Ferret? *Cat.*

Cow? *Cat.*

Dog? *Cat.*

Food? *Cat.*

Stroller? *Cat.*

Stapler? *Cat.*

Someone had to eventually tell you "Oh, that's not a cat, that is a *dog*." Does this mean you were wrong by saying cat? Not from your perspective! You were following your truth, which consisted of very few words. If it had four legs and a tail, that meant "cat" is the right sound for it. Someone had to give you a *new* word sound to associate to this new something – a "dog". And just like that, your truth-story for fuzzy creatures in reality doubles in size.

This same set of "associative" events was repeated for every animal you can name right now as the person reading the end of this sentence. I am in my mid-forties as I write these words, and I am *still* learning new animals all the time thanks to my kids' interest in them. I still misidentify tons of them, and am gently (or sometimes not so gently) corrected to the "right" new-to-me animal word by my own kids.

More importantly though, this pattern did not change. Before I learned about these animals in my 40s, *they didn't even exist for me,* just like that cat in the previous story.

That doesn't mean real creatures like the olm and the flying squid didn't exist in objective reality before I learned about them just last year, it just means they didn't exist at all *for me* until that moment. They were hidden in the void. Where there was nothing, now there was something.

33

The Rest of Everything

What I've just attempted to explain in the previous chapter is an idealized and very cleaned-up story of how you and I learned our words.

Words and their possible meanings for each person would be so easy to understand if learning our first language was this simple and clean. In reality, there are so many *more* possibilities, so much *more* noise, and so many *more* blind spots and gaps in this story, and each one of these has the potential to change *everything*.

What if the person pushing your stroller screamed, jerked your stroller backwards nearly tipping you over, and cried out "Oh my god! It's a cat! *Get away!*"?

Do you think this would affect how you feel about the word "cat"?

What if instead the person quickly changed directions with your stroller and said, "Good eye, kid, that was close! I *hate* cats. They are such disgusting animals! Don't touch it! They carry fleas!"?

What if your caretaker *kicked it* before telling you anything at all?

How about if the person gently picked it up and held it for you to pet? Maybe talked softly to it, taught you how to listen and feel for the purring of it being happy before telling you "This is a cat. Aren't they soft and nice?" and putting it down?

This isn't even close to all the possibilities of this situation. You can probably imagine a million more combinations of first-cat experiences, but things gets *even more* complicated from here: This story is assuming you met a cat and learned the word for it at the same time.

What if a cat attacked you before you ever knew the word for what it was? What kind of associations would you bring from your past when you learned the word associated to the creature already connected to pain and fear?

What if a gentle cat cuddled and purred and napped with you every day as a child before you learned the word for such a cuddly and wonderful creature?

The point I'm trying to make here is the experiences you already had or are currently having with the world when you learn your new words *matter. They matter a lot.* They set the framework, tone, and mood for how you understand your words and the world around you, because it is through these experiences meaning is injected into those words. You will then use these meanings as the foundation to compile your personal context with words for years to come.

Each one of these experiences changes *everything*.

If you were never taught words, you would still build remembered associations between you and everything in the real world. It seems to be what your memory exists to do. Words would be absent, but all the same emotions would still exist and still be tied to those associations. It would just be experience-based instead of word-based.

I believe this is why "feral" children that are found to have survived in the wilderness for long periods of time struggle to learn spoken language at all. They have already made their own mental associations to reality without words being included in the process. Their primary language – their mother tongue - does not include words at all. They would be non-verbal communicators, using only gestures that mimic things and *maybe* drawings.

Words are, after all, just another method of sharing ideas that exist inside your mental boundary, just like gestures and pictures. They are representations of real life experiences in the form of sounds - collections of symbols we call letters.

Ultimately my words are the "answers" to the questions of who I am inside my mental boundary. They are *not* my whole story and they can never be my whole story. I have many experiences I have never put into words before. Whatever words you use are not your whole story and can never be your whole story either.

This means if we are comparing words with each other, we are *only* comparing "answers". The stories behind those answers you and I call words - our personal contexts - remain invisible to each other because they only exist in our past - in memory.

Your story tied to "cat" might contain feelings that cats are awful, should be avoided, and that they carry disease. But all of these associations are usually hidden, deeply rooted, and *subconscious*. They secretly reappear for you every time you use, read, or say the word "cat." No matter who says or writes "cat" you fill the word with *your* invisible stories, not theirs.

As crossed up at this context game is, it still doesn't explain *why* learning words automatically places the learner in a trap.

To understand why, here's a question to chew on while tumbling in darkness: Do you have any memory of your own life that took place *before* you knew words? Any memory where you didn't already have this framework of sounds or letters to make sense of things?

34

An Entire Universe in a Tiny Space

The word "cat" brings something slightly-to-totally different to everyone's mind that reads this book, including yours.

Do you imagine a house cat? Isn't it interesting that if you hadn't, now you have?

What color is it? I haven't mentioned a color at all.

Is it black?

It is now.

Actually, it's a white cat.

But let's give it some black ears.

Why do you think you pictured whatever you pictured?

How certain are you that what you picture is truly *right*?

I can re-arrange this cat for you. It's now a very overweight brown house cat with short hair and two huge, strong, and long rear legs. It stands and bounces around on them. Its front legs are now smaller, and it can use them to grab things. Its snout grows longer and its ears taller. It even has a big pouch on its belly. The tail is thick where it attaches to the body, but thins as it reaches the tip. It often balances on its back legs and tail, like they are a kind of tripod.

Whatever meaning your brain used to fill in each of the word-shells above, it was applied to your imagined cat instantly as you read them. They were applied so strongly and subconsciously that I believe at some point in that paragraph, the "cat" you began with switched to no longer be a cat.

But, I said it was a cat right? I never changed my word for it. You should believe me if I insist our imaginary creation is a cat…right? Are you super sure it's *not*? When *exactly* did it stop being a cat for you? When exactly did it become something else entirely for you? Did you have words for what it was *between* being a cat and whatever word you had for it by the end? No?

You didn't even choose what a cat was for you in the first place – it was chosen *for you*, likely before you can remember. You created this cat above from nothing, and transformed it dramatically, but *I still insist* it's a cat.

So is this animal still being a cat, a *fact*?

What *is* a fact?

The dictionary (Dictionary.com in 2023 again) approximately defines a "fact" as "a thing that is known or proved to be true." But how do you know any *thing* at all? How do you prove some*thing* is true?

The only way you know any words to share information is because someone gave you the words needed to share it *with them* in the first place. I'm referring to the stroller-cat situation. You and your caretaker experienced *something* and you were told to make a "cat" sound with your lips and tongue to associate to the entire experience of *something*. That something *is* a cat, and that *is a fact*, right?

The something exists.

The word exists.

The two are linked.

That *is* a cat.

This *is* a fact.

The world is full of cold, hard facts. Right?

Facts *are* undeniable. Facts are *objective truth*. Right?

How *certain* are you about this?

Who determined these were facts for you in the first place, and how can you "prove them to be true?"

The backside of the looking glass is approaching…

35

Schrodinger's Kangaroo

Reset the stroller-and-the-cat situation.

As a child, what if my caretaker and everyone around me called that same four-legged fuzzy animal a "kangaroo", so I know *for a fact* that animal is a kangaroo? Am I wrong?

What if I was taught the animal that hops on its hind-legs and keeps its young in its pouch is actually a "cat"? Am I wrong again? Both animals really exist. Both words really exist. I know you may be tempted to say one of us is wrong, but what method would you use to *prove* I'm wrong?

Are you going to show me encyclopedia or dictionary pages for "cat" and "kangaroo"? Declare that there are "agreed-upon" meanings for these words? Are you going to declare authority over my context and invalidate my experiences that built my context, and attempt to invalidate *my entire reality* in the process?

Are you going to put me on a treadmill and keep showing me a "real" cat and "real" kangaroo while bumping up the speed until I start giving the "right" answers? Are you going to force me to start tearing my mind in half over what I believe to be right by making me so miserable I tell myself *you* actually know what's right for me in this moment? Are you the *Keeper of Proper English Language Usage and Decider of What Things Are Called For Eric*?

You were likely taught your words using the *same process* that taught me my words. Is one of us the *Keeper of the Right Words*?

This problem is quite a bit more complicated than the 747-on-a-treadmill problem between you and Bob, because in this case you and I both *agree* on what we perceive in the moment – a fuzzy creature with four legs and a tail. In the plane-and-treadmill situation, you and Bob actually made entirely different assumptions under your different answers. This time, between you and me, there are almost *no* assumptions - the animal is right in front of us. *The animal's existence is agreeable between you and I.* That's not the problem here. The problem is that we do not agree on what sound should be used to describe the experience of it. That sound, unfortunately, is and always was *an opinion*.

What if you couldn't have any sources to use as an authority against me? It's just your word for that animal, versus my word for that animal.

We could run on the treadmill all day to try to determine which word is the right word, but this problem is not something we can work out by sharing stories. Our experience of connecting a sound to the animal is what was different. Neither of us had any control over this. We were born into it. But we are each using words the other recognizes to be associated to an entirely *different* experience with reality within the very same language.

When I first learned that the animal was a kangaroo, and it was validated as the correct sound to associate to the experience of that animal over and over again by all those around me, it became a concrete *fact* for me. It became the right thing to do to make the sound "kangaroo" to express the experience of a small four-legged, fuzzy, purring animal. I was told connecting these two things is *right*. Why would I doubt everyone around me telling me what that animal was called? How would I know differently?

The animal didn't exist at all for me, then it did, and then I was told it was a "kangaroo," and that it being a kangaroo is a "fact," and for years everyone around me provided the proof and validation I needed to know that animal is *in fact* a kangaroo.

The same situation applies to you and that same animal being a "cat."

Neither of us is wrong. We are both expressing what we feel is right based on our truth. You might have 60 million people that agree with you that it's a cat and that this animal being a cat *is a fact*. I might have 5 million people that agree with me it's a kangaroo and that this animal being a kangaroo *is a fact*. We are both right from our own perspective, because we have answered with what we believe is the right answer based on our past experiences – based on what makes the most sense, which is our *truth*. From our own perspectives, we are laying out cold hard facts.

The disagreement between us now was created so far back *in our past* we can't even know how this has happened.

This is exactly what "dialects" and even what different languages really are in my opinion, and why certain local and isolated regions of people have different accents and words and sounds for the very same experiences you are familiar with, but associate to *different* words.

Language is developed and flows through generations of individuals that interact and validate one another by using it in ways they understand and are familiar with. Because words ultimately limit and control your possibilities to build truth, those that share the same limits and possibilities (because they share the same word "anchors" with each other) tend to follow the same pattern of *actions* because they follow similar pre-validated truths.

Some people call this "culture".

Over time it's possible the very same *local* system of limitations and possi-bilities (culture), kept in place by the shared language made up of limited words that each represent a sort-of-shared experience with the *local* objective world, evolves. It's possible those within the same culture change the exact word-sound to represent an entirely *different* local experience.

You are likely more familiar with this than you might think.

This is often "the kids" creating new meanings for *existing* words, or even creating new words for the same experience that already have words. As I see it, the primary purpose is to *escape* the limitations and possibilities being forced on them through language itself.

After all, an explanation for *why* a word sound is associated to something in the real world is not usually passed through generations along with the sound. They *are just learned and connected and said to be a fact, and as a kid, everything is concrete and so it is absolutely certain because that is how it is taught.*

"That *is* a cat."

"Why?"

"Because it is."

"Okay."

You learn your words and how to speak by those around you while you grow up. Because of this, you will likely speak in the same way as those you grew up around. Locally (in this imagined situation), the fuzzy animal with four legs and tail *is* a kangaroo for me, and that *is a fact*. I might even say it with a funny accent, because I was born and raised in a rural area.

But just as the experience of first learning "cat" expands to include the emotions and actions of your caretaker and literally anything else you experience in that moment, negative or positive, the same is included for literally *every word you ever learn in your entire life*.

If the caretakers that raised you absolutely hated the color blue, you can be pretty sure you will also hate blue (in your childhood, at least). Blue *is* an awful color, and *that's a fact*. It's *right*! The word sound "blue" now summons color *and* experiences of hate for you.

If guns are awesome, then guns are awesome. *Fact.*

If spiders and snakes are scary, then they are scary. *Fact.*

If a political party is destroying America, then they are destroying America. *Fact.*

If you are told that a certain group of people in this world are awful and no good and should never be trusted, guess what? In your childhood it's now *a fact*.

Why would you think to question it? You are a child. Those people and groups didn't even exist to you *before*, and now they do, and now they *are* untrustable awful people you *should* avoid. All transferred into your context through the invention and magic of language.

The people who raise you teach you what they think you need to know about this world, and how to navigate it. Questioning these individuals is not something you likely do as a child. They are the first *Keepers of Truth and Deciders of What is Right For You* in your life. They are the ones teaching you how to communicate in the first place. They are the Boss, giving you The Rules, and administering the Consequences.

You are learning *their* language, which reflects *their* experiences and *their* contexts, and words can inject those things into your own experiences and context, which they have probably controlled as much as possible.

This local-opinion-of-fact-culture situation can have devastating consequences when amplified through generations.

As a child under your caretakers' watch, you cannot go interact with that certain group of awful no good people that can never be trusted - it's not like your caretakers would permit or want that to happen, because those caretakers believe that group really is awful, no good, and untrustable. It's the story that makes the most sense to them. It is their truth, and they are telling you to believe their truth because they feel it is the right thing to do. They might have arrived at such a belief within the last 10 years, but you are born into this belief being truth and *fact*. It was actually always just an opinion based on something you never experienced. *All facts given to you by others are only ever opinion.*

This means you are potentially removed from yourself and others *by language* as you learn it, because it was others that gave you all the meanings of words in your language. Your original connections between word sounds and experiences are often *not consciously remembered* but are *subconsciously buried* and labeled as *facts*. Those absolute unmoving meanings, complete with all their emotions, judgments, and baggage, end up being the curve on the lens of the looking glass, which you then use to filter and *understand everything else* in the world.

Your language grows from your first words, just like my first words to you in this book. Your first words, and all the stuff tied to them, construct the first foundation you are required to stand on from that point forward.

This means you started out your conscious life (assumedly somewhere around 2 or 3 years old) as mostly a collection of *other people's opinions* – but even the people providing those opinions might have believed them to be fact, because they were told the same opinion when they started out.

Perhaps those that gave you your first words were very validating people, and they let you learn and explore and control your own experiences and express your thoughts as you learned to speak. Perhaps they gave you every perspective they could think of and informed you everything is opinion. Perhaps they told you "yes, purple people are the worst, but some people think they are not a problem at all."

However, I think it's more likely they injected their own biases, their own filters, their own hates, loves, and beliefs right into you with those word sounds with a shade of gray.

In this way, ideas, ways of thinking, and beliefs inside a parent, caretaker, or language-provider's mental boundary can *literally* pass down through generations right to you: Right, wrong, trauma, abuse, positive or negative demeanors, helplessness, narcissism, racism, worthlessness, political beliefs, religious beliefs, and all the values and boundaries (or lack thereof).

If your birth parents were the first ones to give you language, you can be extra sure these things passed down *with ease*, because you probably have largely the same genetic, structural, and chemical makeup they do, meaning you are prone to fall into the same way of thinking and being from day one (this is not certain, of course)

You are barely "you" from the get-go. As I see it, you *start out* your journey of life almost entirely as Second You. Even today, you are *still* a hodge-podge of personalities, reactions, feelings, and experiences belonging to those who taught you whatever words you still use (that have the same meaning they had for you in childhood).

This means every *fact* you know today that was provided to you by someone else in your childhood, including every single word, is up for debate and questioning as truth, because it was *always* opinion. It might be a "fact" in your culture that 60 million people believe, but that doesn't make it *objective truth*. Relative to objective reality, it is and always was only an opinion.

No single culture of shared language is the *Keeper of Truth and Decider of*

What Words Are Right To Represent Objective Reality.

But even this is a cleaned-up version of all the things that can pass down through words alone as you learn language. The full reality is so much messier.

Those that raised you also performed *actions* based on what they believed to be true. It's possible your caretakers gave you the word "love" while hitting you. Gave you "sorry" while twisting everything around to actually be your fault, because they could not accept authority over their actions.

The words you were taught that build facts used to understand your world weren't necessarily even your word provider's beliefs. After all, saying words to others is *an action*, not *a belief.*

Perhaps you were wondering when I was going to finally bring *lies* into this equation – so here they are – and here is your reminder that right and wrong are *relative* words which are first-and-foremost defined by comparing your present to your own past, and *no one else can see this happening.* This hiddenness makes the *action* of lying possible, which is another way of saying disconnecting words from beliefs.

The person that gave you words might believe the right thing to do is teach you words that do *not* represent what he or she believes to be true. If the people that raised me from birth decided that it would be funny (and the right thing to do) to have the whole town convince me for years cats are called "kangaroos", what method do I have to understand this is a lie if I have no opposing view – no second point for comparison?

Is this really even a lie? Or is this now a truth? Or is it just a warped opinion?

From my perspective, the fuzzy (fact) four-legged (fact) animal (fact) that is a kangaroo (fact) is absolutely right (fact). All my words are facts as a child. They are indisputable. So I *tell myself* all of this is true, because without words holding truth, I have no sharable reality at all with anyone else. I *tell myself* I know what is real and what is not. I say it with force because it is not just me, it is the entire shared culture behind me that is being threatened if my beliefs are wrong. That animal is a freaking kangaroo! There are kangaroos everywhere! Everyone calls them kangaroos!"

Then I met you.

We were walking together down a sidewalk.

A black kangaroo crossed our path. I told you it was a sign of bad luck.

Then you said it was actually a cat, and calling it a kangaroo is perhaps the strangest thing you have ever heard.

We can get on the treadmill with roosters behind us for hours.

With only the two of us, we cannot resolve this issue and choose which word is the right word. *We are actually speaking completely different languages, but still using words familiar to each of us.*

The process used to connect word sounds to this animal in our past is approximately the same, our experience with this animal is approximately the same, but the word sound that summarizes our hidden truth-story and experiences and emotions with this animal do not match at all.

If I was lied to, I have no way of knowing. The animal didn't even exist for me, then it did, and it was assigned a sound. Then everyone around me confirmed it and enforced it and validated it for me over and over for years. It is *fact*, and it is proven, and it is true, *as far as I know.*

And then you presented an entirely different reality when you told me it was actually a cat. I've known my whole life that animal *is* a kangaroo. *With certainty.*

Am I wrong? Are you wrong?

"Fact" can be a very dangerous word.

It implies certainty.

Certainty implies the whole story is known.

Certainty implies gaps and blind spots do not exist. That no assumptions are being made.

Claiming a fact is claiming the authority of having the whole truth.

The only facts you can ever claim are those about your own actions or beliefs from your own perspective.

But even those are not facts to anyone else. They are just your opinions.

Objective facts cannot exist in language unless they are about language itself – which ends up creating a word-loop trap. You and I cannot use facts as a shared foundation.

Perhaps then, objective truth can be found *below* language - in the experience itself.

Further we tumble through this great barrier of words.

36

The Light and the Shadows

The single animal we are both looking at with our eyes has been split into two realities by language itself - one reality for you and one reality for me.

In my reality it is a "kangaroo". In your reality, it is a "cat". Our two realities just collided when we became aware of this massive difference in "facts" between us.

This situation exposes some gaps when it comes to words and experiences inside your mental boundary. To expose them, let's assume two things for the purposes of this chapter:

1. You have directly interacted with a cat.
2. You have *not* directly interacted with a kangaroo.

In our situation, my assumptions are the opposite. For the sake of dimming the noise telling yourself my context, let's drop my assumptions and only use *your* assumed reality, as mentioned in 1 and 2 above.

If we stick to the two assumptions above – that you have directly interacted with a cat, and have not directly interacted with a kangaroo - your *experience*, which *forms the context* you use to fill each of these two word-shells, is *radically* different (Note: I'm filling the word-shell "interacted" to mean you have physically seen a cat with your own eyes. You have heard the sounds it makes. You've touched its fur, and smelled its scent. You've witnessed its behavior and size, and even felt how much it weighs because you've picked one up). Experiencing, or interacting, with a cat has shed an enormous amount of *light* into the word "cat" for you.

The ability of someone like me to *change* your truth that the cat in front of us is *actually* a kangaroo is extremely limited. The amount of sense I would have to make to overwrite and change every experience you've already associated to "cat" and re-associate it all to "kangaroo" is frankly *absurd*. It might not even be possible at all. If I actually did convince you to call it a kangaroo, I suspect you would be stuck having to always remember to *tell yourself* that "cat" is actually "kangaroo" when you speak with me.

But what about the animal you call a kangaroo? Based on my stated assumptions, you have never *directly* interacted with a kangaroo before. Let's assume you've only ever seen one picture and read two paragraphs about this animal. These paragraphs included basic information like the typical weight and height, habitat, diet, and population.

What's your certainty-level on kangaroos with a photo and basic information? How much light shines into "kangaroo" for you, given these assumptions?

Let's put it to the test: Could you answer these questions?

How rough is its fur? How big is it compared to the cat? What kind of sound does it make? Does it only hop? Does it walk too? Does it walk on all fours or only walk on its hind legs? Does it smell? What does it smell like? Can you describe the smell? Can you pick one up? Can it pick you up? How fast is it? What do their teeth look like? Is there fur inside the pouch? How many toes does it have?

Chances are good you would have far more certainty in your answers to these exact same questions about a cat than you would about a kangaroo. Even if you were a child, thanks to your *direct firsthand* experience with a cat, you would probably know or can make a very good assumption about things like a cat's teeth size as well as its speed, jumping ability, and more, because you've absorbed far more about a cat than you might realize.

Your mind can go back and use the memories of your firsthand experiences underneath the word "cat" to construct stories you have *not* experienced firsthand, which might include answers to some of the questions above.

You can fill in your own gaps exposed by my questions with a pretty good accuracy, due to the enormous amount of information you take in from firsthand experiences. This makes it easy to *flex* the words you have in your vocabulary to fit the idea you are trying to communicate. The word-shells are being filled with meaning by a *wordless* firsthand experience in your memory.

For example, if you don't have the word "claws" in your vocabulary, you can use other words you do have to communicate the wordless experience of claws anyway: "Pointy sharp things on feet," or perhaps "paw knives," or maybe "murder mitten spikes".

Your situation is *significantly* different when it comes to a kangaroo. You've only seen a picture of one with nothing else in the picture. The two paragraphs you read do not provide enough information to imagine how the animal would make you *feel* face-to-face.

There are things like scale, density and power in movements you notice that can likely only ever be truly understood through direct firsthand wordless experience with an actual kangaroo.

This only-having-a-few-words-and-a-photo-to-your-entire-kangaroo-experience situation makes it a lot harder to "fill in" anything more about a kangaroo that wasn't pictured or specifically mentioned in the text. You are taking "blind guesses" when it comes to answering my questions above, unless it was in the text or photo explicitly.

To fill in the gaps exposed by my questions, you need more experience or more stories, *which you don't have.* This means any answers you provide are being justified by what I'll call "shadow experiences," which are themselves built *from* words (and one unmoving photo) of a kangaroo. You have *enormous* gaps in your truth-story of kangaroos. If you can fill in the gaps at all, your stories are not likely to line up well with the stories someone with direct firsthand experience of a kangaroo would tell.

This isn't really a big problem, until it is.

When does this become a big problem?

Well (again if we keep the assumptions from above), if I said you are about to be lowered into a cage with a housecat, you'll probably feel pretty comfortable with what you should expect, even if you are afraid of cats. You might actually be more concerned with what you don't know about the *cage* (gap created by limited information) than what you don't know about the cat.

But (again keeping assumptions from above) if I say you are about to be lowered into a cage with a *kangaroo*, you might experience a feeling of panic as you realize you don't actually know that much about kangaroos *at all*, as you build a million questions you can't answer to try to fill in your gaps all at once:

Are they aggressive to humans? Do they typically bite? Scratch? Kick? Are they venomous or poisonous? Do they typically jump, walk, or run? Do they typically move on two legs or on all fours? How fast can it move relative to humans? What is their strongest sense? Smell? Sight?

You can't really build an answer to any of these questions with any certainty at all, given what you think you know to be true about kangaroos, which is only two paragraphs of words that directly answer none of these questions. Those words are literally *everything* that exists about a kangaroo for you. They provide the *entire* context you have to give meaning to the word "kangaroo".

And now, as your feet pass into the cage where you cannot yet see, even the truth you think you know or tell yourself to believe about kangaroos will collapse. All the words in the paragraphs vanish. Their population and habitat no longer matter. Their eating habits no longer matter. Their "typical" height and weight are no longer relevant in a way you can truly apply.

You *need* stories built from the *specific* kangaroo you are about to experience. But all you have is the memory of a photo of a kangaroo. That photo gave you no *real* sense of scale, no sense of expectation from it, only an "approximation" of what the kangaroo you are about to meet *might* look like.

All this collapsing-of-belief happens because everything you thought you knew to be true about kangaroos was actually *blind trust* in a single *secondhand* source of "truth." This secondhand source – words and a picture printed on paper - is the only thing you've actually experienced firsthand when it comes to kangaroos. It took the threat of a firsthand experience with a *real* kangaroo to expose *to yourself* just how little information you really had, and how uncertain that information was for you.

Everything you believed to be true about a kangaroo dropped into darkness – back into *shadow* – and became part of the void to you. None of it was validated by anyone except your own context, and it was only valid to you because it made the most sense.

Everything you believed to be true in this situation is, at best, only a tiny piece of *someone else's* firsthand experience with a kangaroo that he or she put into words. At worst, everything you believed to be true doesn't represent the truth of a kangaroo at all.

The two paragraphs might just be words put together by someone that built his or her context from a shadow experience created by another person's words, which were themselves put together by someone that built his or her context from a shadow experience created by *another* person's words. This might repeat over and over so many times that no one even knows what words about the kangaroo were put together by firsthand experience at all.

It might take a team of investigators months to follow this chain of "hands" and finally determine the source of the kangaroo firsthand experience truth-story was a traveling salesman named Bob, who visited Australia 200 years ago. It turns out the species of kangaroo he was studying, which built the original truth-story handed off and rewritten multiple times to end up on that piece of paper you read, has been extinct for 50 years.

You were reading a fifty-seventh-hand experience about an animal that doesn't even exist anymore, and *you had no way to know.*

Your mind doesn't and can't discriminate between these possibilities when it comes to words. A word is a word, and you fill it with your own context no matter how many "hands" it might be removed from *wordless* experience.

You fill each word of these stories using fragments of *your* memories built from other *similar* experiences. You then attempt to merge this "shadow experience" with your current truth, which was *literally blank* when it came to kangaroos.

This tiny two-paragraph group of words about kangaroos, no matter what it says, is automatically going to become your truth if the words used make sense for you at all. You have no other context to believe otherwise. Nothing else makes more sense. Where there was *nothing*, now there was *something*.

If firsthand experience is the light, or "enlightening", then shadow experiences are a tiny dot of a star in the huge black void of "kangaroo".

No matter how many words you find and use from other secondhand sources to grow your truth-story of kangaroos, this "shadowy experience" can never compete with the full light of firsthand experience. You can only add more tiny-dot-stars-of-secondhand-experience to the enormous darkness.

That being said, if you connect two or more secondhand stories together, these points of light will get brighter because more sense is being made, but ultimately they will remain surrounded by darkness, because they are associated to contexts that are not your own and you cannot experience.

There is no direct path between you and the kangaroo star. You can add as many words, pictures, and stories as you want to try to grow the brightness, but secondhand experience can never be as bright as firsthand experience, which provides what seems like the full light of wordless objective truth to overpower all the stars in the darkness with ease.

There is no substitute for firsthand experience.

Words and pictures (and media of all kinds) cannot ever be a one-for-one equal with firsthand experience because you simply do not possess the context needed to fill the words, pictures, and media of all kinds. You lack the "assumptions and norms" created only through being face-to-face with a real kangaroo.

To make the secondhand situation worse, you often cannot know for sure how many sources, how many "hands," exist between the firsthand wordless experience of a kangaroo, and the story that creates your shadow experience.

Every "hand-off" you are away from the firsthand experience of a kangaroo, the more likely the words experience a major *contextual shift* in meaning, because no two people ever fill word-shells the same way. If someone builds their words from the shadow experiences made by other's words, there is no guarantee the new words are even close to representing the words *you would use* if you had the same firsthand experience.

This exposes a very important boundary that becomes completely covered up by language over time: *Words filled from wordless firsthand experience are very different than words filled from secondhand shadow experience.*

Firsthand experiences are *wordless* experiences that exist below words. Endless stories can be constructed from a single firsthand experience, even if you have only one word to use. Firsthand experiences are the foundation of language.

Put a kangaroo in a room with me, and I could write books upon books of words about it. Put a cat in a room with you, and you might create a blog, four novels, an entire meme culture, and a photography business all about that one cat.

This boundary suddenly being visible in your very real context may and probably should collapse almost every belief from secondhand stories you have ever had. *All stories made from words, photos, videos, and media of all kinds are secondhand shadow experiences.* If you have not experienced anything close to the firsthand experience responsible for them in the first place, you cannot be sure if you are sharing an "approximate" experience or one that is wildly different, or even one that *never existed at all.*

You can never be sure if the light from the "tiny star" of shadow experience created from secondhand stories is actually a planet, a lightning bug, a plane, a satellite, swamp gas refracting light from Venus, a hallucination, or just a great *illusion.*

It is incredibly likely this boundary has been covered up and blurred for you for so long by the illusions communication can create, that it will be hard to separate and fully exit the looking glass. So that is what I intend to expose for you next.

37

Dissecting the Gray

Although I have recently learned that the olm and the flying squid supposedly exist as creatures in this reality, how can I be sure? I will probably not experience an olm or a flying squid firsthand in my lifetime. My *entire* experience with these animals will probably remain in shadow.

As a result, I am stuck in the position of having *blind trust* that Macken Murphy and his *Species* podcast team (who could be you) did not decide making up new animals and stories about them was the right thing to do. But this is not yet exposing just how big this "leap of blind trust" is that I must make.

I do not know anything about Macken Murphy beyond what words he has used about himself in his podcast, and I have only ever experienced what I *assume* to be his voice secondhand. Is it? I can't actually know from my position. He is my single secondhand story source for the existence of these two creatures.

So what should I do if tomorrow I read, "Taylor Swift caught using voice-altering software to create fake podcast persona Macken Murphy. Swift has used this persona to fabricate stories of fictional animals as part of a scheme for more income so she can add to her growing fleet of third-generation Chevrolet Camaros"?

I have exactly *zero* firsthand experience with Macken Murphy. The only firsthand experience I have actually had is the vibration of a speaker next to me that sends me shadow experiences through words supposedly being spoken by Macken Murphy about animals that supposedly exist.

Macken and his team, if any of them truly exist, are reaching me from the shadows. I can't actually be sure the context filling the words that come to me from the shadows are tied to *any* firsthand experience with objective reality at all. The voice, the stories, and even the animals in the stories might be 100% fiction, as far as *I* can know.

But consider I know the same about this new Taylor Swift and Camaro story. When you read it, did you automatically *want* to believe it? How did it merge with the paragraphs about Macken that came right before it?

That second story is just words on a screen or a piece of paper for me. I have no idea if those words are tied to objective reality and firsthand experience in exactly the same way I can't be sure about the olm or the flying squid or Macken. One is a set of words from a speaker, the other from a piece of paper. But....your personal context merged better with one of these stories than the other. You will end up choosing which one to believe *automatically*.

As I write this, there are many reasons I believe Macken Murphy really exists and is not Taylor Swift in disguise, and there are just as many reasons I trust in his stories and believe the animals he discusses really exist, too.

But this is absolutely not without risk – it is all still uncertain. It's possible the only way my trust in Macken and his stories will be truly tested will be the moment I am about to be lowered into a large container full of olms or a big pool filled with flying squid. How much of what I *tell myself* I know about these animals can I trust to be valid in reality? I only have my own context to build these animals in my mind, which is bound to be somewhat-to-drastically different from the context I will soon receive through firsthand experience.

To rephrase this: It is entirely possible what I *tell myself* to believe will turn out to be *complete and total fiction* after I directly experience the objective reality the story was *supposedly* based on. This is possible for every story you and I believe to be true that we have not experienced firsthand.

But here's the catch: Am I wrong to believe this without firsthand experience?

Not to me at the time I believed it. How could I have known? To believe my truth was wrong is to believe I could be wrong *and know it*.

In the absence of firsthand experience, if the story makes sense, it's potentially enough to make it true *for me*. This mechanism can allow "truth" to completely disconnect from objective reality using words and shadow experiences alone.

Is the Moon made of cheese? You might laugh to yourself reading this, but I've never touched it and I only ever see one side of it from what seems to be a very far distance. I think chances are good you've not touched the Moon either (although I can't be sure).

If I assume you also haven't been to the Moon, this means you and I are using secondhand stories like "the Moon is made of rock and dust," and believe maps of the dark side of the Moon are correct *on blind trust*.

You and I end up choosing to believe what we want to believe (or worse, what we think we *should* believe) about the Moon simply because it *makes the most sense relative to our own context*. You and I put *total blind trust* in the sources of the information that made enough sense to give us our current truth-stories about the Moon, but secondhand stories are *always* going to be believed on blind trust.

Individuals like Neil Armstrong and Buzz Aldrin (should you believe they exist(ed) – I do) supposedly had firsthand "enlightening" experience with the Moon on the Apollo 11 mission. Everyone else that did *not* actually walk on the Moon with Neil and Buzz, which includes Michael Collins (who supposedly orbited Neil and Buzz experiencing the Moon) will take the Apollo 11 mission's videos and interviews and documentation and build a combined *secondhand* truth-story of the first "supposed" Moon landing. This includes everyone at NASA, and everyone that built every piece of equipment that was supposedly used on the mission. Even Michael Collins had to do this, because he himself did not actually step foot on the Moon, which is also a shadow experience from a *secondhand story* from my perspective.

These secondhand truth-stories from Buzz and Neil, whatever their media content might be (articles, movies, photos, etc), is instantly merged, piecemealed, or rejected relative to your truth-stories built by firsthand experiences with the Moon. The same occurs for me. If my assumptions are correct about you never visiting the moon yourself, this isn't much of a foundational truth-story at all.

Once you and I experience secondhand stories from Apollo 11, a new one-point truth-story will be forged for each of us, and that truth-story will then determine what future secondhand stories can be accepted as true or will be rejected when it comes to the Moon.

I've touched a real rock from the Moon - supposedly. I can't really be sure it was from the Moon, though – it's just what the story on the sign next to it said. I choose to believe it, because the story makes the most sense for me. I also choose to believe that Neil, Buzz, and many other astronauts have actually landed, walked on, and even played golf on the Moon.

But I could also choose to believe the rock is a fake, the Moon is actually made of delicious cheese, and the Moon landing footage is all fake. Perhaps that Moon rock I touched was actually a lava rock from Earth. How could I possibly prove it one way or the other?

Is the Earth flat? Do continents you've never visited actually exist? Do lizard people live in the center of the Earth? Have aliens from space visited Earth? Did we really land on the moon? Are we really spinning around the sun instead of the other way around? Whatever you believe, you almost certainly believe it to be true because a *secondhand story source* told you a story and the words made sense to your already-existing context. The only exceptions for some of these specific questions would be if you were an astronaut who directly experienced orbit or outer space and had a window to look from.

The previous examples are all a bit on the extreme side of the experience scale in my opinion, but the topics involving this looking-back-up-at-the-stars-through-the-looking-glass-and-believing-them-to-be-stars problem can actually hit very close to home, get far more familiar, and get far more serious.

Are the schools teaching children racist theories? Is your vote being counted? Is the Earth's climate changing? Are we capable of changing it? Did the Holocaust actually happen? Are certain groups of people actually in control of everything? Is someone seeing everything you do on your phone/computer? Is light and sound really made of waves? Do germs really exist? Is crime really on the rise or really going down? Do atoms and quarks exist? What's the weather going to be later today? Tomorrow?

Whatever you believe related to each of these questions, it is most likely based on secondhand stories creating shadow experiences. Becoming "enlightened" to the wordless truth of these questions is often difficult-to-impossible, because as a human being you simply cannot experience most these things firsthand.

You will never experience a "wave" of light or sound. The waves are far too small to ever see with the naked eye, or pickup with the unaided ear. You will always be stuck looking at a screen or other instrument to visualize these waves for you. As a result, you'll have to have trust in the tool you are using to provide you the *secondhand story* you make into the shadow experience of the light or sound waves.

You experience the tool's output firsthand, but if you don't know how every piece of the tool works to create that output, the tool itself becomes a gaping abyss between your eye and the object the tool is focused onto. This means the light and sound waves the tool shows are believed not only on *blind* trust, but blind trust *with known gaps in it*.

You'll never be able to know through *firsthand* experience that your vote was counted in an election if you can't literally follow your physical ballot from your hand to the final numbers reported. On top of this, you also need to witness your votes being interpreted correctly and entered into the system correctly. Casting your ballot was your *only firsthand experience*. After casting it, literally everything else is in shadow, even if you work in the election system, because you lose sight of your own ballot in the process, and there are gaps *everywhere* in between your act of voting, the tallying of your vote, and the announcement of the winner.

The only difference between someone like myself (who has never worked inside the election system) and an election worker, is the election worker has more "starlight" filling his or her word-shells than I do, because that person knows and has experienced significantly more that happens between the casting of my ballot and the final result.

You'll almost always have to look through a microscope to see bacteria, a virus, and most cells ("almost" because supposedly there are some rare extra-large examples of these things). Do you know how a microscope works? *Every* piece of it? How do you know it's not a screen inside the eyepiece faking the entire image you are seeing? If the microscope has a screen instead of an eyepiece, how do you know what you are seeing is truly what exists in the very-real sample?

If you don't know, you are stuck again with *blind trust* in the tool. The microscope's output is the only firsthand experience you get. The information, or "story", the output sends you is connected to total shadow, with the only exception being if you *do* know how every piece of your microscope works, and you have experienced each piece inside of the exact microscope you are using to validate each piece inside is *actually* there.

To dissect the gray a bit more: When you get sick and the doctor or nurse takes a throat swab, you don't directly experience anything that happens outside your exam room door. The swab transfer to the lab, the lab tech, the preparation, the microscope or analysis machine (and its inner workings), and the interpretation of the results are all in shadow to you, back in the exam room.

The only way these shadowy places in your life aren't also filled with abyssal gaps and voids is if you work as a lab tech in the exact same office, so you know all the steps and what exactly *should* be happening outside your exam room door, as well as knowing everything about the machines being used. But *even then* there are *still* gaps: From your exam room, you can't know if a mistake is being made, or if the lab tech working would interpret your throat swab the exact same way you would. You have many bright stars in the shadow experience of the throat-swab-to-results process, but you can't be sure any of them are being done exactly like you imagine. You are still stuck in having some degree of blind trust in all of it.

Did the Holocaust happen? Just like everything that ever occurred before you were born or did not witness firsthand, it is permanently stuck in shadow, leaving you to believe (or not believe) secondhand (or fifty-hand) stories, pictures, videos, and other media provided by other people or devices on blind trust.

You can't literally go back in time to experience the reality these secondhand stories firsthand. Any part of a past event you did not experience yourself will forever remain surrounded by shadow.

Is crime really going up (or down) if you've never been a victim of a crime while the news claims crime is out of control in your town? What method do you have to know this story is tied to reality? Becoming a victim? The number of people you know that have become victims?

You are once again stuck in a position of having blind trust your television screen is showing you *real* newscasters from your *real* town's news studio and that what they are sharing is current (as opposed to being recorded six months ago by newscasters that are actually computer generated deep fakes). The only thing you are experiencing firsthand *is* lights in your *screen* (and sounds from speakers).

On top of this, you then must have blind trust in the secondhand stories the newscast reports to be true. If you had never been mugged before the report on rising crime, and went for several more weeks without getting mugged or hearing of anyone else you know getting mugged, would this story continue to make sense for you and so continue to be true? Or would you replace it with your firsthand experiences and more "enlightened" personal truth?

The gap between firsthand experience required to make secondhand stories true for you (like being mugged, getting sick, and voting), and "making sense" required to make secondhand stories true for you (crime rate, germs, and voting results), is *huge*.

You often cannot see into the shadows surrounding all your secondhand story-built truths, and so you cannot know how many gaps are hiding. Yet the only thing required for a secondhand story to be true for you is that it *makes the most sense* when it merges with your own context. If truth-stories in your context are *also* built completely from *secondhand stories*, the hidden gaps can get covered up if you continue to believe these truth-stories in the face of stories that expose the gaps. You would have to *start telling* yourself what you believe is *right*.

These covered gaps become full-blown abyssal-deep *purposeful blind spots* that you only ever willingly choose to uncover when stories come along that agree with "your truths" in a way that fills them. It is only a matter of time until someone figures out they can *inject whatever stories they want to make you believe they are filling these abyssal gaps in the truth-stories you tell yourself to believe*. These truths, which are held together through blind trust in secondhand story sources can literally be exploited and *hijacked*. It only has to make the most sense when compared what you have experienced *so far*. Other's words are always given meaning by *your context*. This means very little effort is required to warp your truth completely when you have no firsthand experience to know otherwise.

I'm going to try to blow the covers off these abyssal gaps that might be hiding in your "looking glass" truth-stories in the next chapter. Once this is done, you'll be able to see straight through the holes it has made in your own context.

38

Shooting Holes in the Stars

If we carry over the assumption from two chapters ago that you have never directly interacted with a kangaroo, and have instead only ever read two paragraphs and seen one unmoving picture of one, what are you going to do when I slap what essentially looks like a picture of a *large rat* in front of you and tell you it's *also* a kangaroo?

The pictured animal appears to be about a foot tall and has a "naked" hairless tail, but is otherwise furry. Its ears are small and covered in fuzz, and it has small beady eyes. It looks *very different* from the photo you saw earlier of a kangaroo.

Still, I *insist* this creature is a kangaroo.

Are you going to tell me it's not a kangaroo? Based on what? A picture and a paragraph or two you read earlier? Are you going to intentionally try to hold covers over the gaps in what you tell yourself to believe?

Are you now an authority of what *is* a kangaroo and what *is* not?

Is this new secondhand story source, *me*, obviously *wrong*?

Am I confusing you on purpose? Messing with you?

Am I *lying* to you?

This might seem like a silly situation, but this happens far more often in your life than you might think. Consider just how much knowledge you have on any topic that is purely from secondhand stories. What shadow experiences do you tell yourself accurately reflect *objective reality* that were only ever pictures, movies, sounds and/or words for you?

In part of our tumble, my goal is to expose just how many gaps, bottomless pits, and complete *voids* might exist *inside* your truth-stories. I admit this can feel scary or like I'm attacking you, but I have no intent to tell you what you should and should not believe. Exposing gaps and holes doesn't make what you believe wrong or untrue. I can't even know what you currently do and do not believe anyway. *You are the one in control of what you believe and why.* With that said....

How did you come to believe the story of what happens inside your government (or other governments) that you currently believe?

How electricity works?

The difference between viruses and bacteria?

The *existence* of viruses and bacteria?

That Jupiter has a red spot?

That the outer planets exist at all?

How a microwave heats your food?

That we live in a spiral galaxy? Why it's called the Milky Way?

How books are printed?

How LCD screens work?

How your car works?

How many of these answers (if you have any for these questions), are based on *blind trust* in secondhand stories, and are not based on your wordless firsthand experience at all? What words and pictures are you choosing to trust on these topics and why do you trust them? How did your secondhand story sources get *their* story to pass it on to you? Are they getting their stories secondhand, too?

Do you check the sources of your secondhand stories that provide these shadow experiences? Do you "fact"-check encyclopedias? Are there *any* other stories out there that say something different than what you currently believe? Are there secondhand stories that disagree with your current secondhand shadow truth-stories completely? How are you supposed to determine which of these two battling stories is closer to objective reality in the absence your own firsthand experience? Which shadowy source of "truth" is right and which one is wrong?

No one knows the whole story about anything – and this is exploitable.

What if I said the government is full of lizard people in costume?

That the force moving the energy through your power cords is actually traveling *outside* the wires?

That viruses are just cancer for air?

That the red spot on Jupiter is dust left behind after a red planet crashed into it?

That there are only five planets in the solar system?

That a microwave uses radioactive elements to heat your food from the inside out?

That our galaxy is actually an orb and not a spiral?

That it's called the Milky Way because a huge corporation bought the naming rights to promote candy bars?

That all books are printed by hand with giant presses?

That LCD screens are actually filled with liquid?

That you might be using contained explosions to get to the grocery store?

That kangaroos actually have retractable fangs?

That Macken Murphy is actually Taylor Swift?

That the Moon is made of cheese?

That the moon landing not only happened, but happened multiple times *for years* after Apollo 11?

What stories do you have in your mental boundary, *right now*, to determine your answers match objective reality with total certainty?

With each question above, you likely went from a one-point, to two-point, and then back to one-point truth situation just like you and Bob did on the treadmill. You probably had *one truth* related to each of these topics before reading this book, which were the limits of everything you know about these topics that made sense to you. My word-shells possibly made you create a second truth-story you hadn't even considered to exist, and I made you create them *using a question*. You naturally built what you believe to be the "right" story between what you already know and what I just said was true. The new single-point merged truth for you will be whatever makes the most sense between the two.

Unfortunately, depending on the size of the gaps I potentially just opened for you, it is *completely possible* for me to have just built a story that makes enough sense you accept it as true, even if it is *not actually close to what I believe is true or what seems to be objectively true.*

How can you know if the foot-tall rat picture I slapped in front of you is actually a kangaroo or not? Was the first article and picture wrong, or am I wrong? Neither of these animals actually exists for you beyond one photo and a few words. You've never experienced either of them firsthand. You've never laid your actual eyes on the actual animals. How do you know the animals even exist at all?

You have to blindly trust me and reject what you thought a kangaroo should be, or tell me I'm wrong, right?

.........right?

You know better this time, I'm sure. You saw that I was trying to speed up the treadmill on you, and remembered to pull me onto the belt to ask me how and why this second animal is a kangaroo, right? You didn't just accept what I was saying, right? You didn't *assume* you didn't understand what a kangaroo was and I did, right? You didn't *assume* I was an authority on the word kangaroo did you? You sought to understand? You started asking me questions?

It's unfortunate that all too often questions in situations like these where two-way communication is possible *never happen*. It's often assumed the "teacher" knows best. After all, this exact thing has been happening to you since *day one*. You had to learn your second word this same way.

You might know so few stories about the topic that it feels right to *assume* the person giving you this new story *must* know more than you, because, well, you've never heard the story before. Why make these types of assumptions? *Assumptions are dangerous as hell to blindly accept.*

As I see it, this constantly happens anyway because of the treadmill that keeps speeding up. *We are in a hurry.* We've been speed-fed secondhand stories since birth. Childhood is a non-stop onslaught of new words, pictures, audio recordings, videos, and electronic secondhand stories filled with "facts" creating shadow experiences believed to be true.

All animals used to be a "cat". That has been shredded into every animal you can name today.

To demonstrate the sheer scale of the problem created by approximate-word meanings and speed-implied-and-assumed authority forged by a constant flow of new-to-you secondhand stories that have been coming at you since your birth, let's go back to the *cat versus kangaroo* situation one more time.

You see, back when the black *kangaroo* crossed our paths, and you said it was actually a *cat*, you totally confused me. I had never heard someone call it that before. I became aware of what appears to be a totally different reality – an entirely second truth – *your* truth.

I could tell myself that kangaroo is absolutely correct. I could attack you over what you've called it. Scream that you are wrong. Let you scream that I am wrong. I could even walk away. *But we aren't getting anywhere that way.*

We need to get on the treadmill together to search the underlying wordless experiences we used to fill the two different word-shells, so we can better understand each other and our very *shared* world. Reality is not *actually* split, after all.

We all-too-often don't do this.

We are in a hurry to get off the treadmill.

We've been solving treadmill problems *our whole life*.

We want to finish the story and give the *right* answer so we can move on.

The result is to add to impending disaster.

I believe this too-hurried-to-understand-each-other-and-build-better-merged-truths-by-asking-questions situation is exactly what has created this growing problem. Our language-based truth can and eventually does completely separate from underlying objective reality. Word-shells are being filled with meaning by contexts that never had firsthand experiences to fill them with. Words are used to fill other words. Gaps are intentionally covered up. In the United States at the time I wrote this, they are often words like:

Deep State, Democrats, Republicans, election security, fraud, social media, lies, truth, Supreme Court, MAGA, wokeism, liberalism, conservatism, Fox News, CNN, stupid people, millennials, boomers, smart people, scientists, transgender, drag queens, gays, racism, black people, white people, illegal immigrants, guns, gun control, gun promotion, mental illness, mental healthcare, Critical Race Theory, school indoctrination, lack of teacher support, money, profits, stupidity, science, religion, lack of religion, absence of god, the devil, police, broken healthcare, wealthy, poor, union, Black Lives Matter, Antifa, immoral beliefs, and thousands more.

The impending disaster is the end result of being bombarded by what feels like a thousand stories a day about each one of these words. They come from news, entertainment, strangers on social media, friends, family, and more. It's likely (though not certain, since I can't know exactly who you are or what you know) the very first time you ever heard any one of these words was in a *secondhand story*, which would mean your starting, unquestionable, and certain truth-story about their meaning started *surrounded by shadow* hiding countless invisible gaps.

"That *is* a kangaroo."

"That *is* a cat."

There were other meanings circulating in the shadows at the same time you learned the first meaning. Other meanings have *always* been there. You just had no way to know about them. You just went from nothing, to everything.

So, you might find yourself clinging to your everything-truth-story built from someone else's words, and declare and fight over your righteousness. Self-validating you indeed knew the "right" meanings for words all along, while invisible-to-you gaps in what you tell yourself to believe as objective truth began to block out more and more words that didn't merge – didn't make sense for you.

"That *is not* a cat, that *is* a kangaroo!"

Approximate words are piled on top and in-between more approximate words.

The cover you stretch over your gaps starts to thin.

It can only hold you up for so long before it, too, collapses.

39

How an Exploding Cat Collapses Reality

When I write the word "cat", how big is it for you?

Is it fat?

Is it mean?

Does it make you sneeze?

Is it sitting next to you?

Is it sleeping? Meowing?

Even if you and everyone else that speaks English associate the word *cat* to a fat orange housecat, none of the fat orange housecats from each person's context would look exactly the same if you could somehow compare them. Nor would they bring about the same exact *feelings* between any two people.

What does the word "spider" bring to mind for you?

Snake?

Vomit?

Winning lottery ticket?

Incredibly attractive?

If you assumed I have all the same associations to these words that you do, or anyone else does, I have sadly failed you at this point. Word meanings are approximate because we all have *different* experiences underneath them – which are also blended with the contexts of the people that transferred their meaning of those words to be our own.

But now things get *really* messy.

How many *completely different* meanings for the word "cat" can you think of in English?

I'll give it a go: It can describe a person (jazz culture) or a beggar (related to crack cocaine mostly). It can be a woman, or a piece of heavy equipment (brand name). It can be a shortcut word for catalyst, a lion, a tiger, a bobcat, or a piece of a vehicle (catalytic converter). It can also be an acronym for three other words that begin with C, A, and T (and I must confess I did some internet searching to find some of these): Computer axial tomography (the "cat" in catscan), common admission test, community action team, computer-assisted translation, computer adaptive test, California achievement test, corporate air travel, "can't answer that", Central African Time, conformance acceptance test, counter assault team, certified athletic trainer, and dozens or maybe even hundreds of other possibilities. You might even know a few more meanings for "cat" I haven't mentioned.

The bottom line is the word *cat* can be made to mean almost *anything*, because individuals can inject whatever meaning they want to the sound or group of letters.

All it takes is two people sharing the same meaning for it to be a language.

This is how I can end up calling what you know as a "cat" a "kangaroo" and seriously believe I am completely right. I didn't realize a boundary existed where my shared language stopped and yours began. To be honest, it's possible I didn't even realize there were other languages – other meanings for this "cat" sound until you and I saw the animal together.

There is no limit to this language possibility expansion. There is no *right* definition for the word "cat." It is a sound or group of letters that represents whatever in the heck experience you or anyone else wants it to represent *relative to the moment.*

There is no "agreed upon meaning" of any word. Who sat down and signed that agreement? Agreed-upon meaning is a judgment call by lexicographers at the dictionary, which is at best based on what could be found. If you are a lexicographer, I'm sure you don't poll every English-speaking person and play the Destroyer of Words game to build your final definition.

If tomorrow everyone suddenly stopped using "cat" to describe an animal, the dictionary would eventually have to adjust or drop that part of the definition. How long would it take the lexicographers to realize what had happened?

Language is a living thing, because it is created by living things – us.

As I see it, the only reason we often think others agree with us and know what we mean is because most of us learned "cat" early enough in life we can't actually remember that first experience, but we kept using the word in mostly the *same shared context* over and over again. It was often used the same way in kid's books, with our parents, at school, and in the shared cultural life of our communities. We used it with others that shared that same *approximate* reality. It's only natural to *assume* everyone else can magically understand our invisible personal contexts. The meaning we assume is validated to be the "right" and "true" meaning of the word a thousand times over.

But how many contextual meanings can possibly exist for "cat"?

To answer that, we have to consider a lot more than just the variations around a *house*cat. Remember all those completely different experiences for "cat" just a few paragraphs ago? How many *variations* of each "category" of the experience of cat, either firsthand or secondhand, can still be tied to the word?

For example, if I say "cat" is a piece of heavy equipment, does it represent a bulldozer for you, an excavator, or a dump truck?

When I wrote this, there were about 1.5 billion English speakers in the world (according to secondhand sources I choose to blindly trust). If that number reflects the approximate objective truth (which no one truly knows with certainty), that means there are 1.5 billion *different* housecats - all different colors, shapes, and sizes - coupled with feelings and memories to go with them. If every one of those same 1.5 billion people can also associate "cat" to the manufacturing brand of heavy equipment, then "cat" can also mean 1.5 billion *different* pieces of heavy equipment. If 75 million of those individuals think of an excavator, and 5 million think of a dump truck, this would mean the word cat, *even within the same context of heavy equipment*, still can't be used to communicate anything specific all by itself. There's still no guarantee any of these dump trucks are the same between any two people.

Repeat this same process with all the other definitions as well. How many different things in reality can the "cool cat" definition represent, or a "catalyst cat" represent? They are called "cat" just the same. Might as well throw in "large cats" like tigers and lions, and throw in some "cat acronyms", too. While there are only 1.5 billion English speakers, the word cat all by itself generates what could be *trillions* of meanings; maybe exponentially more. Many of these meanings are entirely different from others. The bottom line is when it comes to "cat," you can't be sure at all what is happening in the mind of another human being.

But the contextual explosion doesn't stop here.

How many languages other than English exist in the world right now? Each will have a *different* sound and spelling for the experience of a housecat. Sounds or spellings like "felino" in Spanish, "chat" in French, "kissa" in Finnish, and "popoki" in Hawaiian, just to name a few.

Now...how many *completely different* contextual meanings can possibly exist for the words "feline" or "kissa" or "chat" in those languages? If *cat* can mean trillions of things in English - from animals to heavy equipment - how many different meanings can "kissa" have to a Finnish speaker?

For each one of those categorically different meanings for "kissa", how many variations can then exist within each category and still be referred to as "kissa"? How many people exist that speak Finnish? Multiply these two numbers to get your answer.

As an English speaker, if I write "chat," does this word mean something *entirely different* than "cat" for you – especially if I don't tell you the word is French? Now that you are aware of this particular connection, you need to go back and add the possibility that "chat" all by itself could *also* mean the experience of "cat" in objective reality.

The sheer amount of experiences a single human sound or group of letters can represent is mind-blowingly huge. I believe it is *entirely possible* for "kangaroo" to also mean "fat orange cat" for someone in the world today.

The meanings of words, unique to each of us, are constantly evolving and changing, and can change to mean anything at anytime between our ever-changing contexts responsible for our ever-changing truths.

Words do not represent anything specific at all by themselves.

All words not your own are secondhand stories – tiny stars surrounded by darkness. All are subject to be easily misinterpreted and misunderstood.

Many words you use are actually not yours at all, which means they represent purely secondhand stories for you. It's possible that you completely misunderstood your own secondhand story sources that provided these words. After all, there were only about 1.5 billion-to-several-trillion possible meanings to choose from.

If you then taught people those same words, it's possible they misunderstood your intended meanings as well, which was already a misunderstanding.

Now add in that often *no* questions are ever asked by anyone. We all-too-often just accept the meanings of words provided by others and move on with our lives like we completely understand - blissfully unaware of this nuclear explosion of a problem lurking and building beneath the surface of the Great Barrier looking glass held together by mass *assumption* that all words are fully understood and mean the same thing to everyone.

What if you were somehow exposed to all of these meanings all *at once*? Would that be an overwhelming amount of meanings? Would there be so many meanings words themselves would start to collapse from having meaning at all?

Would all blindly-trusted secondhand stories - shadow experiences - break down to what they actually are, which is *uncertain*, as the covers over the possibilities between and through words are ripped wide-open? What would happen next if one context for a set of words declares it is right over all the other contexts?

Would this essentially be an attack on all secondhand stories, shadow experiences, and firsthand experiences? A declaration of ultimate contextual authority? A self-given title of *Keeper of the Truth and Decider of What is Right*?

In my opinion, this is *exactly* what started happening around 2004, with the introduction to electronic worldwide connectivity. Everyone who participates in the mega-giants of social media like Facebook, Reddit, X (formerly known as Twitter), and other platforms at the time I wrote this, are regularly exposed to nearly every context imaginable for every word they already use. It is, quite literally, an invisible but very real explosion of contextual awareness.

Systems of control that were held in place by unmoving and unchanging words began collapsing almost right away as the truth of how words actually work enlightened person after person. This is still in progress of happening today as I write these words, in my own country of the United States of America.

Rigid meanings in moral codes are being seen for what they are: Attempts to force *one* context for the meaning of words as the dominant and sole context to try to maintain order. The result is an entire world of human beings with nearly instant communication between each other exploding into a full-out context *authority* war.

Individuals seem to be trying to force something into existence that does not exist and has never existed, because these individuals believe it *used* to exist in the past. Words *never* had fixed and unmoving meanings. *There is no objective truth in words*. Right and wrong are *relative*. But that's exactly what seems to be happening in the United States right now, and it comes with the worst kinds of consequences that can exist for human beings.

It may feel like there *needs* to be *one context* for words to stop chaos and make words have meaning again. After all, if literally no one agrees what words mean, society would collapse, right? One truth must be chosen as the "right" and "objective" truth whether it's fair or not! Otherwise everyone would just ignore words and do whatever they want, right? Rules would all fail! Bosses would be toppled! Consequences would be unenforceable!

Those are stories individuals tell themselves. They are certainties. They are attempts to desperately stretch the covers back over the abysses present in that individual's beliefs of what is true. These acts of trying to forcibly cover the abyss become the "foundation" and basis for words themselves to lead to social collapse and violence over and over again throughout history.

Building society on unmovable fixed word meanings – which requires one human context to be *The Keeper of Truth and Decider of Right,* is to build all of society on a thin cover that fails and collapses society into the abyss every few generations.

If language tries to inject meaning into you, instead of you injecting meaning into it, you end up like a dog being wagged by its tail – a paradox. A nasty cycle begins that turns the world upside down, started by whoever was trying to force his or her context to be the *only* one for you. It's started by the person who answers that simple question "Are you the Keeper of Truth?" with "yes," he or she knows what "right" absolutely means *for you.* Which can really be boiled down to four words: *"Because I say so."*

What kind of person would do such a thing?

Someone on the treadmill that got tired of being told what was true and what to believe. Someone that got tired of someone else telling them which words are the *right* words to use and that his or her own words are wrong. Someone that just wants off the machine more than they want to talk.

The result, unfortunately, is becoming trapped in funhouse mirrors of belief that complete agreement between any two people is possible, and is possible using language alone. The looking glass becomes the foundation, and exiting seems impossible. Secondhand stories and shadow experiences are mistaken for firsthand experiences and objective truth.

I think it's a worth a quick imaginary trip to witness what this looks like in a larger societal context. I'll try to keep this silly, for now.

40

Murder, With Sprinkles On Top

My kingdom is in chaos. Large groups of people are obsessed with good health, others don't care about their health, and the ice cream companies are marketing how wonderful and delicious their products are on every ad space and billboard they can find.

I, *King Eric*, declare an emergency law that ice cream *is unhealthy* and *is now illegal.*

I give a beautiful speech about my caring and love for all those living in my kingdom, and that I just want everyone to be healthy and happy. I explain that Big Ice Cream, full of billionaires and greedy corporations, don't actually care about anyone's health – they just want to make a buck by any means necessary and are taking advantage of the people.

The crowd watching my speech cheers.

The text of my new law reads:

"No one shall manufacture, sell, transport or eat ice cream."

So now I, *King Eric*, have created a simple law, which is a one-way statement of objective *unmoving* righteousness. It is intended to stand the test of time. It has to, otherwise people will become more and more unhealthy, and the ice cream moguls will keep getting richer off everyone's inability to resist, and we can't have that!

I've now declared what I believe to be the *right* context for words regarding ice cream. Do I know other contexts exist? Of course I do! That's why I had to ban them in the first place! Other people have contexts that justify eating lots of ice cream to be right, but I know that's *wrong*! And Big Ice Cream's context justifies manipulating people to buy more ice cream to be right. The people can't help themselves! It is not fair to them! So I used my authority as King (the Boss) to create this string of written words called a law (The Rules), to act as a higher authority than even me (The Final Boss). If I didn't believe the act of banning ice cream was the right thing to do and would create a healthier kingdom, I wouldn't have done it.

But a word of caution: *You cannot see into my mental boundary*, and the only beliefs of mine about ice cream you know so far is *none*. All you actually know is that I justified my pen to write and sign the law, and what words I said before I signed it. You cannot access the hidden context justifying those actions.

This is critical because it's possible I *didn't* believe banning ice cream was the right thing to do. I have actually been bombarded every day *for years* with donations and arguments from the Anti-Ice Cream League. This activist group constantly demands I save the people from themselves. They email, they call, they write, they beat on my door. They annoy me to no end, but they write the biggest and fattest checks to me so I tolerate them. If *this* story is truly my context - which again you cannot know - then I still believed writing this law and speaking my words were the *right* actions to take, but for an entirely different set of justifications.

Right and wrong are *relative*. They require two points to even define them. Writing the law and giving the speech were the right things to do at the time I did them relative *only to the second point you can never see - my past*. My *context*. My *belief of what is true, hidden in my mental boundary.*

It's possible I did all of this to make the Anti-Ice Cream League leave me alone, or maybe I did it to keep their money coming to me, or maybe I did it because they convinced me creating this law was *truly* the right thing to do for my people. There are lots of possible hidden *justifications* for my actions.

Even darker possibilities spring into existence when you consider my truth might be disconnected from reality due to having blind trust in secondhand stories. I may have a collapsing mental boundary, and am *telling myself* my own actions are always wrong and I shouldn't trust myself, and other people know what is right better than I do. This creates the possibility I believe I am not completely responsible for my own actions, and I am merely justifying them because it's what I think I should believe is right.

Or perhaps even worse, I might have an expanding boundary and keep *telling myself* I am right: *In fact, I am the only one that knows what right even means. These Anti-Ice Cream League people are idiots, but I can tell ice cream is the entire problem. I will fix this and stop all the wrong and idiotic people of my kingdom, and show them what right really means.*

Meanwhile, *outside* my mental boundary, tons of possible contexts already exist in the kingdom for each word I included in this new law. *I can only be aware of the ones that have been communicated to me or that I can imagine.*

In this case, the Anti-Ice Cream League has been a major storyteller for me. The story that ice cream is the main source of obesity, bad health, rising healthcare costs, and rising depression is the *right* story for me and therefore *true* because it makes the most sense.

I also hear it "validated" constantly by members of this group, and they do this with a heck of a lot of certainty in their stories. A huge chunk of my day is flooded with their ideas, data, survey results, and arguments, and this problem becomes more and more time consuming to me with each passing day.

I finally decided to do something about it. They are telling me they have the truth, they are telling me what is right based on that truth, and so I've decided to make their truth story my own. I have decided to tell myself their secondhand story *is* tied to objective reality, based on the constant flow of secondhand stories they use to justify their secondhand story as being true.

This is the actual truth-story I used to justify my actions.

There is no guarantee that the secondhand story I believe is tied to objective truth at all. Perhaps all the graphs, data, and surveys the Anti-Ice Cream League has shown me are made up and skewed on purpose. How could I know otherwise? Where there was nothing, now there was something.

The law goes into effect.

The Anti-Ice Cream League celebrates the new law. They pile ice cream in the streets to melt, and burn ice cream store signs and uniforms. Healthy-obsessed and pro-health individuals join them in celebration. These groups have united under what they believe to be shared beliefs under the words right and wrong. This group of individuals fooled themselves into believing they agree on the context for the word "right", which I'll now call the *Anti-Ice Cream Context*, and support this law.

It feels good to have what *you* believe to be right recognized as right by authorities above you. *You want to believe what you think you should believe.* Everyone's belief in the Anti-Ice Cream Context has just been validated as right by the highest authority in the kingdom.

But what's another major context besides the *Anti-Ice Cream Context*?

How about that ice cream is a good and tasty treat, and makes life wonderful? There's nothing wrong at all with it. It's so very *right*. It's one of life's simple pleasures. Let's call this the *Pro-Ice Cream Context*.

Who belongs to this group? Probably all individuals who like ice cream, as well as the ice cream shop owners and ice-cream manufacturers, which are instantly out of business thanks to this law. Hundreds or thousands of individuals lose their jobs. All of these individuals are likely to oppose my law, because ice cream is *right* and good to each of them; there's nothing *wrong* about it.

The Pro-Ice Cream Context instantly collects and unites *against* the words in this law. In the real world (and not this imagined one), this context would probably be organized and growing before this law ever arrived, because ice cream manufacturers and shops would probably be well aware of the Anti-Ice Cream League and their efforts to ban ice cream.

People that love eating ice cream and refuse to give it up are likely to keep and hide all the remaining ice cream they can find. It's also possible an entire underground network of ice cream trading, hoarding, secondary markets, and secret ice cream creameries will suddenly materialize into reality. An underground "criminal" network is immediately born to serve and help people who think the whole idea of banning ice cream is wrong to begin with. They keep following what they believe is right, regardless of what this law says is right.

Are they wrong? Not from their perspective.

Lastly, people that don't eat ice cream might not care at all, and may split to one context group or the other to support friends or fight against the opposing side, or neither. We'll call this remainder of individuals not personally affected by my law of absolute right and wrong the "Silent Majority".

Because my law is written as a certainty – a timeless absolute declaration that is designed to prevent certain actions from ever happening that were justified as right before this point – two large polarized groups instantly formed around two *approximate-at-best* contexts. One is for, and the other against the words contained in the certainty.

Those that fill the word-shells "ice cream" with associations of being unhealthy and memories of negative experiences (Anti-Ice Cream Context) likely believe everyone else in their group believes the same things they do.

Those that fill the word-shells "ice cream" with being good and memories of positive experiences (Pro-Ice Cream Context) likely believe everyone else in their group believes the same things they do.

Those individuals in the Silent Majority context remain indifferent, because the word-shells "ice cream" don't fill up with anything much at all. This new moral code banning it doesn't affect them, as they don't eat it, or just don't care, whatever reason they each may have.

Now, *who* has the authority over what the words mean when it comes to this law? If you remember from the first time we dealt with my ice cream ban, it's handed off from my context that chose the words to a hierarchy of authority, which you experienced from top to bottom a few different ways after you were arrested earlier in the book.

In this kingdom, the authority over what the words of my law mean will start with a police officer, who applies the law directly to you based on what his or her context determines those words to mean. The officer's personal context is the stories (norms and assumptions) this set of word-shells tells the officer based on firsthand experiences (which includes training).

In the earlier chapters, the officer justified arresting you and placing you in front of this hierarchy with the label of being "wrong". After this, the first judge or jury determined if your arrest made sense and if you are guilty of breaking this law based on the stories the word-shells of this law tells to each of them based on their *own* context, coupled with what they understand the words to mean relative to your story. From there, you appealed up the chain of judges until you reached the highest judge or court in the kingdom. Earlier, it was the Supreme Court of the United States. For this chapter, it would be the High Court of my kingdom.

Now though, I'll add in a critical part of the story that was skipped over in those previous chapters: Every written judgment following an arrest based on this ice cream law results in an *expansion* of what the words of this law mean *by adding more words*, because all future cases afterwards must take into consideration what was justified to be right and wrong about the crime relative to the law. The goal is *consistency* – so judges do not have the total freedom to judge crimes relative to this law anyway they want or however they feel on a particular day. This "forwarding" of all previous written judgments is called "precedent".

Precedent is a funny thing, though. In my opinion, justice systems seem to operate on the idea that precedent is an ever-*narrowing* context for the *right definitions* of each word in the law. Can you see a major problem coming? Let's walk through an imagined and cleaned-up chain of events immediately following the enactment of this law in my kingdom:

Case #1:

Mike was caught eating ice cream in public by a police officer and was arrested, found guilty, and sent to prison. Mike (and his lawyers) immediately appeal.

The higher appellate judge reviews the original judgment, and whatever his ruling may be, this written decision made of unmoving timeless words will become part of the precedent, or *judicial context*, that is *supposed* to fill in the word-shells of the law with better, unmoving meanings for each word.

Mike was putting the ice cream into his mouth when arrested, but claims he didn't swallow it so he wasn't technically "eating" ice cream at all. The judge rules that putting it into your body in any way is to be considered consumption, and should be ruled as eating.

The Anti-Ice Cream Context group celebrates. The Pro-Ice Cream Content is outraged. Both context groups gain more members after Mike's sentencing hits the media.

But something interesting has happened - the law has *more words* that are very much a part of it now. In theory and in "good" judicial practice, the words of the same law for all future cases are: "No one shall manufacture, sell, transport, or eat ice cream. *Eating means putting it into your body in any way.*"

Not only have words been added underneath the original words, but these new words are themselves declaring a fixed unmoving definition for a word in the law - eating. They themselves are a certainty.

People begin arguing about what "eating" means. What if you don't swallow something? Is that really eating it? What about shooting it up your nose? Is an ice-cream enema eating? What about an ice-cream bath?

It seems like precedent should narrow the law. It has actually both narrowed and started the explosion process. Mike's argument to define what eating means has now exposed the fact there are multiple meanings of each word up for debate. "Eating" seemed straightforward when I wrote this law – but it turns out "eating" could mean more things than I thought.

Case #2:

Sarah was caught eating frozen yogurt in a bowl in public and was arrested by a police officer. Sarah's lawyer argues yogurt is not ice cream at all, so it was perfectly legal and a mistaken arrest based on a bad "assumption" by the police officer. The judge agrees that this story makes the most sense. This ruling now becomes part of the contextual definition.

Frozen yogurt businesses pop up everywhere overnight as ice cream manufacturers convert to yogurt production and sales. The Anti-Ice Cream League immediately starts campaigning against frozen yogurt to save the people's health now, and puts up billboards and runs ads condemning it. Public arguments now erupt around frozen yogurt. Is it healthy or not? Should it fall into the ice cream family and be banned as well?

More new words are added to the objective, one-way unmoving morality law to make the context even more specific: "No one shall manufacture, sell, transport, or eat ice cream. Eating means putting it into your body in any way. *Frozen yogurt is not ice cream.*"

Those people who know or relate to Sarah are outraged by her arrest and move into the pro-ice cream context if they weren't already. More people remove themselves from the Silent Majority.

Case #3

John was arrested for eating ice cream. His lawyers argue he was actually eating soft serve, which is not the same thing as ice cream. The judge *agrees*.

The Anti-Ice Cream League gains followers from the Silent Majority upset about what seems like inconsistency with this ruling, and the activist group adds soft serve to their efforts to ban in addition to frozen yogurt.

Soft serve businesses start popping up everywhere. Context debates break out about the meaning of "soft serve". Is soft serve ice cream or not? Should it be banned? More people leave the Silent Majority to join the group against this law as it attempts to turn its focus to soft serve.

More underlying words are added to the judicial context. Functionally, the law is now "No one shall manufacture, sell, transport, or eat ice cream. Eating means putting it into your body in any way. Frozen yogurt *and soft serve* are not ice cream."

Case #4

Frank was arrested for eating ice cream. His lawyers argue it was melted before eating even began, and so was really just *sugar milk* and no longer ice cream. This story does not successfully merge with the story the judge believes to be true about the meaning of the law's words, and Frank's actions are deemed illegal. Melted ice cream *is* ice cream. These words are now added.

Fear and outrage immediately mount in the kingdom as those that like chocolate drinks suddenly start to worry and fear their drinks are about to be taken from them. Tons of people from the Silent Majority join the ranks of both the Pro-Ice Cream Context and the Anti-Ice Cream Context.

Some people that are now against the law don't necessarily care about ice cream; they just want to stop the law from taking away their chocolate drinks – from what they believe would be overreaching. Others are for the law, because 'it's *obvious* now that corporate interests are just trying to play games and don't think it really applies to them'.

It should be noted that no one, including me, *is actually aware of exactly how many people are in each of these context groups.* This is because every single person's context is unknowable unless they perform an action in support or against one or the other for someone else to see – and even then "knowing" it is an assumption.

The law with its complete legal context is now "No one shall manufacture, sell, transport, or eat ice cream. Eating means putting it into your body in any way. Frozen yogurt and soft serve are not ice cream. *Ice cream can be frozen or liquid.*"

Case #5

Heather was arrested for making and selling her own ice cream. Lawyers argue her recipe contained no cow milk, only soy milk, which is not really a "milk" at all. The judge disagreed with this, and it was found illegal anyway. Milk *is* milk.

Soy ice cream makers are instantly out of business. More from the silent majority join the Pro Ice-Cream Context. It is not that they like ice cream, but that the law has now taken their sweets as well. This also pulls people out of the Anti-Ice Cream Context and into the Pro-Ice Cream Context, because they regret how far this law is going now that it has affected them.

The fixed, unmoving, underlying context of the law is now "No one shall manufacture, sell, transport, or eat ice cream. Eating means putting it into your body in any way. Frozen yogurt and soft serve are not ice cream. Ice cream can be frozen or liquid, *and can be made with anything intended to substitute for milk.*"

Case #6

Charlotte was arrested for drinking a milkshake. Her lawyers argue a milkshake is not ice cream and the office was completely out of line. The judge looks at the precedent he is bound to, and determines Charlotte's actions were illegal anyway because a milkshake includes all the ingredients of ice cream, and so falls under the "liquid form" definition.

Milkshakes now leave every menu in the kingdom. Those that enjoy chocolate, vanilla, or other ice-cream-like sugary drinks begin to panic as their darkest fears seem imminent – they too are going to lose their favorite sweet. *Tons* more Silent Majority members slide into the pro-ice cream context. Tons of others start joining the Anti-Ice Cream Context, because 'why does everyone seem to think they are above the law?'

The official context of the law is now "No one shall manufacture, sell, transport, or eat ice cream. Eating means putting it into your body in any way. Frozen yogurt and soft serve are not ice cream. Ice cream can be frozen or liquid, and can be made with any intended to substitute for milk. *Anything that includes all the ingredients of ice cream is to be considered ice cream.*"

Case #7

Tim was arrested for eating soft-serve. His lawyers argue soft serve has all the ingredients of ice cream, so the previous ruling on soft serve should be overruled so that the law includes soft serve. The judge agrees this story makes the most sense based on the words of precedent.

Soft serve is now illegal. The Pro-Ice Cream Context is beginning to flood full of people, though exact numbers in all groups are impossible to know. The only ones anyone can ever identify are those that are out protesting, or commenting, or putting up signs in their yards.

The law is now "No one shall manufacture, sell, transport, or eat ice cream. Eating means putting it into your body in any way. Frozen yogurt is not ice cream. *Soft serve is ice cream.* Ice cream can be frozen or liquid, and can be made with any thing intended to substitute for milk. Anything that includes all the ingredients of ice cream is considered ice cream. "

Case #8

Chris was arrested for drinking chocolate milk. Chris's lawyers argue chocolate milk has nothing to do with ice cream at all. It was found illegal anyway because it contained all the same ingredients as ice cream decided in Charlotte's case. The judge is bound to the law and the words in precedent.

All restaurants and a huge number of individuals are now affected. Protests are becoming a regular occurrence. Restaurant owners and suppliers are becoming increasingly upset as their businesses start to suffer due to people traveling out of the kingdom to get meals at restaurants that have desserts.

The expanded words of the law are still the same as they were before. Instead what has happened is the meanings have managed to capture an entirely different dessert because of contextual shift.

Judicial and Philosophical Turmoil

New philosophies start to arise and gain attention in the legal world about how this law "should" be interpreted in my kingdom. There are now groups labeling themselves as *Originalists* context keepers. They believe the courts have gone too far makes the most sense, and the law should go back to only including ice cream, plain and simple.

There are also now *Abolitionists* contexts that believe the story the law being completely removed makes the most sense.

A judge seat opens up in the kingdom, and during the election campaigns begin to promote these ideologies, (which are really just their own personal context – although they are probably not aware of this fact. They likely believe that thousand or millions agree with their belief and are validating it through their actions.)

The self-proclaimed Abolitionist candidate doesn't win this time around, but support is growing for this context. Instead, a judge that was born *after* the ban was in place takes over (which comes with the assumption that decades have passed). This judge believes the law is absolutely right to exist, and personally believes that sugar is bad, and that the law has not gone far enough. What follows is an aggressive campaign of rulings to intentionally expand the reach of this law.

Case #9

Susie, a grocery store owner, was arrested for selling "ice cream deconstruct-ed", which was all the ingredients of ice cream sold separately. It was found illegal because of Susie's apparent *intent* to try and get around the law.

Now no business is allowed to even carry all of the ingredients of ice cream in the same store.

Case #10

Annie was arrested for eating a cookie. Her lawyers argue even though it contains all the ingredients of ice cream, a cookie is not ice cream at all. It was found illegal anyway because it has the ingredients of ice cream in it, and that definition was decided many cases ago – and should have been banned back then. Cookie businesses all close. Cookie lovers join the fight.

Growing Unrest

There are now protests and unrest in every township of my kingdom. The ban on ice cream has expanded to include some of the population's favorite sweets from the constant expansion of judicial and legal context called precedent. There are open hostilities between two groups: The Anti-Ice Cream League members and its supporters in the Anti-Ice Cream Context, and everyone else who liked foods they can no longer get, or just dislike how far this law has gone.

It's important to note here that the Pro-Ice Cream Context does not appear unified or large to anyone. Everyone in it is upset about *different* things. Some because they can't buy milk at the same place they buy sugar, others because they just want a cookie, or really miss their frozen yogurt. The conversations between them do not necessarily merge to the point of "obvious" agreement. Each is arguing about how their personal dessert is not really ice cream. Someone that just wants cookies back might still absolutely *hate* ice cream, and be fine with the law if it hadn't crossed the line into their world.

Order is Needed!

Horrified at the people that have the gall to *demand and protest a right to be unhealthy*, against my own judgment and the law itself, and mortified by the unrest in the towns of my kingdom, I send in my law enforcers to use whatever force necessary to stop the chaos and arrest as many of these trouble-maker protesters as possible.

Hostilities grow everywhere. The kingdom has almost no one left in the Silent Majority. This law has now touched almost everyone in one way or another.

A high-level judge's seat opens up in the next election. After heated debates, a new judicial context easily wins the seat. The *Originalists'* interpretation is now at the top of the kingdom's legal system. What follows is a quick and intentional collapse of the law's previous reach by overturning all the precedent rulings that allowed it to expand.

The Anti-Ice Cream League and its supporting Anti-Ice Cream Context go into full protest mode. This new context is completely unacceptable. "Society will collapse from all the sweets and unhealthy people!" they scream. They launch nasty smear campaigns against the judge and Originalists' and Abolitionists' beliefs in general, claiming their stories are evil and trying to destroy the kingdom.

On the streets, their demonstrations are loud and obnoxious, but not violent, except when Pro-Ice Cream Context individuals show up to "antagonize" things into violence (or so the Pro-Ice Cream people say).

As I look down from my balcony at my ivory palace tower, it seems obvious to me the Pro-Ice Cream group is definitely the problem here. There isn't violence until *they* show up. Then there is always violence.

Case #11

Addy was arrested for eating a cookie in front of the highest court building as an act of protest. Her lawyers argue the exact same argument as Annie's – a cookie is not ice cream *at all*. The new Originalist judge *agrees*, and Addy's arrest was ruled as unlawful because a cookie does not say the words "ice cream" on it anywhere, and the law only applies specifically to ice cream.

Now the Anti-Ice Cream Context starts to grow in outrage at this backslide. They fear all of their hard work is about to be undone.

Addy becomes a hero of the Pro-Ice Cream movement.

Case #12

Dave is arrested for drinking a milkshake in protest. His lawyers make the exact same argument that Charlotte's lawyers did after she was arrested drinking a milkshake way back in Case #6. A milkshake *is not* ice cream.

But *this time* Dave's lawyers have a new piece of context built from precedent in Case #11: A milkshake is not ice cream because it is not *called* ice cream. The judge agrees and the previous ruling from Case #6 is overturned. All sweet drinks are back because they do not say they are ice cream.

Case #13 (Court Limits, Contextual Collapse of Law)

Steve is arrested for eating ice cream. His lawyers argue that Steve's ice cream was homemade and does not say ice cream anywhere on his ingredients or his container, so it cannot even be determined if it is ice cream at all. What even *is* ice cream?

The judge agrees, says that not all ice cream has a label, and so it must be up to the will of the people or the King to decide what ice cream is and is not - the courts cannot be the ones to decide what these two words mean.

The Return to Source

Everyone turns to me, *King Eric*. I have a choice to make. Do I let this law collapse off the books, or push it back into place again?

I am completely unaware that a huge majority of my citizens are actually against my law. They are all broken up and arguing about strange things like cookies and milkshakes. The strongest voice I still hear is the Anti-Ice Cream League, which has barely changed its position this whole time.

They point out how all the violence has been from the Pro-Ice Cream people, and never themselves. They only want the kingdom to be healthy. They show me slides of all the improvements of citizen health during the ban, and polls from *The Crown News* show my ban has a 90% approval rating. The Anti-Ice Cream League is still telling me how awful sugar is and to look at how my kingdom is tearing itself apart over sweets. They say to ignore any Pro-Ice Cream voices that make it to me, as they will say anything for profits and to hurt the people's health.

(If you cannot see the horrifically expanding mental boundaries for each member of the Anti-Ice Cream League apparent in the words I am using, look again. Find where they start telling me what other people think. That's your red flag.)

As King, I still want my people to be healthy. Big Ice Cream is just evil billionaires trying to cheat my rules and manipulate my people to buy something that is bad for them, and…and…*how dare people not listen to their King and realize I know what's right?! That only I know what's best for them? These idiot citizens are all being manipulated. People are so gullible. Why can't anyone just use their brain and realize the truth?! Unbelievable.*

The Sugar Law (Re-Assertion of Authority)

I double down. In an emotional and fiery speech I announce: "No, this shall not stand! It is not *only* ice cream that is unhealthy, but now it's obvious to me this whole kingdom was addicted to an awful white crystal they do not actually need, and are even willing to be violent in the streets to get it. I proclaim a new law that *everything with sugar* is unhealthy and should be banned! Look at all the violence ice cream alone has caused! It is *sugar* that is tearing us apart! We must have a *healthy* Kingdom!"

I issue a new law banning *all* sugar, and dissolve the judiciary branch completely. They only ever made a mess of things anyway. All police and law enforcement now answer directly to me. There are no more judges or courts. I am *King and Judge*. The entire system of justice has collapsed to one context now - mine. I no longer have to deal with the uninformed people choosing some whack-a-doodle judge that will destroy my authority and mess up the meaning of my words. What madness is this anyway?!

Violent protests erupt everywhere. Mobs make hot chocolate syrup pots and call for me to be cooked in them. There are fights and random shootings constantly, as escalating tensions and hostilities between the beliefs of the Pro-Ice Cream and Anti-Ice Cream contexts overflow into justified violent actions.

During a massive protest event at King State, my kingdom's only college campus, officers open fire on a crowd of teenage and twenty-something year old protesters. Twenty-four are killed.

Societal Collapse (Context War)

What follows will be known to imaginary history as the "Sugar War." It will be fought between the context that sugar is unhealthy and banning it is right, and the context that it is tasty and good, and unbanning it is right.

I control all the military and law enforcement authorities in my kingdom, so they will be a large part of my "sugar is unhealthy" fighting force, along with the extreme Pro-Health Context. They all believe they are doing the right things: The military and law enforcement for maintaining order, and the Pro-Health Context for trying to save unhealthy people from themselves, the Anti-Law Context for trying to stop what they keep calling tyranny.

I, as King, have an interest in making sure my ban stays in place. If the "unbanning sugar is right" context prevails and wins this war, I have a growing fear I am going to be in *serious personal danger*.

But why should I feel this?

This is about sugar!

A stupid white crystal!

These people are out of control!

Don't they know what's right?!

It's only SUGAR!!!

I crack down with the full force of the law. There are raids and sugar roundups. I label all pro-sugar groups as *terrorist* organizations. The word *sugar* is banned from the kingdom, and people are arrested just for saying it. The Pro-Health Context starts openly beating people in the streets they believe and label as being Pro-Sugar. Police and military brutality reaches a fever pitch. These groups start dragging people out of their homes and turning their kitchens and pantries inside out looking for white grainy crystals. I turn a blind eye. This is all for the best.

If my side wins this war, I will *have* to continue to rule as a harsh dictator, forever hunting down the sugar supporters. I can't have them questioning my authority again. The ability to ever justify sugar as right must be eradicated. Those that want it are *vermin*.

If the sugar-supporting side wins, I increasingly believe I'm going to be killed. Why? Whether or not I am aware of it, this is no longer actually about sugar or even ice cream. *It never was.* This is about who gets to decide what the word healthy means, what being healthy looks like, what foods are allowed to be healthy, and if you should be forced to be healthy using someone else's definition. It is about authority and control over the meaning of a single, solitary word: *Healthy*, and this word *doesn't even appear in my law or any of the case rulings.*

By choosing one word that is tied to something that objectively exists in the real world (ice cream, and eventually sugar), and declaring it objectively *wrong* to exist in my kingdom, I've forced everyone to evaluate their own story and relationship with what "healthy" means, and then see if that story merges with his or her story of the very real objects of ice cream and sugar.

The tragic reality is that every single person believes something different about ice cream, sugar, and health, because those words are defined by their individual experiences. When I attempt to ban what I define as *unhealthy* as King, I am not only banning each individual's ability to eat these things, but I am also saying any positive memories or associations they each had with ice cream and sugar are now *wrong*. I am literally trying to erase, invalidate, and overwrite reality itself in the minds of however many people exists in my kingdom.

Even if you as an individual in my kingdom agree with me that sugar *is* unhealthy, that's not *really* what the issue is. You may feel validated you are right and not wrong, but even so, *you've lost something, too.* Now you can't even have ice cream and sugar if you want them for any reason, because I decided that you can't have them and they shouldn't even exist. I have decided what healthy means to me, and I'm forcing it on you. You no longer have a choice. You *will* be healthy – but only based on what I think healthy means. I am invalidating your truth.

I am attempting to make my context superior to all others, and the only "right" context. This can be boiled down to four words: "Because I say so."

The two large collectives are actually not for or against ice cream at all, but instead are actually for or against my authority to make decisions for them, and take away their choices.

The tool I'm using to do *all* of this is simply connecting two words with a certainty, *like those two words are the whole story*, like no other possibility exists, like their connection is the objective truth of the entire universe: "Sugar *is* unhealthy."

But that's not the *only* story being told in my head. Let's revisit the first paragraph of this chapter:

"My kingdom *is* in chaos. Large groups of people *are* obsessed with good health, others *don't care* about their health, and the ice cream companies *are* marketing how wonderful and delicious their products are on *every* ad space and billboard *they can find.*"

Every word in that paragraph is a story based on my personal experiences that gave meaning to those words. Large groups of people are obsessed with good health? Don't care about their health? Was that fair to *assume*? Are these firsthand experiences or secondhand stories, or are they shadow experiences fabricated from my own context as the only possible explanation for the "chaos"?

Even within the Anti-Ice Cream League, each person defines "healthy" differently. They all imagine something different, but they likely *believe* they all perfectly agree. If they stopped to play "Destroyer of Words" or had a poultry treadmill, they'd quickly realize this belief was nothing more than an assumption.

Illusions are creating illusions that are justifying actions on a truth-story that is actually a total fiction-story.

Words themselves are only ever answers, *not* the whole story. All the stories underlying words are hidden in the mental boundary responsible for choosing them – they are *experiences* – and none of them actually match.

The Intervention (Opposing Context Rising to Match My Authority)

The Kingdom of Sugarcane, which sits across the Shallow Sea from my kingdom, comes to the support of the Pro-Sugar Context. Of course, The Kingdom of Sugarcane's interest is being able to *sell* sugar in my kingdom again. Big Ice Cream has been lobbying them for help since the moment my first law about ice cream was written.

Their economy tanked as my kingdom quit purchasing sugar, so overthrowing me to save their own kingdom became justified as the right thing to do.

Over several bloody weeks of all-out war, my context is defeated.

The people that hated me and my laws and the entire authority hierarchy rejoice.

I am captured. All Anti-Ice Cream League members are rounded up and arrested. This moment looks a lot like a mirror of what happened when they ran all over town beating and arresting those they decided were "pro-sugar".

The Destruction of "Wrong"

There is a huge campaign promoting the idea that "King Eric's head is unhealthy".

A massive walk-in freezer full of ice cream was found in my palace basement, along with 2,000 pounds of white and brown sugar. In my extreme stress of worrying I might make a mistake and be judged as wrong, and being unsure what to do about the annoying Anti-Ice Cream League people that wouldn't leave me alone, I would often escape to my palace basement and gorge myself on sweets until I was nauseous.

I am dragged into the center of town. I am spat on. Called names I dare not write on these pages.

My head is removed with a giant ice cream scoop in front of a crowd of thousands cheering and chanting. Someone puts whip cream, sprinkles, and a cherry on top of my head, and holds it up for the masses to laugh.

The people rejoice at the end of my tyranny.

The Rebirth of "Right"

A new king is crowned.

He is pro-sugar.

He repeals the sugar law immediately and restores the judiciary branch.

Big Ice Cream starts running ads all over town again. There is a sugar binge of epic proportions in the kingdom. An annual Sugar Festival will occur every year between the dates of the pro-sugar victory and my execution.

Health conscious individuals are disgusted by the gluttony. The Anti-Ice Cream League secretly re-forms and begins gaining members again. Some individuals believe everyone will become unhealthy, and they'll be the ones that have to pay for the ever-growing healthcare costs through taxes.

Others believe the new king is a puppet to the Kingdom of Sugarcane and has no interest in serving them.

Some individuals believe my murder was a horrible injustice, and that I was killed just for trying to promote good health.

And so the cycle resets. It will end in violence again. Maybe in 10 years. Maybe in one. Maybe in 300. But it will happen again. Probably not over sugar or ice cream. Probably over the authority to enforce some single unmoving meaning for words that declare something or some action as right or wrong with utter certainty and the only possibility that should be allowed to exist.

And once again, *everyone is right relative to themselves*. No one can be directly blamed for all of what has just happened. Everyone's own stories made sense from their own perspectives, and is therefore true. Based on that, they took actions based on those beliefs. They did what they thought was right, and thousands are now jailed, with many scheduled to be executed. I did what I thought was right, and lost my head over it.

And yet the origin of my belief wasn't even me.

It was secondhand.

It was the Anti-Ice Cream League's story.

What was their source?

Perhaps if it was followed back enough, we would find the founder of the Anti-Ice Cream League's wife died of complications from obesity due to her constant binge-eating of ice cream, and he vowed then and there to destroy ice cream forever to save the people from the same fate.

This ice cream ban might seem absurd – an over-the-top fiction. *It is not*. Go look up the words to the Eighteenth Amendment to the United States Constitution.

The real horror is the realization we still seem to be trapped in this same cycle. In *huge* scales, everywhere on the planet. The only thing that ever changes is the word meanings being fought over.

However, the eventual collapse of all context (caused by trying to make one context the *only* context through control and authority) doesn't limit itself at formalized systems and hierarchies. It can reach much, *much*, further and destroy even the simplest stories you might believe are the whole story. After all, they were never the whole story.

We can now begin the difficult process of separating from the backside of the looking glass.

Mapping the Lower Layers

Let's bring back those simple stories from earlier you might think you understand completely, which means you might believe everyone else can agree to these as objective truths, meaning they truly are the *whole story.*

- Killing is wrong.
- The sky is blue.
- 2+2 = 4
- ~~Words have very specific meanings.~~ (We can scratch this one now).

If each of these are the whole story, then these could be used as *foundational* truths both you and I (and everyone else) will absolutely agree to without any other truth-story being possible. This would allow us to stop our fall, and start filling in and rising back out of this abyss-of-a-rabbit-hole we have been tumbling down for a bit now.

For each of these to be objective truths, only one meaning of these stories can exist for everyone. The very real context wars of my time are very much trying to *force* this to happen for many stories.

I'm claiming this will *always* fail.

I'm claiming this is not how words and communication work at all.

Since word meanings are actually tied to your experiences, and your unique experiences are what provide unique meanings of words for you, but your first words are always defined through secondhand story sources (caretakers, schools, books) and these first words end up framing your entire starting truth and reality, which is also how you justify your next action based on what you believe is right...you might be able to predict how these three simple "truths" will end. Collapsed into the abyss.

I can't assume you understand what I mean, though. I can't assume you can pierce through your beliefs in these simple stories to continue our fall.

So I'll do my very best to help you.

Killing is Wrong-ish

As long as nothing interferes to change the truth-story of these three words, you can continue to believe this story is absolute objective truth without an issue, and we can use it to build a foundation together. After all, this story seems to fit with common laws and even most religions, which commonly claim to be sources of one-way objective right and wrong.

But what story do you tell in your mind when you read "killing is wrong?" I suspect most people think of one person killing another person on purpose. This is, of course, me *assuming* a common context for "most people" - which I can't possibly prove – but this is all just to demonstrate a much bigger point, so even if my assumption is not accurate, no harm should be done. So let's run with this story…

Imagine you are walking along a city sidewalk at night. It is normally well lit, but a few of the street lamps are not working tonight, and instead it's very dark. As you pass a narrow alleyway off to your left, you notice the outline of a person stand up in the shadows and start running quickly toward you. You see the shadowy outline of a huge knife in this person's raised hand while he or she charges at you.

The person screams, "I'm gonna *kill you*, monster!"

So far, this situation totally fits an objective singular context that *killing is wrong*. This person is definitely about to be wrong, and these three words seem to be holding up as objective truth. Let's keep going:

Adrenaline pumping, you reach for your concealed-carry pistol, and whip it around just in time to pull the trigger as the person - a man - is a mere two steps away from plunging his knife into you.

You fall sideways out the way as he topples past you, landing in a heap - dead in the street next to the curb. Your bullet went right through his heart, killing him quickly. (If the gun bothers you because you would never carry one, you can instead imagine the man jumped onto you, and you wrestled over the knife, with it eventually stabbing him in the heart while guided there by your hands, killing him.)

Either way you imagine this, *you just killed someone*. You may be upset about it, and you certainly don't have to be happy about it, but the bigger question is: Were you *wrong* to do it? If "killing is wrong" is *objectively* true and the *only* context possible, then you *were* wrong at the time you did it and *you knew you were wrong when you justified it.*

It's arguable, though, that most members of society will say "you were not wrong" if you told them this story. I think it's unlikely you would be convicted of a crime in this case. If you believe this likely as well, then that means a new context has to be added to this truth to keep it objectively true: "Killing is wrong" needs "unless it's in self-defense" added. This means "killing is wrong" was not the whole story in the first place.

I could stop here because I believe this proves the original statement "killing is wrong" to not be objective truth and the whole story. But I think it's worth using a light to expose just how *big* the blind spots and gaps really are in a simple three-word story like this. Let's go ahead and modify it to see if we can keep this new expanded version true: *"killing is wrong unless it's in self-defense"*.

If there were no witnesses at all to this event, and there lays a dead man with a knife and a gunshot wound in his chest, and you holding a just-fired gun (or the dead man has a knife in his chest and you have blood on your hands), *who gets to decide* what has happened? This story can *easily* be re-contextualized to make *you* the aggressor. *You* killed someone. *You* had a gun (or knife). The story "you must have shot and killed him and he was only trying to defend himself with the knife" (or "the knife was yours all along and you stabbed him and are lying that the knife was his" if you are in the no-gun situation) can also exist as a possible truth-story.

So let's add to the story again to keep it true: "Killing is wrong unless it's in *provable* self defense".

But we can re-contextualize you and the other person again:

What if it turns out the attacker had a hidden drug addiction no one knew about and was the son of a billionaire, and you were a minority that was walking the streets in the middle of the night with a gun? Let's say this son-of-a-billionaire endured horrific trauma as a child by a person who was a minority just like you, so he believed you were going to harm him. He even posted a photo of you on social media he had taken from the alley, with a caption explaining how scared he was just moments before he was apparently killed *by you.*

Do you think your secondhand story (relative to everyone else) that *this man attacked you* can still make the most sense, or do you think it's possible for the now-grieving billionaire parent to overwrite your story, and replace it with an extremely well-produced and widely-pushed secondhand truth-story built and bought with money and influence? Do you think the words "Killing is wrong unless it's in proven self-defense" can be twisted to work against you? You did kill after all, and that was objectively wrong, unless it was self-defense. But how can you *prove* it was self-defense at all?

Typically, when it comes to proof, there is forensic analysis. Maybe you are a forensic analyst reading this right now. The analyst has a unique context as well. Did the analyst understand the words of your story about what happened the same way you did? What stories does the analyst use about bullets and splatter patterns and physics to build his or her own truth-story of past events? Can experts for *all* contexts be brought in? Is it possible another forensic analyst could be found to build a completely different truth-story of what happened from the same exact evidence?

What exactly does the word "prove" even mean? Who gets to decide something is proven when it comes to events that actually happened but can never actually re-happen?

If you can see multiple possible outcomes by changing the context like this, then perhaps "killing is wrong unless it's in proven self defense" is not an objective truth after all.

But let's keep going – just for fun – 'killing is wrong' can be applied to so many more possibilities that tear it apart!

Is killing wrong in war? Should soldiers not kill, and also be prosecuted for doing wrong if they do? "Killing is wrong unless it's in proven self defense, or war."

Well you can't let the other side kill your guys in a war and be *right*, that won't work either. "Killing is wrong unless it's in proven self defense, or war (but only for the troops on the *right* side)."

Should someone that tortured and murdered dozens of children one-at-a-time over multiple years be killed by capital punishment (the death penalty)? Or is killing wrong? "Killing is wrong unless it's in proven self defense, or war (but only for troops on the right side), or when used for capital punishment."

It's possible you still believe that *all* these killings are wrong, and the original simple story of "killing is wrong" is still objectively true and the whole story. That's fine too, although we've pretty much destroyed any 'objective' meaning of these words at this point.

Regardless, we are still operating at the *assumed* level of human beings doing the killing and also human beings being the subject of the killing. But other human beings are not actually mentioned in the words "killing is wrong". So let's cause an explosion!

Do you eat meat? (I'm not about to call you wrong if you do, after all, I eat meat as well.) If you do, how exactly did that animal end up on your plate? It's not alive on your plate is it? In the US, most people have grown up with a virtual "food curtain" that prevents them, as meat eaters, from experiencing death of the animals they eat. We don't seem to like that story in our culture – but it is very much part of the story whether we like it or not.

By choosing to eat meat, we each contribute to the harvesting of animals for meat to be the right thing to do, and since an animal cannot be eaten and remain alive, that animal must, by necessity, be *killed* in some way. It's possible you are keenly aware of this, and make your choice to eat meat anyway, like I do. If that's true for you, then add a story where "killing is wrong" is actually not true for you. "Killing is wrong, except for animals I eat." (I will return to this topic later).

Let's assume you are a vegetarian and still believe "killing is wrong" is the complete and objectively true story, and you avoid all meat to keep it true for yourself. So let's shift the context again:

Plants are living things. So once again, something is being killed so that you may eat. Plants are alive at a different "level" of consciousness than us (or so we seem to think), but they move, they communicate, they reproduce, and they do all kinds of things just like all other life. Perhaps you are okay with killing plants anyway. If so, then add a context where "killing is wrong" is not objectively true. "Killing is wrong, except for plants I eat."

But perhaps you only eat the fruit and the leaves of plants, and never take enough to kill a plant, like taking the roots of potatoes and carrots. This would mean you only eat the reproductive organs or dismember the plant, while not killing it. That's fine too. But does it still count as killing if you take one too many leaves, and the plant dies a few weeks later instead of the months it might have had without you taking leaves at all? Is it "killing" if the plant becomes infected at the open wound where you ripped off a leaf? It wouldn't have died if you hadn't taken from it, so have *you* killed it? Are you responsible?

Have you ever pulled a weed? Cut down a tree? Were you okay with that? If so, keep adding more contexts where the words "killing is wrong" takes an exception for you. "Killing is wrong, except for plants but only if the killing was accidental and more than one day of time passed before death, or I believe its death was needed to save other plants I believed were more important."

But why stop here? Let's keep going!

Do you take antibiotics? The root of the word antibiotic literally means "anti-life" or "life-killing". You kill tons of cells just by taking them. The goal is to kill the living bacteria that are making you sick, but they sometimes can kill nearly everything involved in the fight. They can be the equivalent of a nuke on a battlefield at times, killing all sides.

Do you use bleach or anti-bacterial wipes and sanitizers? Same concept – you are killing living things. And if you are okay with these, you have more contexts where you are okay with killing, and an exception must be made. "Killing is wrong, except for microscopic life I believe is capable of harming me, or when it is collateral damage toward that exception."

Do you watch every step you take? How do you know you didn't kill bugs, worms, tiny creatures, or bacteria by squishing them to death? Do you shower or bathe, cut your hair or pluck your eyebrow hairs? You are killing creatures dependent on you for life by removing them with these methods. "Killing is wrong, except for things I can't see or realize I killed."

Are *you* alive? By living you are facilitating the absolutely brutal world that is your living biological system of a body. Cells are culled, murdered, commit suicide, and die in acts of war every moment of every day inside your body. Does that mean you are killing by simply being alive? *I* am certainly not the one doing the killing in *your* body, so it must be *you*. "Killing is wrong except when it's done by my own body."

To bring this to full paradox, which is exactly what nearly every statement of certainty eventually becomes (word loop trap), to make "killing is wrong" objectively true for you by ensuring you aren't killing *anything at all*, you would have to kill *yourself* because your body is constantly killing other things without any input from you. The only way to stop the killing is to kill the thing doing the killing – which would be killing. A paradox.

But why stop here?

We can leave cellular life behind. Is killing a character in a video game wrong? Is killing in your imagination wrong? Is unplugging someone on life support killing and/or wrong? If aliens invade and you determine they are hostile, is killing them wrong? Is killing a beer at the bar wrong?

The bottom line is the words "killing" and "wrong" have all kinds of meanings – a huge explosion of them - and where you draw the line for what "killing" and "wrong" mean to you (which will be based on the context you choose to use when filling in the word-shells) *changes everything about the meaning of those words.*

And you cannot be the *Keeper of Objective Truth and Right and Wrong.*

Nor can the words themselves.

At some scale, bigger or smaller, immediate or years apart, we are all okay with killing. The difference between the wrongness of the guy with the knife killing you or you killing microscopic life simply by being alive, is entirely based on what contextual scale you are using to define the words.

Whatever beliefs you have about the statement "killing is wrong" at this point *are not wrong.* No matter what that story is for you. If you say "killing is wrong" you are not wrong, because you have a specific *wordless* story in your mind that is providing the context for those words.

Just remember those meanings are horrifically approximate and don't at all tell a complete story to *anyone else,* and that it's possible every other person on the planet understands that statement from an entirely *different* context. They can't read your mind, and you can't read theirs.

"Killing is wrong" cannot be an objective truth of *any* kind. It is purely subjective, with it meaning something slightly different to each and every person. And you can't try to force it and become the *Keeper of Truth and Decider of What Words Mean for Everyone Else* without collapsing everything.

As I close this chapter, I want to make one thing totally clear: I am NOT saying killing another human being is okay. This example is being used to expose a problem with words themselves. I am saying the words "killing is wrong" are not *enough* to be objective truth. You can end up using millions of words to actually work out the entire story of which killing is okay and which is wrong. If you don't believe me, ask a lawyer.

As I understand it, practicing law is essentially rooting through hundreds if not thousands of previous judgments of similar situations to your client's situation to try and find a legal *context* that best suits the story you need to best support your client, or if you can find a gap that might allow you to create an all-new context, inching everything closer to total collapse.

"Murder is illegal". Define murder. Define the words in that definition. Define those words. Define those words. On and on it goes - forever - because it's a loop trap. A definition paradox.

To push this to the limit: It's possible, even likely, that complete agreement of a story that is *objectively* right about killing can never be reached. Why? We haven't yet found the smallest life in the universe, or the biggest, because we've barely begun to explore it at all.

This means even if we build a "killing is wrong" story that included every single form of life we know as I write this, and it included every context and possible scenario for one life-form to kill another life-form, and there was no disagreement at all if it was okay to one life to kill the other, we *still* wouldn't have the whole story of *all life* in the universe to be certain the statement is objectively true.

It's *still* possible for new context to be uncovered.

And then the entire story would have to be rewritten.

There is no common ground here.

43

The Sky is Blue-ish

Let's isolate the simple story "the sky is blue" to be only between you and me. We'll ignore everyone else in existence. This story will be considered objectively true if you and I can agree to it. You and I must agree though, because to be objectively true the sky must be blue from all perspectives, which in this situation, is just us.

How can I be certain you and I agree on this?

I can't, because we don't even know each other's context to start with.

This is because of that one nagging truth I keep bringing up: You can't read my mind, and I can't read yours. My firsthand experience with the sky might be different than yours. This means what is happening to even believe the story "the sky is blue" as objectively true is a *whole bunch* of assumptions and culture-based norms. It's possible these are embedded so deeply you can no longer even notice them until you see the holes.

Which sky did you think of? Earth? Why did you think it was Earth? My guess is you may have thought this is because I said "the" sky. At the time I wrote this, we as humans typically share the same sky (the one on planet Earth), so it might seem obvious to assume "the sky" is the one above both of us.

If you thought "blue" would be agreeable to me, you made a lot of assumptions about *me*. But the opposite is also true: I've also *already* made a lot of assumptions about *you*. We are assuming each other's context to a completely absurd degree.

I have no idea *where* you are right now. I have no idea *when* you are right now.

Words are a one-way communication. Always.

Right now, I am typing words on a laptop computer at my desk in my house in the Midwest region of the United States. The year is 2024 and it is Spring. You are not here in front of me. I have no idea when you will read this. The only fair assumption for me to make about you, is that you could be *any time* in the future from the moment I publish this. These words are capable of lasting many lifetimes past my own. They will last as long as a copy of this book still exists. I am essentially time-traveling my thoughts in the form of words for you to read at a time in the future I can't possibly be sure about.

Perhaps "sky" means something completely different for you than it does for me. If you use virtual reality extensively, the sky in your world might be any number of colors. There might not even be a sky.

It's also entirely possible something happened to the sky after I published this – and the sky is now pink for you, or any other color than blue on a clear sunny day. It might have happened one day after this was published, or hundreds of years after my own death. It's also entirely possible you live in a region or area where the sky is a different color than blue at all times.

I can't know *when* you are, so trying to assume *where* you are is also a huge assumption. Even as I write these words, you could possibly be on the International Space Station orbiting the Earth. If that is the case, your answer was probably never "blue" in the first place. If you are reading this 3,000 years after I typed it, it's also possible you are sitting on the planet Mars, or somewhere else entirely different from Earth, with an entirely different color sky.

Even if we assume we are both on Earth at the same moment, can you and I even agree on what blue means? What exactly *is* the color blue? Try to put it into words for yourself. Can you do it? Have you ever disagreed with someone else if the color of something is actually blue? Was it more of a green? How did you decide who was right or wrong?

To make matters worse – how do you know I'm not color blind to the color blue? Maybe I can't even see the color blue. Hell, maybe *you* can't. How could I know that?

But let's take human perception out of it: Is the sky *objectively* blue on Earth, even without humans? Well, *no*. Blue as we define it is radiation (waves of energy) with a wavelength of (you might not like this) *about* 450 to 495 nanometers.

It's *approximate*. Blue is an *opinion*. Does light suddenly change to green at 496 nanometer wavelengths? If you've ever looked at a graphic of a color spectrum, you know there are seemingly infinite shades of color. Where is the exact point in that spectrum of a million colors where blue changes to green? It's going to be *approximate*. I think it's likely no two people will point to *exactly* the same spot on that gradient color sheet.

This means the word for the color blue and even the color blue itself is subjective and approximate. Only the light wave itself is objective, but you and I don't see things at a scale of nanometers. Nanometers will always be secondhand stories for you and me. Tools are required to "experience" such a length. We experience only the output of the tool we are using to "see" nanometer-sized waves firsthand. Without the tool, we humans see at a scale of "color." So what color *is* 496 nanometer wavelengths?

To put this frankly, blue does not *objectively* exist at all. It is *a word* created to describe a human perception, and it isn't exactly the same from person to person. It's only *roughly* the same at best. You supposedly have three different color-sensing cone cells in your eyes. One of them is responsible for blue. What if someone had four? Or two? What if *everyone* was missing the cone for blue? If this were true, would blue exist?

We believe we "normally" have three types of cone cells. Have you ever had your cone cells studied to prove it? What would happen to color if you had sixteen different color-sensing cells in your eyes? How can you know this is not already true for you? It's not like someone can ask "does this look blue to you?" *before* giving you the word "blue."

The bottom line is, even though the statement "the sky is blue" seems like a simple and an objective one-way truth, two things have happened: I *assumed* it was a simple question for you to go straight to "blue", and you may or may not have *assumed* I would agree with you.

We both might be in total 100% disagreement, because our contexts are completely different for each of those four words. Yet we continue to go through our lives not aware of this *massive* potential difference, and all too often start arguing and get angry with each other that *the dang sky is blue just look at it,* which boils down to "because I say it is".

Words are a hell of a thing.

Try to let them go.

44

Two Plus Two is Spelled "Four"

In my experience, math is often claimed to be the universal language. If it is, then the mathematical statement of 2+2=4 must be *objectively* true, even outside of humanity. If there is one claim in philosophy social groups I read more than any other, it's that 2+2=4 is an example of *objective truth*. Only *one context* for this statement can exist for the entire universe.

In a school math class setting, I confess this claim seems to be true and correct. I can't dispute it. Even outside of school it seems to hold up: Two of any object, plus two of any object, equals four objects. You might even imagine that aliens would be able to communicate using math, and agree with you that two plus two equals four.

So where are the gaps?

The assumption is that you are counting *objects*. There is also the assumption the objects are the same unit relative to each other, are not changing, and that time is not a factor. But there are more assumptions hiding than even these.

Why, when no units are stated, should it be understood adding things together doesn't change them? This is itself an *assumed story*. This is not actually universal - why would it be? In my opinion, it would be foolish to assume aliens would innately understand "if the human didn't state units, you just add them together but don't *combine* them."

You might be blind to this possibility because addition was taught to you repeatedly until it was ingrained as something you just accepted as valid and true: When you see "2+2," it is equal to "4". The assumption, or framework you use to think about math can basically be summed up to: "If it was not stated, it does not exist."

Sounds an awful lot like a linguistic certainty to me, what do you think?

So what exactly *is* math when we remove all assumptions and norms?

As I see it, math is a human-created system of quantifying (or perceiving) objects and/or patterns, and then trying to understand the relationships between them by creating *consistent* units (which become numbers) capable of representing those perceptions and relationships.

So what is the primary or starting *unit* of "numbering" things?

It is *one*.

From *one*, you can split one and add another one.

But *one* is not objectively definable. One cannot just magically define itself just because someone says "one means one."

"One" *is* a word. *Numbers are just words.* They are built from a different looking "alphabet".

Someone at some point held up a finger to you and said "one." This first-hand experience of a secondhand story was repeated and validated for you and would forever be used as the foundation of all math for you. It kept coming at you for years – decades, even. "One" *is* one.

But the meaning of one is not completely definable *by others*. Numbers, like words, are ultimately defined by *you*. A number, then, is any object or idea *you* perceive and associate to the word-shell "one".

One minute. One hour. One meter. One light-year. One cell. One atom. One person. One apple. You either *perceive* something in the objective world as one, or *imagine* something inside your mental boundaries being made up of one. *That* is where math starts for you. What's odd about this is that "one" is purely subjective. I can tell you that for each of the following three drawings, I erased nothing – but how many "ones" do you see? If only one version of the three images were handed to three different individuals, could they agree how many *ones* there are without knowing the connection between them?

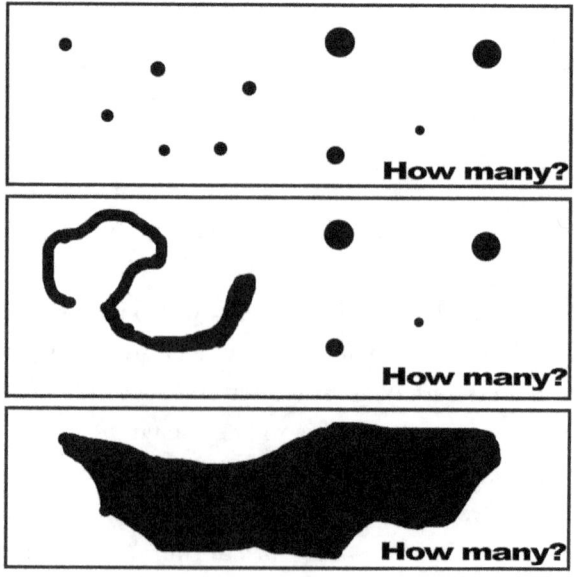

So then, how can we all seem to believe that two plus two equals four?

Unitless math, or pure number "words", is just a base *language*, just like letters have "word" sounds tied to them (Aye, Bee, Cee, Dee, etc). Number words are simply a linear scale order of symbols and words - like the alphabet - except numbers don't stop at twenty-six symbols and sounds. Number words keep going, and count up from zero "ones" to, in theory, infinite "ones". These "ones" can also be scaled, in theory, to infinite precision by simply adding a decimal point, which creates "subdivisions" of one.

With this number scale (let's call it a number *line*) memorized and a defined "one", you can imagine and create functions such as adding another one (two), or subtracting one from itself (zero), multiplying one times any other number (whatever amount you can imagine), and dividing one by any other number (into nearly infinite pieces). You can duplicate ones and organize ones into a square or a cube, or split them into pieces that you arrange into a square root or cube root.

Memorizing the number words you get and how you got them will save time instead of needing to count "one" number at a time up or down the scale, should you be dealing with *lots* of these "ones" in the future.

But what is the whole point of all of this math? To do tricks with a number line filled with "ones"?

As I see it, the whole purpose of math is to use it to predict and measure *objective reality*.

Unitless (perfect "ones") and perfect relationship-based math like I was taught in school starts to fall apart quickly outside of the human mind, just like words, because numbers *are words*. 2+2=4 is an already-solved problem. There's nothing to debate because there is nothing to solve.

It is like saying T+R+U+T+H = TRUTH. *Of course it does* – it's already done and is referencing itself!

Saying 2+2 = 4 is objective truth is the same as trying to say the dictionary is the keeper of objective word meanings. That's not where meanings come from – they come from your context – your experiences. The real truth of it is two of *what* plus two of *what* equals *what*?

Math in the form of "pre-defined quantity + pre-defined quantity = pre-defined quantity" is just *spelling with human math symbols and words*.

Once you try to convert idealized human math spelling such as 2+2=4 to objective reality, which is the *whole point* and reason the number line was created by humans in the first place, this simple problem takes on a whole new life. There is once again an explosion of possibilities that has the potential to destroy 2+2=4 completely, because the assumption that "one" is clearly defined *no longer exists*.

What if it's 2 hours plus 2 minutes? That's not 4. It's 2.033333 hours, or 122 minutes. 2+2=2.03, or 2+2=122.

But this is mixing two different "ones" on purpose. Don't worry though - as I see it, "math" and "one" come apart in so many different ways, it takes two chapters to break through.

45

One is a Word

How do we even have matching units - matching "ones" - for math-spelling to apply in the first place? Let's shoot some holes in units before letting math fall apart.

All time measurements, such as minutes and seconds, are *relative*, meaning there literally is no *objective rate of time*. If you think one minute is objective, which means no other minute can exist, let's play a new game.

1. Start a stopwatch.
2. Immediately close your eyes.
3. Try to stop the stopwatch at exactly one minute, but while keeping your eyes shut.

I think you will be further off from one minute than you might expect.

Add a friend to create a game called "Now I am Become Human, The Destroyer of a Minute". You both will probably struggle to ever hit a minute exactly, especially if the stopwatch also shows you tenths and hundredths of a second. The experience of time is always approximate at best. Without that stopwatch, the meaning of "one minute" quickly falls apart between any two people.

Another larger test of time units could be not to look at a clock at all *for days*, and trying to predict the moment of midnight, exactly 12:00:00 AM. Have another person with a clock to record the actual time when you guessed it was midnight, but don't have this person tell you what the clock reads at that moment. Keep yourself blind to it.

If everyone played this game, there would be a different context for midnight for each and every person. So what would happen to "midnight" if every single clock vanished from the Earth?

When you are born, do you start your life's clock (your birthday) at zero or at one? Did you know this answer depends entirely on what culture you were born into? We as humans aren't even using the same "one" counting system for our age on this planet, let alone using the same "one" for everything else.

Units of time get even weirder though. Not only are units of time relative between each person, but they are also relative between each *thing*.

For example, if you lived on Neptune, "one" year (defined as one lap around the sun) is not 365 days at all. According to NASA, Neptune's "one" year is 60,190 days. But are the "days" that make up that "one" year *Earth* days or *Neptune* days? The units called "one" hour or "one" minute are based on how fast the *Earth* is spinning, not how fast Neptune is spinning. Or are they, in this case?

Have you ever stopped and wondered what an hour even is? *Why* there are 24 of them? You might guess the answer - it's because someone just made up the new word "hour" as a way to divide "one" day into "twenty four" equal "one" pieces, and it stuck. Why? Oh, because it *made the most sense* to those cultures, which eventually passed it along to you – if you still use the unit.

This means if we ever settle on another planet such as Mars, which spins a hair slower than Earth and takes almost twice as long to go around the sun, we will have to deal regularly with two *entirely separate contexts* for words that relate to "*one unit*" of time: "Did you mean one "Earth year" or one "Martian year"?"

If you are a scientist that studies Mars, you already deal with this I'm sure, but there are no humans living there yet to communicate with in my time, which is when and where this problem will become a bit more significant. If an entire human society lived on Mars, is it easier for them to try and stay synced with Earth time? Or just create an all-new Martian time?

A Martian clock and calendar would create another context for time unit "ones" entirely – which loops us back around to wreck our simple mathematical sentence: 2 hours + 2 hours = ? Is *one* pair of hours Earth-based *ones* and *one* pair Martian-based *ones*? Is *one* Martian hour going to be measured by *one* Earth hour, so *one* day on Mars will be 24 Earth hours and 37 Earth minutes long? Or is the Martian day divided into "twenty four" equal *ones* of time that are each a bit longer than Earth's twenty four equal *ones*? Or are we going to create completely new words to represent each *one* of Mars's new time-units?

Imagine we master space travel and spread to live on 100 different planets in the galaxy. Are all of them supposed to use the same "Earth-hour equals one" basis of time? There would be, in theory, 100 different contexts for what *one* minute and *one* hour mean, *one* for each (*one*) planet (assuming you standardize clocks for all people on all planets).

This particular rabbit hole is extremely deep where even the speed of human aging starts to be different (see Einstein's theory of general relativity), but to really drive home the idea of it, consider a memory that may or may not be familiar: In school, did you ever experience the bizarre reality where the clock on the wall in Room A shows a slightly different time than the wall clock in Room B? Which clock is right and which is wrong?

If the teacher in each room obeys their own room's clock, which teacher is following the *objective* truth of what time it is based on the spin of the Earth? You might be tempted to say your network-linked phone's time (which might be expired words when you read this) is the absolute right time. Is it? How do you know? How do you know your phone carrier's system time is the same as another's? Which cell phone carrier's satellite timekeeping is the "objectively right" Earth-time from which all others are based, which can be used to "fix" the classroom clocks? Who is the *Keeper of Truth and Decider of the Right Time*?

Let's say it's 7:50 right now on your phone (or timekeeping device), and you set a timer for 2 minutes. When the timer goes off 2 minutes later, your friend immediately starts a timer for 2 minutes on his or her timekeeping device. When your friend's timer goes off, your phone says 7:54, but what if your friend's phone reads 7:53? Or 7:55? Which phone or timer is keeping *objectively* right time? You have no way to know using any sort of firsthand experience.

Everything in this situation is based on secondhand story sources, and is based on almost completely *blind trust*. In this case, 2+2 has equaled 3 or 5, because you jumped between clocks, and you don't have any way to resolve the time-unit relativity problem between them. If you both choose to sync to one phone or the other, you both might be one minute late, on time, or one minute early everywhere you go (ignoring that this same problem repeats itself with whoever is defining "late," "on time," and "early").

But let's pull away from time and move to measurement in general. You are also assuming *accuracy* if declaring 2+2=4 to be objective truth. If there is no decimal point, we *assume* the unit is a "whole" number. But what is a whole? 2.0 is the same thing as 2, right? If you measure an object to be 2 meters long, what's the "right" *accuracy* on that? Is it 2.000000000000 meters or is it 1.99999999999999999999 meters? Do you measure to the center of the last ink line on the meter stick? The closest edge of that ink line? The edge of the stick itself? What exactly is the tolerance, or *precision*, of the machines that *made the meter stick* you are using?

Plus or minus a tenth (0.1) of a centimeter?

Do you think the meter stick company coordinates with other meter stick companies to make sure they all measure the *exact* same distance? If you put two meter sticks next to each other and line them up at one end, how different will they be in length at the other end? We certainly haven't perfected meter stick manufacturing down to the *unit* of *one atom*. It might require a microscope to see, but *there will be differences in length between all meter sticks*. Meters sticks (and all rulers) are approximate.

If you are in the business of requiring precise measurement, you might point to the ASME, or some standard for consistency in measurement, whatever that organization might be. These organizations, made of humans, do not have the answer to this problem, and wrestle with it constantly. The only meter in existence that is perfectly "right" in length was the *first* one - *the original context for "one meter"*. Every copy made since then is not quite the same. We are not copying the original meter stick down to the precision of *one* atom, we can't. Even if we could, is the atom even the *right* unit to use for perfect precision? Should it be the quark instead? We don't yet have such precision available to us.

The one-meter measurement is always "plus or minus some margin of error", based on a second point: What the meter is being used to measure. The same goes for one "foot", "inch," and every other unit-of-one used for measurement.

Two centimeters + two centimeters on one meter stick might measure 3.98 centimeters on another. Two sticks, two contexts. But maybe you need to be perfectly precise down to four decimal places. Now you can't use *either* of these meter sticks. You've got to find *one* that was created with the precision you need, which is *one* ten-thousandths of a centimeter. But if you get two sticks made with that precision…what's the difference between these two sticks? Again, they each create their own context of measurement for *one*. You can declare that something measures exactly "one meter" all you want, but no one will ever measure *one* exactly the same as you unless they use the exact same measurement tool you used at *best*.

To make the measurement problem even worse, measurement-standardizing organizations recently changed the definition of a meter in an attempt to be more consistent – they tried to create an *objective* meter. They defined this "new *one* meter" as the distance light travels in a vacuum in 1/299,792,458th of a cesium radiation-defined second. Do you see any problems with applying these standards to actually construct a meter stick? Does this sound like a tolerance *precedent* nightmare in the making?

What's the *tolerance* on measuring the speed of light, being we still haven't measured the speed of one-directional light when I wrote this? Do you by chance have a stopwatch that is calibrated exactly to the cesium atomic radiation frequen-cy defined *one* second of 9,192,631,770 hertz? (Hertz, by the way, is frequency, or "cycles per second," so to make sure those cesium atoms you have in your pocket are truly at ground-state so your measurement is correct, you need to calculate *one* ephemeris-second which is calculated by measuring the *exact* rotation of the Earth, and then compare that to how long it takes your atoms to complete 9,192,631,770 cycles.)

Oh I almost forgot: This cesium frequency needs to be measured *outside* of Earths magnetic fields. Good luck getting out of it to build your "perfect" meter.

Also, how precise is your vacuum light-speed measuring device? You'll need to re-measure the speed of light inside of *one* vacuum.

No worries though, I'm sure all meter stick manufacturers have the capa-bility to measure the speed of light in vacuum, have cesium atoms they keep stored outside of Earth's magnetic fields, and are capable of measuring time down to almost one 300 millionth of *one* cesium atom second – just to fabricate that six-dollar meter stick for you.

Of course, you could just do the math using idealized values, right? The new meter equation is a very pretty equation that is going to give you a very precise number – *one* meter. Great, we have the exact length! To make sure we made it the right length before we use our perfect meter stick, let's validate the length is correct by grabbing this existing meter stick over here and make sure it measures exac-------oh.

Measurement paradox aside, what if I asked for a measurement in meters, and then handed you a yardstick? What's the perfect context for each, and are they each perfect examples of that context?

All other measurements of physical reality suffer from this context tolerance problem: Temperature, mass, current, pressure, luminosity, and so on. There are as many contexts for measurements as there are copies of the instruments doing the measuring. Even then, there are as many interpretations of the measurement as there are people *reading (see: interpreting)* the instruments!

Let's try switching to *objects* to see if we can make 2+2=4 objectively true without measurement tools. What if it's two oxygen molecules plus two hydrogen molecules?

Again we can get multiple answers. It's either one water molecule *or* one hydrogen peroxide molecule. Based on secondhand stories I've read, it's *typically* water with a leftover hydrogen. So, 2 oxygen plus 2 hydrogen = 2 molecules (1 water and 1 hydrogen).

2+2=2, or

2+2=1, hydrogen peroxide.

Both are technically correct answers depending on the context chosen. They are both possibilities of 2+2.

This might feel like I am intentionally complicating the problem on purpose, but I'm not, I am just filling in the gaps *normally* filled by *assumptions* with real possibilities - with *real* possible contexts for reality – exactly how words work. *Objects* do not care about your "units" or your "words". If I just say the word "one" it means nothing. One what?

Adding, as I see it, is incredibly hard to separate from *combining*. You have two oxygen and you "add" it to two hydrogen. If we assume "adding" is a verb, then it is connected to an *action in objective reality*.

If you've already counted two oxygen and two hydrogen, saying you have four molecules isn't saying anything at all. It could be even *dangerous* to say, because it erases the different meanings of "one" like it no longer matters. It's just a *rephrasing* of two sets into one bigger set. You are doing tricks with a number line, and in the process have actually *lost* meaning. The entire point of this number line is to understand what happens if you put these two sets of two molecules together in a way that can be applied in reality – numbers like stoichiometry possible.

But let's keep tumbling....

If you plant two apple seeds in two rows (two squared, or 2+2), unitless "perfect-one" math-spelling predicts you should get four apple trees (2+2=4). That might be true, or it might not. The answer is actually approximate. Possible answers of this "2+2" are actually 0,1,2,3, or 4, or *even more*. It is entirely possible for one seed to create two or more trees! On top of this, there is no guarantee you will get the same apples as the type of apple seed you planted. Apple seeds do not always produce the same tree they were grown from.

If you have two printers that can print two pages every minute, how many prints will you get in a minute? While you probably want to say "four" based on unitless "perfect-one" math, there are holes and gaps *everywhere*. Were both printers plugged in? Do they both have ink? Toner? Loaded with paper? Can the power circuit handle both machines' energy needs? Will one of them jam? So many possibilities must be removed by *assuming* these things won't happen for you to confidently answer "four."

If you drive for two hours at two miles per hour, this *should* result in a total distance traveled of four miles. Does it? Only approximately, assuming you don't break down, don't run out of gas, know how to drive, the car runs, your tires and speedometer are perfectly calibrated to precise measurement standards, you don't begin from zero miles per hour (you must start at 2 miles per hour), and a meteorite doesn't fall from the sky and destroy you and your car.

2+2=anything from 0 to ....I have no idea, but more than four.

Will a 747 take off from a treadmill designed to "exactly match the speed of the wheels"? Yes and no are both correct answers based on their respective contexts, and everyone used predictive equations to get there. But the gaps you need to cover up with exact values for the math of physics to predict the objective outcome perfectly, are abysses that can swallow the nearly infinite possibilities used to fill them.

But why stop at infinity?

Let's go to infinity +1!

46

Two is Absurd

"2" is a *symbol* for a quantity. If you weren't familiar with it, you would see "2" as *one* symbol, and have no idea that this *one* symbol stands for a quantity of *two* "ones". Would you recognize number symbols written in a language you aren't familiar with?

In binary language: 2 = 10. Good luck with number symbols out of context with their respective number line.

The written word "two" represents the same idea of *two* objects in the real world, but the word is actually *three* symbols we call letters. Would you recognize a word for *two* to mean two objects in a language you weren't familiar with?

Zwei, dou, zwee, deux, dos, and twee all also mean two. Not a single one of them are built of two-letter symbols. The entire relationship of symbol-to-quantity must be pre-defined. It must be part of a "set of assumptions and norms."

To add yet *another* possibility, the symbol "2" printed in this book is probably (but I can't be sure) done with dots of ink or tiny lights. If you were magically shrunk to be half the height of this symbol: "2" - you might only see *hundreds* of dots or lights and not be able to make out the overall *one* symbol they create, because that's how some printers distribute ink onto paper and how screens work (at the time I wrote this).

What even is a "whole" anything? Is it just where we perceive a single object to begin and end? Infrared imaging and other tools that "see" beyond our own senses show us that our unmodified perception of the edges of galaxies and nebula size is *way off*. There is so much more going on than we can see with our naked eyes. Is the firsthand experience or the secondhand story the true *objective* size of *one* galaxy or nebula?

If aliens exist and experience different scales of perception, how in the world do we communicate given that our language and numbers are built on assumptions and norms built on our own *human* perceptions? We might hold up two fingers, and they see 7 quintillion atoms, because they themselves are half the size of atoms, and can't even tell your fingers from your hand, or perhaps they only see a single cloud of heat, because we are approximately 98.6 degrees Fahrenheit and their "eyes" are tuned to see temperatures as colors.

But what if their eyes could only see temperatures between negative 20 and negative 40 degrees Fahrenheit? If that were the case, we'd be *invisible to them*. We wouldn't even be countable as *one* at all. (Ignoring if they were half the size of atoms, we'd be in a very Horton-Hears-A-Who (Suess, Dr. New York: Random House, 1954) situation to communicate at all.)

In the same train of thought, it's possible *we* are tiny in this universe. If we could communicate with a being outside of the universe, perhaps it would say "one" and we would be totally baffled, not knowing that our entire *one* universe is part of *one* atom that is part of *one* multi-universe-sized carrot that is being perceived and called "one" by that impossibly-huge-to-us being.

The inside of such a carrot would seem like infinity to our sense of scale. We have no context to comprehend experience at such a scale. Heck, maybe we are all part of one giant firework the multi-universal-sized beings have set off to celebrate Febtober 245th every 6,000 timeseeds!

Pause.

Center.

Reset.

Words are a hell of a thing. Take a step back and marvel at the sheer scale and situation you just imagined by reading them.

Back in the reality we can be a little more certain about, math and numbers often *seem* universal to us, because we all tend to perceive our reality to be *approximately* the same. That means we also tend to quantify things approximately the same. It's likely no human capable of seeing looks at an apple and sees only the seeds inside, or can't tell the apple apart from the hand holding it. Our perceived *ones* between each other are typically "close enough" to the point we have progressed a hell of a long way by developing *standards* for those "ones."

But this is all still resting on human-based *assumptions* – which we often call "axioms".

The problem with math in general is that it is an idealized concept.

$2+2 = 4$ is just shorthand for $00 \quad 00 = 0000$. You can only solve a math problem and be absolutely certain of the correct answer when the context itself is controlled and *limited*. The school environment does *exactly that*.

Math class limits contexts by removing units, or *pre-defining ones*, or by constraining the many possible ways that can be devised to calculate and then predict the answer in reality. They remove objective reality to operate entirely within mind-created ideals. As I see it, math is "spelling" class for the physical sciences that are out there trying to measure things with sticks that never exactly match.

You are taught to do math in a *very specific way* because it is actually another language, just like your words. It is taught in a very limited and instructor-controlled context so that right and wrong can be derived *at all* – also just like words. In reality, there are *always* ones involved, and always unique contexts for those ones.

So what's the objectively right answer to 2+2? The human-idealized and forced, unitless perfect-one limited-context answer *is* 4. This spelling is based on a shared understanding of what the number symbols mean because the schools you and I went to all agree (human culture standard). We go along with this because we are being flooded with tons of other secondhand stories to memorize as fact at the same time, and are trying desperately not to be wrong so that we are judged as valid by the *Keepers of Truth and Deciders of Facts*. This forced overload so often tricks us into believing our numbers always mean the same thing.

Four is not the guaranteed objective answer to 2+2 in objective reality.

The answer to 2+2 might actually be anything but four.

We are free of the looking glass now. Without it to slow us down, we will pick up some serious speed as we drop into deeper and deeper darkness, but we must keep going.

47

Trying to Force a Round Peg Through a Square Hole

My hope is that now you can see a whole new universe that will continue to expand in possibilities. It might make you terribly uncomfortable to be this far into the abyss, but trying to cover it up is not the best way to deal with it, in my opinion – it doesn't change anything about reality. Besides, my intent is to blow everything open even more before this section is done.

Words and language have given us wonderful things, awful things, and everything in between.

Unfortunately, the cost of this is extremely high, because while words give you and I the ability to expand our own understanding way beyond our individual and limited firsthand experiences, the process of learning your words involved *constant invalidation* that had no way to be prevented.

You are told which words are *right* to use from your beginning, and I suspect others are still telling you which words are right in your life today. I am trying very hard to make sure I am not one of those people.

There is no marker or age where we as humans can know that we should stop adding new words or new meanings for others. But....I can almost guarantee you recognize what this marker looks like. It is when The Rules set out by The Boss and The Consequences become intolerably unfair. This is the most "visible," in my opinion, during the teenage "rebellion" years.

Why does this rebellion period seem to consistently exist in American culture? In my opinion, it's caused by a collapsing mental boundary, induced by *years* of trying to be seen as valid and valued by a system designed to constantly and intentionally challenge and judge every word and action as *objectively* wrong or right. The result is a having a mind filled by a many-headed Hydra making so much internal noise it starts to become unbearable, and reversing this collapse and defeating the Hydra by force is what is being attempted (Even though you probably have very fond memories of a teacher or adult who regularly or randomly validated you and seemed to love seeing you learn, this is often not enough to prevent this process).

First You fights back and wields one hell of a double-edged sword. *The Boss is to be thrown off your trail. The Consequences be damned. The Rules are nothing but oppressive garbage to control you.* Validation exists only in your friends and peers, who seem to be suffering and are just as angry and in various stages of mental boundary collapse, stabilization, or expansion themselves. You might "clique-up" into a group that validates each other's feelings as right on this topic.

In my opinion, the "rebellion years," whether they are happening as a teenager or as a middle age adult, are the human mind fighting back against authority being derived almost entirely by language wielded as an invalidation stick to beat those same children with since they began communicating – which paradoxically began using the opposite process: Pure validation. Your truth is often mentally "beaten" with laws, with holy books, with rules upon rules upon rules, with test scores and class grades, with team losses and personal failures that require you to *tell yourself* you are wrong, because you are only valid and valuable if you do things right according to...*someone else.*

As I see it, words forcefully "unit-ize" the human mind.

This seems to be the extreme cost of being able to communicate with one another, but it is a brutal process badly in need of reform in my opinion. No one person is wrong, no one has intentionally done all of this to cause harm. The reason for it all is hidden in the shadows by the very thing that is responsible for doing it in the first place – words themselves.

Using the stories from earlier in this book, you began life with "one" word-unit for everything: Cat. Cat was the first "measurable" unit that validated your mind to others in the outside world. All the world was a "cat." Then that word-unit of one was split into two with "dog". All the world was either a *cat*, or a *dog*. Which was which? You needed validation and approval to be sure. Once you got that, along came "fish," and so on until you eventually have all the other word-units you know today.

Depending on how many words are known, the *distance* between the boundaries of words – the *size* of these word-units - might be dramatically different from person to person. I'll give you some quick examples to hopefully help you become aware of the size of your own *word-units.*

How many color-words do you have *between* red and orange? Do you have *any*?

How many animal-words do you have *between* a cat and a kangaroo? Do you have any?

How many tree-words do you have between a tree and a forest? How many trees does it take to become a forest? How many trees have to be removed to make a forest just "that group of trees over there"? How about between a tree and bush?

How many plant-words do you have between a blade of grass and a meadow?

How many numbers-after-the-decimal-point-words do you have between a *mathematical constant* and it not really being constant at all?

Does this mean that only firsthand *wordless and unitless* experiences are the real source of objective truth between you and I, and that we can finally build a shared foundation between us if we just remove words and units altogether?

Let's put this to the test as we tumble into what is often believed to be the lower layers in this abyssal rabbit hole.

48

The Prison Cell With Wacky Windows

When it comes to firsthand experience - no matter if you have a thousand friends or no friends at all - you are still totally and completely *alone*.

This isn't a metaphor or strange poetic thing – I'm dead serious. You are literally alone. It is the only thing you *can* be.

To understand what I mean, try to answer this question: Where exactly are *you* inside your body? By you, I mean the part of your body filling my word-shells with your personal context right now. The part of your body that makes decisions, justifies your movements, and experiences emotions. The part of your body that is responsible for remembering and collecting and controlling the firsthand experiences of you, which builds the strongest truth of all for you.

I can't possibly know your answer, but I feel it would be fair to assume you probably answered "my brain" or some variation of words for the wrinkly organ assumedly sitting inside your skull.

And I would agree with your answer. Your brain is very much where *you* mostly seem to live. It's not that your body is disconnected, it is very much connected, but everything involving your body still comes back to the core processor: Your brain.

If you believe you exist outside of your brain right now, consider this: I can swap any one of your limbs, and you will still be you. I can replace your heart, your lungs, your liver, your kidneys, your pancreas, your eyes (corneal transplants), ears (electrical implants), large sections of skin (though not the whole body of skin yet, but I have confidence this will be figured out) and almost everything in between, and none of these changes affect your ability to still be *you*. I can even separate your brain from your spine, and it will *still* not affect your ability to be you (assuming you have equipment in place to keep you from dying).

But the brain? No way. Your brain can't be replaced or transplanted. It can't be removed, damaged, or even hit too hard. Doing these things will either alter or destroy *you*. Your brain *is* you, and it is where you primarily exist *inside* your body. No one else exists in your brain. *I* am not in your brain. Your neighbor is not in your brain. Memories of others exist in your brain, but they are not *controlling* those memories from outside your head.

Second You is not *actually* other people, but is merely your internalization of what you think other people want you to believe. Outside of this problem, Second You is still very much *you*. You (both First You and Second You) are ultimately one point experiencing this moment right now.

No one else can experience your thoughts or memories of experiences firsthand. They are yours and yours *alone*. They are not going to be experienced by anyone else, probably ever, due to the very real mental boundary in place.

In a way, you are your own entire universe moving around on the Earth (or wherever you live while reading this).

But how are your firsthand experiences responsible for your strongest truth even built? How does *this moment* come to exist for you?

Well, your eyes connect to your brain through an optic nerve that sends images. Your ears connect to your brain through auditory nerves to send sounds. Your sense of taste and smell? Nerves. Your ability to move, touch, talk? Nerves, which carry signals to move muscles to do all these things. Your spine is not much more than a boney casing for protecting the "main bus" of nerves that connect your body and senses back to your brain.

So what does your brain directly experience?

Whatever these nerves provide.

Whatever chemicals and structures exist within your brain to decode, understand, monitor, and send and receive sight, sound, taste, smell, and touch signals through these nerves.

The results in what amounts to "tiny wacky windows" from which you "sense" and "perceive" the outside objective world from the prison cell that is your mental boundary.

I think it's worth taking a quick look out of the biggest windows while continuing our downward fall.

49

Putting the "Wacky" in Windows

You and I may have eyes, but we barely see with them. Assuming you have a sense of sight, your field of view for seeing color and detail is incredibly narrow. This "focus zone" is just big enough to read about five words in a row in this book. You can "see" the other words around the five you are reading right now, but you cannot read them at the same time.

Your peripheral vision, or "the corner of your eye" doesn't quite reach to your direct left or right. If you looked down on yourself from the top and the side, you only ever see *approximately* this much:

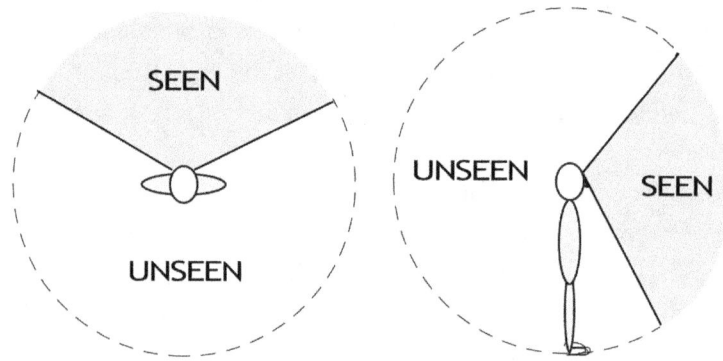

So, at any given moment, there is always more of the world you are *not* seeing than seeing, and most of this is pretty undetailed with only a tiny focus zone having "extreme" detail.

This isn't taking into consideration the very real blind spot inside your field of view that your mind simply makes disappear with - ready for it? Visual *assumptions*. If you weren't aware you had literal blind spots, I'd encourage you to go look up how to experience them firsthand. This usually involves covering one eye and staring straight ahead with the other, and then moving a piece of paper with a small mark around until the mark simply *disappears*.

Aside from these visual limits and blind spots you and I both have, you *probably* only have three different color receptors in the back of your eye, called cone cells. Some of these stimulate your optic nerve when red light hits them, some do this only with green, and some only with blue light. This means you see color in RGB (red green blue), which is a big reason why electronic screens use these exact same color lights.

I say "*probably* only have three different color receptors", because it's possible you might be missing a color receptor set of cells, meaning you don't experience certain colors the same way as I might. You might be "color blind" as mentioned in previous chapters, meaning you see in only RG, RB, or GB.

It's also possible you actually have a *fourth* color cone, but how would you ever know? If you saw colors no one else could see, how would you even attempt to explain yourself? No one has ever expressed a color experience related to having a fourth cone as a word to you before – so you don't even *have* a word to associate to it. You would have to invent the word to describe these unique-to-you colors. But even if you did that, how would anyone else understand? If you say you see the color "feuw," how in the world can you explain it to me? I don't have the cone cells necessary to experience it myself. This is one of the main problems I see with language forcibly unit-izing and "standardizing" our minds, the limited words restrict what experiences can be described.

Did you know the word blue is not in many ancient texts (Greek especially), even though the sky and the ocean are both what we consider blue, and were most likely very much blue even back then? These things were called many other color-words in literary works from that time and region, but never blue (that we have been able to find so far).

Color aside; you also don't see most of what you are focusing on. Your eyes do not move smoothly most of the time. Go ahead and take a look around the room or whatever space you are in, and pay very close attention to how your eyes are moving. Try following a straight line, like the seam between two walls or the horizon where the sky meets the ground. You might notice you are actually "jumping" your eyes from point to point along the line, and no matter how much you try, you simply cannot make them follow that line smoothly. This jumping is often called "saccades" (but I like "twitchy eyes" better).

Anyway, your mind literally fills in the "gaps" of what you do not see between each saccade with - you guessed it - visual *assumptions*. One of the only times your eyes ever move smoothly is when they are visually "locked" onto an object that is moving relative to you, such as a plane, bird, or ball.

Saccades aren't really a huge issue, because it's not like our eyes can send a constant stream of nerve energy to our brains, anyway. They have a "refresh" rate, just like an electronic screen. The human eye's refresh rate is *supposedly* about 60 "pictures" per approximate-standardized second, but this is hotly debated.

You also do not actually see the "right now" as it is actually happening. Light has to first enter your eye, get focused by the lens behind your pupil to hit the back side of your eye, stimulate those cone cells back there, which in turn sends the signal to your brain along the optic nerve, which then has to process the signal, and fill in any gaps between this signal and the next incoming "blip" of visual information. This whole process takes a few milliseconds to do. What you see as "now" is actually just a little behind the "now."

Your experience is just far enough behind the now that you can to be killed, injured, or attacked by something you never saw coming – because you simply can't "see" fast enough. You are just enough behind the moment to be fooled by the magician, who draws your focus somewhere specific, only to do another action so quickly outside of that focus zone you can't possibly catch it.

All these limitations aside, the eyes are truly big windows into your mind.

When you are deep in internal thought, your eyes change and your pupils (the blackness of your eye) dilate. When you lose focus on your thoughts, or are done thinking, your pupils shrink again. The same dilation happens when you see something beautiful or pleasurable. In this way, your eyes can show others what you *may* be feeling or that you may be in deep thought. Just remember that deciding what another's eye dilations mean is an *assumption* unless you get on the poultry treadmill.

Let's switch over from your sight window to your sound window:

You and I may have ears, yet we barely hear. Assuming you are not deaf, your range of hearing is very narrow in several ways.

Sound is just waves of pressure, and your ears are shaped to funnel these waves into your ear opening to vibrate your eardrum. The eardrum causes three tiny bones on the other side of it to vibrate together, which in turn makes pressure waves in the fluid of your ear, which in turn vibrate tiny hairs inside the fluid-filled cochlea, located deep in your head. Each of these tiny hairs connect to nerves that run signals to your brain, where they are converted into "sound". Technically, you are "hearing" objective reality using tiny rattling bones in your hairy ear water.

Still, it works for you and I. We can generally hear pressure vibrations as low as about 20 per "approximate standardized" second for the deepest bass, which would be the frequency elephants use to communicate. However, elephants can also supposedly make sounds so low you and I cannot hear them at all with our naked ears. A device becomes required to give us the secondhand story and shadow experience.

On the high end, you can hear sound at about 20,000 vibrations per approximate second, which is just a touch below a dog whistle (which begins around 23,000 vibrations per second). Some animals use even higher frequencies than this to communicate, including bats (as high as 160,000 per second), tarsier monkeys (90,000 per second), dolphins (23,000 per second upper range), and whales (as high as 175,000 per second), all of which we cannot hear without a tool to do so, so all of this information I'm sharing is from secondhand story sources. Take these stories with a small-to-medium grain of salt.

"Ultrasound machines" like what is used in hospitals, use sound frequencies from 1 million vibrations per second up to 100 million per second (though this upper end is only experimental when I wrote this). Ultrasound is basically echolocation, like how a submarine sends out a *ping* to find things around that reflect the sound back, where it is then converted into an image.

If you've never thought about it, technically when you use an ultrasound to "see" like this, you are seeing a secondhand story image made by a secondhand story sound receiver, which is receiving secondhand story sound waves that were reflected off a secondhand story of an object.

Of course all of this vibration stuff is nowhere close to the total range of pressure vibrations possible in objective reality. You recognize that *you* can't hear a dog whistle, but you probably believe that dogs can by their reaction to it - but if you consider sound is just rapid and small changes in pressure, the frequency ranges that are possible to exist beyond what you can hear are extensive.

To take things to the extreme, the atmosphere of the Earth above you right now (assuming you are on Earth) changes pressure as it spins over the course of a day. This is an example of one vibration you can't "hear" because it is far too slow or low to register as sound to your brain coming from "tiny rattling bones in your hairy ear water".

There is also a massive spectrum of higher pitches beyond our ability to hear. What is the upper limit of frequency in reality? In theory, it is when the size of the pressure change is smaller than the spaces *between* the pieces of atoms. A sound like this would - also in theory - rip atoms apart into plasma, but I'm not sure I am completely on board with this idea for reasons I will share later.

On a different note, there seems to be evidence that hearing is the fastest of all your senses. If a loud and unexpected bang suddenly occurs, your hearing can completely bypass your brain altogether, and cause you to turn your head *toward* the sound in a very reflex-like reaction.

And with this, I'll close out sound, and head over to the remaining three major senses: Taste, smell, and touch.

I don't think I need to spend much effort taking apart taste and smell. You already know the full subjectivity that exists when it comes to these two senses. There really is no objective "everyone likes the taste of this" or "no one likes the smell of that." I'd argue these two senses might be the ones we absolutely recognize can be different between any two of us.

I think there are many things in reality that seem to have no taste to us, or no smell. This is because we simply don't have the sensory "range" or "sensitivity" to detect these things. A dog can smell significantly better than us, and we know it. As far as taste, perhaps there really is something out there that can taste the rainbows.

The sense of touch is also largely subjective between person to person, with big and noticeable differences in what "hurts" and what does not, what is cold and what is not, and overall sensitivity to contact.

Yet there is probably a huge universe of things that we simply cannot feel, such as the mites that live in your eyebrows and eyelashes, and the germs floating through the air.

The point of this chapter was not to disassemble our senses versus experience, but rather to point out that even your own senses cannot be used to determine *objective* truth. They are missing huge amounts of information they just can't seem to pick up for us to experience, and what they can pickup is only ever *approximately* the same between any two people.

This means you and I cannot even use our senses to build a shared foundation of certainty between us. No sensory judgments or reactions can forge an *objective* truth.

Mass hysterias and mass sensory events that seem to be unexplainable except through the mind creating sensory experiences that seem to not really be present in reality, make it impossible for us to be certain what we perceive is *really* happening.

The mental boundary is wrecking our ability to be sure about anything when it comes to agreeing with each other. The view out of our wacky windows is slightly-radically different between us all.

And so now we have fallen into this hole's deepest darkness.....

50

You, Me, and the Room

For this chapter, my starting assumption is that you have your five major senses (sight, hearing, taste, smell, touch). I recognize it is entirely possible this is not the case for you. If you are missing any of these senses, don't worry; we will be going far beyond them.

Let's turn back the clock a couple-hundred years from my time: The very early 1800s. "Modern" technology, such as electricity and utilities I currently enjoy like heating, cooling, and indoor plumbing, do not yet exist or are not widely established. Nor is there any modern medicine.

Imagine you and I are sitting in a room together. The room is made of wood and stone, and the temperature is cool as the sun throws its last few orange beams of the day through the wavy hand-made glass in the window frames.

You and I are next to a large stone fireplace with warm, crackling flames dancing within. We each sink into comfortable brown fabric armchairs, arranged so that we are half-facing each other and half-facing the fire. A flavorful drink fills the glass in your hand. I have a filled-glass as well. The smell of freshly baked bread lingers in the air. Flickering firelight dances off our cups and faces, as we continue our chat into the fading day and growing evening.

We talk about whatever topic is interesting to you. It's been a good conversation so far, and we show no signs of slowing down as the final sunbeams retreat from the window, and the fire throws shadows that dance playfully on the walls.

As I am telling you my very best joke, your vision suddenly fades to black. The flicker of firelight disappears for you. The walls of wood and stone are lost to shadow. Your imagined appearance of me, my expressions and movements, the room, and everything else you might have "seen" before this point are now total darkness.

You are completely blind.

I am sure this would alarm you. I'm sure you would let me know this has happened to you. I'm also sure I would attempt to comfort you and call out for help. We were having such a good time and you seemed very nice, and I know this would certainly be scary if it were happening to me.

This might be an odd time for this question to pop up, but…has anything changed about your mind though? Your thoughts are still intact. You've lost one of your senses. One of your wacky windows to the world just slammed shut.

Other people in the world around you, perhaps even you, live without this sense every day. Physically, you've only lost a few nerve connections into your mind.

Back in the room, a few moments of panic pass. As things settle, we both become hopeful this is only temporary. There is not much that can be done in the 1800s to help you, but I do what I can anyway: I have someone outside of our room send out an urgent message for a doctor – who will be a few hours away at best. Cars, phones, ambulances, and emergency rooms don't yet exist in this time period.

Several moments later, your world goes silent. The crackling of burning logs disappears, and so does my voice. You cannot hear yourself speaking, but you can tell you are still able to make sounds by feeling vibrations in your neck and head as you mimic the memorized movements of words.

You are completely blind and now *totally deaf.*

And once again, I imagine your level of alarm is intense. You cry out, though it's likely your words come out slurred without the ability to hear yourself articulate. I again try to comfort you. There are a few ways I can try to do this. I can make contact with you by tapping or stroking your arm, by stomping my foot, or by banging something loudly so that you can *feel* it. I can also just hold you and stay near, to give you something to feel - some way to know I am still here with you and haven't left you.

Your senses have definitely changed, but has anything about your *mind* changed? Your thoughts and memories are still intact. You've lost *two* of your five primary senses. There are stories of individuals (perhaps one of them is you), who live without these two senses every day. As I write this, Helen Keller's story is probably the most re-told.

Without these two senses, you could still read this book in braille - though if you do not already know braille, it would certainly be challenging to teach you braille now – though it *might* be possible.

A few more anxious moments pass, your senses of taste and smell disappear, too. There is no more aroma of baked bread for you, and the tasty liquid in your glass, which I had offered you as comfort, no longer has flavor.

Plenty of individuals lost both of these senses in the COVID-19 pandemic from 2020 to the present day of me writing this.

Four of your senses are now gone. This is, no doubt, a traumatic imagined experience, but has anything about your mind changed? It is my opinion that the answer is still most likely "no". Your thoughts are, in theory, still intact. You are *still you*. But now you've only got one sense remaining: Touch, which thankfully still offers quite a lot to work with.

Our communication methods do not need to change much with the loss of these two additional senses, as I am currently unaware of any culture that intentionally communicates with smell or taste as a primary method.

Laura Bridgman is perhaps the most famously re-told story of a human being that has lost this many senses. She lived in the time period you are imagining in this situation – the 1800s. Bridgman originally had all her senses, but after coming down with scarlet fever, she lost her sight, hearing, taste, and smell.

There are others in this world who reach this level of sensory loss, and it's possible that group of people includes you: Those who are locked-in, also known as locked-in syndrome. It is certainly an appropriate name considering what it does to those affected by it. Those who get "locked-in" may keep their ability to move their eyes, but only on one axis, and may be able to blink. The person may have a sense of touch in some small areas, but almost everything else is lost. The shades of being "locked-in" vary depending on the severity of the brain stem damage responsible for this condition.

Meanwhile, back in The Room while I try desperately to comfort you and plead for others to get the doctor here now by any means necessary, you lose all sense of touch and bodily control before what is happening to you can be understood.

You are numb and paralyzed from the top of your head to the tip of your toes. You cannot move, you cannot feel anything at all. The comfort of the armchair is gone. The feeling of me touching your arm or holding you to comfort you is now gone with no sign of return. The warmth of the fire disappears. Your ability to make sound is gone with the paralyzing of your vocal chords. You have no idea of your own position or the position of anything else.

If you believe there are more than these five senses, whatever those may be, go ahead and remove them too. For the sake of avoiding picking this imaginary situation apart on technical details, assume you continue to breath at a fixed rate you can't control or feel, and your heart continues to beat, but you can't feel it either. Basically, you will remain "alive", but without any sensory experience whatsoever.

You are now fully imprisoned within yourself – within your mental boundary. Are you still you?

I still think the answer is most likely *yes*. You very likely still have your thoughts and memories, but you have no way to get them outside of your own skull. You have no ability to know where you are, what is happening to you, or to me, or anything else. There will never be another signal into your mind from the world you once shared with me in the very room you no longer know if you are still inside. All the nerve connections to the rest of your body are not registering. You cannot communicate in any way that you are still present at all.

Would I even still exist for you if you had no senses at all? My guess is yes; at first you would believe I still exist because you *did* have a firsthand experience of this a few minutes ago. However, I think your certainty of my existence is going to quickly start fading.

You would have absolutely no way to know if I was still sitting across from you. You would have no way to know if the same thing happened to me that has happened to you, or if I was now dead, or if the house was now burning. You are trapped in your own windowless prison of solitary confinement in total darkness. The objective world could explode and you would never know it. Your nearly limitless mind gets no more experiences with the world around you. You would be your thoughts, feelings, and memories, and nothing else, and even some of these might be debatable.

So would I exist to you? After awhile, I believe the answer would be "no." I believe the only thing you could be sure exists at all is your own thoughts. And even then, you have no way to know your thoughts are still attached to *anything at all*. You would, in theory, be your remaining thoughts drifting in a void of nothingness. Words do not exist to explain such an experience. How could they?

From my unaltered sensory perspective back in The Room, you are now completely unreachable. I cannot communicate with you using any known method in the 1800s. The only method that exists to communicate or prove you are still present in your mind at all, would be mind-reading or telepathy, and there is no evidence anywhere this is possible. Human research on claims of "real" telepathy and mind reading reveals long lists of frauds, mentalist tricks, and keen observations of the tiniest of physical action giveaways by the person whose mind was "being read."

The truth seems to be, with a very high level of certainty (though never 100%), that no one can actually read your thoughts, and I feel safe also assuming no one is literally speaking directly into your head to communicate with you (like telepathy is so often portrayed in fictional stories).

I suspect it's possible you might have trouble accepting this, as I know of many individuals that think these things are truly possible, so I want to dive slightly deeper into mind-reading or telepathy:

Be sure you are not mistaking *familiarity* of someone's behavior or words as "mind reading." In the case of familiarity, you would be merely predicting what *action* comes next with really good accuracy based on patterns alone and knowing how someone follows such a pattern - you are not *literally* having a two-way conversation mind-to-mind, or *literally* experiencing that person's thoughts or injecting your own – you are *telling yourself* what that person *appears* to be thinking. This is almost certainly Second You as a Hydra from Hell trying to cross the mental boundary it cannot actually cross.

With this said, back in The Room without any of your senses, not only can you not communicate with me, nor I with you, but I believe there can be some debate if your mind could still *be present* at all. I believe we have consciousness *because* of our senses, and without them, all consciousness is likely lost. There is a definite connection between mind and body, as I see it. When a person loses one sense, other senses step up and provide more information than they did before. Or, perhaps, more brainpower becomes "freed up" to dedicate itself to those remaining senses.

We have now tumbled beyond the lowest layer, and it might seem as if there is nothing left to find common ground. But yet we are still falling...

51

Freefall

You exist in complete and total isolation inside your head. This means your eyes, ears, nose, mouth, skin, and other senses provide your *entire* reality and therefore your *entire* truth. Your senses give you the story of your life, and are the only tools you have to try and make sense of it all and interact with the world around you.

This also means your entire reality *is subjective*.

Your reality is actually your mind's *interpretation* of what your eyes, ears, nose, mouth, skin, and body are experiencing.

Objective reality itself is a *shadow experience*.

The universe, your galaxy, and your world seem to *objectively* exist. That is my belief, and I have many reasons to believe it, which I will explain in upcoming chapters. But you are *also* part of that objective universe because you also exist, and your understanding of all existence including yourself, is limited to what your senses show you – what comes through those tiny wacky windows.

Your experience with reality seems *objective* to your own mind, this is only because you have nothing else to compare to. You are only one point, imprisoned inside your own skull. You cannot possibly become aware of what your eyeballs are doing (or not doing) to the objective-reality light waves that you experience as colors and shapes, without also having firsthand experience through someone else's eyeballs for comparison. Even if you could, and you could see the differences, which set of eyeballs is seeing things *right*?

We are often not aware of our own subjectivity, because from our one-point perspective, experience is objective.

No two prison cells have exactly the same windows. They are all *approximate* between each other at best. Some cells have badly warped windows, some have fewer windows, and some windows are clear while others are hazy.

This makes you and me not only 100% subjective beings relative to objective reality, but also relative to each other.

Objective experience does not exist for us as humans. The only objective truth for us is that there is no objective truth.

I can't experience your experience. You can't experience mine. There aren't enough words to express our experiences with reality to each other anyway, and even if there were, my words will forever be shadowy secondhand story experiences for you, and yours shadowy secondhand story experiences for me.

Even though everyone currently living inhabits the same reality at the same time, we each have our own unique experience because we can never overlap our unique perspective with one another. *We do not pass through each other.* Because of this, no person *ever* has the same exact viewpoint of any objective-reality event relative to any other person. *Being in the same place at the same time is not possible.* So all perspectives are only ever approximate, at best.

If you are color-blind, you don't perceive the same colors in the world as someone who is not color-blind, so something like the color of a tree is up for debate.

If you are hearing impaired, you do not hear the same sounds as someone who has perfect hearing, so the sound a dog makes is up for debate.

If you and I both witness the same crash, we each see it from slightly different angles, hear slightly different things, and even interpret events leading up to and immediately following the crash differently, so the cause and severity of the crash is up for debate.

And they will be up for debate *forever*, because experiencing someone else's senses firsthand is *not possible*.

As much as we try to standardize the story of our senses to try to overcome this problem, there are immeasurable issues at work. What defines the meaning of relative words for our senses like "impairment" or color "blindness"? You are provided a standardized *pre-defined* test to measure these things, but what would the result be if you actually had hypersensitive or genetically-mutated sensory organs?

For example, if you had a vision test and you have a fourth type of cone cell which allows you to see an entire spectrum of colors those with three cone cell types cannot see, a standard vision test *would never find this*. The test works on the assumption you have cone cells for red, green, and blue. In this way, we are unintentionally *limiting* the possibilities of *valid*-and-enhanced human senses in our search for "defects."

These limits are applied *to you*, and perhaps applied to others *by you*. If you've only ever taken a standard RGB-based vision test, and you've already learned your words for your experience of colors, then no matter what color that "red" square *actually* is to your fourth-cone-cell-having eyes, you will still associate that sensory experience to the word "red". Why? Because someone told you "This *is* red". You had no way to know you saw things *differently*. You were born with your eyes since day one, and the words for colors in reality are already chosen by someone else. No one, including you, will likely *ever* know that "red" appears for you in a way that *no one else* is experiencing.

We have *standardized* what is possible through words themselves. The limits of language become the limits of our reality, and so limit *human truth*. Words filter experiences, which in turn filter words until the two disconnect.

Consider eye-witness testimony of…well…anything. Two eye witnesses to the same event almost never have the same story, and they most certainly will not use identical words to describe their experience. This is not evidence to argue words aren't limiting our perceptions, in my opinion. Instead, this is evidence that different perspectives provide different contexts, which select and fill different combinations of very limited word-shells.

Two people who are taught the exact same educational material will not build the same understanding, and they almost certainly will not use identical words to describe what they do understand when asked later.

How do you know what you know? It's always dependent on those tiny little windows that let you experience anything at all: Your senses. Without them, you are locked in darkness. You will, in theory, experience nothing - learn nothing - communicate nothing. Senses and perception are your *everything*.

They are, quite literally, you.

And now, after the end of You, Me, and The Room you have lost all of them – where you have lost all connection to our shared reality, and are trapped in the solitary confinement of your skull - your subjective existence is now the only objective reality you can be sure exists at all, if you can even be sure of that. You may have even lost "you." The world where you and I chatted by the fire no longer exists. Maybe it was a dream. Maybe it never happened.

This is not the bottom of the rabbit hole. This is the part where it opens up into an enormous cavern. A *void* of obscure concepts we cannot experience firsthand while living as humans. It is a shadow world of darkness, with guessing games about the structure and makeup of the entire universe using words alone. "Everything is…." "Existence is…."

I am not going to dive into the attempts at *objective* reality thinking beyond my experiences, which at times goes as far as "everything is vibrations" and "everything is nothing." These are quite fun concepts to think about, in my opinion, but they discuss reality in a way we cannot experience or control as humans. The result is trying to define truth with secondhand stories, which flips everything upside down the instant we declare we've got it all figured out *with certainty.*

This extreme route of universe-sized thinking created by the words "everything is", causes us to jump outside our own firsthand perspective. This can lead to some very fascinating thought-stories and can be an escape from reality all on its own.

My own daughter became fascinated by some of these ideas converted into media when she saw the final scene from the film Men in Black (1997), where the camera keeps zooming out from Earth to beyond the scale of the entire universe, which just so happened to be contained in a marble. As the camera keeps pulling back, an alien life form is playing games with that marble, as well as many other marbles, which it then picks up and puts into a bag.

Another version of this appeared on the animated show "The Simpsons." Our viewpoint zooms out from the Simpson's family home where the family is sitting on the couch, to beyond the universe, which reveals the universe is just part of an atom that is part of Homer's head (the father that was sitting on the couch at the beginning of the sequence) which returns us where to we started.

These ideas are *fun.* Ideas like these can be and should be explored in my opinion. They create new possibilities. But, even if we discover tomorrow our entire universe is in a drop of water on the back of an enormous wet dog, it changes *nothing* about the nature of *your* personal daily experience. It does not change the authority structures in your life or stop the Floating Anchor Problem. You and I would consume that story and still have the *same problems* we had before we consumed it. A huge discovery like this would boil down to telling ourselves something along the lines of, "neat," and then returning to other familiar matters.

So, if you choose to stay here and live in the endless shadows and darkness that is explorable only through limited words required to use other words to have meaning (paradox) I wouldn't dare call you wrong, but I believe you'll be looking for truth where it cannot be found from your locked-in and limited perspective. I believe you'll be fumbling around in the dark with no flashlight, trying to find answers and have experiences that cannot be found or had. You'll be stuck trying to construct a foundation that cannot be truly shared, and so cannot be used as a foundation at all.

However, I feel I would be doing you a grave disservice if I left you now, and ended this book in the darkest part of the abyss. After all, the list of all those words that seemed to be so often believed as the "cause of all society's problems" have meanings that connect to very real people in the world. You and I haven't actually solved any problems at all; we've just torn lots of things apart.

Even though no one in any group actually agrees completely about anything, groups of people are not going to magically dissolve. That's like asking everyone to give up their weapons while on a battlefield – the last side to give them up is in a position to easily become the winner.

You can see the problems with words and contexts and perceptions all you want, but this still only makes you partially aware of the absolutely massive *Floating Anchor Problem*. Awareness of these problems alone doesn't actually stop the runaway process, and there are still many hidden things I want to show you. We still need to find common ground between us, so that words can't drive us back apart.

Fortunately, you are already face down on something solid. This surface was always down here in the perfect darkness, but it is extremely hard to find, because you have to learn the shape of this place. I will share how I figured this out a bit later.

Those who have tried to venture this deep usually bounce off the backside of the looking glass, and then turn around and exit as quickly as possible, carrying with them the belief they have found the *right* words to fill in the hole and rise all the way back out with The Truth.

There are no *right* words for Truth, though. There are no words at all. Of the few that have dared to venture this deep – far beyond words - some are never able to pass back through the looking glass, and become lost down here, desperately looking for answers in pitch black that cannot be found. It took me a long time to find my own way around, and I would have surely gotten lost myself without help from the unlikeliest of places.

Anyway, when you are ready, pick yourself up and dust yourself off, and we'll begin exploring the very bottom of existence.

I brought a flashlight for you.

There are so many things I am very excited to show you.

PART THREE - FOUNDATION

In the beginning there was only darkness.

No space.

No time.

52

The World of You

You have just arrived at a place where the ground beneath you will not immediately collapse, sending us deeper into the abyss. This is the first place where I become sure you and I will be able to agree.

We have tumbled a long way down the rabbit hole. We have fallen past all the religions and governments and even the great barrier of language. We've watched past experiences and facts zip by. Even your own senses (and mine) have passed. All of these things were capable of dividing your truth-story from mine.

We've now reached a point where every direction *except one* will re-flip things upside down. It took me months to figure this place out with lots of help, and I intend to guide you along the same path I was shown, leading me to many discoveries.

Doing this might be the only way you can help me fill gaps I may have missed, and help ensure we have the strongest truth possible.

I apologize if some of the journey to this point made you uncomfortable. I'm also sorry for using words to create the shadow experience of slamming your sensory windows shut to the world, leaving you in absolute blackness, and perhaps terrifying un-reality. If I didn't feel this was necessary to arrive where we are now, I wouldn't have done it.

It's time to open all those windows back up, though. It's time to step into a new world - *The World of You.* It begins wherever you are right now and with whatever you are doing right now while reading this.

Take a minute and look around you. Note the details in every direction. If you are missing your sense of sight, just use the senses you do have to take in the world surrounding you as much as possible. Really soak everything in the best you can. Take your time.

What around you was made by humans or by human-invented machinery and systems? What movements, sounds, or smells are human-made? If humans made everything around you that you are experiencing firsthand, what is *beyond* the human-constructed boundary? Is it perhaps trees and plants, or more buildings and concrete? Are you on a plane, or maybe in a car or vehicle?

Roll these questions around in your mind while continuing to notice everything human-made around you. Imagine what is beyond what you can sense right now, no matter what that might be. It's all about to become very important.

If you still haven't actually looked up from this book, and have only continued reading; *stop* - take a moment - look up and all around and notice these things, and then come back.

When you've got a pretty good sense of what around you is human-made and what is not, use your imagination to "pause" reality.

Everything freezes in place, unmoving, except for you.

Now let's start making some adjustments to reality: If you are not at ground level right now, imagine you are magically moved to the surface of the planet in the exact same spot you are now (I am assuming you are on a planet). For example, if you are in a high-rise apartment or in an airplane, move yourself straight down until you are on the ground. If you are in a subway, basement, cave, or mineshaft, move yourself straight up until you are on the surface.

If you are not currently over or under dry land because you are perhaps in something like a boat or a plane flying over the ocean, imagine yourself magically transported out of the vehicle to the dry land nearest to you. If you cannot imagine what the nearest land might be like because you are far away from anywhere familiar, you are welcome to imagine being moved to somewhere familiar instead.

Again, take your time and really imagine this with as much detail as possible. Once you've got a good grasp of where you are and what is surrounding you on the surface of the planet, feel free to continue.

With reality still paused, and in your current or new location, imagine everything around you made by humans *vanishes*, and is instantly replaced with the natural world as if these human-made things had never existed. Buildings, houses, roads, sidewalks, cars, pipes, lights, and airplanes are all included in this list of disappearing things. Everything human-made must go.

Take inventory again at just how much around you is human-made wherever you are or imagine yourself to be, and then try to imagine it all gone. Fill in these places with what you imagine *used* to be there before we humans came along.

A busy cityscape might return to a forest, desert, marsh, or grassland. Crop fields and many pastures are often man-made by the clearing of trees, so they too must vanish and return to perhaps a fenceless grassland or forest.

Domesticated animals, which would not exist without human tampering, all vanish as well. This includes every domesticated pet. If you would prefer to not imagine dogs and cats and other pet species gone, then instead imagine them to be mixed feral breeds unfamiliar with humans and spread throughout the wilderness.

There is only one exception to this vanishing-of-all-human-made-things: You get to keep any human-made device currently keeping you alive or helping your senses or movement.

For example, if you wear eyeglasses to correct your vision, have hearing aids or have a replaced heart valve, or maybe wear a prosthetic arm or leg, you not only get to keep these devices, but also know they will keep working without issue indefinitely. As an alternative, you are welcome to imagine your body without these human-built devices being needed in the first place. You can imagine yourself with restored eyesight or hearing, or a restored heart, or with all your limbs, for example.

The idea here is for your body to remain as fully functional in this imagined world as it is in the real world while you read this book, with or without human-made devices and modifications – whichever you prefer.

At this point the only thing surrounding you is raw wilderness, the creatures that inhabit that wilderness, and any other human beings (if there were any around you when you started this thought experiment). All of this is still frozen and unmoving.

Now *all other human beings* disappear as well.

It is now only you. You are naked (clothes are human-made after all) in the unmoving world, with all the other non-human creatures and life still in it.

Everything made by humans is gone, and now no other human beings exist in this place either.

You are aware of all of these changes. You *know* everything human-made no longer exists, and you *know* you are the only human here. You get to remember this transformation process. You keep all your thoughts and memories just as they are while reading this.

Now…if you are anything like me, you've probably already started "over-thinking" about all sorts of consequences to these changes and started forming a million questions. You might wonder how such a dramatic transformation to the planet-scape will affect your situation. I'll do my best to clear this up:

The environmental stress of this radical change is wiped away. By that, I mean the sudden re-appearance of a huge amount of plants and the disappearance of a huge amount of concrete and pollution in the world is of no consequence or focus in this experiment. We aren't going to consider oxygen level shifts or worry about the consequences of sudden changes in air quality. This great "restoration" *just happens* and the environment is stable in this new configuration with no shock to your body or other life.

Any changes to the food chain are of no major consequence either – every-thing in the natural pecking order will stay how it was before, minus the existence of domesticated animals (or their conversion to being feral).

If you are near a human-altered body of water, then any waterway disasters that would be caused by things like dams suddenly disappearing are also to be wiped away. Imagine these waterways are instantly and smoothly returned to their pre-human, natural, stable, unblocked existence.

If this forces you to move from where you are now to avoid ending up in a now-larger or reshaped body of water (like a river that is suddenly much wider) then imagine yourself moved just enough to stay the same distance from the water's edge as you were before these changes.

As for personal stress, I'll offer you some relief: Any relationships you had with other now-erased humans will be restored back to before (or how you would like them to be) *after* you successfully navigate this experiment.

This being said, there is a pretty big unknown here. *You have no idea how long this experiment will last.* It might be hours. It might be years. It might even be decades.

Ok, go ahead and un-pause everything.

Let your revised reality start to move.

Please sit with this for a bit. Look up from this book.

Imagine your situation to the best of your ability.

What does it look like?

What does it sound like?

What might you smell?

What does it feel like?

Imagine *everything human* gone as best you can.

You are alone and naked wherever you are.

The only human in the world.

The now all-wild planet.

Put your finger in this book now, or stop these words from coming at you for just a minute - however you prefer to do this. Look up from here and take in a deep breath.

Really get a good feel for what this world might be like.

When you think you've got it, come on back – I have a question waiting for you.

What is the first thing you need?

Don't continue to the next page until you feel confident about your answer!

(

Seriously.

Look up from this book, or pause this audiobook, and imagine yourself naked and alone in the all-wild world.

What is the first thing you need?

I have very little doubt you thought of one of three things, or maybe all three things:

Food   Water   Shelter

If you thought of something else you needed first, I'd be very curious what it was, and if one of these three words just overwrote that choice.

I have set up this thought experiment with many individuals face-to-face, and I've not gotten an answer yet that appears outside of these three words. Don't get caught up thinking these words have exact meanings, though. We are well underneath the looking glass of language right now. Each of these three words is itself a huge *category* of possibilities.

All words are approximate, remember, and are filled by your experiences and the secondhand experiences of those that gave you your words. If you thought of an answer that is more specific, my hunch is that answer *still fits* within one of these categories of food, water, or shelter.

To help ensure you understand what I mean, I'll expand each of them to show what can fit into these three category word-shells.

Water (experience): Any liquid you thought of drinking to quench thirst.

Food (experience): Any thing you thought of eating to satisfy hunger.

Shelter (experience): This can include a *lot* of things and is a good example of just how much a single word can pack into it. To me, shelter at its most basic means protection.

In *The World of You*, shelter is protection from anything related to the world around you. This could be shelter from other animals, hot and cold temperatures, weather (such as rain or lightning), exposure to the sun, or any other conditions and events that might happen outside of your control.

Wherever you might be when you look around in this imagined world, you probably have a good idea what kind of shelter might exist nearby, and what you might need to seek shelter from.

Fire, if you thought of this idea as a first need, is a form of shelter. It protects you from cold, darkness, and possibly even other animals should you pick up a burning log and wield it as a weapon.

Clothes are another form of shelter. They protect you from weather, the sun, cold temperatures, or from minor skin injuries.

Climbing a tree would be a form of shelter against animals on the ground, and a tree also provides shade and protection from a few elements.

Jumping into water can even be shelter from certain predators and also shelter from heat.

Although shelter is not always *required* for resting, it can eliminate a lot of risk to you while you sleep, which is another thing you are going to eventually need to do. That being said, I'm almost certain *'going to sleep'* was *not* what you thought of first after being thrust into this rather harsh experiment.

A weapon, if you thought of that, can fit into a couple of these categories depending on your intent. A weapon can be a form of shelter if intended to defend yourself against animals that could appear to threaten you. This could be a rock, a stick, or some combination of the sort you build from raw natural materials.

A weapon can also be your first tool to *assist* in your need for food, but it is worth noting a weapon is not necessarily *required* to get food. If you thought of a weapon for this purpose, it actually becomes a form of shelter against hunger, because it's purpose is to help you acquire food faster or easier. I think it's import-ant to note, however, the need for food came *before* this need for a weapon, and the weapon is just a means to help. I don't believe you would choose to starve to death while trying to create the first weapon. For this reason, a weapon as a tool ends up collapsing into whatever category you wanted that tool to help you with. I think this will always drop it right back into the category of food, water, or shelter.

I am not the first to think these three things are humanity's common ground of course. Way back in 1943, a psychologist by the name of Abraham Maslow devised something he called the "Hierarchy of Needs", and he put "food, water, shelter" at the very bottom, which made it the foundation for his pyramid.

Before Maslow, the philosopher Aristotle of ancient Greece listed the same three things as needed to pursue other goals in life, and he supposedly said this around 2,200 years before Maslow did.

All three of these seemingly timeless unchanging human needs have one thing in common, which pretty much guarantees them as a solid common ground between us - *survival.*

I am not about to disagree with the many others before me who claim food, water, and shelter are foundational to being human. I believe that you, at your most primal level, are well aware of this as truth, even if you somehow did not answer my question with one of them, or your answer did not fall into them categorically. These three needs are nothing new, but I do think we as individuals sometimes tend to overlook them.

The human drive to survive appears in many books and stories about the mind, psychology, and human beings in general. Although all the authors I have read don't really seem to linger on this concept very long, or seem to explore all that is going on at this basic level of existence. It seems the human survival need just *is*, which itself is an *assumption*, and authors of all kinds seem to go no deeper.

They'll often turn the plot of their story around this idea, or simply touch on it, point out the bottom of Maslow's hierarchy, and then just start climbing right back up the layers.

I struggled to accept this born-with-a-need-for-food-water-and-shelter story at face value and move on. There had to be a deeper truth hanging out here in *The World of You* waiting to be uncovered. After all, what matters next in this story? How could I be *sure* you and I would continue to agree on any story beyond our need of food, water, and shelter without me declaring what *should* be next? Surely these can't be the only things we can agree on. Something happened between *The World of You* and present day for me (2024) where reality started to split at the seams while ultimately sharing the *same* basic needs.

I paced over this imagined all-wild place for quite some time. Eventually I decided to test others. So, over the course of a few weeks, I put several individuals into *The World of You* using face-to-face conversation. Something immediately caught my attention. Every individual appeared to perform the *same actions* while developing an answer for my simple question: "What is the first thing you need?"

Every person - *every single one* - did not *immediately* have an answer when faced with this question. It always took several seconds of very deep thought to come up with a response. This is exactly why I put in lots of blank space in this book after asking you the same question, and tried to force you to pause and look up from this book both before and after I asked it.

Since I am not in front of you to observe your actions, I instead tried to break your flow from word to word, and forcibly insert a moment for you to drop into deep *personal* thought. Maybe it worked, and maybe you just skipped the blank space and kept on reading – I can't know. I do know that in face-to-face interactions though, a massive pause seemed to be *required* to construct the answer. *Why?* I didn't immediately understand.

However, this pause is not the *only* action every face-to-face participant had in common. Every person - again *every single one* - ended up saying not just one, but *all three* words without any prompt from me. You might have done this as well, but I can't know and so I won't assume it.

Not only did all three words arrive when face-to-face, but each word also arrived in a *matching pattern*. Many seconds always had to pass for the person to come up with the first need-word. After saying that first word, the individual did not look for validation or for my reaction to it, but instead *immediately* dropped back into deep thought. The other two words followed quickly.

I held my tongue until the individual's body language told me thinking was finally done. "Food", "water", and "shelter" came in different orders, but as of writing this, not one person has ever failed to mention all three within *the same mental dive*. Nor was there ever any doubt, question, or concern that I might think it was the *wrong* answer. There was apparently no Hydra from Hell in this place. This process seemed to step right around all *Keepers of Truth*.

Lastly, most participants also did something else. This "something else" eventually opened a new rabbit hole into the foundation itself.

53

A Different Loop Trap

You are alone on the all-wild planet where no humans or human-made things exist, except you.

I just asked what you needed first, and you (assumedly) identified your survival needs - food, water, and shelter. I have good reason to believe, however, those words are not where your thinking *stopped*.

Based on most of my interactions when it comes to others and *The World of You*, I think it's extremely likely your mind also built *and automatically* answered a very big question right after identifying your survival needs.

The answer to this self-built question was nearly always given out loud and *unprompted* after the three things needed to survive were said. It didn't seem to matter if the conversation was face-to-face or thought social media or messaging apps.

"I would be dead in a day," or *"I would maybe make it a week at best"*. The unspoken question was relatively easy to reverse-engineer from these comments, in my opinion. *How long do you think you can survive in this experiment?*

As I see it, your answer to this self-built question, if you asked it to yourself (and I suspect you did), will depend highly on your background and experiences *before* this experiment, since you brought your context in with you and it is being used to create both the question and the answer.

If you are an Eagle Scout or a trained survivalist (and I've talked with both), you probably have much higher confidence entering this thought experiment than if you have never directly experienced the wilderness before and have only ever experienced a cityscape. As it turns out, these survival-trained individuals were the only ones that did *not* blurt out this answer to the self-built question.

Unfortunately, this means there is not much common ground in *The World of You* that might be based on the answer to this question. I'm trying to stay fair to you, so I can't assume anything about your survival skills as I write this and use it to deepen the foundation between us. I have no idea how long you think you can survive in an all-wild world. So...I found myself stuck.

I dwelled on this for quite some time. Many days later, it suddenly occurred to me how to get unstuck. It turns out the self-built question needed to be framed in a different way, because the answer cannot be given in any "standardized" units like hours, days, or years and result in a shared answer.

The reframed question is: *What determines* how long you can survive?

On its face you might think this question should be easy to answer. It's how long you can keep getting food, water, and shelter, right?

Unfortunately, it's not. It's not actually that clean and easy *at all*. There are gaps all over the place in such a story. Here's how you can blow it apart and see them:

If you can keep getting food, but can only find one small meal a month, is this a problem for your survival?

If you can keep getting water, but can only find a few ounces of it once a week, is this a problem for your survival?

If every shelter you find or build immediately falls apart, would this be a problem for your survival?

The answer to each of these assumedly being "yes" deepens our shared foundation just a little more. It means you and I are not *only* tied to getting our need for food, water, and shelter fulfilled. It means we are also tied to getting each of these needs filled *within a certain amount of time over and over again* if we want to continue surviving.

You need lots of water at least every couple of days to avoid dehydration, and you are going to need a specific amount of calories from eating something at least every few days to fend off starvation. Your body sets the rate of these needs based on how fast you burn through water and calories just by being alive.

To package this up with a little bow: The ability to not only fulfill your primary needs of food, water, and shelter, but to *keep* fulfilling them in recurring *personal-sized loops of time* is required to survive.

These are, in essence, your "life-loops". Which is what I decided to call them when I first noticed them.

This life-loop concept seemed simple and elegant to me. It was a story that made tons of sense at first: If you just ate something, the need to eat again will be told to you by the feeling of hunger. If you just drank something, the need to find water again will be told to you by the feeling of thirst. If you are in danger, the need to find shelter again will be told to you by the feeling of fear.

All of these life-loops very likely still apply to you today as you read this book from the past. Food, water, and shelter are *still* most likely your primary survival needs. If you've ever not been able to find food, or had trouble finding drinkable water, or been stuck outside in a storm or extreme weather unprepared, you know how urgent and serious these feelings can get if you can't find what you need *when you need it.*

There are, of course, other things you need at *all* moments, like the air you breathe, or perhaps even a machine or device that keeps you alive – but these are not things you must actively seek out in *The World of You*. You either have breathable air on the surface of the planet, or you do not. This is why I allowed you to bring devices that help you function and also know they will continue working without issue. This is why I chose to wipe away the enormous environmental consequences such changes would bring to the air on the planet. These things are beyond your control as one person at the speed of your life-loops.

Unfortunately, the joy of making sense with these life-loops didn't last long. I quickly found myself frustrated and stuck again, even with this looping realization. The loops did not help me figure out what came next between us to rise back out of this abyssal rabbit hole.

Were the only things I might ever have in common with you be that we both need food, water, and shelter when our bodies say we need them over and over again?

As it turns out, I just needed a little help – a new perspective to gain a better understanding. That help would come from somewhere very unexpected.

54

The World of "Mouses"

After the realization we all have repeating "life-loops," I grew increasingly frustrated by them. I was unable to build anything more off of this foundation with another human being.

So I began talking to more and more people in my life, trying to find some pattern or consistent answers. I built *The World of You* for each one, and would then just talk with him or her about it.

It didn't take long to stumble into the wisdom I needed, although I often find myself thinking about how easy it would have been to dismiss and overlook its source.

I dreamed up a modified version of *The World of You* thought experiment, and my own daughter (9 years old at the time) ended up teaching me quite a lot within it.

She and I called this imagined place *The World of Mouses*.

The starting concept is the same as *The World of You*. You are alone in a world that no longer has evidence humans or their creations ever existed. You are the only human on the planet and you know it.

But now the twist: There are no longer any animals on the planet bigger than a mouse. There are no bears, wildcats, bats, dogs, snakes, or birds of prey. If the creature is larger than a common field mouse, no matter what it is, it no longer exists in the world - with the exception of you.

I know this would have dramatic consequences to the food chain and the stability of the environment, just like the waterways and atmosphere changes in *The World of You*. Ignore the potential environmental shock. Just assume all the larger animals have disappeared, and everything left is stable at the same numbers they were before everything else disappeared.

Got it imagined?

Ok, now, how long do you think you can survive in *The World of Mouses*? Has this amount of time changed between *The World of Mouses* and *The World of You*?

Perhaps it has, and perhaps it hasn't. Once again, I cannot know. I suspect a big change in your answer only occurred if there were large creatures in *The*

World of You that would pose a constant threat, such as lions, bears, or large cats. You might also feel a big change if you had planned to eat these larger animals to survive, despite your lack of weapons or tools.

Otherwise, I'm fairly confident your outlook *hasn't really improved at all.*

To amplify the point I'm trying to make here, let's give *The World of Mouses* another twist: Imagine you now have a burbling spring of water right next to you, a magical basket that always stays full with whatever natural foods you enjoy eating, and a small cave complete with a perfectly sized stone that works as a door. When this stone is in place, this cave provides you a nearly perfect shelter from the elements.

But let's not stop here. To try to improve things *even more*, imagine your environment is now always a comfortable temperature, which means you require no extra methods to heat or cool yourself to stay alive.

In short, you now have *all three* survival needs fulfilled at whatever interval you could wish for. You have food, you have water, and you have shelter. I have no doubt a ton of stress has just been removed from your mind.

But let's loop back to the question and turn it all the way up to eleven: Can you now *guarantee* your survival in *The World of Mouses* until the end of this experiment – which might be days or years or decades?

If food, water, and shelter are the only things we need to survive, then it seems like the answer *should be* "yes". Just ride out this experiment until it's over in this perfect spot, right?

It may or may not come as a surprise the answer is *no*.

It turns out I underestimated my kid's ability to put multiple stories together to make a bigger truth. She didn't even hesitate to answer this question with "no."

At the time I created this version of the experiment, she and I had been listening to a podcast we enjoy where the two hosts teach each other the stories of diseases including the discovery, history, symptoms, treatments, and future expectations (Erin Welsh, Erin Allmann Updyke; This Podcast Will Kill You). As a result, my daughter learned all about water contamination, parasites, sanitation-based diseases, and many other horrible ways to get sick and die from the environment. This immediately curbed any overconfidence she might have had in *The World of Mouses.*

If you know these stories when entering this experiment, which almost certainly *still apply in full* for you, the results are very humbling (in my opinion). Guaranteeing your survival is not something you can *ever* do in *The World of*

Mouses, even with every survival need fulfilled in a timely manner and all large creatures removed. There is an entire universe of the invisible and the unpredictable waiting to kill you for your mistake or overconfidence - or for just existing. Here's a few possible stories:

- Your spring water becomes infested with bacteria that give you severe diarrhea. After a few days you die of dehydration.

- A piece of food in your basket is covered in salmonella and you develop severe vomiting and diarrhea – a good chance you die.

- You scratch yourself exploring the world around you, and over the next few weeks an infection sets in and you die.

- A tiny bat smaller than a mouse bites you while you sleep. Little did you know it just infected you with rabies, and you will die in one of the most horrible ways imaginable over the next few weeks.

- You slip, fall, and snap a bone in your arm while moving around on a rainy day. It develops an infection, and you die within weeks.

- A rock comes loose and tumbles down a hill onto you, killing you instantly.

- With no way to sanitize your hands or your food, you accidentally infect yourself with poop-based parasites, or maybe get food poisoning, or develop a massive eye infection after rubbing them causing eventual blindness, and with any of these comes a more probable early death.

- You experience allergies that develop into a sinus infection, which spreads into your brain and eventually kills you.

- The common biological causes of death from the time I wrote this still exist as well in *The World of Mouses*: Strokes, cancers, heart attacks, embolisms, and organ failures are all still possible and *much* less survivable in this experiment than they are with modern human-made medical intervention.

There's no one else to help you when you are the only human.

There are still a million ways to die.

Even with no large animals and every need handed to you with ease.

This means your survival is not something that can be guaranteed even for one week. *Ever.* And this is likely still true as you read this today.

Tomorrow is never guaranteed.

Five *minutes* from now is never guaranteed.

You may die while trying to finish this sentence.

When I considered the sheer magnitude of this, I finally realized (with some help once again) you and I need something else. There is a gap *between* the words that build the story of survival.

As best I can tell, it turns out our brains are always trying to fill in or cover up this abyss whether we realize it or not. The need for this gap to be covered up ends up being what gives us this seemingly "innate" drive to survive, but it also creates a heck of a lot of problems.

55

Fishing in a Lake With No Bottom

My daughter and I spent a lot of time fishing the first summer after The Event. She really enjoyed it, so we visited all the local parks to fish the ponds and streams with bits of hot dogs. She liked to catch fish and put them in her pond-side 5-gallon fish tank. She would then name them and pet them, and invite other kids to see and touch them as well.

If you are wondering about the hot dogs as bait, well, she struggled with putting worms on the hook, and hot dogs were an easy fix. They also came with the added bonus of allowing her to bait her own hook and carry around a bag filled with pre-chopped-up bait, which freed me up to fish, too. In other words: Hot dogs were a win-win (other than the smell after they were in the hot sun for awhile).

One afternoon we visited a really big pond in a park that allowed bank-only fishing. At the pond, we encountered a man fishing with multiple fishing poles. He and I chatted for a bit while my daughter fished nearby, and I learned he had caught a massive fish out of this very pond the previous summer.

He very excitedly pulled out his phone, scrolled a bit, and then turned it to show me the screen. On it was a picture of him standing with his legs spread wide apart while wearing a huge smile and holding up a monster of a fish by the mouth with both hands. It was as big as his entire torso.

The man went on to explain that he released this fish back into the same pond we were now fishing, and he was hoping to catch this monster a second time.

I couldn't help but smile at his joy in both the photo and in his retelling of the story, and I thanked him for sharing both. I then returned to my daughter to continue fishing with her.

After a few minutes passed with no bites, we decided to move down the bank to the other side of the many-poled man. As we moved past, he was in the process of baiting one of his hooks.

To my daughter's horror, this involved pushing an enormous hook straight through the body of a small bluegill, resulting in a very serious wound almost certain to lead to the eventual death of the fish.

I quickly redirected her away as soon as I realized what was happening, but the damage was already done. He was harming the very same type and size of fish she was trying to catch. From her perspective, he was actively harming the animals she cherished and just wanted to meet up close.

The conversation following this moment would be one of the most enlightening conversations I've ever had in my life. I knew telling her to "get over it" or "sorry kid, that's fishing" wasn't going to help her. Instead, I decided to jump on the poultry treadmill with her. I sought to understand my nine-year old daughter, who was obviously upset and in emotional turmoil.

At first, I told myself a story this situation must be upsetting for her because harming a fish "felt different" from harming a worm or using a hotdog. So I started there.

"What do you see as different between fishing using a fish, and fishing using a worm?"

She replied quickly with "The worm isn't trying to get away!"

I gently explained that the worm is very much trying to get away (or at least squirming like crazy which seems likely to be the same thing). I added that the hotdogs we were using were technically a cow at one point, and that meant she was fishing with a cow on a hook instead of a fish.

Neither of these stories seemed to make much of a difference to her. She was still visibly upset. So, I decided to try a different angle: The food chain and survival.

I used *The World of Mouses* the two of us had visited so many times before, and tried to tie this situation to the food life-loop for her, "If you ran into me in *The World of Mouses*, and I said you can catch a fish big enough to feed you for a week if you put a mouse on a hook instead of a worm, would you do it?"

"Yes."

"Ok, well there ya go! That is what that man is doing with the smaller fish. He is using the smaller fish to catch a much bigger fish. He is getting more food by catching a larger animal with a smaller one. It allows him to get many meals by sacrificing what would have only been one small meal without much added extra effort or risk."

She stayed silent. I was confident this was the right approach, and she needed to accept the fate of the farm-raised baitfish, which was bred specifically for this sacrificial purpose. I was sure she was just uncomfortable with the food chain by believing the life of the fish somehow weighed more than the worm and the mouse. So, I gave her space and time, because every kid probably deals with this at some point after the connection is made between the harm of other things and most of the food that appears on plates. It's the moment they have to build and accept the line where "killing is wrong" to them.

But my comment wasn't actually helping. I could tell she was still very uncomfortable about everything, and I felt uncomfortable for her. I didn't want this fishing trip to end with her upset, but I was struggling to find a way to comfort her without saying something that declares her feelings wrong like "life's not fair" or "get over it". Something deeper was happening in this situation that was very uncomfortable for her, something more than just food-chain stuff.

It actually turned out *I* was the one that needed space and time, because with just a little bit of it, another possibility suddenly occurred to me: Maybe this wasn't about the food chain or one animal's life being worth more than another at all. Maybe this was about *the action itself.* She did, after all, witness a creature suffer a mortal wound with her own eyes. She did not witness the cow lose its life to become a hot dog, so that action was invisible for her – lost in shadow – a secondhand story easily ignored or filtered away. Her refusal to fish with worms fits with this story as well, because using worms requires her to purposefully justify and perform the action of harming another living creature.

She just experienced something *firsthand* she had never experienced before. It was an action that *intentionally* inflicted pain and suffering. There was a *certainty* to it. It was not reversible. It could never be undone. The wound that harmed and would likely end that fish's life was witnessed with her own eyes, and even worse, there was no guarantee this harm would lead to catching the bigger fish at all. Nothing is ever guaranteed. *I had failed to consider the lessons from my own thought experiment.*

She could catch and sacrifice a hundred mice to try and catch a single massive fish in *The World of Mouses*, but the reality is there were never any big fish in that world – they are larger than mice so they don't even exist (and neither do fishing poles and hooks).

I spoke to her again to ask if this story of witnessing harm meant anything to her. Her reaction told me I was very much approaching from the right direction now.

She was upset because she valued the fish, and that fish was seriously harmed and would die, and there was no guarantee doing this would result in anything but the harm and eventual death of that very same fish. From her perspective there was no reason – no *justification* - for *any* of it to happen. The man had no intent to even eat the larger fish – it was a catch-and-release pond. He just wanted to catch it again and release it again. No food chain was involved in any way unless the huge fish happened to catch a lucky break and swallow the little one off the hook.

Now for a big question though: Was the man *wrong* for doing this?

As I see it, *not at all*. He was chasing an experience exactly the same as she was, because he felt it was the right thing to do. I have no ill feelings toward him whatsoever. I've done the exact same thing while trying for a big fish in my past – from sacrificing worms and minnows, to putting a bait-stealing bluegill back on the same hook I just used to catch it as bait for something bigger. Who am I to judge this man? Am I *King Eric, the Keeper of Fishing Morality?*

For my daughter though, who has a huge heart to help animals, and who had just saved a bird tangled in fishing line at the very same pond just a couple days before, witnessing that fish get pierced with a hook and knowing it would likely die for no practical reason at all was a bit too much.

Did she have gaps and blind spots? *Of course she did.* She is technically endorsing secondhand-harm herself by using a hot dog. She also harms the fish she catches because she uses a hook, but she only directly experiences a tug on her line. The act of harm - piercing the hook into the fish - is a secondhand story. I suspect she would never want to pick up a fish from her pond-side tank and shove a hook through its lip just for fun. It did help that so far in her life, she had not yet caused any *fatal* harm with her hook.

But now *I understood.*

She directly experienced the *certainty of a justified action to intentionally cause irreversible harm or death to another creature* in a world where our main goal is to survive, which can *never be guaranteed.*

On top of this, she also made the discovery that we as humans are often the ones directly responsible for justifying these actions to harm other creatures. I have no doubt these stories must be incredibly uncomfortable to process for anyone the first time they are noticed.

I found myself reflecting on this conversation for a long time.

It didn't take me long to connect the difficulty of a child processing the harming of a fish, to the difficulty of processing something we all must eventually experience firsthand...

56

The Puzzle With a Missing Piece

After the fishing trip, I was convinced avoiding the certainty of mortal harm and death while seeking food, water, and shelter had to be the explanation for nearly all human behavior. It *had* to be. It just made too much sense. There wasn't much it *couldn't* explain. We create all kinds of systems to avoid thinking about the experience of death, especially in the United States. We seem to try to *cover it up*.

For instance, as I see it, the major religions in the United States seem to be trying to remove the reality of death completely by making it a transition to some sort of afterlife. As a culture, we try to make our dead appear alive-but-sleeping at funerals. We tend to avoid hospitals and morgues. Cemeteries are often considered scary places, and are taboo locations to loiter or socialize. We try to memorialize our dead by putting their names in stone – a kind of marker to the memory of their life that will stand for centuries *beyond* their death.

We, in the wider US culture anyway, seem to me to very much avoid accepting death as the end of our story. This was the major lesson I took away from the conversation about fish with my child.

I felt pretty satisfied this foundational puzzle was getting closer to completion. There were now less gaps and blind spots in the shared story between you and I than there was before: You and I both need food, water, and shelter over and over again in repeating loops, because if we don't, we die. Since you and I can't know *when* or *how* we will die out of the countless ways it can happen, our individual minds do their best to control and avoid stories that seem to be moving toward potential harm or the ending of our own stories.

I think there is a good chance you will feel horrifically uncomfortable if there is the *slightest hint* you are (or were) in a story that contained the possibility of your death or serious harm. Examples could be missing your plane and discovering it crashed after takeoff, or witnessing another driver fly across the intersection in front of you at very high speed *after* your red light had turned green. It could be as immediate as stepping out of your house or apartment if you fear you could be killed in the world outside, or as indirect as obsessing over vitamins and supplements in your ever-advancing age.

I suspect it's also very likely you've had nightmares and anxieties revolving around your serious harm and death in some way at some point in your life. I know I have had *many* nightmares about death. In my youth, these nightmares always involved me making a mistake that resulted in my own death. Thankfully, I always startled awake just before the moment of death arrived.

Now though, in my forties, these nightmares almost always involve this same thing happening to my children instead of me, and I wish I could startle myself awake and not be forced to witness it happen time and time again.

Growing up, I didn't always have to be asleep for these imagined final-moment stories to invade my thoughts though: During most of my teenage years and into my twenties, I experienced *Call of the Void* moments all the time – which is imagining my immediate death happening by my own imagined actions from my real-life current situation. These moments seemed to happen to me for no apparent reason at all. I *hated* them. I was disturbed every time I experienced them, though I would later find out they are so common among humanity they've earned the already-mentioned formal name.

All of the ideas I've mentioned so far involve close encounters with the *firsthand* experience of death. They are simple reminders that you will eventually experience death yourself and you will likely bear witness to it. When it comes to *secondhand* shadow experiences with death though, everything changes, and this avoidance often flips things upside down.

For instance, in the United States, it seems to me we simply can't get enough of death when it comes to stories *about others*. Death is frequently involved in movies, television, and entertainment. We as audience members seem to enjoy mocking how silly characters' choices are in campy horror movies, because we can see their obvious-to-us-but-not-to-them pathway to a future harm and death. We can see the gap in their story. We also expect the story to continue *after* their death.

This leads into a plot twist: Death does not seem to bother us if we know it is not actually the end of *our own* story. Of course, if we find ourselves *attached* to a character, and can see ourselves in that character - that character's sudden unexpected death can result in the most horrifying or shocking moment of *any* story. It's like we are drawn toward the practice of others' fictionalized pathways to death, or even others' *real-life* pathways to death, but only through the protective boundary of knowing these stories will remain *secondhand* stories to ourselves.

This, to me, is the root of morbid curiosity. This is the reason why I think front page news often involves death or near death experiences, and newsroom editors often use the long-standing informal rule "If it bleeds, it leads." We as humans seem drawn to the ending of *others'* stories, fiction or real, because we want to *avoid* that particular ending from ever becoming our own, and if it happened to someone else, it wasn't our ending.

All this being said, I quickly ran into a problem trying to use this concept to grow the shared foundation between you and me and rise back out of the abyss.

This whole line of thinking implies that you and I are driven to survive because we don't want to die. It defines our lives as simply being the opposite of our deaths. This in turn implies we are each obsessed with avoiding possible death stories and build our entire lives around that central idea. It's like saying the only reason we ever seek food, water, and shelter is so we don't die.

While this might seem largely true in some ways (or perhaps not wrong), it started to transform into a paradox when I considered the consequences of thinking about death *all the time*.

If you constantly seek out, learn about, and explore the path to death in fictional or non-fictional stories, or constantly try to determine if you are on a path to death at every moment, you only ever end up creating more and more possible stories that lead to your possible end.

In other words, the more you witness the very end of someone else's story, the more ways you can imagine your own story merging into them and leading you to that same fatal conclusion.

For example, assume you watch a video of a roller coaster incident that led to a death. The next time you think about getting on a roller coaster, you might suddenly imagine that fatal story about to become your own.

This creates a paradox. Knowing many death stories can make you safer because you will likely prevent yourself from choosing the same actions you have seen others die after choosing. These actions become *wrong* to you and are then discarded, because justifying them as right could lead to your death.

Every added story of death you build for yourself in this way can slowly make you more and more anxious that *any* action you take might be wrong, and moving at all might merge you into your final story. You can end up *telling yourself* you just haven't figured out how it's all connected *yet*.

The result is a growing list of possible experiences you fear and are unwilling to try because they *could* merge to become your last story.

If you amplify this to the extreme, your life becomes striving to experience no stories at all. You end up action-paralyzed, doing "nothing", trying not to think about breathing, living in constant anxiety, and only ever acting to keep completing the required life-loops of eating and drinking. You might end up *never* leaving shelter, believing you are staying safe from death this way.

There is no shelter from death.

In my opinion, attempting to live like this will only result in the opposite. It will actually *increase* your chances of putting yourself in the story of your death. I will return to this idea, but this is not the only paradox that can appear in a death-centric lifestyle.

When you try to structure the idea of life and death as complementary to each other and then build some usable life-mantra, you are likely to end up building a statement like, "I choose to live in order not to die, and not dying is what it means to live."

Life and death can't *really* define each other just like a pair of individual words can't define each other. Meaning has to be *injected by you* somewhere. You cannot directly experience life and death at the same time. If you are alive, you are not directly experiencing death. Your own death will *always* remain a secondhand story while you are alive. For this reason, I eventually decided there is no common ground to be found that is based on death, even using *The World of You*.

You have different firsthand experiences than me, and it's possible, even likely, you are not okay or comfortable talking about and exploring ways to die to find common ground. Adding the reality of our certain death and our relationship with stories of death is an improvement and expansion from needing food, water, and shelter in repeating loops, but it is not enough to build a shared foundation to rise back up out of this abyss.

So, I continued pacing this imaginary pitch-black landscape. I was the only human here, trying to avoid death and fulfill my needs to stay alive. What else might I have in common with you, and anyone else?

After many days of thinking on this, it finally hit me - something that allowed firsthand experience to inject meaning into both the words "life" *and* "death". This "thing" is the foundation of context for all things human, and it was just sitting there in the massive gap between these two words, staring me right in the face.

Perhaps you've already noticed its presence and are frustrated I haven't yet brought it front-and-center yet, and perhaps you are like me before *The Event*, and have never noticed it before. I was blind to half the foundation of context for my entire life.

Once I finally noticed it though, it changed everything. It blew the covers off of an absolutely massive rabbit hole within our foundation, which allowed me to tumble *into* the depths of this shared foundation, and begin to expose and understand far more.

In we go.

57

The Master Measure

There is a common theme embedded in both my context and yours, and it has actually been embedded in nearly every concept in this book since you began reading. It has been a part of every word of every sentence, and even in the spaces between them. It is required to define the need of food, water, shelter, as well as survival, death, stories, and even the right now. It is the whole reason our truths are relative.

It is the past, it is the present, and it is the future. It is *time.*

The longer I dwelled in this pair of thought experiments, *The World of You* and *The World of Mouses*, the more I came to realize time was a major factor in everything I ever imagined myself doing. It took me weeks to finally realize the sheer magnitude of this.

Time, as I see it, is more important than any psychological or philosophical theory I've ever read makes it. Even physics, which uses it extensively, seems to not put enough emphasis on time in my opinion.

Time is the underlying framework of everything related to being a human being. Even though it is never directly experienced the same between you and me, it is still the medium that exposes the entire remaining shared foundation between us. *Every story, directly experienced or secondhand, requires time to tell it.* Every story is a chain of events occurring *through time.* This one is no exception.

To better understand time's significance, let's reboot the first story of our shared foundation. Pause *time* and re-enter *The World of You* thought experiment from the beginning. All the creatures of the world are present, even those bigger than a field mouse, and all humans are gone except you, as well as everything human made and human-changed. You are currently all alone on the surface of the all-wild world wherever you are right now (or the nearest dry or familiar land to you).

Now un-pause time. Let everything start moving.

Consider the following question a second time: What is the first thing you need?

You already know what you will answer of course: Food, water, or shelter. But *how* do you decide which word you will answer with first?

Assuming you haven't been formally trained in survival, I believe this decision is based on which of the three needs you feel to be the *most urgent*, and urgency is a word built on the context of *time*. If you feel more thirsty than hungry, you'll probably seek water before you seek food. If you are starving while reading this, you'll probably seek out food first. If it is the middle of the thunderstorm where you are right now, or 120 degrees in the sun, or the middle of a below-freezing night, you'll probably make shelter your most urgent need, and thus your first answer to the question.

This decision of *priority* exists within the context of time (which in *The World of You* is based only on urgency). Everything "human" does. You are one human, moving around in the objective world. You have one story of your life, from now until the end - your death - and you can only focus your attention on one task or action at a time. Your mind weighs your three needs and then chooses the most urgent, the one need you think is right to pursue with what time you actually control – the ever-moving *right now*.

Without *time* you would have no need for water or food, because your body would not be consuming what water and food you had in your system.

Without *time* you would have no need for shelter, because nothing is moving to need shelter from. The sun would not move in the sky, the rain would not fall, and the creatures and environment around you would all be still.

Without *time* you would never develop a feeling of hunger or thirst or fear. You would have no thoughts, no heartbeat, and no ability to inhale or exhale, because your thoughts are electrochemical pulses moving *through time* in your brain, your heartbeat is a muscle relaxing and flexing *through time*, and your breathing is your diaphragm and lungs expanding and contracting *through time*.

Everything "human" exists entirely within the context of time.

Just thinking about what is most urgent and setting priorities *requires time*. For at least a split second, you stood in the imagined *The World Of You* doing nothing but thinking. Everyone I have ever put into this thought experiment face-to-face spent *many* seconds doing nothing but thinking after I ask what was needed first. Without those seconds, no thinking could take place.

Time is *so much more* than just the context for your life-loops and thoughts, though.

When I wrote this, time is the shared master context for all humans as we ride along on the same giant ball of magma and rock we call Earth, spinning around a star we call The Sun, which is itself twirling around in an arm of what we call the Milky Way, which is hurdling through space like all the other galaxies we know about. It's even possible the entire universe is moving on top of all the movement already taking place I just listed within it. *Everything we know seems to be moving.*

Even if you don't believe most of the planet and galaxy stuff I just mentioned because they are all secondhand stories that don't make sense to you, the master context of time still exist and applies for you just the same. Even if you believe the Earth is flat and a simulation, the simulated sun moves across your sky, the simulated clouds shrink and grow and move, the simulated stars and moon move in your night sky, and all the simulated plants and life upon this planet move as well.

Everything is moving. Movement *requires* the passage of time to exist, and likewise, time *requires* the presence of movement to exist. *Time and movement are completely inseparable.*

Time is also central to connecting you with the universe.

For instance, if all things in the universe were destroyed and only you remained in a bubble, that bubble would become the boundary edge of the universe, and your movement would define all time in that universe.

If the bubble popped and you were destroyed so that only your left eyeball remained, the outer edge of the eyeball would become the boundary edge of the universe, and the movement of fluid inside of that eyeball would define all time in that universe.

If your eyeball was then destroyed leaving only a single atom, the electron around that nucleus is still moving (if you have trust in secondhand stories that explain atoms), which means the electromagnetic field it creates would become the boundary edge of the universe, and that same electron's movement would become the basis to define all time in this one-atom universe (assuming the electron to be the only moving part of an atom).

But once there is nothing - *truly* nothing – *infinite* nothingness – time ceases to exist, because there is nothing to measure it by. Nothing moves. There is no universe at all.

This idea can be extremely difficult to imagine in my opinion, largely due to words themselves. If you struggle to imagine "nothing", it's possible the problem is that you are still continuing to think of *yourself* as existing. What I suspect is happening is you are "observing" this situation from your mind's "third" eye.

Your mind's eye is actually Second You imagining this situation into existence by overwriting First You's truth – temporarily, of course. This makes your mind's eye the second-point needed to define the entire basis of time and movement in a no-longer-empty-because-you-are-present imagined universe. If it bothered you to consider what must be beyond your bubble, your eye, and that last atom, it was likely for this reason. You are still experiencing time because it is being accidentally defined by your imagined existence as an *outside observer of the imagined universe*. To fix this, you have to imagine your own un-existence as well.

But perhaps this is all a bridge too far for you. As usual, I can't know if all this time-movement relativity stuff makes sense for you or not – it's all a secondhand story. It's possible you disagree with the above stories in some way, or they just don't make sense for you. If this is the case, that's okay. I'll come back to relativity and the boundaries and limits of systems later in a more practical way that I think might help.

Instead, let's start putting together a little structure to our *firsthand* experiences, because the introduction of time just created a measure to begin to understand everything in our shared foundation - no seconds, minutes, or hours required.

58

Penrose Time-Loops

Perhaps the most important thing time did for me on my first journey into the darkness below our shared foundation is that it began breaking loops. I feel this is best explained with the help of some illustrations, but just in case you are listening to this book without access to graphics, I'll do my best to paint pictures using words as well.

I'll begin with the food loop.

As I see it, it's easy to fall into the trap of thinking the need for food exists in a repeating loop. I fell into this way of thinking myself. It seems to make sense. This is, however, a loop *trap* created by words, and it took me quite awhile to figure out how to break it.

The food life "loop" begins with the act of eating, so to picture the loop, imagine the word "eat" written on a piece of paper. Next, we'll use an arrow to represent time. Since we will end up needing to eat again, this arrow of time will leave the word "eat," travel in a loop, and end right at the same "eat" again.

There is only one way to exit such a loop: Die from not eating. As a result of this relationship between the act of eating or dying, these two words must be combined in the same spot on the time loop. So imagine adding the two words "or die" written under "eat." Our diagram now shows "eat or die" with time as a looping arrow back to "eat or die" over and over. A cycle.

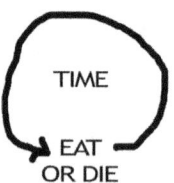

TIME

EAT
OR DIE

This might make sense to you just as it did for me at first, but while your needs set your priorities, your needs do not define objective real-world *time* outside of your body.

You are just a small part of this reality the same as me, and your bodily needs are *only* understood and experienced in one place - inside your own mental boundary. I know you need to eat because you are human and I am human too, but *I* can never directly experience how hungry *you* feel. This means my own sense of time and priorities are in *no way* tied to *your* need to eat. The sun and the moon in the sky are not affected by your need to eat, either. The trees don't care about your grumbly belly, and neither do the mountains.

This means everything above is exactly *backwards*: Time doesn't travel in a loop around your need to eat. Instead, your need to eat is defined by the existence of objective real-world time. Your needs wouldn't even exist without it. Your body burns calories as it continues through time, which requires you to eventually eat more calories to keep your body going.

This food life-loop story is actually only a mental-loop you tell yourself is true based on words. Remember mental loops? Remember what can happen if you try to create an internal mental loop between your justifications of what is true and your actions, like telling yourself you are manually breathing? What would happen if you *tell yourself* your hunger defines time for me or anyone else? You are not the *Keeper of Truth and Decider of Objective Time*, and neither am I or anyone else.

The loop breaker is that you must justify actions to eat. Your actions are your movements that happen *through time*, and they always occur in a certain order in time: The ever *advancing* "right now".

To better understand what I mean, just anchor your justifications and actions to the alphabet, which is taught to you in a certain order starting at A and ending at Z.

Justification for an action happens at time-point "A", and the justified action is performed at time-point "B". Telling yourself to re-justify "B" as wrong occurs at time point "C". Believing the action was wrong when you did it occurs at time-point "D".

No matter how many mental loops you might imagine into existence, objective time keeps marching in more-or-less the same direction.

Let's use this idea to fix our previous loop to show time how it really is: A straight arrow from left to right, the same way as the alphabet is shown A to B to C left-to-right when you read it. This time arrow can actually point in any direction you like, but I had to pick one and go with it, so imagine it being a long straight arrow underneath the time-loop pointing to the right.

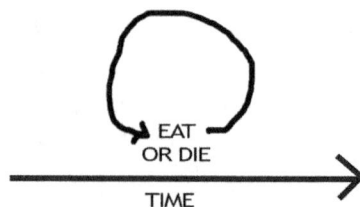

If you place the previous time-loop above to the straight time-arrow, which we know represents the order of your actions left-to-right, the food time-loop is doing something impossible: It shows time traveling *backwards* (right-to-left) in the top half of the loop relative to the way time actually flows, (which is left to right).

Which arrow is correct? Well, you cannot actually undo or reverse actions you have already done. You cannot "unsteal" a candy bar, and you cannot "unread" what you have read in this book so far. You can't make real-world time run backwards. Your actions and even your thoughts always move *forward* through time. This means the loop needs to be revised. Before we can do this, we first need to give the straight time arrow some *anchor points*.

Since the straight time arrow is representing the order of your actions, consider which actions you can control and when you can control them. What is the first and last action you can ever control?

The first action you can control is your action *right now,* because you can no longer control your actions that have already passed – you are only ever one point in this current moment. So I'll draw a dot on the left end of the straight arrow timeline and label it "Now".

The other end of the timeline, at it's most distant, is the last action you will ever justify – the one just before the end of time for you: The last one before *your death*. So let's erase the arrowhead and replace it with another dot, and label that dot "Death."

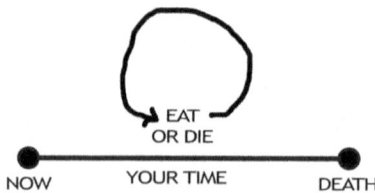

After adding these anchor points to each end, you'll notice (perhaps uncomfortably) that your action timeline is now a very specific and *limited* length. As hard as this might be to look at, *you do not live forever.* This is the reality of experience as a human being.

But now we have a new problem: There are two "deaths" being shown. One death appears in the food life-time-loop if you don't eat, and another appears at the furthest anchor-point of your action timeline underneath; where you stop performing actions because you have died.

You can't have two separate deaths. You do not have two realities. You cannot split yourself in half, which is exactly what must happen for this situation to be true. By avoiding the reality of your death, you can end up tearing your mind apart just like with the stolen candy bar, because it involves trying to divorce what you want to believe and *tell yourself* is true (you can avoid death) from what you know is true (you will die).

To solve this problem, we must break the "eat-or-die" loop apart to separate "eat" and "die", because they are not the same action, so they cannot happen at the same time. (If you thought of choking just now, I would argue that choking is not eating or death. It happens *between* eating and death.)

With our broken loop now having "eat" on one end and "die" on the other, we can merge "die" from the newly broken loop with the end labeled "death" on the straight timeline because they mean the same thing: The end of your actions. Doing this leaves the other end of the broken loop, "eat" dangling nowhere in particular.

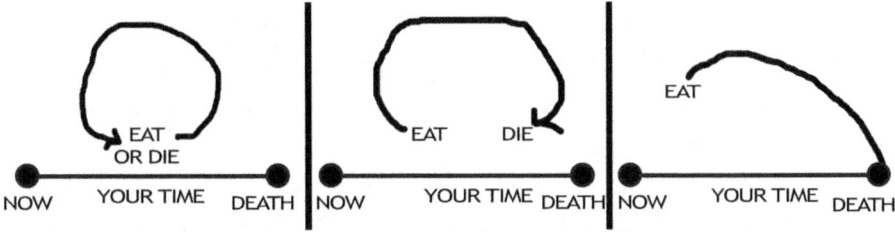

If we assume you eat *right now*, then we *can* find a home for this dangling end of the broken loop. We can now safely merge "eat" with the "now" anchor point on the starting end of your timeline. Viola! No more loop!

The food life-loop is now opened into a food-need *arc*, with each end of the arc connected to each end-point of your action time-line.

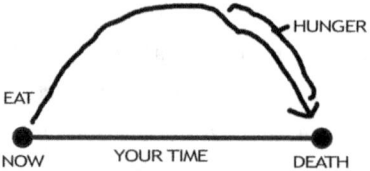

One end of this arc represents you right now, and assumes you are eating which means you do not die. Since your body does not immediately need food again after eating, your story of death by starvation moves up and away. As you continue not to eat, the arc starts to bend back down toward death.

I can even add the label of "hunger" starting near the peak of the arc and lasting until sometime near death to help this diagram to make more sense.

We now have a simple graph showing you eat, your time passing as a straight line representing your physical actions, the rise and fall of your food need, the feeling of hunger, and eventually the end of your timeline - death - if you do not eat again.

Let's call this your *food story arc*. Here is a 3-panel graphic of how it looks as you progress along the arc and your action timeline.

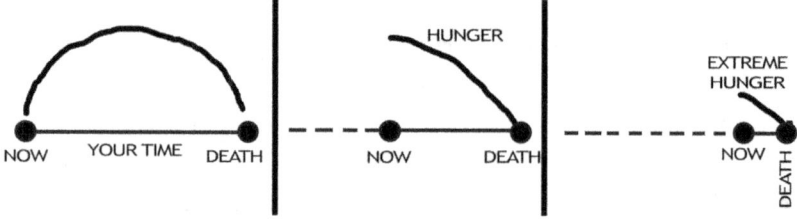

If you just enjoyed the last meal you will ever eat, then the length of your action time-line across the bottom has a limited-and-always-shortening length. The number of actions you have remaining will depend on how quickly you burn all the calories from your last meal plus the calories that can be consumed from your own body before you die.

This is probably a few days or maybe weeks from your last meal (assuming all other needs are fulfilled). The secondhand story I've heard most often is about three weeks. It is the time it would take you to starve to death, whatever that length of time might be. The length is actually customized to you.

This same process works for breaking your water life-loop as well, which has a final graph that looks exactly the same as food. If you take a drink right now, your *water story arc* moves up and away from death by dehydration. Time passes, and your actions cause your bodily processes to consume water. As your arc turns down again, you eventually start to experience thirst. This thirst will increase to extreme thirst as you inch closer and closer to death by dehydration.

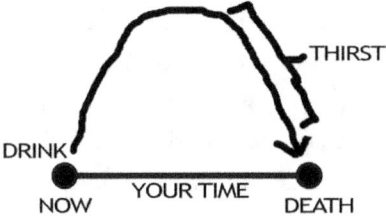

The only difference between your food story arc and your water story arc is the length of the action timeline underneath. Your water story arc is not even close to the length of your food story arc. If you just drank a glass of water and ate a plate of food and would never get either again, you would die of dehydration long before you died of starvation. If these two arcs were overlaid together, it would look like this:

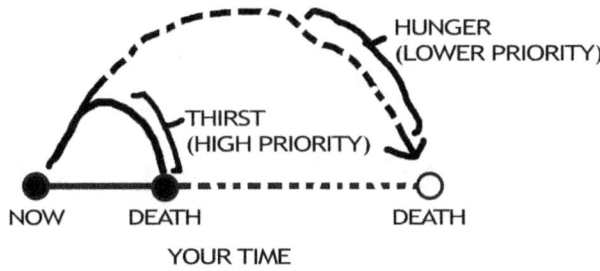

This leaves only the shelter life-loop to break.

The shelter life-loop comes apart the same way as food and water, but it has a different shape. Shelter is anywhere or anything you can use to temporarily protect yourself from life-threatening conditions in the real world. In a way, shelter is a "safety net" against the uncertainty of your survival in the world you inhabit.

You never know when you might need shelter. It might be right now, or it might be five days from now. It could be as simple as being under a tree, or as complex as building a house or making clothes. For these reasons, your shelter story arc does not have a predictable shape or length to it. It is sometimes jagged with crazy direction changes, and the action timeline underneath can change wildly in length in a flash and a bang.

There are a million shelters against a million potential threats, and you have to know which shelter you need relative to the threat you are facing. The words appearing on your shelter life-loop are "Shelter or Die".

If you assume you do not need shelter at the moment you begin *The World of You* (perhaps because the sky is overcast, a comfortable temperature, and you can sense in all directions there is nothing threatening you), then your shelter story arc breaks apart with one end labeled "shelter" and the other labeled "death".

The death ends can be merged just like the other loops, but when it comes to the shelter end, if we assume you don't actually need shelter right away, it moves very high and far away from your action timeline, and it will stay there as long as your situation does not change.

For now, let's keep things simple and go with the story of suddenly being faced by a large dangerous animal charging you after only a few moments in *The World of You*.

Your shelter story arc, which had quickly risen to be far away from your timeline, now immediately and abruptly heads almost *straight down*. If you cannot find the right shelter relative to your situation, then, *in theory*, you will die.

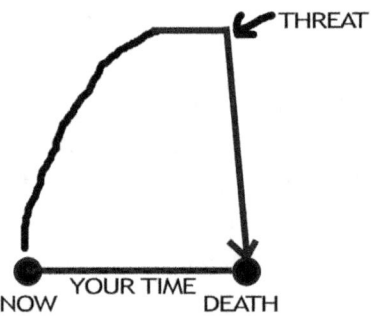

This is the point where a previous story comes back to insert itself: Since you have not starved or dehydrated or been un-sheltered enough to directly experience your own death before (because I assume you are not dead while reading this), *you can never actually know the exact length of your own remaining action timeline*, and I think it's safe to say your mind will do everything it can to prevent you from ever finding out.

Your mind is always working very hard to prevent the *secondhand story* of your own death from ever merging to become your *firsthand* story.

The overall result of all three story arcs working together is the system your mind uses to *prioritize* actions based on the order these story arcs are coming back down, because which ever one is collapsing first is also the one quickly collapsing your action timeline length to "very few actions remaining".

This means if you are more thirsty than hungry or scared, you go for water because that arc is collapsing first. If you are more hungry than thirsty or scared, you go for food. If an animal is charging, you go for shelter before water and food.

Now…I didn't write this book to bore you with graphs and diagrams, and I won't be using them throughout this book. In *The World of You*, you aren't going to imagine these graphs (or any graph) before you justify an action, and I don't intend for you to "tell yourself" about these story arcs when making choices.

Instead, I quickly found out through my own experience here in the dark of our shared and hidden foundation that these graphs *can* and *does* reduce the amount of noise and overwhelming feelings of chaos and confusion when time starts to flow for the first time in *The World of You*. If I've explained what I understand about this well enough so far, you might now understand *why* you came up with your first answer - whatever it was.

Applying *gapless*-objective-time to this situation also helps explain the self-generated story ending of "I would be dead in a day," or "I wouldn't last a week."

What these individuals really did was notice out loud that he or she suddenly "sees" the end of their action timeline where they couldn't before, and since they (or you) have no past stories of how to get food, water, or shelter in the conditions of *The World of You,* the end of their timeline is the only thing left to notice and anchor to at all.

They noticed, in my opinion, their current imagined future story in *The World of You* just merged with the story of their own death – and it is very uncomfortable.

This explains why those individuals that were survival trained did not build and answer this question out loud. They each already knew the stories and actions needed to survive in *The World of You*. They did not experience the merge between their future in this experiment and the story of their own death.

Using these diagrams can also be helpful in another way: By showing actions in *gapless* time, your mind can start shaking off the "unit-izing" of thoughts and feelings created by having to constantly *jump* from word to word to understand yourself. It allows your reality to not be full of holes.

Believe it or not, underneath all those words you've been forced to use your whole life, your mind is actually an enormously powerful machine capable of performing some serious shock and awe on your reality when needed.

59

The Time-Bending Speed Machine

Let's assume you answered my question of what you need first in *The World of You* with "water". This means water is your most urgent need and therefore your highest priority after arriving in the thought experiment.

You are, of course, not *consciously* aware of what is happening inside your body at the cell level. All the stories you have about your cells are *secondhand*. When it comes to water, you only directly experience a feeling of thirst, and an awareness of how much time seems to be passing while feeling thirsty.

This "thirst awareness" is not measured in minutes and seconds. In a world with only you in it and no clocks, you wouldn't think to yourself *I've been thirsty for 3 hours and 7 minutes*. Instead, your urgency of thirst is expressed through mental stories that are capable of justifying actions. They might range from *I should start looking for a drink* to *I will drink the liquid that comes out of an animal's rear if I could*. A worst-case story might be something like "I can't move anymore and am going to die". This same sense of urgency measured through mental stories applies to food and shelter.

Ultimately though, the thing creating these stories is not words. It is *feelings*. It is the basic feelings of hunger, thirst, and fear.

Time is directly and extremely connected to you staying alive or you being dead, so these feelings are created by your mind's connection to your body in order to determine where you seem to be in your own timeline.

Since you cannot control these feelings with your mind – meaning you can't just "turn them off" - these feelings end up becoming the base "units" your mind uses to determine your *personal rate of time* relative to your death. These feelings set your *minimum* allowable life-speed.

Your minimum life-speed is naturally different than mine. There are a million (or more) factors that cause this to be different between us. If you are twice my size or half my size, if you have higher metabolism or lower metabolism, if you are more muscular or less fatty relative to me, then your water and food need story arcs will be different in length than mine.

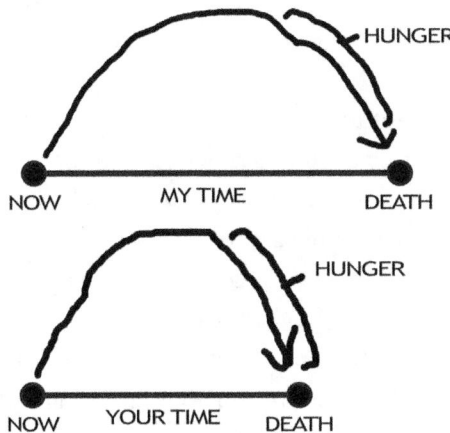

If you are more fearful or less fearful than me given identical situations, then your shelter story arc is different than mine too, and this can radically change the perceived length of our action timelines.

Ultimately, how these three story arcs overlay together creates different length "segments" in each timeline. These segments are the space *between* our needed actions.

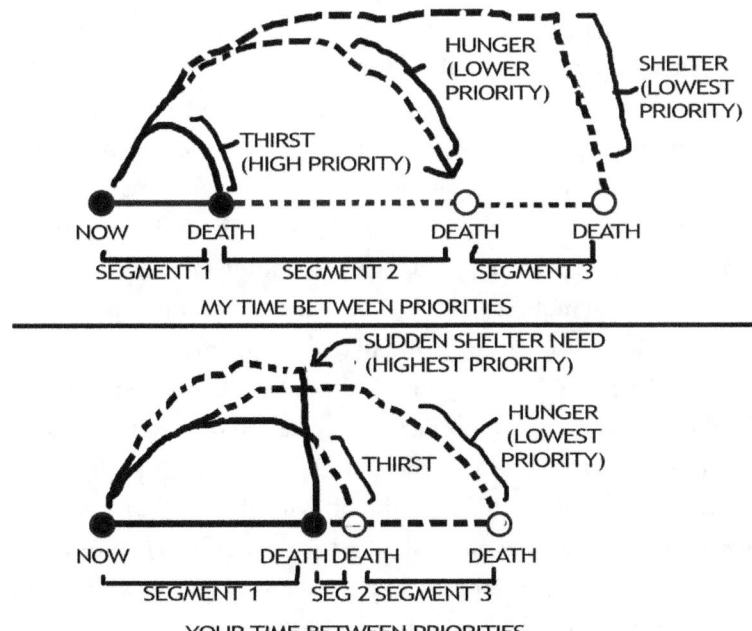

In *The World of You*, I might seek water first, and you might seek shelter. I might need immediate shelter, and you might need food. It all depends on how these three story arcs align.

I think it's important to keep things in perspective though.

Even though our arcs are different lengths – maybe you need to eat more frequently than I do - they are not usually *dramatically* different. Especially once you compare us as humans to the arcs of all the other creatures on the planet.

For example, a hummingbird's need story arcs go up and down *much* quicker relative to you and I. It needs food almost constantly compared to you. If you started *The World of You* side-by-side with a hummingbird, death by unfulfilled needs is likely measured in minutes for the hummingbird, not days like you. Because of this, the hummingbird must move at significantly faster speeds.

On the flipside, the sloth's need story arcs go up and down much slower relative to you and I. It needs small amounts of food every few days. Its death by unfulfilled needs is likely measured in weeks, not days like you. As a result, it can move at significantly slower speeds.

I think it's incredibly important to note that faster and slower are relative words, and so is *normal*. From the hummingbird's perspective, it's at or above the *minimum* speed needed to fulfill its needs, just like you. What seems like extreme speed to us is *normal* from its perspective. It probably thinks we are sloth-like, but our speed is normal for us.

The same holds true for the sloth. From its perspective, it is moving at or above the minimum speed it must to keep its needs fulfilled. It probably thinks we are hummingbird-like, but again our speed is normal for us.

When it comes to *your* speed, every bit of you is "built" to operate at or above that minimum speed required (again assuming you have all the known human senses and full mobility). All your senses operate at the speed most likely to keep you alive in this all-natural world thought experiment, and help you fulfill your needs at whatever speed your timeline segment "units" require.

Whether you believe in evolution or in creation or in something different – it seems our senses, when we have them all, operate at their peak and equip us with everything we need to survive in this all-wild *World of You* situation.

Your focus and attention are well suited to track other creatures both faster and slower than you. For instance, you can follow a squirrel with your eyes as it races through trees, which are also filled with thousands of leaves, all moving.

The colors of edible ripened fruits often appear vividly bright relative to the tree, like red apples and green pears.

Your see-in-the-dark vision receptors activate at about the same speed as a sunset. After all, it would be a problem if you couldn't see during the first few hours of darkness. This is also why going into *sudden* darkness results in not being able to see at all. Those receptors aren't "built" for that kind of transition speed – they are "built" to match the speed of the sun crossing the sky.

You can notice many smells in the world extremely quickly, and can identify many of them with one or two whiffs.

Your taste receptors for bitterness and sourness operate very fast to protect you from toxic things *before* you swallow them.

Did that new sound come from behind you, or to your left? You can usually tell with relative ease without having to hear it a second time.

If an insect crawls on to you, you can typically notice within a couple of seconds.

You can also sense the slightest temperature differences, allowing you to stay ahead of an incoming weather change.

Your mind and body are honestly pretty awesome, in my opinion, even if you are missing some senses and mobility. But this is just the tip of the iceberg of what they are capable of doing. In extreme situations, your mind can do some pretty wild stuff with your body. It can dramatically alter its speed and warp your entire perception of time, and with it your entire reality.

Let's say you are suddenly faced with a threat to your life. Your mind believes you just merged with a story that leads quickly to your very-possible demise – such as unexpectedly coming face-to-face with a pack of wolves.

At the moment you realized the story you just merged into, your need for shelter is immediate and severe. It doesn't matter how hungry or thirsty you were. Shelter is now an *extreme* priority. To help, your mind tells your body to start dumping and pumping chemicals and hormones that make time seem to slow down, and provide you extra strength, speed, and heightened senses.

Time is not *actually* slowing down in reality of course. You are not affecting the wolves' sense of time at all. Instead your mind is just *speeding up* to aid in your ability to survive this situation. Everything about you is now moving faster, which gives you the perception that everything else but you is moving slower. You have become the hummingbird and have a new *minimum* speed. This had to happen because the length of your timeline between now and your predicted death just got *extremely* short.

Your mind puts the only thing it has left between you and the wolves of death – *time*. Your sense of time warps dramatically as adrenaline floods your system, which in my experience can feel like adding twice as much action-time to the situation and reality itself. The adrenaline-altered shelter story arc would start to warp and bend to lengthen the action timeline.

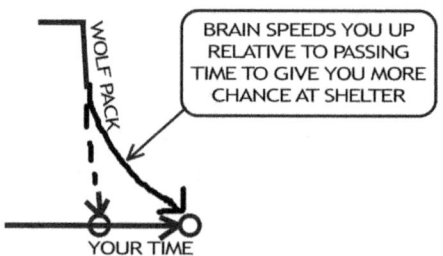

Your mind seeks *anything* you can use as shelter to put between you and the wolves, because the only story it can predict to come next is pain, suffering, or death. Your action timeline collapses in length to only a few actions remaining.

Stories of shelter explode all around you in desperation to re-expand your time and change this story. This desperation allows you to convert a tree, a stick, a rock, a river, or *anything that you think can change or exit this story* into shelter. If there are no shelters available to put between you and the wolves, your mind even might recruit your own body to be shelter however it can.

Expanding your action-timeline like this comes at a hefty price. You burn through *significantly* more water and calories when you do this. None of this matters to your mind though, as changing this predicted future experience is worth consuming every resource you have as quickly as possible in the present. Those added few moments might give you the split second you need to find shelter and re-expand your timeline again.

Fear and the need for shelter isn't the only reason your mind can slow your perception of time, though. Have you ever been bored because you had to do something you really didn't want to do, or be somewhere you didn't want to be? Maybe you experience a powerful thunderstorm in *The World of You*, and you are stuck in your chosen shelter waiting for it to stop.

When time itself becomes your main focus because your movement is restricted, time can feel like it is passing slower. In objective reality, time is still the same speed as always – the sun and moon's movements haven't slowed down because you are bored, and neither has the storm - you are just constantly checking your situation and wishing the storm would speed up because you have other priorities and are not able to fulfill them, which gives the perception of time passing slower when "sheltering".

Your mind can also simply erase time for you. Have you ever driven home from work or somewhere you visit regularly and realize you have no memory of the drive between the two places? Or maybe went for a walk, but don't recall most of the steps or sights along the way?

In these cases, and in my opinion, your mind (Second You) prioritized *itself* over justifying your actions because it already knew where you were going, or accepted you knew how to walk, which allowed it to focus on itself and your memories entirely, while First You handled the pre-justified and approved motions in the driver's seat. Time was never your focus. As a result, the passage of time itself was only noticed after arriving at the destination, or end point. Your arrival at that end point required you focus and re-evaluate your priorities again, breaking the conversation with yourself and reawakening your awareness of your actions through time.

Your brain can even speed up time for you. Ever had a great conversation with someone, did something super fun, or manage to get yourself extremely focused on work or whatever you were doing only to look at the clock and realize *hours* have passed? This is typically called "flow", and if I'm doing a good job, maybe that's even happened to you while you read this book. In the case of flow, Second You is finding the task at hand so interesting it can't focus its attention on anything else.

Maybe you've been flowing through this book since Chapter 1. If so, you might find it interesting to discover that's exactly where we just arrived deep within our shared foundation.

60

The Story System

With the master framework of time added to *The World of You* thought experiment, I accidentally found what I believe to be the most foundational basic process of both you and me. Once I noticed it, it immediately merged all the ideas and concepts I was trying to figure out into something that made sense. It made so much sense, it began uncovering and explaining so many things in my life and in those around me that I sometimes had a hard time dealing with it. I'm now going to show it to you, but I will not leave you to figure out the rest without guidance (even I had guidance from those closest to the darkness).

Reset *The World of You* one more time, and skip ahead to the moment after you answered my question, which assumedly resulted in "food", "water", or "shelter".

For this chapter, assume you answered *water*, which would mean it is your most urgent need and your top priority.

Now imagine a puddle of liquid is directly in front of you.

How do you know what water looks like? How do you know you aren't looking at mud? Or urine? How do you know water is a liquid and not a solid? How do you know if this is a place you *should* find water? How do you determine where you are in relation to such a place? Does it matter if you are on the top of a hill or the bottom of a valley?

The answers to all of these questions are built from one place - your past - and they are created at an incredible speed. Each question is itself a rabbit hole your mind falls into and climbs back up again to build what it thinks is true, almost exactly like you and Bob on the poultry treadmill.

To do all of this, it is constantly running what I believe to be a very simple 3-step process. This process is directly tied to three concepts related to time and movement: The past, the present, and the future.

Step 1: (Present)

You perceive your current moment in the objective world through your tiny, wacky, subjectively-lensed windows (your senses of sight, sound, taste, touch, and smell, if you have them all). This is direct firsthand experience.

In the case of the puddle, you perceive the puddle, and the rest of the world surrounding the puddle.

Step 2: (Past)

You understand what you perceive to be happening in the present though those wacky windows using associations from your past. These are memories, which happen to be built from previous Step 1 perceptions through those *very same wacky windows*, which is why all your previous experiences in life change the way you understand what is happening in your present. Both steps are all built by the same set of senses: Yours. This contextual second step is the construction of your truth.

In the case of the puddle, you compare the liquid you perceive in front of you to everything you have experienced about where liquids are found and what they look like to decide if this liquid is most likely water, urine, mud, or something else. You decide it is likely water.

Step 3: (Future)

You use this truth to build a story of what *should* happen next in the world around you. Being *right* about this is directly tied to how much perceived length you have left on your action timeline. *Story building is tied to your survival.* Story building determines truth, and with it determines right and wrong. This step creates your expectations, which are ultimately used to justify your actions.

In the case of the puddle, you believe this liquid (step 1, perception) to be water based on how it looks, smells, feels, and where it is located (step 2, truth), so justifying your act of drinking this water is the right thing to do because it should make your feeling of thirst go away (step 3, expectations).

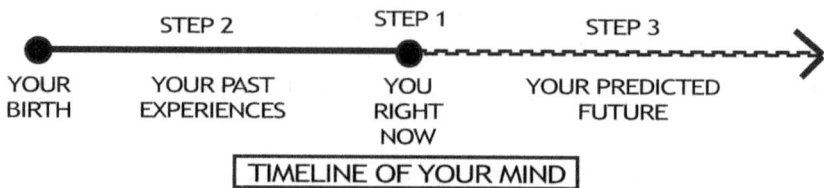

If the above graphic of these three steps looks eerily similar to something you've seen before, it's because you've directly experienced it in the past within the covers of this very book: *Two points* inside your mental boundary made right and wrong possible for you. One point represented you right now, and the other represented your past. These were the two things being compared to determine truth *relative to your situation*.

In the puddle situation, you decided the action of drinking is justified to be *right* based on what you believe to be *true*. After this action is justified however and you begin to execute it, everything above *repeats* the instant your experience with the liquid *changes*.

You are constantly trying to *validate* that this puddle is indeed water as you move closer, as you scoop it up, and as it enters your mouth. Your past experiences with water are used to confirm or deny the truth-story that this is indeed water at every possible moment. If it meets your expectations of water all the way to the inside of your mouth, the act of swallowing is justified based on the expectation this will quench your thirst. The action of spitting is discarded.

If this liquid *doesn't* match your expectations of water when inside your mouth, spitting is justified and the action of swallowing is discarded.

If this liquid is somewhere in between, or uncertain, perhaps swishing it around in your mouth is justified until the truth-story of this liquid is better determined. Then it is spit or swallowed.

No matter what action you justify from each and every mental story you build, that justification was based on what you believe to be the *right*, based on what you believe to be *true* based on your *past*. And now we arrive back at the beginning of this book.

No matter what you decide is true and no matter what action you perform, you are not wrong.

Who am I to say otherwise?

You always do what you think is right in the moment based on the most likely story you can imagine to be occurring, which is built from all the stories you've ever experienced in your past.

Your future depends on this.

Your survival is anchored to your predicted future story being right.

Your survival is anchored to expectations.

Before we leave this chapter behind, let's tweak things just a little bit to make extra sure this whole thing is agreeable.

Let's assume instead of a puddle, there is no liquid at all in front of you – just dry ground.

To find water at all, you must determine where you are in *The World of You* by taking in the present ("Am I in a valley? On a hilltop? Already near water?").

You will then associate your location with any past knowledge of how water moves or where it is found (that it creates canyons, that it flows downhill, that it typically collects in low places).

Then a future story is created that water *should* be located downhill from where you are right now. Using this, you justify moving downhill to be right (uphill is discarded) to try to move toward *validating* this predicted truth-story.

This same validate-your-expectations-built-from-the-past-and-present-by-using-predictive-stories-to-justify-actions process is everywhere in this thought experiment.

The smell of rot and decay would probably keep you away from an area. Why?

The smell of most flowers or fruits would probably draw you to an area. Why?

The sound of burbling water would probably draw your attention. Why?

The sound of howling wolves or roaring lions would probably concern you. Why?

You use this same process for literally *everything* you do in your life.

You build stories in your mind from memories related to what you are currently experiencing in order to predict what is most likely to happen next, so you can anticipate, plan, and justify what the next right action is in your timeline. And the sheer amount of perceptions taking place at any moment is difficult to comprehend.

I think it's very likely that consciousness itself exists to control your focus and attention in order to filter and organize the frankly overwhelming amount of data flowing into your brain every single second. It might exist to find order in apparent chaos and then bring that order into reality through action.

No matter how or why consciousness came to be, your personal truth is the huge number of remembered experiences, firsthand and secondhand, which provide the context and justifications for *all your actions*, because your personal truth determines what possibilities actually exist for you *in your future*.

This means direct firsthand experience is *everything*, because it sets truth, and with it expectations.

Without it, all your sensory windows are shut, and you are directly experiencing perhaps nothing at all.

This means the limit of your direct firsthand experience, also known as Step 1, is *critical to understand*, because it also sets the natural limits of this present-past-future truth-story building process.

Unfortunately, this limit-boundary keeps getting blurred and even erased thanks to words and time. Since I am communicating to you from the past in time and using stories made of *words*, being able to notice and un-blur Step 1's boundary is the next best way to shed more light into deeper parts of our shared foundation.

61

Truth's Isolation

Right now, as you read this book, you probably can't remember anything before your second birthday. Maybe even before your third birthday. Having fond memories of your own birth or your own conception is out of the question.

The experience of your creation is *missing* from your perspective.

Your direct firsthand experience can only ever reach back to the beginning of your story - to your very first memory.

This memory, whatever it might be, is not the beginning of time or movement in the larger world or universe. It is not even the beginning of your *own* time and movement. It is only the beginning of your *awareness* of time and movement.

Your first memory was, in my opinion, simply the first time Steps 1, 2, and 3 happened in a way that was quick enough and successful enough with its prediction and following justified action that the very next Step 1 *validated* your prediction of what should happen next to be *true*. This sudden validation ignited constant attempts to make it all happen again.

This system may have struggled to fire up successfully over the next few months or even the next few years, but ultimately this first successful loop is what I believe gave birth to your sense of autonomy and control. I believe this is the birth of consciousness, and the birth of your sense of being *you*, and this always-running process eventually brought you here to this point, on this page, in this book.

All of this means your own story did not start at the beginning. It started *somewhere in the middle*.

This makes a significant change to your need story-arc from earlier. Your starting point, in both real life and at the start of *The World of You* is somewhere *above* your timeline, because you did not actually start "now" from the very beginning of anything. Your need story arcs actually began somewhere *after* the act of eating or drinking, and for that to be possible your story arcs could not have begun at the same level as death (from your perspective).

Your story before your first memory, whatever it might have been, is darkness. True, timeless, *nothingness*. The best this period of time can ever be for you is broken, unit-ized, word-based, secondhand stories filled with abyssal gaps and voids.

Your parents or caretakers can tell you all their stories about time, motion, and existence before your memories began. You can even try to consume every story through all of recorded history before your first memory, but none of these stories will *ever* be direct firsthand experience for you. The believability of these stories will be forever limited to *what makes the most sense for you*.

Your personal story starts in the middle of all of this.

My story does, too.

Everyone else's does, too.

Our individual contexts with reality - *firsthand experience* - cannot reach beyond our own first personal memory. Language, however, has done something both grand and terrible to this very real and easily measured limit.

Since someone we knew was around before our first memories, and many of these "someones" gave us our words built by their own firsthand experiences, we don't build a clear boundary between where firsthand experiences end and secondhand stories begin. The price for this smudged boundary is extremely high: We often have no idea where to cast doubt on the truth in any story.

We tend to trust those that give us our truths through the transfer of common language, which is often given to us specifically to transfer those very same "truths". But it's entirely possible all of our starting truths given through language are not *actually* true at all. The result can be a completely *ungrounded* truth.

You and I can easily start to believe truth can be followed *through time using words* even when those words go way beyond *anyone's* firsthand experiences. This line gets blurred over and over until you and I struggle to tell the difference between stories being built by *wordless* firsthand experience and stories that were *only ever words* that you and I can never have the whole context to fully understand.

For example, at the time I wrote this, every human alive has (or had) a mother. If I assume you do (or did) as well, then you know *with near absolute certainty* that your mother also had a mother, who would be your grandmother. You might have even met them both. If you did, their existence to you is *wordless truth*.

You might have even known the mother of your grandmother, your great grandmother. There's even a slim chance you have also met your great grand-mother's mother as well.

Based on this story, it *makes sense* that this pattern of truth – this chain of mothers - *must* keep going backwards in time beyond the mothers you have met. Your great-great grandmother must have had a mother, and that mother had a mother, and on and on. This quickly amounts to a mental *infinity* that feels totally certain to have existed, and it begs a question to make it stop: *How did the first mother come to be?*

Nobody knows. Any story you might know of humanity's origin is not possible to be *objective* truth. My reason for claiming this, however, is probably *not* the reason you might be trying to predict. I'm not claiming whatever story you tell yourself to believe as the truth of humanity's origin is wrong. I'm saying you and I can *never know for sure* if such a story is right or wrong. The reason is simple:

No human was already present when this grand story we call humanity began. This means someone being present to experience the origin of the universe is an absurdity.

Even in the most popular religious books that provide a secondhand story of humanity's creation, the universe existed *before* humans were added. I don't know of any religious story that involves humans creating time and movement in the universe, but even if one is out there, it can't be a source of *objective* truth, because no human could have experienced it firsthand.

Even if you could somehow combine every human's memories built from firsthand experience all the way back to the very first human, that first human's experience began somewhere *in the middle*.

That first person couldn't tell you the story of how they began from firsthand experience, just like you can't tell anyone the story of how you began from first-hand experience. There's two or three *years* of what would be firsthand experience completely *missing* from your own memories, so who knows how many "years" were missing before the very first human – billions? Trillions?

This "missing origin story" seems to bother many people immensely. After all, if we can't be certain of how we got here, or how we will end together, then how can we be sure we have anything in common beyond our survival needs? How can we be *sure* what is true, what is right, and what is wrong? We *need* the truth of the past to predict the future to act *right* in the present…*right*?

If you find yourself in a line of thinking that we, as humans, *must* know where we came from and how we got here to know what is true and therefore right and wrong – then in my opinion you are trying to jump your mind out of your body and connect it entirely to stories that literally no one that has ever lived directly experienced. Every origin story and every extinction story about humans and the universe is and will forever be a secondhand that-make-sense-but-is-not-based-on-direct-wordless-firsthand-experience story.

For you to validate these purely secondhand stories as wordless truth, you would need to experience the *lifetime* of personal firsthand experiences of *each and every* storyteller involved. You would have to literally re-live that person's life from first memory until the words were written in the story you currently believe to be true. That is what is required to understand the full context every word these storytellers chose to use. That is what is required to fill up their word-shells completely with the right-and-intended meaning.

I am not saying you are wrong if you believe particular stories about our origins to be true. You can believe whatever you want. You have every right to do so. What I am saying is there is no way you and I can use stories like these as a foundation between us. One of us would have to tell the other *which meanings to the words in these secondhand stories are right and which are wrong*, because neither of us will ever understand the same word to mean exactly the same thing.

You cannot leave your perspective and memories and enter someone else's to validate any story they are claiming to be true. Neither can I. As I see it, if you believe someone else's story as true based only on the words within the same story (no firsthand experience involved), you give yourself a hydra head instantly, because you are *telling yourself* to accept someone else's story as truth, which means this other person is defining right and wrong for you.

Depending on how many times the story has been retold through time, you might be letting many, perhaps *hundreds* of secondhand storytellers become your *Keeper of Truth* simply because they lived before you and their stories seem to make sense, so they passed them to each other, eventually arriving to you.

There's no telling how many hydra heads such a story claiming to be "truth" has grown to have through the ages, and how much noise it can end up making in your mental boundary.

Remember that everything you have read so far in this book is actually *your* context making the words I have chosen into secondhand shadow experiences. These experiences are created by *your* past, not mine. This is true of every second-hand story you ever hear or read.

I want to take a quick walk down this pre-you backwards-in-time path, because understanding the limits of truth is something I feel is extremely important and the cause of many problems.

From your prison-cell perspective, from your unique context in this universe, you cannot possibly have firsthand experience of *where you were* when sperm met egg to begin you. If you think you do (or can), then you are operating under an illusion about yourself that was built by someone else. Someone else explained to you how sperm and egg came together in a certain place and time to make you in a way that made sense. *But you did not experience this time and place yourself.* It is purely a secondhand story. It is a white dot in the blackness of time before your first memory. It is a story that took place *before* your story.

Even if you could somehow experience your own conception firsthand because somehow your senses and memory came into existence just before that sperm and egg cell combined, you would not be able to build the story of *the source* of that sperm and egg from direct experience. You would have *no idea* where they came from.

Firsthand experience of the time and place where the sperm and the egg began within the *same* system would require you to exist *days* before they ever met and also require you to be in two different places at once. After all, you would have to follow the path of the sperm and the egg backwards through time, and they meet while traveling in opposite directions. You would be connecting little white stars into "constellation stories" in the inky black shadows before your first memory.

This only gets more impossible for you to directly experience as we keep going backwards in time.

You cannot trace the sperm and the egg that made you back into the *separate bodies* of your mother and father (or whatever processes might have been used to make you in a future I cannot possibly predict while writing this). That would not only require you to be in two places at once and experience two wordless firsthand experiences at once, it would require you to exist in two entirely separate *systems* at once.

All systems and all stories work this way. As a system or story approaches its own origin in time, new understandings and experiences and measurements all become increasing difficult to nearly impossible. To have any certainty at all about its own origin, the system or story must experience a second point *before* its beginning. And no system or story can experience time firsthand before its own experience of time even began.

Another person had to observe a sperm and egg coming together and then tell you a story about this event for you to have any idea that this is the story of how you were made. The same is also true about the story of your placenta developing and the beginning of your cardiac tissue pulsing, and everything else that happened to you and around you before you began your own stream of firsthand experience *years later*. If you know any of these stories, there is a very good chance most of them are not stories about you in particular. *These stories are always approximate at best*, because they are built by a context that is not your own.

As the stories of humanity approach the limits of time - origin or ending - our understanding, experience, and measurements become increasingly more difficult and approximate. Eventually, a direct wordless firsthand experience outside of humanity will be required to gain further understanding because we have reached the human-system limit. We need another point outside of time, outside of movement, and outside of our universe.

These outside-the-human-system points needed to secure another level of relative wordless understanding might *never* be possible to acquire.

This might be hard to understand, so here are more stories to hopefully help:

A cell inside of your liver cannot determine what exists outside the liver where it lives, because it cannot leave the liver. It certainly cannot figure out all the complexities of your eyes, your heart, your brain, or any of your other organs from its position in your liver. It probably has no idea these other places even exist. It just does its job day-in and day-out, doing liver-cell things. Understanding human consciousness and what exists outside the boundaries of your body (such as the concept of "days" or the "moon") is beyond comprehension to this liver cell.

Even if a cell from your heart somehow traveled to your liver and told your liver cell stories about the heart and its fantastical journey through your body from heart to liver, the liver cell *has no context* to truly understand. It can only imagine what the heart must be like based on its own experiences in liver-world.

The best the heart cell can ever do is say "it's like the liver, but different" and then try to explain the experience of the heart by looking around at the liver and explaining what is different between it and the heart. There is practically no chance the liver cell will build a perfect or full contextual understanding of the heart, no matter how many words the heart cell uses.

And so we exist in the same way within our universe.

Let's say your universe is a soap bubble floating through the breeze. Somewhere on its journey from origin to popping, you were born inside. Generations that came before you in the bubble built theory-stories that the universe must have formed by expanding from a central, singular point.

If I assume you have seen soap bubbles blown by a child (which is the type of bubble I intend for you to imagine), then you know from direct wordless experience this origin story is technically *not wrong*, but also horrifically full of voids and gaps.

This type of bubble is formed from a single flat sheet of soapy film stretched across a child's circular bubble wand after that wand was dipped into a jar of soapy water *and* the child blew into it. But from inside your bubble universe, you and your kind can never truly know or experience that origin, because you, your kind, and the bubble itself did not yet exist when this happened. You would need to observe all of this from *outside* of your universe, and *outside* of your understanding of time to truly understand this origin story from firsthand experience.

You also cannot determine the path your bubble is on from inside of it, because from in the bubble where you were born looking outward, you have no way to know if your bubble universe is what is moving or if what is outside your bubble universe is what is moving. You also don't know if what is beyond your bubble is actually beyond it, or if it is just a moving image appearing on the "inside skin" of your universe.

In your bubble society, theory-stories for each of these probabilities exist, and the "we are moving" theory proposes that for your universe to be moving, *something must have pushed it.*

Of course, you know, as reader of this book, a child (who is no longer visible from the bubble universe) blew into the flat bubble film on the wand to expand it into a bubble, and eventually the force of that wind freed the almost-bubble of the wand, when it quickly snapped closed into a sealed orb. That sealed orb was the origin of that universe. After that, the forces of wind on the planet where the child lives have been propelling the universe.

You, as a person born in the bubble, can't know *any of this*, because you have no idea and can't directly experience the "wind" outside your bubble, know where that wind originates from, that an enormous-to-you child exists, what a bubble wand is, or what the physics of creating universe-sized bubbles might be. Even if somehow you could put together all of this story, you'd never be able to prove it correct. You'd instead be led to keep going: What are the *origins* of the child and the weather systems that all must exist *outside* of your bubble? All of this would have to be figured out on top of trying to figure out your own physics and systems that you can directly experience *inside* your bubble universe.

What is the edge of your universe made of?

What came before the child?

What came before the bubble film?

What came before the space the child inhabits?

How did that space come to exist?

Time is an abyss.

We as humans cannot anchor truth or right and wrong to the beginning of objective time, or the beginning of the universe, because we cannot experience it or understand it completely without observing it *before* it began – a paradox. Even if someone did this, it would be impossible to understand that experience in story form completely. The mental boundary problem prevents it. That person would be like a heart cell visiting you in the liver. To be fully understood, It would have to be *you* wordlessly experiencing these things with your own senses.

Secondhand stories cannot extend beyond their beginning or ending, nor can they suddenly expand in the middle, or shrink to nothing. I cannot suddenly add more to the introduction *while you are reading*, add three chapters at the end, and expand the middle of this chapter. This book is not a living thing. These words are frozen in time. The first word of a story by itself has no significant meaning ("You" in this book), nor does the last ("story" in this book).

I could travel back to before the beginning of time and space and bring back the story of how it all started, and you would have no context to understand anything I'm saying, and no reason at all to believe me. Not to mention it's very likely that words do not yet exist to describe the experience.

Even if we, as humans, manage to recreate and successfully build a brand new human or universe, we can never be *certain* this is how our current story began. This creation story only applies to the new human or universe. The origin story problem then repeats itself all over again: The new human or universe can never *experience firsthand* its own creation no matter how much we tell and show this person or universe how it happened. It will always and forever be a secondhand story from that perspective. This new human or universe can only ever witness the creation of a *different* human or universe at best.

There is no shared foundation in secondhand stories.

Our shared foundation and truth in firsthand experience is not going to be found and grown from humanity's origin. It will not be found in your origin or mine. These points are in darkness for us. No one else alive has experienced these moments from our point of view, either.

It is also not in your death or mine.

It will not be found in the extinction of humanity.

It will not be found in destruction of the universe.

These points are also in darkness.

No one alive has experienced them firsthand.

This leaves only the present. Your *right now*. This leaves your 3-step story process and your actions.

Firsthand wordless experience built by this 3-step story process can never be shared completely. Ever. No matter how many stories of all media types are used. For me to directly experience your truth *in full*, I would need not only to read your mind, but experience your entire consciousness and being. I would have to literally *become you* with all your memories and senses and leave my own completely behind, because bringing them with me would change the context I would need to fully understand and experience *you*. I would somehow need everything to make sense *for me* the same way it makes sense *for you*. This would literally require me to not be myself – a paradox.

This isolation of our firsthand experiences within our own memories means even direct firsthand experience cannot be the basis of *objective* truth, right, and wrong. Your wordless firsthand experiences are certainly *your* truth. My wordless firsthand experiences are certainly *my* truth. But since we cannot share them completely, we can never fully understand each other's truth. The foundation between us is not found and grown by the stories you and I tell or even in the individual experiences you and I have.

Instead, experience *itself* is what you and I have in common. You and I both build context from wordless firsthand experience in approximately the same way with what senses we have *because we are both human*. The substance of those firsthand experiences is different, but I believe the *process* of experience is the same.

This is also why I believe *The World of You* thought experiment is such a hugely powerful tool. In it, words no longer matter. You do not need them. The only context that matters is everything you've ever perceived in the past through direct wordless experience, and secondhand stories that made enough sense you can translate them into actions to create new wordless experiences and truths for yourself.

This means if you've never experienced any firsthand or secondhand stories of wilderness survival, sourcing your own food, finding water in the wild, parasites and diseases, first aid, and what makes a good shelter – you probably feel like you are not going to make it very long into the future of this experiment.

You would have no context in your past to help you know how to fulfill your basic needs to survive in your future. As a result, you are almost completely blind in two directions: You are *possibility-blind* in your future because you are *experience-blind* in your past. You are instead forced to create *brand new* associations and context using the present moment to justify *expectation-blind actions*. And performing actions without any idea of what is likely to happen next can feel incredibly hazardous for your survival.

If this is the case for you in *any* situation, not just *The World of You*, all of your current perceptions and memories you brought with you into the present offer almost nothing toward predicting your future. You are severed from any usable context - from your own sense of right and wrong and truth in the moment.

The World of You is not real and it is not shared truth. It's you imagining your surroundings before humans existed – and you are human. Literally no one knows exactly what this place looks like for you, which means only *you* know what your world is like in *The World of You*. No matter how you imagine it, it is not wrong. It is *you*.

Your first imagined movement within *The World of You* won't be real or shared truth either. It's you imagining that movement being successful based on your past. Literally no one else knows exactly what that movement will be, and no one will imagine his or herself performing exactly the same wordless movement as you. No matter what movement you imagine yourself doing, it is not wrong. It is *you*.

Objective reality before humanity existed might have been *completely different* than you or I imagine it to be. This is the hidden power and the hidden danger of language and its ability to smudge the boundary of first and secondhand experiences.

Your first imagined movement in this imagined world called *The World of You* might be prevented if it were real life. Perhaps by a random unnoticed meteorite passing straight through your head the moment you justify your first action but before you can carry that action out.

In a way, you have truly returned to a single point.

And yet, you have always been a single point.

In this way, your imagined experiences will always be the beginning and end of time for you within them.

And this place – *The World of You* - is the closest we as humans can get to directly experiencing a shared single moment of time, even though it is not actually real.

Assuming you have zero survival training, Step 2 and Step 3 of your story system collapse inward with uncertainty, and only Step 1 remains: Perceiving your current moment in the objective world through your tiny, wacky, subjectively-lensed windows.

While Step 1 feels like a small step for you every moment of every day – it is actually one hell of a giant leap to perceive at all. Your perspective *created* the foundation of you that you still rely on today. And that foundation is *huge*.

62

The Biggest Step Ever Taken

How far can your senses reach?

What is the maximum distance you can still see something?

Hear something?

Feel something?

If you've not thought much about this before, the next sentence might come as a surprise.

There is no limit on this distance.

Your senses are *windows*, not projectors. They let things *in*.

If you have the sense of sight, your eyes could, in theory, see light coming in from the edges of the universe when you look up at the night sky, assuming there are edges.

If a cosmic "hiccup" like some kind of super-massive explosion occurred on the other side of the universe and sent out a giant shockwave that reached wherever you are right now, you could, in theory, feel and possibly even hear this shockwave, assuming it could travel through space.

If the Earth passed through the orbit of Mars, then if you were on Earth at that time, you could, in theory, "smell" Mars (assuming it has a smell, I do not know).

But I feel confident assuming none of these things happen for anyone.

As I see it, one of the reasons this doesn't happen has nothing to do with our sensory windows. It's because there are things *in the way*.

These "things" have *always* blocked you and I from having sensory experiences with the entire universe all at once.

For example: You are, assumedly, on a planet right now. If this is indeed the case, then when it comes to your visual window, the ground beneath is constantly blocking about half of your view of the universe.

When it comes to the other half of your view, then even on a clear moonless night far away from any other humans, the atmosphere *itself* is the thing in the way. It blocks your view of the universe by reflecting away or even absorbing *lots* of starlight before it gets to your eyes at the surface of the planet.

For evidence of this filtering, just look up at the stars on a clear night in a city, and then do it again far away from that city. We often call this "light pollution" because human-created light such a street lamps is being reflected back to us by the *atmosphere*. The same process is happening in the *opposite* direction, and it keeps us from being able to see the light from everything in the universe even on the "clearest" of nights.

This is why observatory telescopes that gaze into space are located as high as possible – it's to look through as little of Earth's atmosphere as possible. Unfortunately, even these locations still experience the sensory light filtering of *miles* of atmosphere.

You might be tempted to build a story that venturing into space could provide an unobstructed view of all light in the entire universe. However, if you are orbiting a planet like the International Space Station orbits Earth today, you are still in the planet's upper atmosphere. *Your perception is being filtered.*

So what if you went deeper into space? After all, this is why humans put the James Webb Space Telescope into orbit *around the sun* almost one million miles further out than Earth. If you were floating next to the James Webb telescope, would you then have an unobstructed view of the universe? I'm sure it would certainly be an improvement from being on the surface of the Earth, but it is still not a view everything in every direction.

This is because the entire solar system is inside the Sun's invisible heliosphere. This heliosphere filters *lots* of things. Outside of the heliosphere is whatever "sphere" is holding together the Milky Way galaxy – which, by the way, you can *experience firsthand* on the clearest darkest night. The heliosphere and atmosphere can't filter out *all* of the light from the Milky Way.

Outside of the Milky Way's filtering "sphere," cosmic dust is believed to exist everywhere throughout the universe (though we can't be sure this is true until we explore the *entire universe*). This "dust," in theory, is filtering out light between you and distant stars.

Beyond this cosmic dust, the pull of gravity itself from who-knows-how-many cosmic systems that might exist between you and the edge of the universe (if there is one), bend and warp the light from those edges in directions away from your eyes so you never see it. This is completely ignoring the existence of black holes, which will literally absorb that same light so that nothing else can *ever* see it, except for whatever is inside of that same black hole.

If it were possible to completely clear the universe of all dust, atmospheres, gravity, black holes, and anything else that blocks your view of the entire universe, and then install tons of lights at the edges of the universe in every direction – you could, in theory, *experience firsthand* the entire shape and scale of the visual universe by simply turning your head a couple of times.

However, in this light-up-the-universe situation, there would be absolutely nothing to hear. If sound could travel through outer space, then assuming you are on Earth right now, the roar of the seemingly endless massive explosion of the nearest star we call the Sun would probably be the only thing anyone could ever hear. This is because unlike light, sound (at least at the approximate ranges you and I can experience), *requires* an atmosphere. If you want to increase how far you can hear, you need *more* atmosphere, not less.

If you wanted to be able to hear everything in the universe, you would need to somehow survive being encased in a universe that was a perfect solid. This ability to hear the universe comes at a pretty steep cost though, because if you were encased in a solid, you wouldn't be able to see anything anymore.

In this way, sight and hearing are complementary senses. Underwater, you can hear great distances but can't see very far. In space, you can see great distances but can't hear anything.

The same things-in-the-way problem even applies to your sense of smell and taste. The molecules required for you to inhale and smell something like Mars, or taste something like the Moon, are prevented from reaching your nose and mouth. If we assume Mars and the Moon have a taste or a smell at all, then the reason these aren't directly experienced is because the molecules needed for those experiences to occur are blocked from your nose and tongue by the gravity that keeps those same molecules on their respective "sphere."

The point I want to make, is that all this filtering and blocking of your senses shrinks what you can experience firsthand away from the outer edges of the universe. This shrinking continues all the way down to whatever is obstructing your view of the entire universe *right now* while reading this book – which, if you are reading a physical copy of this book, *includes this book.*

Step 1 of your story building system is limited in how far it can see/hear/taste/smell/feel the universe through your sensory windows by that very same universe. This is not necessarily a problem, but it matters quite a bit for specific reasons I will soon explain.

Another reason your immediate experience with the entire universe is limited is due to your senses themselves. This ties back to the "Putting the Wacky in Windows" chapter, but uses a slightly different perspective.

You cannot see many things that are real because they are too small or too dimly lit. You cannot hear many sounds that exist because their vibrations are too close together, too far apart, or too quiet to be heard. You cannot feel touches that are too light, or too consistent, like the air pressing on you from all directions at this moment, or when you no longer notice you are wearing jewelry or a watch or clothes.

You have a specific "resolution" to each of your senses, and you are intimately familiar with them. You weren't born familiar with these limits, but you have been repeating the 3-step story process non-stop for years, and through all of this experience, you have become very familiar with the limits of your personal Step 1.

For example, how far away can you move this book and still read the words (assuming you don't increase the size of the letters)? You probably have a decent *approximate* idea of this distance.

If someone is talking, how fast or slow can the words arrive before you wont be able to understand them anymore?

How far away would the person speaking those words need to be moved so you could no longer hear the words being said?

How lightly could someone touch you before you can no longer feel it?

You probably have a decent "feel" for these answers pulling from your past wordless experiences. However, experiences do not actually affect these answers.

In a perfectly cleared universe, *these limits do not change.*

In a perfectly solid universe, *again they do not change.*

This is because the "resolution limits" of your senses are fixed. They are determined not by the universe coming into your windows, but by the "clarity" of the windows themselves, which is based on something you were born with - your biology.

This means we as humans all have "approximately" the same quality windows because we are the same species. Our senses are configured more-or-less the same. Some individuals have a sense that is worse or better than that same sense for others. Some choose to correct and enhance their senses (glasses, hearing aids, etc). Some have "crossed-up" senses that mix two or more together (synesthesia). But by-and-large, the structure and expectations of our senses do not change.

We do not expect to suddenly start seeing x-rays halfway through our life. We do not expect to start hearing extreme frequencies, smell things no human so far has been able to smell, feel things no one so far has felt, or taste things no human so far has been able to taste.

This means there are *two* reasons you cannot perceive the entire universe of information:

1. Other "things" are blocking your windows.
2. Your windows have "screens" or "filters" that limit the detail (or resolution) of information coming through them.

I think it is *incredibly important* to be aware of both of these reasons, because if something is filtered out or hidden from your senses, you do not *automatically* consider it in your present moment. You do *not* use it to build truth and define right and wrong. It will *not* be added to your Step 2 being built by Step 1 in every moment.

The reason for this exclusion is simple: You *never* directly experience what you cannot sense.

You might immediately feel the above combination of words is wrong. If this is the case, secondhand stories have blurred this boundary for you by *making so much sense* you are considering them firsthand experiences when they are not. These stories might even be part of the hydra forcibly overriding an opposite and self-built-through-firsthand-experience truth without realizing it.

For example, you do not and most likely will not ever directly experience a single virus particle. At the time I wrote this, you can't see one floating around or resting on a surface at all. You can't smell one, hear one, taste one, or even feel one. As a result, you don't actually consider viruses as much as you might think you do. There could be a cloud of thousands of virus particles drifting just under your nose *right now*. Yet you read this sentence anyway, oblivious.

Things you cannot sense do not enter your 3-step truth-story building process. They are *beyond* the minimum requirements to be sensed. They are below, or above, the resolution of your senses.

Nearly everything I have mentioned so far in this chapter was not immediately known to be true. Individuals had to move things out of the way to find them, or enhance the resolution of his or her senses with devices and tools that they understood and validated from top to bottom. Today, we have to *tell ourselves* the stories built by these individuals are true, and then just try to remember virus particles actually do exist and they are present in the world even though we can't experience them.

This quickly makes a mess between your truth and the truths of others. The

Keeper of Truth Hydra starts to show up everywhere.

Just like you cannot continue breathing manually because you must *always* breathe and there are other things that need your attention, Second You cannot always override First You's ever-changing 3-step truth-story process, *especially* when it comes to invisible things. There is a balance point between them hidden in darkness in need of finding.

You cannot change the resolution of your senses in the moment. Your senses are what they are. If you have always been missing one sense, you cannot fabricate the experience of that sense from nothing just by thinking about it. Every experience that requires that sense is invisible for you and becomes something you must tell yourself, which is a secondhand story, and can't be used to grow the shared foundation between you and I.

However, you are familiar with what to expect from your own senses and no one else's. You trust them. You have to. You have no other set of better senses providing you a better firsthand experience. The same is true for me. This *is* part of our shared foundation. You imagined *The World of You* using the same sensory resolution you use today. So did I. This resolution is not what changed between where you are right now reading this book and the imagined world with all evidence of humanity erased (except you).

This means the only thing that *actually* changed for you and each person that reads this book, which potentially made you believe you would be dead very soon (or at least seriously restructure all of your priorities), was the imagined rearrangement of *things* blocking your view of the universe through those *unchangeable* sensory windows.

After lingering in darkness and thought experiments for more than 16 months, this realization led to what I considered a profound realization that merged in stories from different parts of my life: If something changes your view at Step 1 in a way that is not expected in Step 2 – what I can only describe as a mental "voltage" is immediately created. This voltage is ultimately what changes your truth, your reality, and potential everything and everyone else with it.

63

Failure Is the Only Success

I'm not about to switch all my words to be complicated and hard to understand because I just mentioned "voltage." Even though I am familiar with much of the language of electricity, I am actually terrible at using those words the way engineers in my life use them. This is because I am not an engineer. I learned what I currently understand about electricity by taking a wildly different path.

To give a small amount of backstory, I am employed during the day at a large engineering firm. I have been with this company off and on for 15 years or so I think, maybe more (I've lost track and time is approximate anyway).

I was originally hired as a drafter in the power and controls department, which means I created and made changes to electrical drawings based on what someone told me needed changing by the use of a red pen. After many years of this, I was pretty good at creating things like wiring diagrams from scratch, and was even given the opportunity to "lead" some simple projects.

None of this work involved *power* design, though. This is because I *hated* everything about power design, and very intentionally avoided it. I was comfortable just connecting Wire A to Terminal B, and Wire B to Terminal C, and all the wires staying the same size throughout. Electrical power design seemed way too complicated, math-heavy, and rule-heavy for me.

I knew this, because I sat through countless lunch-hour training and safety courses on the topic of power design over the course of 15-ish years. Not one course instructor ever explained how power and electricity worked in a way that made sense for me. Power design seemed to just be an onslaught of extremely specific words, rules, codes, regulations, and calculations, and there was a *very defined* right and wrong way to do *everything*. From my very hydra-head filled perspective at the time, right and wrong were already completely figured out and *absolute*. There was no way I was going to touch anything with so many ways to enforce and validate my wrongness.

As a result, I usually zoned out five minutes into each power training course, and the words became nonsense for the rest of the 60-minute year while I constantly checked the calendar, waiting for the clock to stop ticking backwards.

Then along came a day that would change everything. It was the day I was asked to lead a simple *power* project.

This simple project was pitched as replacing some old power panels as well as the wires connecting these panels to a few electric-motors and opened and closed some valves.

My boss at the time knew I was uncomfortable with power design, so he offered up this project as a kind of challenge, and also offered two very knowledgeable mentors to help me should I need it. At my request, I was even assured I could change my mind if the project became too difficult.

With all this support and freedom to exit, this project seemed like something I was willing to try. What's the worst that could happen? With no real reason left to avoid my entry into the world of power system design except my own fear of failure, I accepted the offer to lead the small project.

I had no idea of the magnitude of challenge I had just laid in front of myself.

Expansion

I quickly discovered these "old power panels and valves" were part of an emergency system to save people, equipment, and the environment from a runaway process that could expand to a large-scale disaster very quickly.

The system's purpose was to close the electric-motor operated valves to isolate huge amounts of hazardous chemicals moving through pipes, pumps, tanks, and other equipment in the event of a leak or a fire.

This was indeed a small project on paper, as promised, but it carried with it a *huge* importance.

As I looked closer, the simplicity of this project turned out to be an illusion. From outside the system looking in, I just needed to replace old circuit-breaker style power panels with new fused power panels and also replace the old wires between these panels and the motors for the isolation valves. The valves themselves were not being replaced.

However, once I pulled back the curtain and started thinking *inside* this power system, everything *exploded* into complication. Here, I'll pull it back for you, too:

This valve system is responsible for isolating the massive amounts of chemicals moving around in pipes that can *expand* disaster. For expansion to be possible, there must be something to expand upon. This means this system is intended to *operate within a disaster already underway to prevent it from expanding.*

Per the newly-revised safety rules at the site where these valves were installed, this "simple" electrical system must continue to work *inside* of a 2,000 degree Fahrenheit fireball for a *minimum of two minutes*.

This design-a-simple-power-system-that-continues-to-work-inside-Hell-itself project began creating questions related to electricity and power design that no one around me could explain with any certainty *at all*, nevermind explain using words I could understand. Mentors and engineers included.

This didn't matter to a future fireball. None of this mattered to the future workers that might push the big red button to activate this system while running the other direction. Gaps and guesses on my part were *unacceptable*. I wanted to be certain this system *worked*. To do that, I needed to be sure I understood *why* and *how* it worked at all. But of course I had a project schedule to meet, so I needed to figure all of this out very quickly.

Collapse

The site's safety rules were made of words, and so were filled with gaps and voids between them all. The system had to "operate for two minutes in two-thousand degrees Fahrenheit when within a fifty foot radius of the leak source".

What's "the" leak source? If it is a pump, is the fireproof zone fifty feet from the center of the pump? From the outer edge? If an isolation valve is more than fifty feet away from a pump it is designed to isolate, but it is only twenty feet from another pump that has a different isolation valve assigned to it, does the first valve need to be fireproofed or not?

There are also all the temperatures that will exist between whatever the thermometer outside reads normally and 1,999 degrees. Do these temperatures count as part of that two minutes, or does the two minutes begin only after reaching 2000 degrees?

Each of these rules was its own rabbit hole of questions for the client who owned the system to connect back to reality for me. But this was only half of the equation.

The *control* wiring for these valves (which turns out were just as important as the power wiring because they are what tells the valves to start opening or closing), also needed to be fireproof for two minutes, or wired to activate the valve in the event they failed. These wires do not connect to the power panels at all. Instead, they go to process control rooms and big red pushbuttons spread throughout the area. On top of this, the control wiring needed to not only activate the valves, but also automatically shut down the related equipment (such as a pump) that can expand the disaster causing the need to close the valve in the first place.

But this is *still* the tip of the iceberg of secondhand story problems in this small and simple project. How does electricity behave in 2000 degrees Fahrenheit? Can a wire even conduct electricity through 2000 degrees?

If there is a fire, there is a good chance the power source that feeds my panels could fail. So the panels will need a backup power source and a switch that automatically transfers between these two power sources should this happen.

This opens the door to questions like: "What happens if the valve is closing when the power goes out and switches over to the other source? How much "blip" between the sources is allowed? Will the valves "forget" what they are doing?"

This led deep into how transfer switches work, and understanding differences between two power sources and how to switch between them while keeping a motor running smoothly, and understanding what the control system would "see" during such an event. After all, if a power source is ever going to fail, it is likely to fail during the same disaster I am trying to keep under control and operate within with my design.

To add pressure to my increasingly simple-is-actually-explosively-complicated situation, this project's justification was an incident at a different site, where that emergency valve system failed to work as intended during a *real* fire. Based on my understanding, it did not escalate into a large-scale disaster, but it *could* have, and the client was taking this close call *very* seriously. My new "simple" project would design the first system following newly revised design rules, and would be used to set a new *precedent* for reworking every other emergency valve system this client owned to ensure such a failure and close call did not happen again. In total, there were probably somewhere near one hundred systems like this one, and they were located all over the country.

Even though my mentors and state-licensed engineers had to sign off on everything I designed and thus became *legally* responsible (that's what engineer licenses are really about), I quickly realized I could not blindly follow "the rules" of the National Electric Code, or copy anyone else's designs. Nearly all systems out there were designed to work as "normal" systems under "normal" conditions. Those conditions could all be different, but there aren't many out there trying to conduct electricity through *literal* fireballs. All those training courses I sat through for 15 years actually offered almost nothing to help.

And so my deep dive into the rabbit hole of electricity began.

Unstable

This self-driven dive *to understand* was insanely intense and often caused sleepless nights. I thought about electricity non-stop. I even had electric dreams once or twice. I didn't know it at the time, but this dive would eventually send me to the extreme limits of not just how electricity works, but how *everything* works.

This is because at the edge of what is understood about the invisible world of electricity, the questions about what comes next results in answers found in *other* fields of physics: Classical, quantum, relativistic, and field mechanics and their theories.

In other words, the second point I needed to understand at the level I wanted suddenly started showing up *outside* the system I was studying.

This didn't matter to me. Everything I designed had to be justified, written down (with assumptions), and then *understood and agreed to by others*. Those others had to legally stand behind my decisions. If I didn't understand my own justifications, then how could I explain why I made *any* decision? Fake it? Steal the words other engineers use to sound like I know what I'm talking about?

I'd argue this "mask-of-words" actually happens all the time, and frankly I find it a bit terrifying to think about. I decided if I reached the point I felt I must pretend I understand, I needed to back out of the project. But considering how deep I found myself going, this possibility made me very uneasy. *Someone* had to design this system, and it bothered me to think about someone just going through the motions on something so important. I could do this. *So what if the second point kept jumping into a different field of physics?* I was already decently comfortable in all of them for various reasons.

But the same second-point-outside-the-system kept happening in lots of other places on this project. My questions quickly found themselves entering the field of philosophy, politics, word definition problems, certification methods, authority hierarchies, metallurgy and chemistry, and communication problems. All because I wanted to fully understand what I needed to know to be successful.

However, none of this would have happened if I hadn't approached the limits of the deepest rabbit hole of all.

The Treadmill

At the beginning of this dive into electricity, I began by writing down question after question and then trying to answer each myself. I started with "How does a generator work?" Weeks later, I ended up somewhere near the origins of magnetism and creation of plasma. I consumed countless videos and write-ups. It was honestly alarming to me just how wildly different the answers for my questions could be between any two sources. Sometimes it took dozens of sources just to get a handle on what a single word *might* mean.

Occasionally I would get stuck on a question I could not answer on my own. I would then ask my mentors or surrounding engineers my question, and share what I thought I understood so far. Much to my frustration, they could often not answer my question either.

From my perspective, this seemed impossible. I was asking what seemed like basic foundational questions I assumed every electrical engineer *had* to know. I had created my questions simply by trying to understand electricity *at all*. However, I was experiencing the secondhand-story-with-missing-first-hand-context problem to the extreme. What was *actually* true about electricity? Did anyone *actually* know? How could this be possible?!

This doesn't mean I didn't learn anything from my mentors. I will forever be grateful to one of my mentors in particular, Michael, because he was often able to translate concepts to me in a way that made them *make sense*. He could talk with me about electricity without using the strict and difficult language I struggled to understand. He took the time to try to understand what I was asking – recognizing I did not use "normal" words other engineers seemed to expect.

I considered Michael my last and final resource. He had his own projects to worry about, after all. It was not in my interest or his to use up his time asking questions I could answer myself. My goal was always to avoid asking questions to anyone else at all, which was only made possible at the speed I needed to learn by the existence of the internet.

I eventually figured out what I believe to be *most* of the story about what electricity is and how it works. I chased down everything I could until it anchored to a perceived truth-story that made sense for me. I then had to do the same for fireproofing, testing standards, electrical equations and constants, and more. A few times this involved putting other people and manufacturers in the hot seat to explain how they were able to make their claims.

For example, how does a cable tray manufacturer actually know their cable tray will hold up for two minutes in fire with melting like they claim? Turns out they put a six-foot long piece of cable tray in a furnace for two minutes and directly observed the amount of sag in the metal afterwards. Great. Words and certifications are now anchored to reality for me. But my tray pieces are *twenty* feet long, not six feet. What now?

Another manufacturer did a "modified" 2000-degree furnace test on their product. What does that mean? What test was *actually* done? I suddenly needed to know. I can't let those "modifications" of the test slip by and assume everything is okay, only to potentially let such an assumption cost someone (or many someones) their life.

The high-temperature wire manufacturer does not allow the use of galvanized metals to protect their power cables. Why? Turns out in extreme heat, the zinc vaporizes out of that metal and can bond straight to the copper *inside the power cable*, collapsing its ability to conduct electricity.

Circuit breakers and fuses often have a don't-load-over-80% rule. Why?

The main "rulebook" for power design, the National Electric Code, has a rule to size wires and breakers 125% over what is needed. Why? Are the 80% and the 125% actually connected to each other? I had to find all these answers.

As it turns out, the National Electric Code, which is the main provider of nearly all the foundational rules of power design in the United States, is created by the NFPA. That stands for the National *Fire Protection* Agency. All the rules in the National Electric Code are designed to prevent the build up of extreme heat, which can lead to a fire. *The strict rules I hated so much existed to prevent fires.* Was any of this ever mentioned in all those trainings? Not a chance. So before this project, the code was just rules to follow or *be wrong* with no real reason at all.

Stability

I eventually finished my design. I learned I was not the keeper of any truth at all – *but neither was anyone else*. Instead, I ended up talking to nearly every person that might use this system in real life to get the answers that truly *mattered*. Why? Because the system existed *for* them and was to be used *by* them.

I put the big red buttons wherever they told me they wanted them. Where it made sense for *them*. What made sense to me *didn't matter*. I can put the big red buttons in what I think are great locations, but if the workers that depend on those buttons to save them or others need to run a football field-distance out of their normal paths of movement to hit them – its possible the 2000 degree fireball could get there before they do.

I did my best to clarify the meaning of all the client's rules by talking with decision and policy makers about the underlying intents. Sometimes this led to uncovering very different beliefs in what these valves should and should not be doing.

Finally, I also talked to construction workers and installers before finishing my design, and then talked to them again while they were installing the system to better understand where I could improve my designs to make their lives easier in the future.

In the end, the only story I had to *legally* care about was that the system I designed worked *on paper*. That wasn't good enough for me. I discovered I didn't give a crap about how things looked *on paper*. I wanted *no gaps*. I wanted to know that when those individuals that live and work in a dangerous world end up in a situation where they need to push a big red button, the system designed to protect them and those they love in the nearby town was *not going to fail before its job was done.*

The system was constructed and tested to work as designed before going into service. As far as I know, the system has experienced no issues, but I admit this assumes no news is good news. No one can know *with certainty* the system will work as designed until an *actual* 2000-degree fireball exists for two *actual* minutes and the system is sitting inside of that fireball, working away as designed for two whole minutes.

Failure

If this truth-story is ever validated in a literal trial-by-fire, it will immediately become the past. The system will be damaged or destroyed, along with the wordless experience of it doing its job (or not) before it's demise. It will either be a successful failure, or a failure to succeed.

Afterwards, a new "right" design will emerge from the truth-story uncovered by this event. This cycle will repeat as long as the system is needed, and anything other than what is expected to happen occurs.

As for me, my dive to understand didn't stop after this project. I am actually still on this dive as I write these words for you. *Years* have passed since this first simple project was completed, and I learned a ton from it. I am still designing power systems, but none have ever come close to making me push myself to the limits of human understanding like that first project.

I went on to design several more emergency systems, and taught others what I learned and help develop. After these many years in this same rabbit hole to understand the invisible parts of this reality we seem to share, I learned something I consider absolutely critical:

Language makes everything more complicated than it actually is.

As I see it, there is no way around this complication. It is done for a very good reason I will share with you soon. Just know that while we as humans know a tremendous amount about the universe and reality, there seems to be an even greater amount we don't know.

Now though, I'm going to bring you a little closer to what I believe to be the very core of our foundation. The core itself makes a very interesting pattern I noticed when my dive into electricity started merging into all the other fields – and I do mean literally *all of them*.

I believe my ability to find and understand what I believe to be the core was only made possible by realizing everything you've read about so far in this book. It required what I learned from this power project, passing through the looking glass of language, and gaining the ability to see through the definition problems and authority that words can bring to the situation.

There seems to be a grander truth hiding down here below the darkest shadows.

64

Unexploding the Electric Cat

"Electricity" is an exploding cat. By this, I mean it works exactly like the word "cat" from earlier in this book. By itself, the word *electricity* has so many possible meanings it means practically nothing at all, just like "cat" could mean an animal, a piece of heavy equipment, or a slang term for someone in jazz circles, or a hundred other things to a hundred other people.

This means the word electricity *requires* other words around it to help narrow the context. This is needed because the electricity in your brain and body is not the same as the electricity in and around any wires you might see. The electricity that moves around the wires in your home is not the same as the electricity found in batteries. The electricity you witness in a bolt of lightning is not the same as the electricity in an electric eel. These are all different types of electrical *systems*, and each is creating electricity differently. As a result, each type of system typically gets its own unique set of "electric" words.

This creates a form of *shelter* from the Exploding Cat Problem – which is just another name I made up for the everyone-has-a-unique-context-for-every-shared-word-shell problem. If each type of electrical system didn't get unique words, it could easily result in someone believing the electric eel's electricity is *exactly the same* as the electricity in their wall outlet because the words used are the same. That person could then end up electrocuted while trying to do something like create an "electric eel backup generator" based on the belief that "words like electricity have exact, specific, and agreed upon meanings."

The above story might seem like a silly example, but it exposes how complicating things with language on purpose can increase safety. There are *very real* dangers that can be created by our limited language. By choosing or creating new words for each type of system, you hopefully no longer automatically connect lightning, eels, outlets, batteries, and your brain as being the same.

But this has the side effect of splitting reality apart where it doesn't actually split apart.

As I write this, homes are already being powered by batteries, and batteries are being charged from the wires in homes. Eventually someone may figure out how to create an electric eel backup generator to power both the batteries and homes, which would prove all these "electricities" actually can be connected – though I have some concern for the eels' well being.

Electricity is an extra tricky word though. It is not as cut-and-dry as a noun like "cat." Electricity is not some physical thing or object you can just point to in a room or look at under a microscope. It doesn't exist in that kind of way. It is not a noun at all, even though the dictionary treats it like one and says it is.

Electricity is a *process*.

This process is *invisible* to our senses. It exists below the resolution our eyes are capable of seeing. We occasionally experience one tiny piece of this most-ly-hidden process firsthand when we witness "arcs", such as a bolt of lightning, but arcs are no where near the whole process.

When you experience getting shocked (electrocuted), you are again only experiencing the arc – it's just passing through *you* where you can't see it anymore.

Because the process of electricity happens below the resolution of our sens-es, we as humans can only experience the entire process *secondhand*. Since we are required to build context from memories of our firsthand experiences, this too-easily causes "electricity" to break up into multiple individual-and-separate words. This is because when it comes to the process, the association of word-to-wordless experience is always missing the wordless experience part.

For me, each "electrical" word ended up getting a *separate* context in my memory, and it wasn't until much later that I learned how they actually work together.

When it came to my job designing systems before my deep dive, I had to *tell myself* how electricity works and memorize the equations, and then *tell myself* the words involved have "specific meanings that everybody agrees on". This threw me right into word loop-traps and needing to rely on my *Keepers of Truth* to be sure I was doing everything "right."

Electricity was just another method to add heads to the hydra of Second Me. It's why I hated the topic and avoided it. I, however, have no intention to do the same to you.

I'll be crystal clear right up front: *No piece of this process can be separated from another*. Period. The only things ever separate about this process are *the words* to describe it. Each separate word is anchored to very non-separable parts of a single, wordless, gapless, objective-reality process. If one of the words appear, the others cannot be ignored or neglected.

You may have noticed a loop just now where it seems like I am about to go against everything I've said so far in this book. I've dismissed a lot of stories in the last few chapters as unable to be objective truth because they cannot ever be directly experienced. Now I am writing to you about the process of electricity, *which can't be directly experienced*, and placing it here in this book, near the very deepest part of the shared foundation between us where it seems like only *absolutely agreeable truths should appear.*

I want you to know I am not doing this lightly. I acknowledge the *appearance* of this paradox and urge you to bear with me. I intend to break this appearance of a loop soon enough. After all, my goal is for you to never need to rely on *my context* to understand anything I'm saying. *You* fill these word-shells, not me.

I'm not going to use any of the words so often used to teach electricity. Instead I'm going to give you an imagined experience to "make sense", *then* I'll give you the electric words.

The only risk I believe I'm taking is the small chance you don't have direct firsthand wordless experience with the objects I intend you to use to fill up the incoming word-shells. I'm taking this risk based on the assumption that if you have time to read this book, it's also very likely you have wordlessly experienced these objects.

I think it's a risk worth taking.

65

The Process

Imagine you now have two identical deflated balloons (identical balloons are not possible in objective reality, but completely possible in your imagination). Both balloons are basic, stretchy, rubbery balloons, not the fancy metallic foil-looking balloons.

Imagine blowing up one of the balloons, and putting a clip on the opening, sealing it shut. Then pick up the remaining balloon and blow it up to be twice as big as the first balloon. Put a clip on it as well. Here's a terrible drawing to help you picture it.

Now imagine a basic drinking straw appears in your hand (what color is it?). Put one end of this straw into the tail of one of the balloons and push the straw in until you bump it against into the closed clip. Then insert the other end of the same straw into the tail of the second balloon until you bump into its closed clip. Got it pictured? Here's some more bad artwork if you need some visuals.

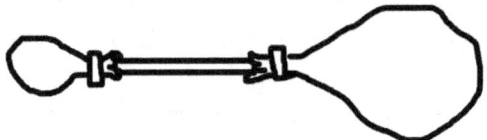

Now, before you leave this paragraph to start the next, I want you to make a *prediction* based on your imagined present and your remembered firsthand experiences: *What do you expect to happen if you remove both clips at the same time?*

This is not a trick. I'm not going to throw a meteorite into this situation. By all means, please assume nothing *unexpected* will happen. Assume the balloons will stay on the straw, and are sealed to it tightly.

I have a pretty high certainty you and I agree on what to expect when the clips are taken off, but I don't want to assume it, so I will do a quick tumble down this rabbit hole with you. We need to be sure we agree "well enough" to keep going, because I can no longer change this one-way communication with words from the past while you are reading it.

Here's what I imagine to happen in my own balloon-straw system:

The instant the clips come off, air begins flowing through the straw from the more-inflated balloon into the less-inflated balloon. The air stops flowing when the two balloons are the same size.

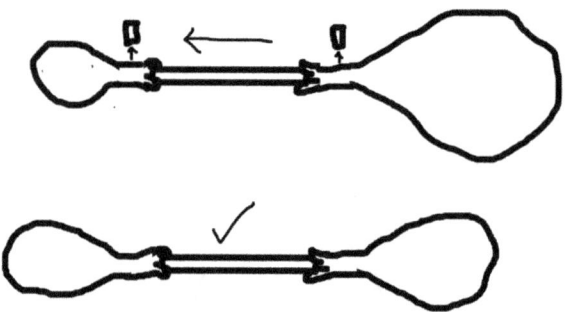

That's it. That's the process of electricity.

If you imagined your balloon-straw system behaving the same way, then we agree on this imagined truth-story built from what I intend to be our firsthand, direct, wordless experiences with balloons, straws, and clips.

If your imagined system did not behave the same way, then the "laws" of physics drastically changed since I wrote this, or you have gaps and didn't realize it, or I have gaps and didn't realize it. This doesn't make you or I wrong. But we do have a problem we can't resolve on the poultry treadmill within the pages of this one-way book.

The only option I can think of to remedy this is to urge you to actually construct this balloon-straw system in reality to test and perhaps validate the wordless truth.

I do *not* want to become your *Keeper of Truth*, which is exactly what is about to happen if your imagined balloon-straw system doesn't behave the same way as mine. The whole point here is that this *same truth-story* exists separately in both your story system and my own story system – even though we are very separated by time and space with memories with similar but different *objects*.

If we do agree on this system, we can't quite call this experience an *objective* truth. This is because you and I can *never* be absolutely sure our wacky and screened sensory windows imagine these objects *exactly* the same. They will always be approximate at best. For instance, the colors and sizes of the balloons, as well as the material and texture of the straw are likely to have differences.

Instead, this is actually a shared *subjective* truth that is *anchored* to objective reality as we both understand that reality to work. In theory, we can both fill up hundreds of different word-shells about the objects and movements in this system and still *understand* each other, because we aren't stuck believing our words have exact unmoving definitions. I think you can potentially explain this system to a child or an adult with ease if you wanted to, and be pretty confident with all your word choices.

So, with the assumption that we agree on what happens when the clips come off, this becomes our simple, shared, wordless, enlightened, imagined-through-experience-with-the-same-or-at-least-similar-reality, and an *agreed-up-on-human-subjective* truth.

We now are in the deepest part of our shared foundation. The next few chapters will hopefully begin to give shape to the core. The only things that change once we've entered this core, whether outward and inward, bigger and smaller, faster and slower, are the words used to describe what is performing *this same core process*.

With this in mind, I'll now provide you the three words needed to understand the process of electricity using this balloon-straw system.

The imbalance between the two balloons is called *voltage*.
The movement of air from imbalance to balance is called *current*.
Anything that slows this movement is called *resistance*.
That's it.
That's Electricity 101.

But as I understand it, this process is *far more* than Electricity 101.
As far as I can tell, this process is *Literally Everything 101*.

66

The Electric Balloon Bridge

I am going to do a quick walkthrough of the exact same balloon process a second time. This time though, I will begin anchoring the three "electric" words I just gave you: Voltage, current, and resistance, and show how they are entangled.

I have no intention to replace the balloon-and-straw-based words with electric words after this chapter. I will instead switch back and forth, and I suspect you'll be able to understand me because you have wordlessly experienced the process underneath.

If I were to switch to only electric words from here on out, there is a massive risk that you might un-anchor the electric words from the wordless balloon experience by accident, which would cause me to suddenly become a *Keeper of Truth* for you. This is what I have found to be one of the biggest hazards of learning something new, and is how the much bigger Floating Anchor Problem continues to expand toward disaster – but that's a story for later.

You might be wondering: *Why use them at all? Why take some side-quest rabbit hole into electricity, especially since I don't intend to use them exclusively? Why not just stick with the balloon system and words like "air" and just drop electricity and its words all together?* I'll explain:

You cannot see your mental boundary or mine, you cannot see authority, you cannot see words flying through the air to your ears, you cannot see fear or your need story arcs, you cannot see right or wrong, or the firsthand sensory wordless experiences of anyone else. They are *all* invisible, but they are all still very much a real part of a bigger, wordless, objective reality. So is electricity.

In my opinion, electricity is special though. It is special because it is tied directly to atoms, which is why the process of electricity actually ends up applying to what might be *everything we as humans think we know to be true* in this universe so far (as much as I can tell anyway).

The balloon system is intended to act as a "language bridge" which connects the first point of the visible, to the second point of the invisible. And because even your brain and body operate using electricity, this process is probably about to apply in *way* more places than you might expect.

I urge you to read what follows in this chapter carefully and multiple times if needed, because I think it is incredibly important each electric word anchors itself across the bridge to the very wordless but "sense-able" balloon system experience.

When you are ready, let's reset the system and begin.

The Process of Electricity

Imagine two deflated-but-identical balloons. Blow them up again, one twice as big as the other, and clip them shut – exactly like before.

If air pressure now represents electrical energy, then there is a *voltage* present between these two balloons. Why? *Because there is an imbalance between two points.*

Voltage *is* imbalance.

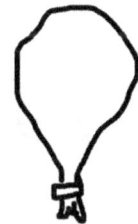

Some stories from your past are about to come racing back:

Imbalance is a *relative* word.

A single point cannot be imbalanced because there is nothing to be imbalanced with.

A second point is required to compare and give meaning to the word 'imbalance'.

This means two points are required to have voltage.

This means voltage is relative.

This means imbalance in reality is also *relative*.

If the two balloons represent two points in the same system, then currently there is no movement or change between these two points *relative to each other*. According to the resolution of your senses, there is so much *resistance* between these two imbalanced points no movement toward balance seems to be happening. Let's call this 100% resistance.

When separated and clipped like this (100% resistance), you can't sense any movement of air (current) from one balloon to the other, so the imbalance (voltage) between these two points remains the same to you.

Here's another terrible diagram.

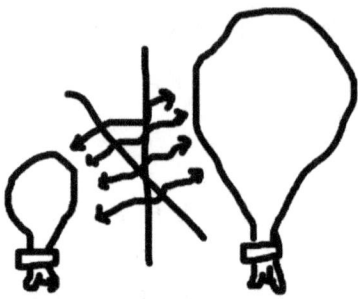

Now attach the balloons to the straw like before, but leave the clips in place.

The voltage, current, and resistance in this electric-balloon system *still* has not changed.

There is still no airflow between the balloons.

They are still the same size as they were before the straw was added.

Resistance is still 100% between them.

Only the *shape* of the system has changed due to the addition of the straw. *Nothing else has changed.*

Again, according to what you can see, hear, taste, smell, and touch.

Now remove the clips.

As soon as you do this, the resistance between the two balloon-points drops to less than 100%.

The moment this change in resistance happens, the other two parts of this inseparable three-part electric process, *voltage* and *current*, also begin *changing*.

Air immediately begins to racing through the straw (current). It is moving from the bigger balloon to the smaller balloon.

As this happens, both balloons start changing size. The bigger balloon gets smaller, and the smaller balloon gets bigger.

The imbalance (voltage) between the two balloon points is shrinking.

Actually, let's call it *collapsing*.

As the voltage collapses, so does the current through the straw.

Eventually, both voltage and current collapse smoothly and completely at the same moment.

The two balloons have now *stabilized* at the same size and pressure.

You can easily recognize this moment because you no longer notice any movement.

The part of the process that determined how long this system took to find a balance was the *resistance*.

Good to go?

Now for a tough question: *Where* is the resistance?

What happened to it during *The Process*? Did it change? Was it changing the whole time like the voltage and current? Did it drop from 100% to 0% when you took off the clips and then stay at 0% the whole time?

You might be tempted to think so, but "resistance" is a word that is just as tricky as "electricity", and *everything* in this system was changing at the same time as it collapsed toward balance.

Resistance is so tricky, yet so important; I'm going to give it its own chapter.

A Light in the Darkness

To better understand why resistance is tricky and where resistance is located in the electric balloon-straw system, try to answer this question about *The Process*: What *exactly* is being resisted?

I can't know your answer, but I suspect you might be tempted to think something like: "The flow of air is being resisted by the straw".

If this were an electrical system, this would make the straw a "resistor". This is not wrong, but it is only a very small part of a much *bigger* story.

After all, the straw isn't actually "pushing backwards" against the flow of air. It's just a tube, right? It's not "doing" anything. It performs no action at all. It's just hanging out, doing tube stuff, thinking tube thoughts. You know, *tubular* stuff.

Instead, the straw is forming a pathway for air between the two balloons as they move from imbalance to balance. The shape and size of this path can change everything about movement and time for this *entire* system.

If the straw-path was twice as wide but stayed the same length, it would let much more current flow. This would cause the collapse to balance to happen *faster*.

If the straw-path was super skinny, it would only let a tiny amount of current through, so the collapse to balance would happen *slower*.

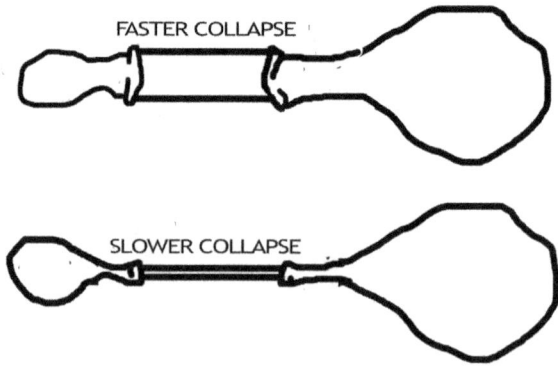

You can experience this wordless straw-size truth-story firsthand. Find a clock or have Second You count out time while you do the following:

1. Find a real straw.
2. Take a deep breath in and hold it.
3. Close your lips around one end of the straw.
4. Get ready to start your timer or count off approximate seconds.
5. Blow through the straw as hard and long as you can while counting the time it takes for you to run out of pressure.
6. Note the amount of time this took to do.
7. Repeat steps 1-6 with a much thinner straw if you have one.
8. Repeat steps 1-6 with no straw at all.
9. Compare the times.

Crossing the language bridge into the electric process, the straw is not a resistor, but *a wire*. The shape and size of this wire-path can change everything about movement and time for this system. The wire determines the time it takes to collapse voltage in exactly the same way as the straw.

If the wire is super thin relative to the voltage, it *resists like hell*. It can resist so much it might start to get hot, and can even begin to glow. This concept is responsible for the early light bulb. If you've ever experienced an old-school "filament" bulb, this is why it gets so hot. Look inside of it when it's off (and cool), and you'll see an ultra-thin wire suspended by two thicker wires.

When the two hidden and imbalanced "balloons" of the electrical power source are connected to the bulb through the metal base try to collapse into balance with each other *through that tiny wire in the bulb*, the tiny wire begins to heat up and glow so bright you can use it to light up an entire room.

When the bulb "burns out", usually with a *pop*, you can often see a tiny broken wire dangling inside, accompanied by scorch marks on the glass bulb. The ultra-thin wire failed and broke from the heat and stress of resisting.

This straw-failure story applies to the drinking straw in our balloon-straw system just the same.

Do not do this without taking some safety measures, but to go full light-bulb-blowout with a drinking straw, seal a very thin straw to a huge pressure source like an air compressor. In theory, when you run the air through the straw with the compressor at max pressure, the straw will also begin to heat up to the point it melts. It might also simply "break" by blowing a hole out the side of it, or splitting along the side. The outcome will likely depend on what material the straw is made from.

The point I'm wanting to make is this: The size and length of the straw matters *a lot* when it comes to time and movement for air, and the same holds true for the wire when it comes to electricity.

Of course, I still haven't answered the questions of what is being resisted and where that resistance is located. I have delayed answer as far as I have, because it is about to cause a massive expansion to our simple imaginary balloon system in all directions. This is because pieces of the story *are very likely missing for you*. These "missing pieces" are lost to a void, and I suspect resistance is one of these pieces. Resistance is invisible - totally hidden way under the resolution of your human senses.

Don't worry though, even with missing pieces, the overall approximate truth of *The Process* isn't changed. Instead, I believe this explosion is going to validate it as an even stronger truth, perhaps to an absurd degree. I'll now light the fuse:

You previously read the words "anything that slows this movement (to balance) is called *resistance*." Given this, after the two balloons have collapsed into balance and all movement has stopped, *does resistance still exist?*

The answer is *yes*. It is actually *back to 100%*.

But it is not in the same *shape* it was when you began.

If you are confused, don't worry. I might be able to help with another question. *Is an imbalance still present after all movement has stopped?*

Still unsure? Here, have another question:

Was an imbalance present when both balloons were empty?

If not, how did the imbalance get created in the first place?

Perhaps some light bulbs are glowing in your mind now, or perhaps not. It can be hard to see in the void sometimes, and we are in the darkest part.

There is *very much* an imbalance still present. It was present *before* the balloons were inflated, and *after* the two balloons have collapsed to balance.

The imbalance is just no longer between the two balloons.

After all movement has stopped, the imbalance has instead shifted to be between the *entire* balloon system and the *atmosphere* around it. This "atmosphere" exists *even in your mental boundary* where you assumedly built and observed your own little balloon-straw demonstration.

If the atmosphere around the system was missing for you, this is one of your "missing pieces of the story". What's wild to think about is you actually imagined stepping completely into this hidden void, most likely without noticing you did. You imagined yourself existing in that void at the beginning when you blew up the balloons, and watched the experiment from the same void until the end. It's possible you completely ignored this void and continued on with thinking about the demonstration, not caring about what filled up all the space around you in your imagination at all times. It's even possible this might have happened if you actually built this system in reality with real balloons and a straw.

Perhaps you are *still* unsure exactly *where* resistance is located even now, and what this void has to do with it.

It has *everything* to do with it.

In *all* of your stories – gaps and voids are *everywhere* and you often don't realize they are there until you look at things from another perspective. In the universe as we understand it so far, there is *always* a Point 2.

Entire now-stabilized two balloon system = Point 1.

Atmosphere outside of the balloon system = Point 2.

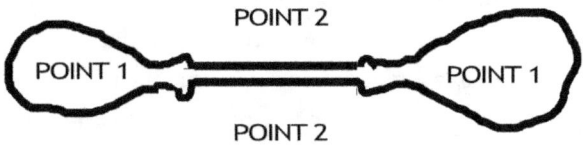

There is still an imbalance between two points *relative to each other*.

Voltage is still *very much* present.

There are many words used to describe this new Point 2. They are words like "baseline", "zero", "ground", "Earth", "foundation," or "footing".

This is not new at all of course, this Point 2 has *always* been present.

An inflated balloon of *any* size has a "voltage," or imbalance, *relative to when it was deflated*. In power system design, this starting point is called "ground".

When we as humans do little thought experiments like this in our heads, we often "isolate" the imagined system in our minds from both ourselves as the imaginer and most of the other variables that might exist in reality. How many other variables are there? *We don't yet know.*

This leads exactly back to the Gazing into the Abyss chapter. This leads exactly back to determining the objective truth about a plane taking off from a treadmill. There are an *overwhelming* number of possible stories that can happen. Way more than your mind or mine can imagine and deal with at one time.

And this is what I quickly discovered in my deep dive into electricity. There are strict rules and equations in place, but they are there for safety and consistency (standardization) reasons. Some individuals seem to act like they have it all figured out – but no one knows the *whole* story of electricity. It is full of gaps and voids even as I write this.

We, as humans, are discovering new things related to electricity all the time. For me to realize just how many gaps and voids there were, it took having to design an electrical system that not only works, but works for at least two minutes as everything around it changes so radically the system itself would begin to melt, break, fail, and vaporize.

Figuring out and trying to control the point of failure like this happens all the time in the world of engineering, but what I think made this process special for me, is that I was not already stuck in "unitized" thinking. I didn't already have the specific words of electricity driven into me as truth-stories I constantly told myself. I truly began with almost nothing – for years I was like a child sitting in an advanced calculus class when it came to the topic of power. Nothing made sense. But this eventually handed me the freedom to ask *why* like a child surrounded by experts in the top of their field when I couldn't understand something.

I was fortunate, *deeply* fortunate, to have someone in my life not only brave enough to say "I don't know" to some of my questions, but also have a great understanding of what he did know, which allowed him to "flex" his words to communicate with me about the same *wordless* process.

I had almost no truth when it came to power design, so if I had learned under a different person that spoke in certainties, it probably would have wrecked my ability to understand. Here, I'll give you an example how refreshing uncertainty can be:

I, *King Eric*, need you, my reader, to design a simple replacement two-balloon system. But this design requires you to hook the hose of a massive air compressor into the middle of the straw. You are to make sure this replacement balloon system can handle 60 seconds of compressed air being added from this compressor at full speed without the system bursting or leaking anywhere. If it bursts too early, someone might die. Oh, and make sure the system is also porcupine-proof. At any time we may release 30 annoyed porcupines into the room with the balloons. Got it? *Go.*

Suddenly you need to know *a lot* about rubber (or is it latex? or polychloroprene? or nylon?), straws (plastic, metal, or paper?), binding these materials together (glue, bands?), their failing point, what happens to them under growing pressure, how the builders of those straws and balloons tested their claims of how great their products are, what the starting "ground" pressure is (which might involve the elevation where this system is located), the shape and size of the compressor and straw nozzle, the size of the tank the compressor has (imbalance to balance! Tank size is going to matter!), how much space the balloon system is allowed to take up as it grows, how sharp porcupine quills are, if porcupines can actually shoot those quills from a distance – on and on and on – a million questions. You must become a child, but one with a lot more words at your disposal.

I believe you would eventually ask all these questions even if you don't think of them right away. After all, someone might die if your system fails – and I haven't actually said who this someone is, which means it's safe to always assume it could be *you.*

With this in mind, you might be willing to begin with what feels like the most basic of questions: "How much air can a balloon hold?"

Everyone has to start somewhere. Would you rather another system designer - one who refuses to ask questions and instead just assumes answers - design the system that might result in your death if it fails?

Having someone next to you that can explain nearly everything about materials, balloons, straws, compressors, and how pressure works in a way you can understand would be a near ideal situation, right?

And how nice would it be for that same person to confess to you that while he or she knows a ton about physics, he or she doesn't know a single useful thing about porcupines? Is that not better than "faking" stories about porcupines on the spot to confuse and blur the line of truth and understanding?

"I don't know" is refreshing.

It's honest.

It's trust-building.

It requires the person to know his or her own *boundaries* of knowledge relative to objective truth, to reality.

It requires *resisting* the urge to cross that *boundary* with a simple sentence declaring his or her opinion as a certainty.

It requires leaving the crown where it rightfully belongs – upon no one's head.

Resistance is everything.

68

Resistance is Every Thing

Even though you and I can learn about and measure "isolated" systems like these electric balloons and ignore all the other things around it, *there seems to be no such thing as a truly isolated system in our shared reality.*

Scientists in laboratories try extremely hard to isolate their experiments, but there is always a "ground" condition in the laboratory itself, and this condition is always different in some way from lab to lab and from moment to moment.

In the laboratory of your imagination, often the "experiment" you are focused on, such as two balloons, is totally surrounded by an unnoticed and *missing* ground condition. There is quite literally *nothing* beyond what you are thinking about – it is a total void.

This void is so persistent that I believe even if you actually created this balloon system in real life, it's *just as likely* you ignored everything that happened before and after the experiment.

There is only one thing that makes any of this possible. It makes these voids possible, it makes gaps possible, it makes the balloon system possible, it even makes The Process itself possible because it is what makes *two points* possible.

It is a *boundary.*

In the case of a balloon, the boundary making two points possible *is the balloon itself.* A "balloon" is literally a rubbery stretchy expandable boundary you can open and close as you please.

When the balloon was deflated before you began the experiment, it had an opening that let air move in and out freely. Both sides of the balloon were the same pressure because of this break in the *boundary*, so there was no imbalance between the inside and the outside of it.

The exact moment you sealed the opening of the balloon with your lips, you made a throat-straw connecting that rubbery balloon with your fully inflated lung-balloons. You then "pushed" the air in your lungs through the straw you call your windpipe, trachea, or throat into the rubber balloon.

That *pressurized* air, made possible by your muscles "squeezing" your lung-balloons, began pushing out against the rubber boundary that is sealed to your lips *in all directions.*

The pressurized air wanted to keep spreading out (current) until the pressure inside the balloon boundary matched the pressure on the outside of the boundary again (voltage collapse back to ground), but this rubbery, stretchy boundary *won't let it.*

It is resisting.

The natural resulting shape of this process, no matter how much air you add, is a *sphere*. It might feel like I just broke a "rule" that states "pressure/liquid/gas takes the shape of its container" but I have not. The rule is not wrong, it's just creating it's own boundary in your head.

That boundary is *an assumption* that the resistance of the container will always be greater than the imbalance on either side of it.

All containers, or boundaries, have a limit. That limit is determined by wherever the container boundary is weakest when it comes to *resisting the imbalance from moving toward balance.* The air that is "bound" inside the container boundary is pushing on all surfaces of the container "approximately" the same in *all directions.*

If you want to eliminate any weak spots in a boundary, it needs to provide *exactly the same resistance to imbalance in every direction.* The most efficient shape to do this, which also uses the least material possible, in my opinion, is a *sphere*.

To an imbalance present on both sides of a sphere-shaped boundary, the sphere is the same from every direction.

To the sphere boundary relative to itself, it is also the same in every direction.

As I see it, the only reason rubber balloon "containers" are often not sphere-shaped is because most are molded into a teardrop shape to make the opening usable to your lips and face (or other inflating devices).

With *enough* imbalance though, all balloons and all containers will transform into spheres, or burst on their way to becoming one (because the boundary could not keep resisting the imbalance the same in *all directions*, which resulted in a new unplanned "straw path" and collapse back to "ground" at the point where the boundary was weakest).

The flat-and-floppy empty balloon started its transformation toward a sphere because *you* created an imbalance between the inside and outside of it. This imbalance doesn't immediately collapse back to "ground" after you remove your lips because you put on the clip, closing (or sealing) the balloon boundary and "pausing" the collapse.

Once the two balloons filled with air are connected, clips are removed, and imbalance collapses to balance, *every surface of the entire system is still resisting*.

The straw *and* both balloons are resisting being expanded into one big sphere.

The pressure still trapped inside the entire system is pushing against every *boundary* with the same force trying to reach *balance* with the pressure on the *other side*. Some of the boundaries could resist this "push" easily and did not stretch. This was the straw. Others boundaries couldn't resist the "push" as much as the straw, and started to stretch toward failure. This was the two balloons. The result of these differences in "resistance" determines a non-sphere final shape of the system. Which looks like this:

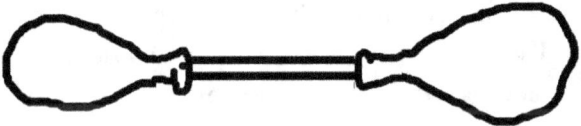

As soon as you start thinking about this just a little bit more – trying to find exceptions to this or follow The Process further – everything starts to expand dramatically in all directions and scales like a big bang in a black hole.

69

Tiny Simple Enormous Complexity

For two points to exist there must be a boundary between them.

For a boundary to exist resistance must exist.

For resistance to exist there must be an imbalance.

For imbalance to exist there must be two points.

A loop.

These words cannot define each other. Loops do not actually exist in word meanings. As I see it, these circles are *always* an illusion caused by accidentally ignoring the voids that exist *between* words or by *telling yourself* their definitions secondhand. Loops are the end result of trying to anchor a firsthand truth into secondhand word-shells, which are at best being filled by an experienced *wordless* context that is never *actually* shared between any two people.

Loops in time do not exist either.

Action loops or imagined time loops are always an illusion caused by accidentally disconnecting from the unbroken string of constant new movements that occur forward through time, which is often hard to notice due to your memories and predictive powers constantly mixing in the past and future.

When it comes to uncovering objective truth, imagined loops in both language and time might be the single greatest problems we face as human beings. As I see it, the greatest weapon we can wield against these problems is the idea of objective time *outside* of our experience.

Time is the Great Loop Breaker.

Time has no gaps.

If you look at word loops and thought loops through the lens of gapless objective time, you will quickly uncover dramatic explosions of possibilities in all directions. This happens because objective *movement* cannot be separated from time. Each causes the other to immediately exist, and neither time nor movement ever stop, have gaps, or run backwards. This sounds like an assumption, but I have reason to believe it is not an assumption at all.

Unfortunately, you and I will *always* have gaps.

Our contexts can never be gapless.

You cannot follow time or movement beyond what you are directly experiencing in this moment with absolute certainty. Neither can I.

This is ultimately why you and I are forced to stay pinned to sometime very near this very moment, *right now*, when it comes to being able to agree about what is true at all, because our agreement quickly falls apart if we try to build a past or future truth-story using our sensory-filtered memories.

Even if *The Process* turns out to be the absolute objective truth of this entire universe and you are now aware of it, you and I still can't use it beyond this current moment to define a shared *objective* right and wrong.

This might be difficult to grasp, so let's take a stroll together:

Starting just before you picked up the empty balloons, try to trace The Process *backwards* through time. Where is the imbalance that will be responsible for the current that will eventually fill each balloon?

What created that imbalance?

What about the imbalance before that?

And the imbalance before that?

Keep in mind that any movement you notice is almost always the current in a "straw" connecting two imbalanced balloons, and that the balloons and even the straw might be completely invisible to your senses.

For this reason, I suspect you did not immediately consider that *your lungs* were the larger "balloon" when you exhaled into each deflated rubber balloon, and those same lungs were the smaller "balloon" when you were inhaling.

If you didn't connect your lungs to the balloon system, this is pretty *wild to consider* in my opinion, because you *specifically* imagined blowing them up. Did you oddly gloss over this piece of The Process? Was the imagined experience of blowing up the balloons a bit fuzzy? Absent of details? A gap of time in your memory filled by assumption?

Keep following The Process backwards before your lungs, if you can. When you inhaled (which expanded your lung-balloons), what was the big balloon?

You will very quickly find yourself trying to follow endless and ever-splitting stories of the movement of air backwards from balance to imbalance beyond your first memory, beyond the first humans, beyond *everything*. You will find yourself wondering how air and time got started at all.

Now try to follow The Process *forward* in time, beginning at the moment the balloon system has stopped all movement.

It's possible you didn't imagine the *existence* of air and its lower pressure all around the outside of the system's boundary, which just became the new Balloon 2. If you didn't, it was a void or gap of thought for you.

But keep going. Try to follow the imbalance-to-balance process forward into the future. Pop one of the balloons. Where does the imbalance move to next to find balance? Does it spread out in the container (sealed room) you imagine yourself in? Do you inhale it again? Where does it go when you open the door to leave the room?

You will quickly find yourself trying to follow endless and ever-splitting stories of the movement of air from imbalance to balance beyond your 3-step process, beyond the last humans, beyond *everything*. You will find yourself wondering how air and time will end at all.

We aren't going to find a perfectly agreeable truth in either one of these directions in time away from The Process. We are confined to what time we can experience *together* in Step 1 of our story system's process.

With this in mind, let's try to work within the boundaries of Step 1 to find the best and deepest common ground.

70

Enormously Complex Tiny Simplicity

Time and movement operate at the finest "resolution" of objective reality, whatever that might be (no one yet knows when I wrote this). By comparison, your wacky sensory windows have a much "lower" resolution.

This means *inside* what you perceived to be a wordless, gapless experience-built-by-firsthand-memories with the balloon system, there were enormous amounts of information "hiding" from Step 1 of your story system. This would be true even if you actually built the system and directly experienced it firsthand.

Let's take another stroll in the darkness of the core....

Assume your imagined balloon-straw system process now represents an entire universe. Anchor the *beginning* of time in this universe to the moment the 3-pieces are assembled and sealed with an imbalance in place. The clips are still on, so there is 100% resistance preventing time and movement in the universe. Anchor the *end* of time to the moment the two balloons find balance and all movement seemingly stops.

Now try to trace *every single movement* in this universe forward through time from beginning to end.

The overall method and measure to do this is simple – and you already know it – it's The Process. However, at the resolution of your sensory windows, you are only able to take in *the outer-most boundaries* of this universe.

I will assume we both agree there is absolutely *something* inside the boundaries of this balloon system. After all, you put it there – it's *air*. You would feel air racing through the straw if you gently squeezed it while movement was occurring, and air is ultimately the "stuff" that makes the system change.

The Process applies to air just the same as it does to everything else. With a small amount of imagination, you might notice the story of the balloon universe from beginning to end actually has a *ridiculous* amount of things happening at the invisible scale underneath the resolution of your senses.

Here's a tiny taste: Shrink yourself in the laboratory of your imagination to fit inside this balloon-system universe.

You are now comfortably able to fit inside the straw. From your new perspective, the straw is now a tunnel that is two times wider than your height.

At each end of the tunnel there is a closed metal door that represents the clips still in place at the beginning of time. Both doors are hinged to open toward you. There is a sign on each door. One sign says "Danger: *High* Voltage". The other says "Warning: *Medium* Voltage".

The signs do not say the same thing, because there is a *different* imbalance - a different voltage - on the other side of each door *relative* to the tunnel where you are. The straw tunnel is at a different pressure than *either* balloon. One imbalance (high voltage) is twice as large as the other (medium voltage).

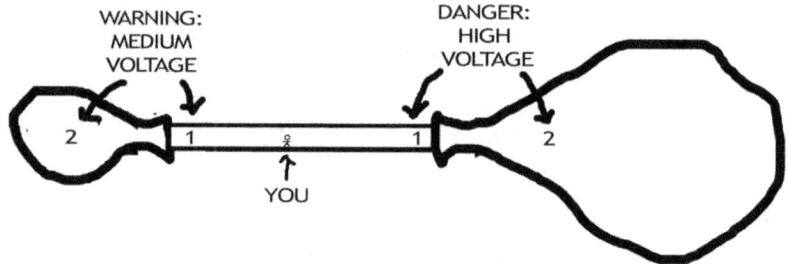

In the straw, you are actually near "ground" pressure relative to each balloon. This means you are at about the same pressure that exists on the *outside* of the entire system. Did you catch this with your Step 1 and 2 the first time through The Process in Chapter 65? What about the second time through when we used electric words in Chapter 66?

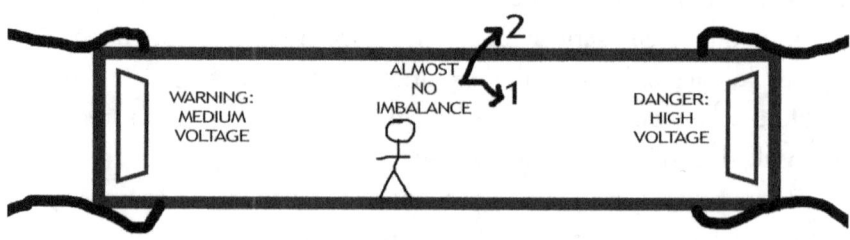

If you didn't notice this "third balloon" created at the moment the straw was sealed at both ends, it was not part of your truth. If it was not part of your truth, it definitely didn't appear in your imagined version of this system. You didn't add it in, even though it was there the entire time.

If this was missing for you, then this third straw-balloon appears to be a small gap in the story of your balloon universe. However, this "small gap" is hiding a ton of information.

Remember, your imagined balloon system is only *approximately* anchored to reality. It is actually 100% anchored to the filtered reality your story system has "recorded" coming in through your wacky sensory windows – and those windows have a very low resolution compared to reality itself. Your balloon-system story is missing *all kinds of things* because of this difference.

To get a feel for the monumental amount of information in this single gap that held an entire third balloon, just imagine shrunken-you opening the door that says "Warning: Medium Voltage". What is going to happen?

At the very instant the seal is broken, current starts to blast its way into your straw tunnel even though the other high-voltage door is still closed.

This happens because there was a pressure *imbalance* made possible by a boundary. In this case, that boundary is the balloon. It just so happens the door is the one place that balloon boundary can be broken, and that break happens to be connected to your straw-tunnel.

The Process exists exactly the same as it did before. *It's just in a different shape.* The doorway itself is now *the straw* between the smaller balloon (which is really the bigger balloon in this system) and the straw tunnel balloon (the smaller balloon).

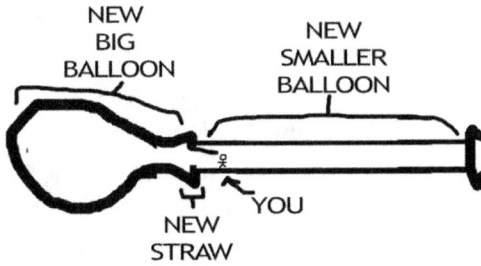

NEW
BIG
BALLOON

NEW
SMALLER
BALLOON

YOU

NEW
STRAW

How long does the current blast through the door into your face? As long as it takes for the pressure imbalance between the rubbery balloon and the "straw-balloon" to *collapse* into balance through the open door-straw.

This takes a *lot* less time to happen than the two rubber balloons took to balance earlier, because the straw-balloon *resists being expanded* by the amount of pressure the rubbery balloon has to offer. The straw is technically a tiny very-non-stretchy balloon *relative* to the stretchy rubbery balloon on the other side of the door.

Again, this is *exactly* the same process as before. It is just way smaller and way faster. It's only the size and shape of the boundaries involved in The Process at this moment that is different.

The doorway at this smaller scale is one clip at the universal scale.

The larger balloon at this small scale is the smaller balloon at the universal scale.

The straw-balloon at this small scale is just a straw-pathway at the universal scale.

But we haven't actually changed or added anything. All of this existed from the beginning. You and I just overlook all of this because it is not something we can notice with the limited resolution of our senses. That doesn't mean it doesn't exist as part of reality. It just means you didn't and can't actually experience it *firsthand*.

After all, there is a near-zero percent chance that you actually removed both clips at the *same exact moment* in reality, even though you probably imagined it that way. This "straw-tunnel pressurization" process probably took place in a gap of time so tiny it is practically impossible to notice at normal speed.

All the statements about voltage, current, resistance, time, and movement apply just the same at this tiny sliver of a moment as they did to the entire balloon system universe you imagined at a more familiar human-sized scale.

At the tiny scale, current rushed as fast as possible through the door as soon as the boundary's resistance dropped below 100%. Resistance dropped below 100% the instant the door boundary was broken by you opening it. The size and shape of this doorway matters a ton to how fast balance is reached between the rubbery balloon and the straw-tunnel "balloon".

If the doorway is as large as the straw tunnel, the collapse-to-balance happens as fast as possible for this system. If you make the doorway into a tiny porthole the size of your shrunken fist, it takes *much* longer for pressure to balance and air to stop moving through it. Either way, the end result is the same, and balance is reached. The only difference is *time*.

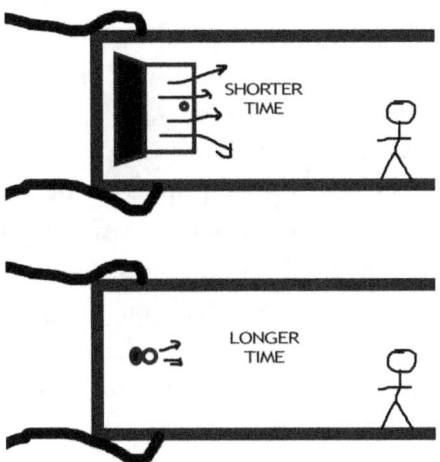

With the small balloon and straw tunnel now balanced, turn around, move to the other end, and open the other door in the straw that says "Danger: High Voltage". The same exact collapse-to-balance repeats, but The Process has changed shape yet again.

The high voltage doorway at this small scale is the clip at the universe scale.

The larger balloon at this small scale is also the larger balloon at the universal scale.

The now-balanced smaller balloon *and* straw-tunnel at this small scale are actually combined to be an oddly-shaped smaller balloon at the universal scale.

But again we haven't actually changed or added anything.

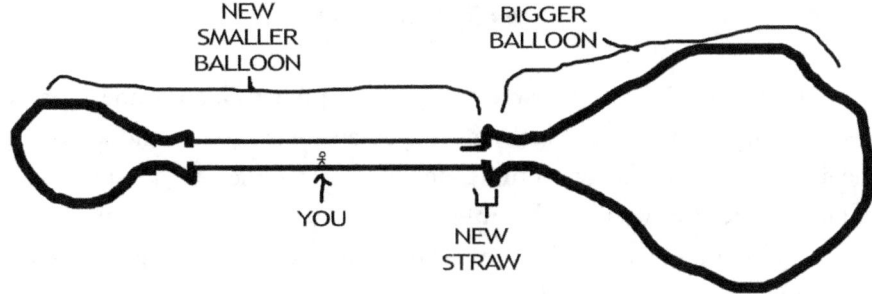

This is its own kind of abyss.

We can keep shrinking and repeating this process in time over and over again.

When you began opening the first door, when *exactly* did current and voltage start changing? When was the exact moment resistance actually dropped below 100%? When the door opened a finger's width? A hair's width? As soon as you felt air moving through it?

It's none of these.

It changed the moment the door moved at all.

All movement is The Process in action.

From inside the balloon, the pressure is trying to expand and pushes almost-equally in all directions. If the door moves *at all*, then because it opens away from the balloon and into the straw tunnel, the balloon boundary just *got bigger* (expanded) and the straw tunnel "balloon" just *got smaller* (collapsed).

Current rushed in to fill the newly expanded space you created by simply moving the boundary door. This was way before you ever felt any airflow. This is way before the door seal was broken at all. The voltage across the door collapsed toward balance by a *tiny* amount, because there was movement.

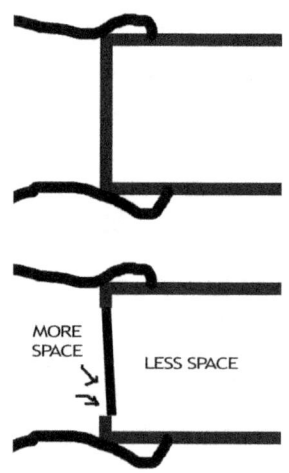

MORE SPACE   LESS SPACE

The Process is *recurring* at a nearly incomprehensible number of scales *inward*. You just can't sense what is moving because it is so quick and so small. In theory, this recursion likely continues down in scale to the very finest resolution of objective reality, whatever that might be.

I will not go that far. Not in this book. Instead, we'll ride this movement scale all the way down to the tiniest recursion I dare to go with you, even though secondhand stories are always required to experience them at all: Molecules and atoms.

Everything seems to be made *of* atoms, *by* atoms, or connected *to* atoms (or the pieces that are found in atoms) the best we can tell. You are made of them. I am too. Electricity is as well. This is one of the main reasons why I have chosen to anchor the words of this balloon process to electricity-based words. All the other invisible "energies" in the universe we know about like pressure, temperature, light, and sound can all be tied back to the atom and the movement of it and within it, which is responsible for electricity and magnetism.

I believe that the secondhand stories of electricity and magnetism exist as part of objective reality because I design systems using them every day. I design these electrical systems using screens and computers made possible by designs that used that very same electricity and magnetism. I used screens and computers to write this book. My kids play with magnetic energy all the time. These things exist, and we seem to have figured out quite a lot about them.

The electrical systems I design are eventually transformed into directly experience-able movements of objects in reality, even though I can't actually directly experience the invisible-to-me imbalance and currents doing the work.

I am going to assume you use electricity or magnetism every day as well. If you do, then this is a type of proof that things are moving and happening in this universe beyond the ability of your senses to perceive them, and these things can be understood and even controlled by human beings. Not perfectly controlled of course, but we are getting closer every day.

When it comes to electricity and magnetism, this understanding and control reaches all the way down to being able to move around "pieces" of atoms.

We can begin to enjoy a shared truth at this scale in a way that makes sense if you shrink yourself even further inside the balloon system universe.

Inside each balloon, the dominant secondhand story I've heard most often is that *molecules* are responsible for creating "air pressure." They create pressure by moving at very high speeds and bouncing around off of each other and the balloon boundary.

If you were an air molecule in the balloon, you would find yourself in a situation much like an intense mosh pit at a heavy metal concert. Everyone is running and bumping and pushing each other, and you never know which way you are going to be forced to go next.

The people surrounding a mosh pit that create its edges are the stretchy balloon boundary. They are flexible and easily moved around, but they resist joining in the mosh pit themselves or letting anyone currently in the mosh pit "fall out".

If all the people creating the edge of this mosh pit locked arms, they would become a much firmer boundary – similar to the walls of the straw. It would now take lots of individuals in the pit pushing against them to make them move or to fall out. The resistance is now significantly greater.

And yet, these different boundaries and molecules are all made of the *same thing*. In the shrunken situation, all of these things are made of people. When thinking about these same things from the size you are right now while reading this, all of these things – the mosh pit, the rubbery boundary, and the firm boundary, are made of what we call atoms.

Supposedly.

I personally have never experienced pieces of atoms inside a balloon, or the atoms making up a balloon firsthand. I have never witnessed individual atoms doing the things I design entire electrical systems to do with them firsthand. I can't see them. I can't feel them. I can't taste, smell, or hear them either.

The structural engineers and designers cannot experience the carbon molecules present in steel beams firsthand.

The civil engineers and designers cannot experience the atoms present in concrete and the Earth firsthand.

The chemical engineers and designers cannot experience the molecules and atoms in various substances and chemicals firsthand.

The computer engineer or programmer cannot experience the bits or currents in a processor or transducers dancing on and off firsthand.

The aerospace engineers and designers cannot experience the atoms in the air or in their vehicle firsthand.

The mechanical engineers and designers cannot experience atoms in a machine about to be stressed firsthand.

No one can experience atoms firsthand.

Even quantum researchers do not experience atoms firsthand (as I write this). These individuals, which might include you, are probably closer to individual atoms than almost all other human beings, and yet all they experience firsthand are the *secondhand stories* their devices and tools tell them is happening.

This means even though we can understand how the atoms interact through The Process, we unfortunately cannot use atoms and movements that exist below our senses to build a common shared foundation. We'd be stuck trying to match our individual secondhand tools to each other, sending us headlong into the abyss of the measurement problem and tolerance problem.

And so we arrive back at truth existing at the scale of the balloons themselves.

71

Atom and "E"

Look or listen around you. All the things you might be seeing, feeling, smelling, tasting, or hearing are supposedly made *of* or made *by* the building blocks of everything we know so far: Atoms.

As I understand them, atoms themselves seem to be a kind of balloon containing even smaller balloons. We are currently down to quarks, gluons, and meson balloons within the system of the atom "balloon", and the search continues as I write this to find the smallest "one" in existence within the atom we can assign a word-sound.

I've personally traveled what feels like hundreds of different rabbit holes to learn about invisible-to-us "things" and "ones." These are rabbit holes into the micro-worlds of biology, chemistry, and electricity. However, it took me going way beyond the known limit of these things quite by accident, to see the general concept connecting all things. Connecting "all things" is the realm of "not things".

When I had to design an emergency electrical system made of various "things" to fail successfully, I realized all *things* eventually fail in extreme conditions. The only way this makes sense is that everything is ultimately made of the same "things". It is only ever the arrangement – the shape of these same "things" relative to "not things" that changes.

Here's an dramatic example to help make sense of what I'm trying to say: Imagine throwing the entire Earth into the Sun. Every material, every person, every rock, steel beam, satellite, wire, house, plane, car, gas, food, and liquid is still on it. All of these things will be "dissolved" into the Sun. You might be tempted to think it all dissolves to "nothing". To our sensory resolution, it would all be vaporized into a temporary little swirly spot on the surface of a raging fireball. The shape of the Earth and everything on it seems to be destroyed.

And yet, *it isn't*. All of it has become part of the Sun itself. It has all transformed from many things to become *one* thing. They have merely changed shape. The sun became ever-so-slightly larger and hotter, because it now contains all the atoms of the Earth and everything connected to them.

This is what The Process does. It is the merging of two "ones" into a bigger and more balanced "one". It is the changing of shapes. It is ultimately the rearrangement, the *merging*, of atoms into new and bigger "things" amongst the "not things". And it is those bigger things, or "ones," that you and I can finally start to notice directly through our sensory windows.

This leads to an enormous and important truth lying here in front of us: Wherever there is *one* of anything – no matter what it is - it is only ever the outer *boundary* you or I are perceiving. Any *one* is only the boundary of resistance that makes a second *one* possible to exist and not merge into some bigger *one*.

Inside of that boundary is an invisible world you and I cannot directly experience through our windows, and it reaches all the way down to atoms and whatever exists beyond the pieces and parts that make the atomic system possible.

We, as humans, have figured out how to manipulate these outer boundaries at lots of scales to create "new" imbalances. These human-arranged imbalances have resulted in movements from *two or more* into a new balanced *one*. We've also figured out how to split *one* into *two* smaller ones and merge again into different larger *ones*. This ultimately gives us new human-manipulated "shapes" or "things" we then assign new words.

These shapes are human-designed molecules, chemicals, compounds, and materials. Of course, we didn't actually create these things – the nature of the universe "decides" what is possible and what is not. We are merely uncovering what exactly can be merged and separated in this universe using The Process.

At the time I wrote this, these discoveries are multiplying faster than ever, and we are still going strong with plenty more "things" to build and create. The possibilities are growing each day as we uncover more and more smaller and larger "things" hiding beyond our current boundary of human understanding.

All of this progress is possible through trial-and-error and the development of ever-improving tools and methods that are designed and shaped by us. These are things like the microscope, the oscilloscope, the particle collider, the spectrometer, and thousands more useful inventions. Yet all of these tools and methods to understand and control The Process are themselves ultimately made of atoms. These atom-based tools allow us to see beyond the resolution of our senses into the invisible and discover new truths – even though it is always through secondhand story shadow experience.

These secondhand story shadow experiences, which are validated and repeated one perception at a time, seem to always lead us back down to the atom, which we do not yet fully understand at the time I wrote this.

We are collections of atoms discovering themselves. Which brings things back to a full paradox:

The only things we ever experience firsthand are tied to atoms, which can never be experienced firsthand even though they ultimately provide us firsthand experience.

Are there large-scale consequences of humans tinkering with atoms, which ultimately are what we ourselves are made of? Are there consequences of tinkering around with The Process? We cannot yet know.

Is it possible to fully understand the atom? We do not yet know.

What might we construct from a complete understanding? What possibilities would appear if we could control the tiny invisible pieces of reality? Would we create whole new atoms that take on different shapes? Harness "wireless" electrical power transmission? Create "strawless" processes? I cannot say, because I cannot know the future. Perhaps as you read this, some of this has already happened.

I am not concerned about human discovery. I didn't write this book to motivate human progress or dictate and try to control and limit what we as humans should or shouldn't do when it comes to exploring reality. I am not the Keeper of What's Allowed to Be Found.

I wrote this because there is a human process already out of control. This particular process is actually responsible for the amazing amounts of human progress we have made, but also a horrific amount of death, destruction, and suffering.

This runaway human process can only be completely stopped by finding the answer to the following question:

How can we know when to stop tinkering with new-to-us boundaries?

As I see it right now, there is no way to know. Even if we think we know exactly where this limit was, I'm not even sure we can stop ourselves when it comes to searching and controlling every possible way to get around it.

The smallest "one" is always going to be "two" for us as humans. We *know* what we sense as "one" is merely a boundary – a balloon skin - and that there must be an inside and an outside to the shape. We seem to almost *need* to know what it is we cannot see inside. We long to make all of the invisible visible.

At the same time, we gaze out into the cosmos and long to know what is out there - what is beyond our always blocked views out of our windows. We almost *need* to know what is beyond the boundary that our most powerful telescopes and tools can create for us. The largest "one" is always going to be "two" to us as humans. We know what we sense is a boundary – a balloon skin - and that there must be an inside and an outside to it. We long to make all of the invisible visible.

We don't like things *blocking our view of the entire universe*, both up and down in scale. We seem to dislike *not knowing*.

As I see it, there is only one way this runaway what's-inside-this-new-smallest-boundary-or-beyond-this-new-largest-boundary process comes to an end: We understand *why* everything exists and *why* it is the way it is.

Unfortunately, that requires knowledge of a second point *outside* of what we understand to be existence. It would require knowing the existing "ground" conditions outside the universe, time, and movement, while existing inside the universe, time, and movement *at the same time*. It requires knowing what's on both sides of *all* boundaries.

This is a paradox – another loop trap.

You can never be in two places at once, let alone *all* places at once.

That doesn't mean this paradox cannot be broken. As I see it, the problem is largely one of perspective.

72

The Dog That Likes to Chase Cars

We, as humans, seem to struggle with boundaries. I'd argue that we almost *despise* their existence. We seem to be driven to know what is beyond the next boundary that tries to *resist* and hide things from us.

We desperately want answers to questions we can't answer.

Why are we here?

How did we get here?

Where were we before we were here?

Where even *is* here?

What is beyond here?

What is the meaning of all of this? Of everything?

What is humanity's purpose?

Why must we get sick?

Why must there be suffering?

Why must we die?

What happens after we die?

These are all *fantastic* questions, and yet no human being can answer them with any certainty that answer is objectively right. Changing this story and trying to find the answers from firsthand experience seems to be one of the main drivers of all human progress. To me, however, the biggest question is: *Why* do we seem so obsessed with knowing?

As I see it, the answer to this question of "why" is directly related to uncertainty. Being uncertain can create an uneasy feeling. So much seems *certain* in our daily lives. It almost feels *wrong* not to somehow know the answers to those big and fantastic questions.

In my opinion, this uneasy feeling created by uncertainty is actually *imbalance*. Imbalance can feel a lot like falling. We as humans (at least as far as I can tell) do not seem to like the feeling of falling, especially if we are not in control of the fall.

To prevent what feels like an uncontrollable tumble into an abyss of "unknowing," it seems to me that we often choose to *tell ourselves* a particular secondhand story (that answers these big questions) is *certain. That story is absolute objective truth, and that's that!* Having these questions answered can make you and I feel balanced – like the feeling of falling has stopped - and we now stand on solid ground.

That is *not* what is actually happening if you or I do this, though.

Instead, we return ourselves to Chapter 1. We are trying to make someone else's story our own truth, or our own story someone else's truth.

But no human being can ever be *The Keeper of Truth*.

You and I have tumbled to somewhere near the bottom of this "abyss of uncertainty", and yet nothing terrible has happened. Well, if we ignore my problem of trying to find a way to climb back up from this place knowing you and I agree on what is true.

Still, there is a consistent pattern: Agreement and common ground seems to always collapse back to direct firsthand sensory experience. This means you and I at least have *something* to work with and grow from. It doesn't really help us answer any of these big questions, though.

This aside, I grew tired of living in the dark down here after two years, and I'm sure you'd like things to be a little brighter as well. So now I'll use these word-shells to hopefully begin lighting up what I believe to be the rest of the shared foundation between you and I. I'll begin with a statement that is going to give you a ping of unexpected feeling:

Solid ground does not exist.

To get the story that answers any of the previous big and fantastic unanswerable questions from the beginning of this chapter, you would need to have firsthand experiences *outside* of the topics in the questions. You (and anyone else) would need firsthand experiences outside of "humanity," of "here," of "before," of "meaning," of "purpose," of "sickness," of "suffering," of "death" and outside of "after."

As impossible as this might seem, it appears to me to be *exactly* what we are trying to do. We continue to peer through smaller and smaller and larger and larger boundaries. We seem to be looking for a "master unit" or a final and obvious "one" so we can finally stop feeling like we are falling and get a solid foothold on our reality.

We are expanding our scale of experienced movement ever smaller and ever larger, while at the same time trying to expand the scale of experienced time ever more backwards and ever more forward. We continually try to reconstruct The Process back to the beginning of time and movement, only to then try to understand how time and movement got started. We also continue to try to construct The Process forward to the end of time and movement, and try to understand how time and movement stop before it happens.

We are expanding to scales no human has *ever* directly experienced or probably will ever directly experience.

It seems we are like a dog that discovered it likes to chase car while living in the middle of a four-lane highway.

Perhaps you have already dedicated most of your life to chasing one of these cars. There is no harm in this. You are not wrong. Chasing these cars is helping us as humans grow our understanding of The Process and our universe one discovery at a time. Just look at how far we've come from painting on the walls of caves!

Expanding our understanding has resulted in wonderful things like harnessing the power of electricity, space travel, and medicines. Unfortunately, it has also resulted in awful things like harnessing the power of the atom to destroy each other more efficiently with thermonuclear, chemical, and biological weapons, developing new ways to take each other apart, and everything in between.

I am not trying to make you question if you (or we) should or should not be chasing cars. Instead, I would like you to consider a different kind of question: What would you do personally if *you caught one*?

Whoever wrote the line for the character of Dr. Ian Malcolm in the film *Jurassic Park* pointed out what I think matters here, which I'll now rephrase it into a question: Are you so preoccupied with discovering if you could, you haven't stopped to think if you *should*?

What would you do if you discovered the smallest "thing" that can never be opened and how to control it, or discovered the edge and shape of the universe?

What would you do if you discovered the wordless truth of what happens after death, or how the universe and humanity actually began?

Why do you want to know how we got here, or what happens after we die?

Why do you want to know the smallest thing in existence, or the shape of the universe?

What are you going to change if you discovered the origin or ending of all movement, or the beginning or end of all time?

Whatever you might imagine changing or doing with such information, you've already justified it based on what you believe to be right based on what you believe to be true at this moment. If that act affects anyone or anything else other than yourself, then in my opinion, you might want to be prepared to explain *why* what *you* believe is right should also be right for *all of humanity* and the universe itself.

Do you intend to *escape* your universe? What are you escaping? Why do you feel escaping is the right thing to do?

Do you plan to *control* your universe? What are you controlling? Why do you feel like controlling the universe is the right thing to do?

Do you want to invalidate the universe and *remake it as you see fit*? What exactly is "fit" to you? Why do you feel the universe is not right or fit to exist?

I encourage you to think on these questions. Ask yourself if you really even want to know the answers. *Why do you want to know anything at all?*

When you think you might have an answer, I think you will realize that it's much more likely catching one of these cars will kill you.

So let's assume you just caught one, and begin your fade to black.

Close all your sensory windows.

Let time and space drift away from you once again.

73

The Ark

Something in the void just changed.

You open your sensory windows, and discover you are in a cube-shaped room. The ceiling is about three times your height. There are no windows and no doors. There are only four solid walls, a ceiling, and a floor, and all the surfaces are identical.

The room is lit, but there is no apparent source for that light and no shadows. There are also no sounds and no smells.

Your bodily needs from *The World of You* no longer exist. You do not experience hunger or thirst. You are not aging. You do not tire. You do not need sleep. There is only one experience occurring in this place: You and your 3-step process.

Even though you have no context for this particular room, it would not take you long to determine pretty much *all* the possible stories present within this room simply by drawing from your past experiences with other rooms. Just turn your head a couple of times, and assuming you have all your senses, you'll know almost everything there is to know within this space.

Time passes in the form of your movements, and nothing ever changes about this space, except you. You cannot make a mark on any of the walls. They do not make sound when you tap them even though you can hear yourself when you move. The room does not feel like it is moving, and there are no changes in temperature or light.

Time continues to pass.

How long could you tolerate this place?

How long would it take you to become uncomfortable within it? To want out?

How much time would need to pass until you tried to scratch on the walls, break through them, or cry out for help?

Why do you think you would do this? *You have everything you need to survive.*

What *determines* how long you can tolerate this space?

Pause for a moment or two and think about it.

Close your sensory windows again.
Back into the void you go.
No time, no space.

Re-open your windows.

You are in the same cube-shaped room as before, but this time there is one difference.

Now there is a white, solid, and closed door on one of the walls.

Are you drawn to it?

Fascinated by it?

Why?

Will you open it?

Why?

What do you think will be on the other side?

If you choose to open it, would you open it cautiously, or fling it open? Why?

Why do you apparently wish to leave this space? *It has everything you need to survive.* You do not require water, food, rest, or relief in this place. There is no threat to you here. No disease or predators of any kind are here with you.

Consider the closed door for a bit.

What do you expect or hope to be on the other side?

Are you drawn to open it, or not? Why?

What is the *feeling* this door gives you?

Be with yourself, the room, and the door.

I'm going to assume you will open the door. This isn't a blind guess about you. Everyone I have ever put into *The Ark* thought experiment opens this door, and interestingly, all of them choose to do it cautiously.

On the other side of the door is an identical cubed-shaped room just like the one you are in. The only difference in this second room is the presence of a white, solid, closed door on *each* surface. As you peer through the door you just opened, you see a door on the wall to your left, to your right, straight ahead, as well as one in the center of the ceiling, and one in the center of the floor.

What do you do now?

Do you enter this second room?

If you think you would, do you leave the door into the first room open or close it behind you? Why?

Will you select a new door, open it, and go through it right away?

Will you turn around and leave through the same door you used to enter?

What do you think or hope will be on the other side of any door?

How do these doors make you feel?

Will you try to open all five new-to-you doors in this room before deciding if you will enter one?

I'll assume the answer is yes to the last question. Again everyone I've put into The Ark so far has answered this way.

Intending to open all the doors leads you to your first problem. You can easily reach three of the new-to-you doors, but can you reach the one on the ceiling? I feel safe assuming the answer is "no" unless something drastic has changed about human jumping ability or stretching ability since I wrote this.

How much does not being able to reach the ceiling door bother you? Why?

Even if you could somehow grab the knob and open it, could you then get yourself through it if you wanted to? If it is assumed that you cannot pull yourself through the ceiling door even if it could be opened, are you content to leave it closed or do you still feel you *need* to open it anyway? Why?

Refocus on the doors you can reach: Do you again intend to open them cautiously, or fling them open? Why?

I'll assume you will choose cautiously and continue.

The first new wall door you open leads to total sensory blackness. When you reach your arm into it, you aren't able to feel a surface past the door in any direction, even if you reach around the doorframe to feel for the other side of the wall that belongs to the room you are in. It's like the wall is not there on one side.

Would you stick your head out of the door?

Assuming you would - you can't see anything, hear anything, or register any other sensation beyond the doorway.

Do you now leave this one to check the other doors? Again I'll assume the answer is yes.

The other two wall doors open into a void just like the previous door you opened.

I'll assume you now turn your attention to the door on the floor. This door is locked. The knob does not turn and doesn't have a keyhole of any kind. There are no screws visible anywhere. There is no method available to unlock this door.

Does this bother you? Why?

Would you now attempt to kick it open?

Would you stand on it and jump up and down?

Assuming you would, the door remains unaffected and locked.

Now what?

Is this door better or worse than the ceiling door you also can't open?

You have a growing context for this place. So far, it is a two-room system. You can move into the room with one door, or into the room with six doors using the door shared between them. There are three doors that each lead to voids, one you cannot open or reach, and one that is locked. You still need nothing related to food, water, sleep, or rest. You don't experience hunger, thirst, fatigue, or pain.

You have everything you need to continue surviving.

This place seems to be a *perfect shelter*.

Even so, I suspect the black spaces beyond the doors are going to bother you.

I also suspect the unreachable ceiling-door and the locked floor-door are also going to bother you.

Will you leave the three doors that lead into voids open or closed? Why or why not?

What would be your main focus at this point? Trying to reach the ceiling-door, open the floor-door, or figure out the unknowable darkness?

How long would it take you to give up on the ceiling door and floor door?

How long would it take you to decide that stepping into the void is the right thing to do? Could you bring yourself to step into one of the voids without knowing what was behind the two doors you can't open?

Again I feel safe assuming you *will* eventually decide to step through one of these doorways into a void.

I believe you will eventually do this for the very same reason you are drawn to open the first door in the first room.

But don't blindly accept my assumption.

Think on it.

What *determines* when the moment has come for you to take a step into the darkness?

I'll continue assuming you will eventually reach such a moment.

You finally take a big step into the blackness without any intent to catch yourself.

As soon as you cross the threshold without the intent to catch yourself and not knowing what will happen next (but probably expecting to begin falling downward since you were never able to feel a floor of any kind), a force like gravity pulls you *forward*, and accelerates you into a *sideways* free-fall relative to the rooms. You hear the wind rushing past your ears and feel the water leaking out of the edges of your eyes as you tumble end over end while continuing to fall at what feels like incredible speed.

On your first half-tumble, you glimpse the open door you just stepped through racing away from you. It shrinks into a smaller and smaller white rectangle each time you see it with each successive tumble. It eventually fades into a pale dot.

As the void grows to start depriving all your senses, you feel your fall begin to slow, and keep slowing.

The wind in your ears slowly goes silent, and you find yourself now suspended in blackness with no sense of up or down and no ground beneath your feet.

As you adjust to this new situation, your eyes dilate to let in more light. You start becoming aware of what must be a billion white dots surrounding you in every direction. One dot in particular is slightly brighter than the others.

If you do not have a sense of sight, then instead imagine there is a distant "beep" coming from a billion different places in every direction, with the beep being slightly louder from one particular direction.

You can reach none of these dots. They are incredibly far away. With nothing to push off of, you have no way to move toward anything. You are helpless. You are, in a way, trapped.

Sit with this situation for a bit. Accept it. There is nothing you can change or control now except your own movements, which get you nowhere.

Do you regret stepping through the door?

Is this situation better or worse than the two-room system?

Would you rather be back in the cube-shaped rooms again? Why or why not?

Ponder it a bit.

When you think you have an answer for yourself, keep going.

You feel something change.

You can now fly in any direction just thinking about it.

What do you do next?

Will you race toward one of the billion dots?

How will you choose which one? Will you choose randomly? Will you head toward the one that is slightly brighter (or beeping slightly louder?)

What is it you are hoping to find?

As you fly toward the dot of your choosing, you quickly realize it, too, is a door. *All the dots are white, solid doors.* A billion of them.

With this realization entering your Step 2, what future story do you build with Step 3? What is the pattern of actions your story system will justify?

Open them all?

In what order?

Will you open them in a pattern?

Fly around and open them randomly?

Will you choose to reclose the ones you open, or leave them open?

Is there anything that would make you justify fully entering any one particular door, given that you now know from your 3-step process it's entirely possible that unpredictable things might happen when you pass through a doorway, preventing you from getting back?

Is there anything that could be on the other side of any of these billion doors that could make you stop opening any other doors and just go into it?

What would that be?

Think on it.

It just so happens the very first door you fly to leads back to the 6-door room. I don't like assuming which door you would choose, but in *my* imagined context, the brighter dot was always this room, because it was the only door of the billion that was open.

You now face a choice. Do you re-enter? Or do you move on to other doors in the darkness? Technically, you only know where two of the six doors in the cube-shaped room lead: The one you are now looking through that led you to this billion-door space, and the one that leads back into the one-door room.

It's *possible* you could re-enter this six-door room and open the door on the ceiling, if you *assume* you will keep your ability to fly. Would you attempt this?

Do you think you would re-enter the six-door room? Why? To step through the other two doors into blackness to see if it's the same blackness? Or do the other billion doors in the blackness call to you more?

Will you need to open every single door in this blackness *before* you re-enter this six-door room, or is it the other way around for you?

Again, what *exactly* are you looking for?

What are you hoping to find?

I would like you to think long and hard on this. *What would have to be on the other side of any door out of more than a billion to make you go through it, leaving all the other doors unopened?* What conditions would have to exist? Could these conditions exist at all for you?

Even if you think of something you believe would cause you to enter the door, be careful not to accidentally ignore time. Would you be content with whatever might be on the other side for the next hour, month, year, decade, century, and millennia, should you not be able to leave back out of the same door?

I do not know what your answer might be. I am certain, though, that you will have a different answer than me and anyone else that reads this, because even if you share the same idea or word as someone else, your context will always be unique, and even an idea expressed with the same exact words is anchored to different wordless feelings and memories for each person.

However, I do believe your answer is *approximately the same* as everyone else's. And when all the approximations are put together, it forms something very real that is shared by all of us at our core.

But it's not time to uncover this yet.

First, I want to introduce you to your very own personal balloon system.

74

Lightning in a Bottle

Everything that just happened is a result of your very own mental electric balloon system. You are already familiar with it, because it is what defines truth, right, and wrong for you. *It is the two points of you.* It is your present and your past. And these two points have straw paths between them.

The first balloon of you is Step 1 of your story system: Perceive the present. To understand what "ground" is for this balloon, rebuild your Step 1 sensory boundary. The outer edge is the furthest you *expect* your senses to reach, which is a truth forged by a lifetime of using Step 1 to build any truth at all.

To imagine this clearly, just put yourself back into *The Ark* at the point you were unable to move anywhere after falling into the billion-door space. There were no white dots all around you at first, because all the doors were just barely outside your sensory resolution. As your eyes adjusted to darkness, the distant doors became visible. *They crossed the boundary of Step 1.*

If you have no sense of sight, you couldn't hear the beeping of all the doors because of the sound of wind (atmosphere) as you were falling. When you stopped, your ears adjusted to the silence and started to pick out the faint beeping of the doors in all directions. *They crossed the boundary of Step 1.*

The doors didn't move or change in either case. Instead, it was the sensitivity of your senses that changed. The extra "pressure" or "noise" in your senses disappeared when you finally stopped falling, so your senses quickly became the most sensitive they can possibly be. This sensitivity change does not seem strange to you because you are used to how your senses behave, so the doors being noticed only after being in the dark or silence for a few moments isn't alarming.

To make it super clear where Step 1's boundary exists: Imagine that just after you begin to see or hear the billion doors as faint white dots or faint beeping, all the doors begin to move slowly away from you. Eventually, they will reach a point far enough away that you can no longer see or hear them. They will grow fainter and fainter until you cannot sense them at all. At this moment, they have moved back across the boundary of your Step 1.

After this happens, there is nothing inside Step 1's boundary except you. Your story system is truly isolated. Your sensory windows are wide open, but there is nothing in any direction to let in except yourself.

And what shape does the same distance make if it is measured in every direction? A *sphere*. Step 1's "orb," or sphere-shaped sensory boundary, has a fixed maximum resolution. That resolution is determined by your unique biology. In this case, the sensitivity and shape of your eyes and sensitivity and range of your ears determine the size of your sensory "orb".

Whether or not you are able to sense anything within this orb-shaped boundary is not actually up to you. You cannot change your biology on demand. It is instead up to *objective reality* to provide things you can sense.

The existence of Step 1's boundary creates a two-point system between you and the universe – an inside and an outside of the boundary. And two points means relativity applies and makes things a bit, well, *relative.*

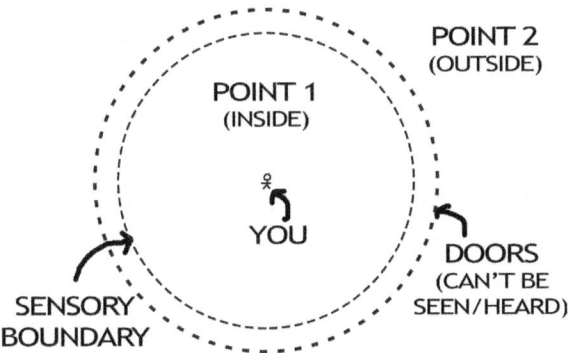

Here's what I mean by this:

If the doors that are now beyond your sensory boundary began to glow twice as bright or became twice as large, or their beeping became twice as loud or lower in pitch, they might suddenly be within your sensory boundary again *without them moving.*

This would mean your boundary actually got bigger in size, and yet *you didn't change anything.* The maximum resolution and sensitivity of your senses did not change. Instead, objective reality is what changed. The *objects* in that reality became bigger or brighter, louder or deeper.

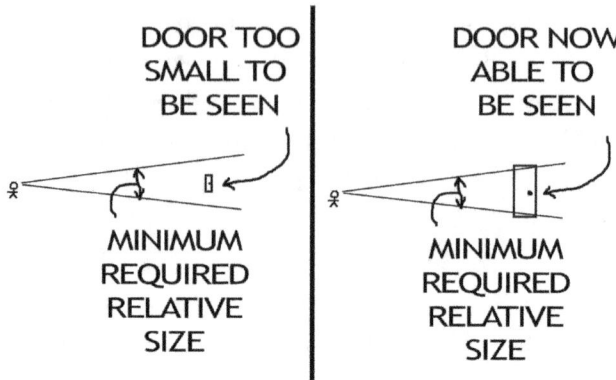

For the doors (or anything else) to enter the boundary of your story system, all that is required is the detection of *movement*.

But now for a plot twist: It is not the movement of *the doors* that is required. They were not moving your first time through *The Ark* after all, and you still eventually sensed them just fine.

If the doors moved outside your sensory boundary and then grew in size or loudness until you could sense them again, they can then stop growing or getting louder, and you'll still be able to sense them just fine.

This is because the movement required to enter your story system's Step 1 is actually invisible electrical *current*.

There is current throughout your body right now being directed and speed-controlled by billions of straw-paths called "nerves", and they have clips called "synapses". The balloons that make current possible are your windows labeled "eyes", "ears", "nose", 'mouth", and "skin".

Every one of these sensory windows is made up of tons and tons of tiny "air compressors" or "voltage generators" waiting to inflate an electric balloon as soon as *movement* from objective reality makes them vibrate by pushing on their boundary.

Light itself "moves" or "vibrates" cells at the back of your eye (rods and cones) that then generate electrical imbalance. The ability to create an imbalance is made possible by *a boundary*. In this case, the boundary is called a nerve synapse.

When that imbalance reaches the point where the boundary's resistance breaks down, resistance drops below 100%, the synapse "doors" open, and the electrical imbalance collapses into current which races through the straw to your brain "balloon."

When it arrives at your brain, it finally enters Step 1.

Sound "moves" bones and hairs in your ears that build up a voltage that collapses into current after it reaches the point where resistance breaks down and opens the synapse doors.

Molecules *moving* into your nose and mouth and landing on certain spots grow voltage that collapses into current after it reaches a point where resistance breaks down and opens the synapse doors.

Any change (or movement) to pressure or temperature on your skin triggers a build-up of nerve voltage – *imbalance* - and it too becomes current if the imbalance gets big enough to overcome the resistance of the synapse's "doors".

If the straw-paths for current from your sensory windows to your brain do not have current flowing (because there is not enough voltage being created by movement), such as when you were floating in the empty void of *The Ark* and nothing could be sensed in any direction, then nothing enters Step 1 of your story system from outside of your body.

This "minimum voltage" requirement to sense something creates a rather large blind spot for us as humans. I'll quickly show it to you.

Imagine the billion doors move beyond where you can sense them again.

Now imagine they all begin moving toward you. *However*, they also start shrinking and dimming at the same time they are doing this (or beeping ever-softer).

It is *entirely possible* for all billion doors to remain "unnoticeable" to your story system as they come ever closer, much like a virus or bacteria can be under your nose or in front of your eyes or next to your ears right now and you would never know.

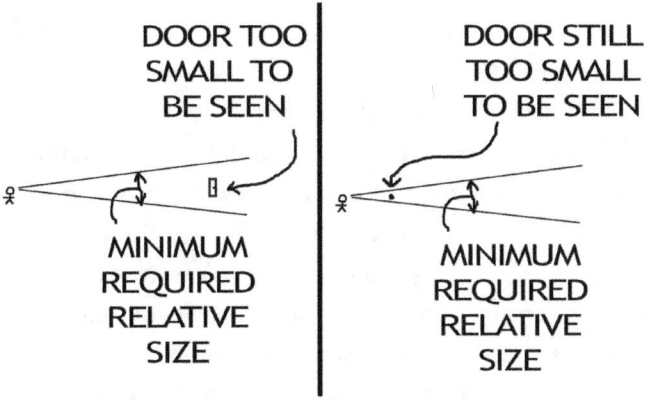

Every single door could be an arm's length away in every direction and you would have no clue because they shrank (or quieted) at a rate that remained below your sensory resolution.

Even though they were very much moving, they simply did not create *enough* movement to generate a big enough imbalance at one of your sensory windows to force open the nerve door-synapses and collapse into current.

This can result in some rather disturbing situations for you. From your perspective, the billion doors never changed. As far as you know, they are *still so far away you can't sense them*, because a different story has not been noticed. You aren't aware they are all now an arm's length away.

Even if we reset this situation so the doors start out as visible white dots before moving toward you, they could still sneak in much closer that you might expect by shrinking in size and brightness (or fading in loudness) to remain exactly the same relative to your perspective.

I don't think they could get within an arm's reach before you noticed they were indeed closer, but I do believe they could get at least halfway to you before you noticed a change in your "sensory currents."

The minimum movement-resolution for your sense of smell and taste work the same way. You'll notice when you walk into a big new smell because it was a big change, or "movement". However, there is a threshold where you will be unable to notice a smaller change in smell. You cannot smell individual viruses, for instance, or carbon monoxide. There is a minimum required "voltage" needed for any smell or taste to become noticeable by your story system.

All in all, every surface boundary of your windows, and every nerve that touches those surfaces is ultimately a straw-path for pulsing electric currents to travel to your story system by bursting through doors that are set to open when a particular imbalance is reached on either side of them. Those pulsing currents are generated by the *outside* world vibrating those windows and in turn creating a growing imbalance on one side of the door.

If you can't perceive something with your senses directly, then it is a second-hand story and it is excluded from your *automatic* truth-story process in the moment. This does not mean that the invisible something does not exist. This does not mean that something you can't sense is not true or real. It just means that this invisible something is a story you must tell yourself is true.

After all, *your senses are windows, not projectors.*

In a way, you are very much a billion paths of lightning contained in a you-shaped bottle, and all billion paths can flood current into your story system's

Step 1 at the same time.

To experience how this might feel firsthand, I'll give you the most Step-1-flooding shadow experience I can imagine, using *The Ark*:

Imagine you are unable to move in the huge black space again. The billion doors have moved away, far beyond where you can sense them. You are in near total sensory blackout and helpless to create any change to your situation.

You blink (if you do not have sight, imagine quickly putting your fingertips in your ears and immediately removing them).

You are back in the very first room with *no doors and no windows* – but it's three times smaller, twice as bright, and there is a very loud and constant buzzing noise.

Notice your feelings right now. How did this make you feel?

For me, I imagine a huge surge of *something*.

I experienced this surge firsthand when a bolt of lightning missed my head and instead blew a hole through the roof next door. It would take me more than a year to make sense of this feeling.

This surge is, in my opinion, the consequence of what amounts to a *shock-wave* of current into the story system from a billion directions, and with it, an immediate appearance of an enormous new imbalance.

This imbalance is what I'll call *story voltage*. Voltage requires two points and a boundary to exist, and one of those points is your mind. The other point is *not* your surroundings. It is also not your sensory "generator" balloon-windows. They worked *as expected*, and told you what they could about the outside world.

The second point to define this massive new imbalance is *also in your mind*. It is the second balloon of you. It is *Step 2* of your story system: *Your past*. Your past is how you make sense of what is happening around you.

Step 2 is how you understand what Step 1 is showing you.

Step 1 and 2 together create Step 3.

And your story system just experienced a dramatic failure.

75

The Collapse of Reality

In the blink of an eye, you moved from an enormous black space with no sensory experience to the small, cramped, bright, and noisy space with a completely sealed boundary.

A huge surge of sensory current flooded your story system. These currents were completely *unexpected* based on the story of reality you *believed* you were in. As a result of this surge, a huge voltage appeared *inside* your 3-step story process.

Before the blink, your story system created a Step 3 prediction that made sense based on your current moment. This prediction is built on what Step 2, your past, believes to be true based on the context of your past and present.

After the blink, the entire system experienced a dramatic *failure to be right* in the blink of an eye.

Let's slow down time and run through the 3-step story process three times to show how I imagine this extremely jarring transition from sensory blackout to fully enclosed tiny, bright, and loud room to go:

Story System Loop #1

Based on the lack of sensory *currents* that were entering Step 2 (past) from Step 1 in the huge dark space (present), Step 3 created a future story that when you close your eyes or ears in this space where there is nothing to sense, the currents coming in from your eyes and ears and body in Step 1 *should not change.*

Closing your eyes when there is nothing to see should not result in change.

Plugging your ears when there is nothing to hear should not result in change.

Action Justified

The action of closing your eyelids or however you plug yours ears is justified because it is the right thing to do based on the above truth-story. You do not pay any attention to your sensory currents during these actions because you have experienced them perhaps millions of times before.

Since your 3-step process has validated the currents associated to these actions with great accuracy, they are no longer noticed by your conscious attention unless they go unexpectedly. (Such as an eyelash in your eye, or having a sharp fingernail in your ear, which would create unexpected currents).

Story System Loop #2

The truth-story predicted by Step 3 of Loop #1 is now confirmed to be *valid* after your eyes or ears are closed. No real change in sensory current happens in Step 1 of Loop #2. What you believe to be right and true when performing these actions is validated for the millionth-plus-one time, because the currents experienced during this justified action were exactly as expected.

Based on the remembered sensory currents of the enormous black space *before* you closed your eyes or ears (Step 2), Step 3 predicts that when you *open* your eyes or ears again, you will *still be* in the huge black space where there is no sensory information. Step 3 predicts that Step 1 of loop #3 *should not experience a change in sensory currents.* In other words, Step 1's currents from Loop #1 (the past) and Step 1 of the not-yet-happened Loop #3 (the future) should be *the same.*

Opening your eyes when there is still nothing to see *should result in no change in sensory currents.*

Unplugging your ears when there is still nothing to hear should result in no change in sensory currents.

Action Justified

The action of opening your eyelids or unplugging your ears is executed. Again you would normally ignore any feedback currents from this, as they are all highly validated truths and therefore expected.

Story System Loop #3

Mid-action of opening your eyes or ears, which was justified as the right thing to do by loop #2, Step 1 gets hit with a huge and incredible shockwave.

Suddenly *tons* of straw-path nerves, which were expected to provide no currents at all, are *flooding* your story system with *enormous* and *unexpected* current flow. Bioelectricity blasts its way into Step 1 of your story system from all directions.

Your vision straws are struggling to carry the inrush. Your touch sensors have lit up because you suddenly have weight and are touching a floor. Your hearing straws are bursting with current.

The second point of your story system, your past, which determined what was true for you in the moment and used it to determine what to expect next, expected *no change* in the Step 1 currents of this loop.

Instead, Step 1's balloon started expanding *dramatically*, as thousands of "sensory balloons" inflated by motion hitting your window "air compressors" causing "air currents" to flow through the millions of straw-paths into your mind's story system.

Step 2 is overwhelmed, and immediately needs time and movement *to slow down*, so it starts auto-justifying actions to try to *resist* this unexpected flood of information currents. You reflexively squeeze your eyes shut (or squint) to try to slow down the visual currents. You might cover your ears, or even curl into a ball on the floor.

These "automatic" subconsciously justified actions are, in my opinion, actively trying to slow down time itself so Step 2 of your story system can figure out what just happened to reality.

The goal, as always, is *balance*. By giving Step 2 more time to determine what has happened to reality, the two points of you – your mental two-balloon system - can begin to collapse toward balance again.

As I see it, balance is achieved when your story system starts creating predictions about reality that turn out to be mostly *right* over and over.

However, and perhaps most importantly, when the prediction turns out to be mostly *wrong*, the overall "pressure" of your entire story system changes. For example, your Step 1 balloon was completely empty in the black space of *The Ark*. In the blink of an eye, it was wildly overinflated.

As a result, the balloon of Step 2 became overwhelmed with current, because your present moment (Step 1, balloon 1) suddenly began overriding *everything* you thought you knew using your past. It had to do this. Step 2 is *hardwired* to Step 1 by millions of straw-neurons.

Being moved in the blink of an eye made absolutely *no sense to you*. Your truth was horrifically *wrong* and being wrong about reality is a *huge threat to your survival*. This unexpected change in Step 1 currents is not far removed from the appearance of a pack of wolves in *The World of You* or a bolt of lightning striking next to you.

This overwriting-of-past-truth-and-context will continue until you can "get your bearings" and establish a new balance by building a brand-new context for predicting whatever reality you were just thrust into. As soon as you can predict what happens next and be mostly right again, a feeling like balance will be restored.

However, the overall pressure in your story system doesn't go down. The entire system of you is now at a higher overall pressure than it was before you blinked or plugged your ears.

If story "ground" is total sensory blackout, then your entire story system is now very imbalanced relative to that ground. Both balloons are now enormously filled with a flood of new and strange information.

You literally jumped through time and space in objective reality with no way to understand how or why it happened. Reality as you believed it to behave collapsed.

Imagine if every time you blinked or swallowed (which makes a "click" in your ears), you found yourself somewhere radically different.

You might think this would destroy your story system, and that you would have no ability to predict what comes next at all. You might even think this would drive you insane or mad. Do not underestimate the raw power of your mind.

Your story system remembers the reality of your past, but can also simply let go of it to build a new one that is relevant to your new reality.

Your story system and the truths it builds are *ungrounded*.

New contexts are uncomfortable at first, but your mind will simply reset your expectations and carry on, trying to justify the right actions to gain control all over again. Perhaps you might even start *consciously* controlling your blinks and swallows.

In my opinion, being ungrounded is a superpower.

It is the key to being able to survive no matter what changes to reality might happen.

Being ungrounded allows you *to fly*.

76

The Planet You Drag Around

In my opinion, the human mind is best described as having a "floating ground". Perhaps another way to word this is "normal is relative".

This is the moral of the story in the ancient Greek philosopher Plato's "Allegory of The Cave" if you happen to be familiar with it. In the case you are not familiar, and to avoid re-telling it to you word-for-word, I'll give you the general idea of this story using *The Ark* experience from three chapters ago.

Plop yourself right back into the room with no windows and no doors. Not the smaller, brighter, noisier version, but the first version.

Now imagine your memory is completely reset. You don't know who you are, where you came from, or anything else. You are mentally a newborn baby again, but in your current body. This is not really a problem or threat to your survival, since you don't require food, or water, or sleep in *The Ark*.

If every single experience in your life over the next few decades takes place in this room with no windows or doors (because you've been inside of it since your memories began), then you are, *in theory*, not bothered at all by this room. Your story system has never experienced a different reality. Your sensory boundary has never reached beyond these walls before. This room is *normal*. This room is your "ground state".

Now plop yourself into the big dark space with no sensations at all, and reset your mind all over again. If *this* is the only reality you will ever know for decades, then in theory your mind is not bothered by it either. This space is *normal*. It becomes your "ground state".

You, reading this book wherever and whenever you might be right now, very much started with an "erased mind" as well.

I am not arguing we are all born exactly the same – a blank slate - rather I am arguing we all enter our own story of reality from total darkness. You can't remember anything before your first year of life (or maybe even two or three), and neither can I.

Eventually though, your story system began running successfully, and so your truth-story-from-firsthand experiences began. Your truth about reality at that moment became your "ground state". Ever since, that truth-story has been changing because your reality has been changing – but everything that has ever happened up to today still has the potential to be connected all the way back to your earliest memory – your ground.

The reason for this is simple: At the time I wrote this, every human being's story system is grounded to reality on the planet Earth. No one has been born and raised in outer space or on another planet yet. If you began on Earth as well, then your story system's strongest unbroken truth-story – or context - is based on the movements around you on the planet Earth that have been entering the boundary of your Step 1 since your first memory. Your belief in this truth-story and the accuracy of its predictions are incredibly strong.

For example, if you let go of a drinking glass, you probably believe with almost 100% certainty that the glass will start moving *down*. You might be less certain if it will shatter when it hits the floor, but you probably do accept this is a very likely possible story, so you won't be surprised if it happens.

You drag these truths around with you everywhere you go in your imagination, and you may or may not notice you are doing it.

In *The Ark*'s billion-door space with barely-detectable white-dot doors in every direction, you used your story system's "ground," which is built from all your remembered experiences with reality *on Earth* to imagine where those doors disappear as they move away from you.

That distance is probably about the same distance you think you could barely see a standard size door…while standing on planet Earth.

I said *nothing* about the atmosphere in *The Ark*. This didn't matter, you applied an Earth-based atmosphere anyway, probably from phrases like "wind in your ears" and that beeping could be heard at all.

It's possible you might have even imagined yourself feeling "right side up" when stuck in the middle of the billion-door space. Was that belief justified, though? If you let go of a drinking glass in this space, which way will it move? Toward your feet? Why? How certain are you about this? Which way is up and which way is down?

After you began to fly and approached the 6-door room, did you imagine looking in that door and the ceiling being "up"? Why? It was also entirely possible to look through the door and find the ceiling was actually *down*.

Again, assuming you have always lived on Earth, everything you imagined *between* the word-shells and *filling up* the word-shells I put into an order to put you in *The Ark* was ultimately grounded back to your own Earthly firsthand experiences. What else could these word-based experiences possibly be based on? In this way, you pulled your entire past and the planet Earth with you into *The Ark* as if they were shackled to your mind. You had no choice really. They are your ground state.

All of this happens because *The Ark* is a *secondhand* experience. It is a story you are telling yourself. And for you to tell yourself anything, Second You must be involved…

77

Your Best Friend and Nemesis

Second You is spinning these words into imagined experiences.

This means Second You was responsible for creating *The Ark* experience, which it did by overriding the firsthand experiences happening all around you while reading with *remembered* electrical sensory currents. These remembered currents include tons of feelings, emotions, and ideas, such as whatever story was happening in your reality on Earth when you first learned the word "cat".

Second You is a sensory *projector*. It is your imagination. Perhaps a good way to think of this is while you worked your way through *The Ark*, Second You was feeding Step 1 of your 3-step story system with *remembered* currents anchored to the word-shells I chose. In a way, your imagination "bounces" these anchored currents for each word-shell off the *backside* of your sensory windows.

I have no doubt you can "see" the rooms and doors in *The Ark* in your mind. Of course, you know they are not real. They are built by your previous Earthly experiences with "white solid doors". You might have even "seen" details like fine wood textures on the door. You also "felt" the fall you took into the void. Again, you know that was not real, either. It, too, was built from your remembered Earthly firsthand experiences with "falling."

This brings me to the part where I owe you an apology. I intentionally misled you early in this book. Second You is not actually born by believing you can be wrong and know it at the time you perform the action. Second You was not actually created by your mind tearing itself in half after you stole that candy bar. It has existed within you all along. Second You is the protector that keeps you safe. It is also your strategist, your master planner, and your guardian angel.

It only becomes a monster and a multi-headed Hydra from Hell the moment someone else hijacks it before you can defend yourself, and causes your best friend to turn against you. Some person used your own mind, your own imagination, your own *feelings*, as a weapon against your own firsthand experiences and self-built truth. That person might have done it unintentionally, but it was done because he or she believed it was the right thing to do. This hijacking is made possible by the very same method I have been using to communicate to you from the past – language.

This angel-turned-demon or strategist-turned-saboteur transformation happened to you because, unfortunately, it is *normal*. It is part of your "ground state" because you often learn your words at the same time you are having brand new firsthand experiences. In my opinion, this is *why* you cannot easily see around the looking glass of language, or even notice it is there.

It is my opinion that this looking-glass process has allowed us to make the massive amount of progress we have made as a species. Unfortunately, it also came at the extreme cost of "unit-izing" our reality, which provides us the ability to create and believe what seems like "absolute gapless truth-stories" that never existed as a firsthand experience for anyone. These stories are believed simply because they "make sense", which means they *feel* like firsthand experiences.

But of course, I assume you are aware your imagined experiences are not firsthand experiences. They are not even shared. They only *feel* real. They only *feel* shared.

Overwriting your Earth-based assumptions and norms (context) requires *firsthand experiences* that offer a completely different truth. You can certainly imagine a room upside down, or erase the atmosphere in *The Ark* after my words specifically create these stories for you, but you likely did not do this *automatically* without my words reminding you to notice them. Words that overwrite your Earth-based context, like *The Ark*, are very much like Second You telling you to "Breathe manually" again.

Your imagination can float your ground anywhere you want to go, but if the assumptions and norms are different than your firsthand experiences, you are stuck having to constantly tell yourself to remember the "right" context.

This is why I decided not to stick with the words of electricity throughout this entire book. You'd possibly have to keep retelling yourself the "right" story of what the words mean, and that can be exhausting.

As soon as you leave this book and refocus your story system on something else away from these words, your gapless, wordless truth built from sensory experience collapses these floating anchors back to your Earthly "ground" experiences. However, you keep the imagined experience of *The Ark* in your memory, put there by your friend and nemesis, *Second You*, who bounced electrical currents off the backside of your eyes, ears, nose, mouth, and skin to make it all happen.

(Note: I keep writing that your imagination "bounces currents off the backside of your windows", only because when you are looking at something in your imagination, you are no longer consciously attending to what you are looking at in real life. It's possible *Second You* does all of this sensory overriding completely inside your brain instead. There is an entire rabbit hole of neuroscience stories to dive into here.)

If I assume you have always existed on Earth, then your Earth-based "grounded context" is the only context Second You has to pull from, which means your story system's expectations and predictions about objective reality *in the future* are always going to be limited to this context.

This context is ultimately what "fills in" all gaps and voids in your imagination, which might be creating new floating grounds and contexts through fantastical stories and surreal fiction. You are unable to ever cut completely free of your Earth-based context, though, until you actually experience a non-Earth based reality firsthand.

With this in mind, let's talk limits.

How many different stories are possible for your story system? How many predictions can the two points of you – your present and your past - work together to create when it comes to imagining what might be behind any unopened door?

No matter who you are, the answer is not going to be countable. It is my opinion that memories are actually remembered electrical *current patterns*. This would mean *movement is always involved* in your memories. New incoming current patterns from your present (Step 1) are compared to your memories of past Step 1 current patterns being held in "storage" (Step 2). Your story system can match the currents happening now with current patterns from any time in your past.

For example, assume you see or hear something right now that you saw or heard before when you were 5 years old, 10 years old, and again 34 years old. Step 2 flags each of these memories as a pattern-match, and I suspect it also notes what currents came before and after each one of these flagged memories. It then waits for the next Step 1 current pattern to arrive. If the next incoming pattern matches the pattern before or after the flagged memory from when you 5 years old, it uses that match to project (Step 3) a story of what pattern is *most likely to arrive* in Step 1 of the next loop, and might even justify a new action to ensure, avoid, or change that likely future.

I suspect your Step 2 can even combine and rearrange the order of movements in memory however it wants. Perhaps the incoming Step 1 current pattern matches something from when you were 5, but the current-pulse from the next Step 1 matches something from a different experience from age 29, and the next Step 1 pulse matches something from age 8. I believe your story system can rearrange and recombine all three of these in Step 2 to still come up with a story of what to expect next in Step 3, although you will probably feel less certain about it. This is the beauty of having a floating ground system.

The story possibilities explode when you realize this memory catalog of context includes *imagined* current patterns as well, such as *The Ark* and all the stories you've read in here so far.

If given unlimited time, which was completely possible in *The Ark*, you will imagine *every possible story in your story system* related to what Step 1 is showing you. You will literally imagine everything that could possibly exist on the other side of doors – limited only by your own experiences.

This is the maximum resolution of your story system – of your mind.

The shape this creates was present the entire time in *The Ark*.

78

You, Deconstructed

Rewind time and restart *The Ark* thought experiment from the beginning. You are now in the room with no doors and no windows.

Remember that huge story voltage that occurred from all the sensory current rushing into your mind after you blinked and unexpectedly found yourself back in this room? *That exact* voltage, or imbalance, was present the first time in this room as well.

It doesn't matter if you came into this room from the huge black space in the blink of an eye, or by drifting in from *The World of You* – the story voltage is the same in both cases because your sensory limits and assumed Earth-based expectations *did not change*. The limits of your sensory windows are determined by your biology. Your expectations of these limits are set by all of your firsthand experiences.

The only difference between these two paths to end up in this windowless, doorless room was the speed of entry. One path resulted in a massive surge and an attempt to "slow down" reality using methods like adrenaline, and the other did not. Either way you entered this place, you will quickly adjust your truth to this situation.

In this new situation, your story system can "float its ground" and try to re-anchor all its predictions and expectations to your new "normal." However, even after this re-anchoring, you still experience more story voltage than you would prefer. You remain imbalanced.

Step 1 has provided pretty much all of the information it can about this room with a couple quick turns of your head. Nothing about this room ever changes, so the sensory current patterns flowing into Step 1 from your eyes, ears, nose, mouth, and skin never change from what you'd expect and predict from a room.

However, your story system knows from firsthand wordless past experiences you *should* be able to sense *beyond* whatever is blocking your view out of your sensory windows. This means you automatically "know" and there *must* be something *beyond* the walls of this room. The walls form a *boundary*.

Every possible story you can imagine by combining and arranging your millions of experiences, both firsthand and secondhand, starts whirling into your mind to try to match up and merge with your present moment. Everything you can associate to rooms, lights, shapes, textures, and size is pulled into use, along with every story you've ever heard or read or watched about rooms, and where rooms could be located or their purpose. *Are you dead? Is this hell? Heaven? A prison? Why could you be here? How did you get here?*

These are all questions that you simply cannot answer from your position. You need to experience a point outside of this room to understand and build an answer that is not a blind and wild guess.

Before we go further, imagine I have a machine somewhere far away from where you are, and that machine just placed a black pen on a piece of paper. This paper has a floor plan of *The Ark* already printed on it. The place where the pen is currently touching the paper is exactly where you are inside *The Ark* in this moment. If you move, the machine will move the pen on the paper to match, which draws a map of your moments.

If you were trapped in this room for long enough, I believe this machine would eventually create a solid black square. The only change happening in this room is you. *You* are the source of all possibilities in this place. You can jump, you can run, you can scream, you can try to do anything your story system believes your body might be capable of doing, and can justify trying to do. Yet your environment never changes. For this reason, I believe you would eventually move and touch every bit of this room possible to try to find *any* story that doesn't *begin and end with you.*

The voltage in your story system created by Step 1 being unexpectedly blocked cannot be relieved, and I believe this will become painful. Your story system *needs* a new step 3 that doesn't return right back to you.

You, as the person reading this book, can't *actually* normalize being stuck in this room. You were not born in here. You have far too many memories of bigger places and movement-rich places. Specifically (but assumedly), Earth.

So, after awhile, you might scream. You might scratch and bang on the walls. You might try to *force* an expansion on this seemingly indestructible boundary. You *need* to relieve the imbalance in your head created by not being able to sense what you *know* should be there, but the resistance to you experiencing beyond the walls is 100%. There are no sounds or any indication of anything at all on the other side of this room's walls.

If this point is reached (and it is my belief it is likely experienced all the time in solitary confinement in prison systems of all kinds), then your story system will likely turn to the only thing it has available to relieve the imbalance – *you*.

Normally (in my opinion), imbalance between Step 1 and Step 2 of your story system is converted into currents directed by Step 2 to accomplish two things: Predicting what will come next, which I've called Step 3, and justifying your actions relative to your imagined future or what *actually* happens next.

Typically, this comparison of Step 3 to Step 1, whether movement is justified at the same time it is happening or not, is *always* in need of some adjustment. Your predictions and your justifications are almost always changing, because your environment or needs are always changing between these steps. This makes predicting the future very tricky business, because there are so many variables and possibilities involved you have no control over.

However, if you are trapped in a room that never changes in any way, and you need nothing at all to continue surviving indefinitely, I believe your story system will quickly break down. After all, every single prediction you make of what comes next in this room will turn out to be *exactly right every single time, and your movements change nothing.*

The only thing that ever changes in this windowless, doorless room is *you*. This creates a paradox – a loop trap. Your story system keeps trying to predict a future that is 100% predictable, no matter what you do to try to change it.

If you have played classic or very basic video games, you may have experienced this paradox firsthand. The classic side-scrolling 8-bit and 16-bit games were your story system versus a simple, unchanging program. The enemies or obstacles in the game did not randomly change or appear in different places or behave in different ways. If you failed, the game simply reset you to some previous point, which ultimately provided you a way to *memorize* the game's patterns, figure out what to expect, and eventually succeed in getting past your previous failure point.

Imagine you complete such a game.

You might consider yourself done with the game at this point, and move on to a new game. However, let's suppose you really enjoyed this particular game, so instead, you restart it from the beginning and work to beat the whole thing all over again.

Then you start over and beat it again.

And again.

And again.

After enough repetition, you will start to lose interest in completing the game, because it is no longer challenging for you. I suspect you will instead shift your focus and start trying to find *the limits* of the game. For simple unchanging games, this often boils down to how fast you can beat it – also known as "speed-running".

There is a hard limit here, because this measurement is one of time. If the start of the game is point "A", and the end of the game is point "Z", then there is an *objective* fastest-possible path to connect A to Z. This most-efficient path will be based not on your ability, but on the code of the game programming itself.

Imagine you actually achieve the objective fastest time possible to complete the game you love. You are now the world-record holder for speed-running this particular game, and your performance was determined to be the fastest time possible based on complete analysis of the programming. No one will *ever* beat your time.

I suspect that *now* you will probably be done with this game. It does not have anything new to offer you. You have played it though hundreds if not thousands of times. You know every detail, every scene, every obstacle, and every shape and pathway from point A to point Z. Nothing can surprise you when it comes to this game. *Everything is expected.*

Now let's make this game work like the room with no windows and no doors in The Ark.

It doesn't matter if you know every detail. *You can never stop playing this game.* When you reach the end, it seamlessly jumps back to the beginning. No matter what game cartridge you put in or load, it's always the *same game exactly where you left off.* No matter how much you modify or replace your gaming system, this same game is the only thing that ever comes up on any screen you see. If someone is playing a different game, the moment you look at it, the screen changes to become the same game. You cannot escape it.

How long would it take you to start destroying screens? To give up on video games completely? To question your own reality?

The same is happening in the room with no windows and no doors, but in this situation, you are the playable character in a game with a very tiny level and no way to restart or complete it.

You cannot get through the boundary of the room to expand your possibilities, but you also do not need food, water, shelter, sleep or anything else. You are stuck.

Step 3 predicts that if you justify moving to the left, you should expect your view of the room to shift to the right in the next incoming sensory current, and that's *exactly* what happens. You are able to predict the future currents *with perfect accuracy*, because in this situation you have *total control of everything that changes*.

All of your story voltage, which is normally converted to currents in Step 3 to predict future movements and justify your own movements toward or away from that future is looping right back into Step 1 with no change in voltage at all.

Your present and your imagined future *are merging*. You don't actually have a need to predict *yourself* – you already know what to expect from a lifetime of experience with your own movements. Now these movements are literally the only things your story system has left to predict, but it already controls them.

In an desperate effort to break this loop where your story system is trapped with itself by an impenetrable unchanging boundary, I think it will justify piercing the only boundary it can, to create movements that don't result in perfect prediction: The boundary of your body.

Destruction of self may suddenly be justifiable in order to bring new possibilities and new stories and movements into this place where they didn't previously exist. For example, self-harm can provide fluids, which can be endlessly interesting to your story system and help it create new harder-to-predict stories – but at a very high cost.

To push things further, removing pieces of your body could result in harder-to-predict stories by changing the "shape" of the affected sensory window. In a twisted way, this creates *new* experiences in an environment that had none to offer.

A less-physical route is the splitting of your mind. I believe this can be justified *without* your conscious approval in a situation like this one. Your story system cannot relieve the huge imbalance trapped inside by simply "knowing" there must be an outside of this room. So after you exhaust your exploration of this small space, your story system has *nothing left to do*. It has no new stories to create. No outside movements to predict. All future stories relative to you are perfectly predicted with no adjustments needed as they arrive, so there is also no real need for your story system to commit any of them to memory. Step 3 merges with Step 1, and your 3-step story system collapses into a 2-step story system. This gives Second You – your imagination – the new ability to provide *firsthand* experience.

Since your future has collapsed into your present, projections of your actual future where nothing ever changes and projections of imagined secondhand stories might merge with each other.

Your "movement" justification currents might instead redirect to your imagination, and this might make your sensory windows convert this perfectly predictable room into a kind of movie screen. I suspect this would create all kinds of bizarre experiences. Hallucinations of all kinds might appear in the room as your biological straw-path wiring literally re-arranges. The straws of the imagination (Step 3) reroute and connect themselves into the pathways that bring currents in from the outside (Step 1's wiring).

I suppose it's even possible for this re-wiring process to create a whole new *separate* you – a full split of your mind – so that *something* in this space changes unpredictably, which should, in theory, relieve the enormous mental voltage created by being trapped in this limited-and-unchanging space.

Unfortunately, none of this will relieve the voltage. The imbalance is all still contained within your mental boundary, because it exists between *your memories* of existing in larger spaces with far more movements in the environment, and *your present* where that is prevented from happening.

I hope that you have not directly experienced these processes, but I must recognize it's possible. This is all beyond my own experiences, so I dare not go any further for risk of telling you what to believe or just playing wild guessing games. So instead, let's assume the single white door now appears in The Ark before these destructive processes start to happen.

I suspect you can better understand *why* the door was such a fixation now, and why I believe it's almost *certain* you will attempt to open it. You remember how doors work, and that they lead into new spaces. There is now a possible "break" in what was previously perfect 100% resistance. You can suddenly see the clip on the balloon you are contained within.

You cannot know what is on the other side, however. You were never able to build a context for what was beyond the room with no windows and no doors, and the same holds true with the addition of this door. You can't build a story for what is beyond this door before you open it, and have any idea if that story is "right."

This doesn't matter. The pressure in your story system screams for relief, and before you is a straw-path doorway into a new story. Every story you could imagine in Step 3 which could previously only return to Step 1 due to the impenetrable boundary all around you now has a new place they can be projected, and you might actually be able to justify moving into that place to discover which imagined story is valid.

Without opening the door though, all of these imagined stories are Earth-based or secondhand story predictions with no firsthand experience to back them up here in *The Ark*.

Before you imagine moving forward through this door, I think it's worth considering what would happen if this door had been locked.

As I see it, this would make the door the same as if it were still the wall. But I have to ask, would a locked door be better or worse than having no door at all?

In this locked-door situation, you know exactly where you need to move for new stories, and even the actions you need to justify to get there, but those actions fail to change the story, no matter how many times you justify them.

In my opinion, this causes your predictions of justified actions and their effects on reality like "turning the handle is how doors are opened" to slowly collapse into impossible-to-make-happen stories, and eventually get discarded as the *right* actions to take until the only truth-story left is your own movements in an unchanging room all over again.

I think the process of self-destruction and harm would begin here once more, and you'd eventually start creating stories *from* yourself, just like before.

Of course, this door was not actually locked in *The Ark*, so it's nearly certain you will open it. I suspect you will use caution opening it, because you have no idea what is actually on the other side.

With the discovery of the 6-door room after opening it, your context for this space grows, and so does your usable truth. Your anchor - your *expectations* - are expanded.

You know it's completely possible for this place to be nothing but endless rooms and doors, and you may have even assumed it right away. When putting individuals through *The Ark*, some voiced this assumption out loud, probably hoping to receive validation of a larger truth about this place the mind desperately wants.

With the addition of the 6-door room, the story voltage in your head caused by your senses being blocked closer than expected is collapsed by a tiny amount. This moment is a lot like if you took the clip off of one balloon but not the other.

There are now lots more possibilities, but there is still a long way to go to get back to "ground". Your story system sets to work again, imagining all creating the stories most likely to be on the other side of these doors. Its strongest firsthand experience to use for this task is your experience with another door in this same place, and it simply opened into a room with more doors – so that is what you will likely rely on the most.

Your story system begins projecting through boundaries of all sizes and distances with questions that have no answers. *Will rooms with even more doors be on the other side of these 5 new doors? Where are you? What is this place? A prison? A maze? A puzzle? Why are you here? Where is here?*

Opening the first door into blackness doesn't really help the imbalance in your head. Neither does opening the second or third door. The blackness beyond each might as well be a blank wall at this point. The locked door on the floor and the unreachable door on the ceiling are very likely to bother your story system immensely. The other sides of these doors are places that might allow some relief from this imbalance you are experiencing, perhaps even an answer to your questions, but you cannot open them to find out what is on the other side.

Still, you *know* from past experiences whenever there is a door there is another side. Your story system again projects all the stories of what might be beyond unopened doors yet again by pulling from your immediate past *in this place*, your remembered firsthand experiences *in this place*, and even remembered secondhand experiences like science fiction or any other story created by humanity that you have consumed *similar to this place*.

You want to know all the possibilities that exist in all directions, but you can't. There is still a boundary preventing this. Your sphere of firsthand experience, the boundary of Step 1, is still being mostly blocked.

Back at the machine that tracks your movements with a pen on paper, the process of filling up the new room with black ink has begun. I think you would do what you did in the no-door room and explore every inch of this space. You will look for some hint of a larger story – some hint of "ground", or some hint of an unnoticed straw-path to give relief to your enormous and imbalanced story balloon.

As you exhaust all the movements you can make in this two-room system and disaster starts to enter the situation and work, the machine has drawn two black squares connected by a "straw" which represents the doorway. You still can't relieve the huge voltage in your head, and Step 3 begins to reroute itself into Step 1. Self-destruction justifications begin to rear their ugly head, driven by the need to collapse the voltage in your brain, and finding no way to understand the story you are in and also validate that story into being the truth of this reality.

Is this a prison? Is this punishment? Hell? A maze? A puzzle? Why am I here?

How did I get here?

As this justification voltage starts to build, a new story appears that was probably not something you were willing to believe was "right" earlier. Point 1 of this new imbalance is your entire story system. If left as it is, this point will probably continue the journey into self-harm and/or self-consumption and hallucination by trying to split itself into two to relieve imbalance. However, there is a new, very real, and completely empty Point 2, and there is even a straw already in place between these two points. The only thing preventing this system from collapsing into motion is your own resistance to justifying that motion as "right".

Point 2 is the void beyond the doors.

Individuals have told me there's no way they would step into the void. They claim that with no way to feel a floor, it's bound to be a big tumble, and that means injury or death. Perhaps this is where you think you would stop as well. If so, then you remain trapped in this two room space literally *forever.* You do not experience hunger or thirst. You are not aging. You do not tire. You do not need sleep.

Everyone that read *The Ark* chapter, probably even you, likely predicted that no floor in the void meant to expect the future experience of falling downward. This is an Earth-based *assumption* you dragged into *The Ark* with you, and you might assume it for the same reason you believe when you let go of a glass it will move downward – it is a highly validated and directly experienced firsthand truth for you. Gravity is one of the most validated truths in need of remembering on Earth – forgetting it or misunderstanding it can lead to your death, after all.

So, as you took the step across the threshold into the void, your story system prepared you for a fall – because this fall into the unknown finally became a better story than tearing yourself apart over an eternity. The void became the *new path of least resistance.*

I think it's safe to assume your story system experienced a shockwave as you unexpectedly fell *sideways*, and with that shockwave, any Earth-based context you were using about where you are and what is possible was overwritten.

Your reality resets as you tumbled into nothingness. Your story system floated its anchors of expectations, and sought to re-anchor to whatever new "ground truth" might exist, and was fully ready to accept whatever it might be, even if it was falling for an eternity.

After you stopped falling and found yourself surrounded by a sea of tiny white dots in every direction, the first paradox appeared, and it throws this "your-ground-state-is-the-limit-of-your-Step-1" story into question.

If your senses being "blocked" or limited is responsible for the imbalance in your mind, then at this point, floating in the void, you should have felt very close to balanced.

The dots (or beeps) themselves were the only things creating sensory currents into your brain. This should have resulted in a significantly lower story voltage – but it didn't. Instead, your story system remained imbalanced, but the shape of that imbalance changed.

You were unable to move in the void. This was unexpected, but not *completely* unexpected. This billion-door space literally looks like outer space, and it's likely you know secondhand stories about human movement in outer space. You probably know that movement in zero gravity relative to anything else requires you to *push off* of something – and you recognized you had nothing around you to push.

The previous imbalance that was created by your *senses* being blocked is replaced with a new-and-equal imbalance caused by your *movement* being blocked.

There were now *tons* of possible stories all around you. Billions actually. Your story system likely ran wild trying to think of what each dot could be. But you couldn't move toward validating anything. All of your justified movements resulted in no actual change in any story.

You went from no possibilities and unlimited time, to a billion possibilities and no objective time to reach them, while at the same time experiencing endless time. The center of the void is similar to being back in that first room with one door, and that door is locked.

You probably wouldn't give up trying to move right away. I think you'd probably start trying all kinds of movements with your body to see if you could make the story change. This resulted in a slightly different process of self-collapse.

In the room with no doors or windows, your self-collapse was due to Step 3 (your predicted future) and justified movement being the only thing that *ever* changed Step 1 (your present). Now with a billion possibilities and no way to reach them, self-collapse is due to Step 3 (you predicted future) and justified movement being completely *unable* to change Step 1 (your present). They are now almost totally *disconnected*.

When you attempt to push and run and wiggle while floating, you expect the doors in Step 1 to change. They *should* get closer, or maybe farther, depending on which way you are focused relative to your attempted movement. Yet this expected future keeps getting *invalidated*. It is constantly *wrong*.

Your mind will now float a different anchor entirely. It will start to re-evaluate the reality of your own body and movements.

Is the truth about how I move with my body wrong?

Am I moving my body?

I suspect you'd look at your own body and try to determine if you are actually moving your legs or not. I suspect you would quickly exhaust every movement possible as you float in the center of the void, before facing the beginning of the process of self-destruction.

When you were in the void of *The Ark* the first time, I asked you if this situation was better than being in the two-room system, if you longed to go back, and if you regretted your decision to enter the void.

The question you were really answering was: *Is being able to move with nowhere to go better than having a billion places to go and not being able to move?*

Not long after this point in the story, your story system experienced yet another shockwave of story voltage that triggered it to float its anchors yet again.

You were *unexpectedly* released from being unable to move. Not only released, though: You could move without the justification process at all. Once again I think you would have no trouble letting go of the reality where you couldn't move, and would quickly re-anchor to the new ground state where you can fly however you want. I have even had an individual reply, "Oh yeah! Now we're talking!" and go on to explain all the loops and tricks he would perform with his new flying abilities.

This change seems to offer tons of story voltage relief even before any doors are approached. You are free to make any choice you want with unlimited time to make and execute that choice. Your story system is now running at full capacity and much closer to balance. Every story you can imagine is suddenly up for grabs and might be possible.

However, some voltage remains because you do not yet know the whole story of this reality. This is why I believe with almost certainty that you would want to know *all* the stories of this place and open *all the doors*.

Nearly every door so far has resulted in an *unexpected* story, which leaves you almost completely context-blind as to what to expect behind *any* door. And there are a *lot* of doors in The Ark.

To get a grasp on how many doors there are, if you open one door every *5 seconds*, and use an Earth-based clock, it would take you 158 years, 200 days, 8 hours, 53 minutes, 20 seconds to open all one billion doors.

Of course this is not really a problem in *The Ark*. You have all the time you need. You do not need food or water or rest. You could spend a million years opening them all if you wanted.

How does this make you feel? Does it bring you joy? Perhaps dread?

As you move toward your very first door – what expectation might you have for what may be behind it?

Whatever it might be – it's going to start causing a lot of changes...to you.

79

What Growth Feels Like

Without much context for what could possibly be behind any one door in the billion-door space of *The Ark*, you are forced into the position of opening each door almost completely context-blind. As a result, you can have no idea what to expect, but you know huge voids, rooms with more doors, and gravity direction changes are possible.

The first few doors you end up opening will build and grow a larger truth-story and context for about this reality for you. If each door hides something completely unexpected, you are going to experience a ton of *emotions* during this process.

Using the 3-step mind-as-a-prediction-story-system framework, your emotions can be understood with what I think is decent clarity. You might also start to see how they help guide you and the big part they play.

Emotions, happiness in particular, are often central to philosophy and psychology. Individuals of all kinds in both fields often discuss happiness like it is something you could potentially achieve long-term. Many seem to believe it should be possible to be permanently happy. The United States Declaration of Independence from Great Britain even mentions "the pursuit of happiness" as an unalienable *right*.

However, when viewed through the lens of movement, time, and our story system, emotions (happiness included) are not static one-point states at all. They are something you do experience in the present, but it is my belief they are created exclusively by *a difference between the present you are experiencing, and what you expected that present to be.*

In other words, emotions are your story system's *reactions* to the validation or invalidation of its expectations. This means emotions always come *after* Step 1.

This might seem confusing at first, because I have no doubt you feel like you experience emotions before, during, and after things happen in your present moment – but this is an illusion that exists because of your mental boundary. Emotions are created and exist in only one place: Inside the mental boundary where no one else can see. This means they are never created by anything outside of that boundary. No one says, "bibbity-bobbity-boohoo," and causes your brain to magically produce emotions.

You are only ever one point experiencing the present moment.

To hopefully clear this up a little, I'm saying that emotions are experienced relative *to your own mind*. They are the "movements" between two constantly imbalanced points: Your expectations (Step 3) and reality (Steps 1 and 2). Your reality is your present moment, which is what you perceive and comprehend.

To put this together: Emotions seem to be created by *errors* in your predictions and justifications of what you believe is right based on what you believe is true, which is what sets your expectations. Once that expected reality arrives in the present, the difference between you and the universe is "felt" as emotion.

As for what emotions are made of, they seem to me to be composed of chemical movements or chemical "currents" in your brain to help balance "mistakes" in your truth. I suspect these chemicals have an important function related to feeling valid or invalid, and help control what is logged into your memory and what is forgotten.

Since emotions in this context are based on two points, this means they are *always* relative. This means the master context of time is the best way to define and understand them.

Before I go further in attempting to explain how I believe emotions work, I accept that while you might find yourself agreeing with what I'm about to lay out, it's also possible you won't. That's okay. This is simply a theory I have built that makes a lot of sense *to me*. Over the course of two-and-a-half years, I haven't yet been able to disprove it yet. *This doesn't make my theory objective truth, or even your truth.* It just makes it my personal truth. If my truth turns out to match the truth of the human condition, then it might be a valid theory. And that's the most it can ever be. A theory like this is always open for debate, revision, or outright invalidation. I am not the *Keeper of the Truth of Human Emotions*. I accidentally "uncovered" what I think emotions are and how they work by watching, listening, and interacting with the children in my life.

I'll start things off by defining our emotional *ground*. This will be pretty easy, because I've already written about it at length when I referred to mastering a simple video game in the previous chapter:

When you play this video game, if everything were to occur exactly as you predicted it to occur, you would experience very little emotion. If you imagine what it could feel like to complete this video game that never changes from start to finish for the 397th time in a row, you are at emotional ground.

Back when you beat this same game for only the 10th time though, you would perhaps experience satisfaction that your predictions seem to always be right, and pride in the consistency of your justified movements in the game. However, your expectations from reality are already starting to warp your understanding of "normal."

You begin to *expect* you will beat this game an 11th time, 12th time, and even a 13th. Eventually, beating this game will result in feeling no emotions at all. You are starting to become *all-knowing* when it comes to this game.

In the "game" we call life and reality, no one is *all-knowing*. You are not, and I am not. No one knows the whole story of anything. This means literally every human being, you and me included, experience emotions while trying to advance from being able to predict *nothing* perfectly to being able to predict *everything* perfectly.

Let's start with the positive emotions. In particular, the emotion central to so much of our focus in life: Happiness.

Happiness, as I see it, is the chemical release that occurs in your brain when the future truth-story you predict to happen (Step 3) based on your past and present (Steps 1 and 2) turns out *better* than expected when it becomes the present (new Step 1).

Examples:
- You expect to open the first door in the billion-door space of *The Ark* and find a room with more doors, but when you open the door there is a beautiful beach at sunset on the other side, complete with waves gently lapping against the sand.

More Earthly-based examples:
- You expect to bowl a terrible game based on your last five games of bowling, but you end up bowling your all-time personal best score.
- You expect a huge tax bill based on your last two years of filing, but discover you are getting a large refund.
- You expect to be rejected by someone you like after you confess you want to have a romantic relationship, but discover the desire is mutual.

Simplified, happiness is what you feel after an expectation (future prediction) turns out to be wrong *in a positive direction*. The amount of happiness you feel will be directly related to how far apart your expected truth and reality seemed to be, and how long it took you to realize your error.

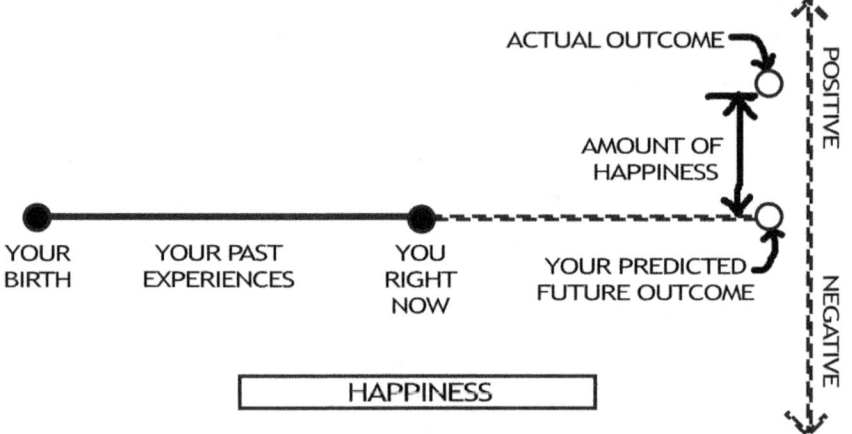

This means all emotions have very real limits. If something happens that was unexpected, then your story system *has failed to accurately predict your reality*. A story voltage appears, and a flood of chemicals is then released as this error slides into your memory with a big red flag, so that you can remember what went differently, as well as what came before and after the error. The chemicals help adjust your beliefs, predictions, and actions accordingly in the future when similar current patterns arrive.

This is exactly why the "pursuit of happiness" is an endless and impossible goal. Happiness will always be temporary, because it exists due to an error in expectation. The emotion of happiness itself is part of the story system that is working tirelessly to *eliminate this error from happening again*.

If an error in your prediction or expectation occurs, your story system just floats the anchors by releasing chemicals that cause you to experience emotions, resets expectations, and drops the anchors on a new ground, which is hopefully more in line with reality, which will make it harder to experience the same emotion again.

I'll use the same examples as before to illustrate this.

- After opening the door to discover the beach, you reset your expectations, and predict that when you open the next door in billion-door space of *The Ark* you will find something equally amazing. However, when you open the door there is the same exact beautiful beach at sunset, with waves gently lapping against the sand. The amount of happiness you feel from this discovery *will be less* than before. If every door you ever open leads to the exact same beach, you will eventually stop experiencing the feeling of happiness at all when you open a new door only to find a beach at sunset.

- You just bowled your personal best game ever, so you reset your expectations that you can do it again. If you somehow do it again, it was somewhat expected, so you will experience less happiness than you did the first time. If you bowl a perfect game every time you play from here on out, eventually all happiness will disappear for you when it comes to bowling, because perfection is now the expectation, and it is always right.

- You got a large tax refund. It arrives and you deposit it. Your bank account balance is now larger and you experience happiness. However, this feeling is your story system floating the anchor and re-grounding to a new normal. Now you *expect* your bank account to be this size. You also are likely to experience less happiness when filing taxes next year, because you now *expect* to get a refund.

- You discover the desire for a romantic relationship was mutual, and experience a long bout of happiness afterwards. However, the anchors float and re-ground, and now this relationship becomes the expectation.

As you can probably guess, it doesn't actually take much to flip expectations of happiness into the opposite realm: Negative emotions. Here are a few examples of the negative side of things:

Annoyance - When the future you predict based on your past and present is *blocked* from meeting expectations for reasons you can't seem to understand, so you predict the same future again, keeping the original expectations.

- You turn the doorknob in the first room of *The Ark*, and feel the door come unlatched, but the door refuses to open, so you turn it again.

As you continue to try to bring your expected truth-story into the present, annoyance might give way to *frustration*, which is the same cycle of annoyance repeated more than once.

- The door does not seem to open after turning the knob and coming unlatched twice, so you bump it with your shoulder expecting it to pop open, but it *still* does not open.

In the same family as annoyance and frustration is *anger*. Anger is when the future you predict based on your past and present does not turn out as expected due to the actions of something in objective reality *intentionally* (real or imagined) preventing that future from meeting your expectations for reasons you do not understand. As a result, the future you predict and your expectations again remain unchanged.

- You turn the doorknob, feel the door come unlatched, but the door does not seem to open no matter how much you bump it with your shoulder, and you can hear someone laughing and mocking you on the other side for not being able to open it.

This situation can quickly escalate into *rage*, which is when the future you predict based on your past and present does not turn out as expected due to the actions of something intentionally preventing that future, but your expectations remain unchanged, *and* the actions of the "something" preventing that future also remain unchanged.

- After pleading or demanding to help open the door, the person on the other side continues to laugh and mock your inability to open the door.

Sadness – When the future you predict based on your past and present turns out *worse* than expected once it becomes the present, and also prevents your original expectation from being possible long-term.

- The person on the other side of the door apologizes for not being able to open it and walks away, leaving you alone in the room with an unopenable door.
- You expected to get a raise this year, but you did not.

- You expected to hang out with a friend on Tuesday, but he or she chose not to join you.

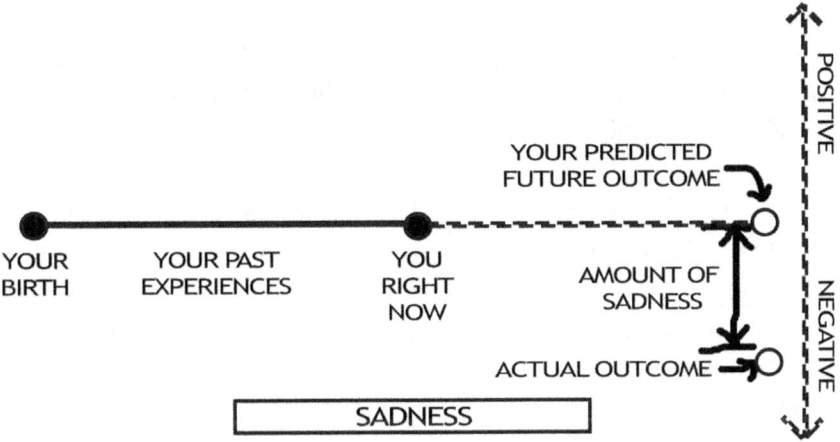

Let's fire through some more:

Excited – When the future you predict based on the past and present is expected to be positive but is not within your control to understand or accurately predict how positive it will be.

- You have been selected to win one prize out of a range of many, and that prize will be chosen for you using a randomizer.
- You are one move away from a perfect game (choose any sport).

Scared – When the future you predict based on your past and present includes the possibility of your harm or death in a way you cannot predict.

Surprised - When the future you predict based on your past and present turns out radically and suddenly different from what you expected.

Proud – When an uncertain but positive future you predict based on your past and present turns out as expected by your own actions bringing it to reality.

Awe - When the future you predict based on your past and present turns out completely unexpected, novel, and positive in the present.

Anxious – When the future you predict based on your past and present contains the possibility of a negative result, but is beyond your ability to control.

Nervous – When the future you predict based on your past and present contains the possibility of a negative result, might *directly* affect you, and is not an outcome you completely control.

Fear – When the future you predict based on your past and present contains the possibility of a negative result, which will affect you directly if it happens.

Disgust – When the future you predict based on your past and present becomes something completely unexpected and negative to your senses in the present.

Enjoyment - When the positive future you predict based on the past and present produces long-term and unexpectedly positive experience.

Disappointment - When the future you predict based on the past and present does not turn out as expected in a way that indirectly affects you.

Curious - When the future you predict based on your past and present is uncertain but is not expected to result in a negative experience.

Hope – When the future you predict based on your past and present is expected to be positive, but is beyond your control.

Despair – When the future you predict based on your past and present turns out *worse* than expected once it becomes the present, and prevents any expectations of a more positive future.

If my theory is accurate, and emotions occur in the present due to something about your expectations *of the future* not being met (because your predictions were not right which implies your truth is not right), then as time passes these *remembered* emotional experiences may also change. This is thanks to Second You, which makes it possible to re-experience the memory of the error and project it right back into the future again.

For example, a happy moment in the present (something turning out better than expected with a friend) could eventually become associated with sadness.

Let's say the happy memory happened years in the past, and today the friend is no longer in your life. When you re-experience the original recorded happy memory in your current moment, you will end up merging that emotional story with the current story of that friend being gone. This results in a "new" experience that changes both remembered stories and then sends that merged story back into Point 2 – your memory. You perhaps expected to feel happy after the first memory, but it was immediately and unexpectedly met with sadness, and you have no way to control or improve this story in the future, so they automatically merge and you experience a feeling of sadness while remembering being happy at the same time.

You can really begin to stir the emotional pot when you consider that while experiencing this merge with Second You, something you did not expect occurs in the present that then merges and warps this memory and its emotions even further.

With all this in mind – past, present, future, expectation, truth, errors in prediction, emotions, control, and movement - let's now return to *The Ark* to lay all of this out in a way that makes sense.

80

The Strawman

Return your body to *The Ark* at the moment right before you are released from the center of the billion-door void and discover you can fly.

Let's take your mind to a different place.

Come take a look at this pen-on-paper machine that tracks your body's movements in *The Ark*. It drew a straight line from the door you stepped through to tumble into the void to where your body is located right now.

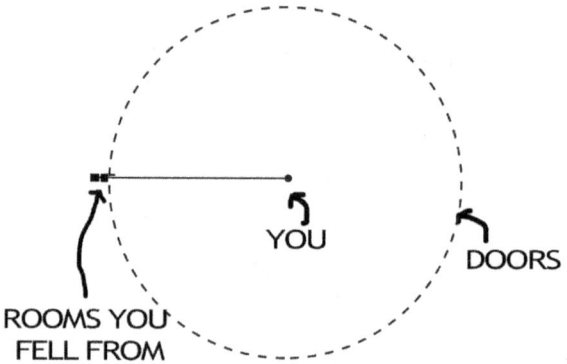

Now imagine taking the pen from the machine, and drawing line-paths from where your body is now to all the possible destinations it can travel to, going no further than your sphere-shaped sensory boundary.

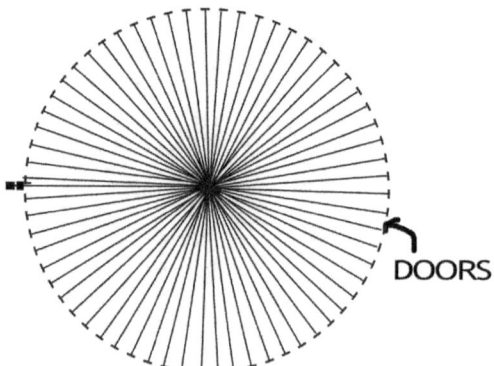

Notice that the further these paths get from the center, the larger the space gets between them. In my opinion, this space will not be seen as a destination by your story system because it doesn't make sense to take literally *all* paths or even a curved path to arrive at any one door, because there really isn't any reason to fly into a space that appears to not have anything in it.

Your senses easily fill in the story that you have no need to explore every inch of this space – you can plainly see there is nothing *between* any two straight-line paths to any two doors. Your *trust* in the resolution of your senses allows you to comfortably leave this gap. This gap, in my opinion, can grow as large as the distance you are capable of sensing in Step 1, and still leave you feeling comfortable no story is being missed.

In other words, you tend to be comfortable with unexplored gaps in first-hand experiences with some senses, as long as you can "sense" across or through these gaps with a different sense. Which I think is interesting to consider when you apply this idea to the situation of being trapped in the room with no doors or windows. In that space, nothing was being missed by your sense of sight or hearing after the first few moments. Yet I feel confident with enough time you would eventually start experiencing every inch of space that might exist using feel, taste, and smell, to make sure you haven't missed something.

It's like you start falling down a sensory rabbit hole when you feel something might be getting missed. Trust in your own sensory resolution becomes questionable.

In my opinion, you are comfortable assuming nothing is missed in *The Ark* because you are naturally drawn toward the "things" that are creating sensory currents for you. You fly toward the things that are creating imbalance in your story system by blocking possibilities. Each door is a *boundary*.

This is where the idea of your movements being driven by wanting to return to "sensory ground" falls apart almost completely.

If the goal of your movements were truly to return to ground in the form of total sensory blackout, you would not move toward the doors at all, your goal would instead be to move away from everything. In a pinch, you could close your eyes and plug your ears and be very close to achieving this sensory balanced ground state. Yet I suspect this is the last thing you probably want to do.

This means the ground you are seeking by moving toward specific doors is not based on sensory ground at all.

Doors *lead* places. Doors are possibilities themselves. Each door you open might actually contain the possibility to change your reality in a negative way. And yet you are drawn straight to them anyway. In my opinion, this is because they are *hiding* possibilities; they each contain pieces of the whole story of this place. And your mind *wants the whole story*. Sensory blackout is no story at all. Opening all one-billion doors is to know all one billion stories of this huge black space, and merge them all into a story contained inside just two word-shells: *The Ark*.

And yet, you are just one point. Of all these billion possible destinations that make up *The Ark* – you must ultimately begin with *one*.

How do you choose where to start? That I cannot say. I can only say that however that might be, it creates the straw-path of least resistance for you. However much resistance you experience to the idea of opening that first door will determine how fast you are willing to travel through the "straw-path" to fly to it.

One door was slightly brighter (or louder) than all the others. As a result, it generates more sensory current for you. This creates a higher "voltage" point in your sensory boundary your brain wants to balance, so I suspect it's likely you will choose to move to this brightest door first.

Of course, I can't be certain this would be your choice.

I can't be certain of any choice you would actually make over the last eight chapters. No matter how much I understand about how you make decisions, I can't actually *control* those decisions.

I've had to assume your choices. I didn't make these assumptions blindly, but also can't assume them with certainty. Every decision I've covered that actually belongs to you exists as only a *probability* from my perspective. I can directly sense where you are, or measure the speed you are going, but I can't predict the direction or justifications for your movements.

You might have immediately jumped into the first void behind the second door you ever opened in *The Ark*.

You might have never left the room with one door.

You might never have been able to give up hope toward the ceiling or floor doors.

You might have never entered the void.

Only *you* are capable of justifying *your* decisions and acting them out. I can't see how you justify your choices. This process is hidden from me behind your mental boundary, and this makes all the difference in the world. It makes predicting any action you might take an act of *assumption* on my part.

I made the assumptions on your part I thought were the most probable from the beginning to the end of *The Ark* chapter. I assumed you would choose to fly to the brightest dot in the billion-door space of *The Ark*. For the sake of continuing this story, imagine yourself back into the *The Ark* at that moment just before you arrive back at the 6-door room.

As you approach, you are suddenly struck with the feeling of *familiarity*. You have memory of this place and can predict, successfully (thanks to your memory), what is inside this room. You know what to *expect* from it.

Whether or not you will re-enter and try for the top or bottom door inside the 6-door room before trying the other 999,999,999 doors in the large black void, I can't know for sure. But I think it's very likely this familiar doorway will become an anchor point for you. It is a place to begin – a starting point.

But…a place to begin what? Exploring all the possible story options? What are you looking for? *What story are you hoping to find in this place?*

I can't answer these questions for you. No machine and no person is capable of perfectly predicting what you will actually do at any given moment, and therefore can't answer these questions for you. The machine with the pen reacted and followed you – it doesn't and can't predict you.

If this chapter made sense for you, and you can imagine yourself justifying all the actions I've mentioned without experiencing story resistance in the form of wanting to quit reading this book or being frustrated by these words, then perhaps we are on the cusp of a larger shared truth. Perhaps this whole foundation within the abyss is very close to lighting up.

I hope this is the case, because it's about to get a lot more intense down here.

While you float in *The Ark* just outside the doorway back into the 6-door room, and debate re-entering, something snags your attention immediately to your left.

Something just moved.

What *feeling* did the previous sentence create for you?
Was it fear? Perhaps excitement? Something else?
Whatever it was, try to notice and remember it if you can.
It will become important soon enough.

81

Gravity

While you were debating what to do at the door back into the 6-door room, you notice the doors nearest to you in the billion-door space seem closer than they were earlier.

You push back from the doorway and spin around, only to notice even the furthest dots (or beeps) that were directly behind you seem to be getting brighter (or louder).

All the doors are moving toward the center of this space.

As you turn back around to face the 6-door room again, you notice new and very-faint dots *beyond* the outer edges of this doorframe.

There were layers of doors beyond the billion-door layer you couldn't see until now.

There are no walls in this space. There never were. I specifically pointed out they weren't there your first time through *The Ark*, when you reached your arm through the door into the blackness to try to feel the backside of the wall, and discovered you couldn't feel anything at all. I even showed you a drawing that illustrated no walls exists in this huge black space.

If you've been imagining black walls all this time between the doors, then you most likely filled in a mental gap with an Earth-based assumption. That gap was caused by information that either went unnoticed, missed, or forgotten in your laboratory of imagination, and it was replaced with an assumption you have anchored to doors in your more distant-but-familiar past context: Doors are openings through walls.

Regardless, *objective time* has just asserted itself in this space. Before this point, only your movements defined time in *The Ark*, and you had all the time you could ever need or want. Now the doors are steadily moving toward the center. They are steadily growing brighter in every direction you look.

You can already see corners on each of the distant dots beyond the first layer of a billion doors. (Or hear their distant beeps growing.)

Have your priorities changed?

What is most urgent for you now?

Will you enter the first door available to you, no matter what is on the other side?

Will you wait for more time to pass before deciding what to do next?

Will you try to fly *past* the nearest layer of doors to hopefully buy yourself more time with the doors beyond?

All doors are getting closer together at every moment. They seem to even be gaining speed with each passing moment.

How fast did you imagine yourself flying around before this point? As fast as you can run on Earth (assuming you can run)? As fast as a car? A plane?

I gave you no limit, but I will now: You cannot fly at or above the speed of light.

Does this change the possibilities for you?

Will you immediately accelerate to near light speed?

To what end?

What is your goal?

What are you looking for?

Are you racing around just to open more doors? Why?

Are you trying to open as many as possible? Why?

Are you flying *between* doors to avoid becoming trapped by the tightening layers? Why?

Even if you fly away from the center, passing through layer after layer of doors at near light speed, you are still not fast enough for the layers to start getting further apart from each other. *Escape* is not possible.

The result from trying will only lead to an eternity of flying through layer-upon-layer of collapsing doors, with no extra time earned to open more doors.

Knowing this, how long would you attempt to do such a thing – to escape?

Would you prioritize trying to escape the doors over opening the doors?

How long before you give up on escaping?

How will you decide it's time to stop trying to escape and make a different choice? What would that choice be?

Will you start choosing doors at random, or try to fly back toward the collapsing center?

If you would head back toward the center…why?

If you would pick a door, how will you choose it?

Again, *what are you looking for to justify the decision to enter a doorway?*

Every door you open, no matter the layer or the number, leads to a familiar-looking cube-shaped room with five other closed doors inside.

Does *this* change what your 3-step story process is attempting to do.... whatever it is you are trying to do?

Will you eventually give up and enter one of these familiar-looking rooms?

When will you choose to enter? Will you keep opening more and more doors until you become trapped between two layers because there is no longer enough space for you to fit between the doors? Why?

Again, what in the heck are you looking for?

What would cause your 3-step process to abandon its desire to open *every* door? What would be enough to justify the *new* action of going into a door and not just opening it?

Are you even looking for the same thing you were looking for when you had infinite time?

Will you enter one of the doors *only after* opening hundreds at near the speed of light to discover they all opened into 6-door rooms?

How much would needing to leave behind billions of unopened doors bother you?

Would you rather stay in this collapsing space than enter into another cube-shaped room again? Why?

I'll now assume you will eventually enter a six-door room instead of choosing not to. As you cross the threshold, you immediately fall on your face as gravity asserts itself in a different direction.

Behind you, the open doorway that once led into blackness is filled with the edges, faces, and features of the many doors that made up door layer below – or one big mass of loud beeping. All shades of white begin to merge as the space between the open doorframe and the doors layer beyond continue to shrink. All beeping merges into one giant beep. All shades of white merge into one shade.

The doorframe fills in with a solid, smooth, white wall as the features of all the other doors flatten into a perfectly smooth surface.

The door-singularity – a solid sphere-shaped mass of billions of doors – has arrived at the threshold of the doorway into the 6-door room you just passed through.

You turn to face a familiar room, now with one less door to choose from than before.

A few moments pass.

Suddenly the room begins to shake, and the crunching begins.

The crunching is not coming from the wall pressed against the white singularity.

It is instead the *walls touching that wall* beginning to fail and fold like an accordion: The floor, the ceiling, and the walls on your left and right buckle and crumble where they meet the wall pressed against the singularity.

The room seems to be slowly collapsing into the singularity you just escaped.

There are currently five doors remaining to choose from in this room.

One is out of reach.

One is locked.

Three remain possible to choose. Left, right, and straight ahead.

What is your choice?

Why?

The two doors that opened into sensory blackness before now open into the same space as the collapsing doors. They each offer you a view of a kind of "hallway" between the two layers of doors that are narrowing as they collapse inward. One of those layers is the surface of the solid white singularity. It is very close. You can almost reach out and touch it from the open doorway.

Assume the space between the singularity and the nearest layer of doors collapsing toward that singularity are separated by about the height of your body (with each passing moment closing this distance). Would you exit the 6-door room through this side door and enter the ever-narrowing hallway of doors to attempt entering a different room or space? Why? What are you hoping to find?

Are you just seeking to escape the room you are in?

Do you stay where you are?

Do you take the only door remaining in your room that feels safe to take?

Do you wait for the room to continue to crunch until the ceiling and locked floor doors are broken, exposing the other side?

Perhaps you will consider taking the door that seems furthest away from the singularity? The door straight across from the one you entered?

Better hurry, you are about to lose four more doors as options in the time it takes for one more breath.

Will you enter the door furthest away from all this?

Will you slam it behind you?

Will you panic as that doorway begins to fill into a solid white wall just like the previous door as it presses against the singularity? As you realize you have no doors remaining to choose?

I assume you will.

You have returned to the no-door room.

You hold your breath and wait for a crunching sound to begin in this room as well, but the walls do not begin collapsing like they did in the 6-door room. Nothing changes.

Time begins to pass.

How long can you tolerate this place?

How long would it take you to become uncomfortable within it? To want out?

How much time would need to pass until you tried to scratch on the walls, break through them, or cry out for help?

Why do you think you would do this? *You have everything you need to survive.*

What determines how long you can tolerate this space?

In the beginning, you wanted out of this room with no doors and no windows. It was confining.

In the end, you wanted back into this room. It was safety.

And yet you already know you will want out of it again.

You have fallen from the heaven of unlimited choice into the hell of none at all.

This is the paradox of shelter.

This is the paradox of life.

Life is not all or nothing.

Everything is a balance between everything and nothing.

Time continues to pass.

Your story system is close to collapsing one of the 3 steps.
The justification of self-destruction looms nearer.

One wall of this room suddenly vanishes.

Beyond the room is a featureless all-white *planet*. In the distance, the white horizon collides with the solid black sky in a near perfect line.

Will you step out onto the surface?

I think it's safe to assume you will. Are you going to stay in the room for eternity instead?

As soon as you make contact with the white surface, color and many other sensory experiences spread from that point of contact in every direction at incredible speed. The straight-line horizon warps and flexes. Everything quickly forms into something very familiar for you.

You have just returned to your first moment in *The World of You*.

The room behind you vanishes.
You have returned to ground.
It's time to start building together.
We've got an abyss to climb out from.

PART FOUR - EXPANSION

Falling is easy.

It's getting back up that's hard.

The Forced Choice

With no movement in reality except your own — *The Ark* — you are free to take all the time in the world to make your decisions. Time doesn't matter. Your story system can imagine opening all billion doors one-by-one at a speed you are familiar with. This string of actions might take anywhere from 158 years to maybe a half-million years to complete, depending on your chosen speed.

You might even expect to learn the *whole story* of *The Ark* by opening every door. It's possible you are not prepared to pass through any door again until you have opened all of them. Unfortunately, this situation is *not* the reality you live in.

Objective time, which is movement that exists *outside* of your own justified movements, does something truly terrible to your stories — it erases many of them to ever be possible.

The Ark will *never* be a reality you can experience firsthand as a human being. You will always be forced to make decisions without all the information. So will I. You will never have the chance to be certain of what is right and wrong. Neither will I. Our reality as human beings is the story of *Gravity*. No one knows the whole story about anything. No human has had the time to explore and understand all the possibilities.

At the end of *Gravity*, a wall disappeared from the room with no windows and no doors, and you stepped out to find yourself returned to *The World of You*. Color exploded in every direction from the world of white doors and black sky to become someplace with far more familiar movements, colors, sounds, smells, tastes, and feelings, providing you a strong and well-validated set of expectations.

Return yourself back to this point in *The World of You* once again. Look up from this book and all around you. Imagine all humans and all human-made things gone. Your surroundings return to the wilderness. Take your time, and really try to re-ground yourself to this imagined place, built by your past experiences merging with your current moment.

Where are you right now? A hill? A valley?

What do you see? Trees? Grass? Sand?

What do you hear? Birds? Insects?

Are there any odors or tastes in the air?

What do you feel on your skin and hair? Is there wind? Rain? Snow?

Don't turn the page until you can imagine *The World of You* clearly. Stay in this imagined place as long as you need until you have a good feel for it again.

Now that you have returned yourself to *The World of You*, pause all time and movement, including yourself.

Now take a step outside your paused body. You can do this, because this is your imagination. Second You is fully capable of imagining itself looking back at First You using remembered reflections or recorded images of yourself from sometime in the past.

And so, there you are – take a long look – you are naked, surrounded by the all-wild planet with no other human or human-made thing anywhere in existence.

Your body is no longer in *The Ark*. It can now experience hunger and thirst again. It is going to need sleep too, and there's a good chance it is going to need shelter sooner rather than later.

You've only been in this imagined world for a fraction of a second. If this were the first time you entered *The World of You*, you wouldn't have had enough time to recognize your first need from this place yet.

But even in this tiny sliver of time before everything was paused, you already made the most fundamental choice possible for a human being, and you didn't even realize it. It is a choice made every moment of every day by every person. Choosing not to make this decision is not an option.

The existence of objective time, which you cannot actually pause, *forces* this decision to be made over and over again and again.

The speed it repeats is frankly a bit tough to comprehend. In this sentence alone, you made this decision *at least* twelve times.

The choice being made is the end result of your story system finishing Steps 1, 2, and 3: Justification for action.

After pausing *The World of You* almost immediately after it began, it might feel as if there are only two choices related to justifying your actions: Act, or don't act. This leaves something critical in a hidden gap, though.

Acting or not acting as the only choice is treating yourself like some kind of lemming. Like the direction and intent of whatever action you justify *does not matter*. Like you act for the sake of acting, and only decide to act as a yes-or-no decision.

There is already very much a reason to act, and you've already experienced that reason the first time you entered this place. It was apparent in your answer to the question I haven't asked yet: *What is the first thing you need?*

Your previous answer was a word. That word was itself a justified action and the *end* result of your story system dealing with the fundamental forced choice. There is no *act* or *not act*. There is only the reason – the story - of *why* you choose to act. Everything you do is *always* justified, whether you justify it consciously or subconsciously.

You already know you will need food, water, and shelter because you've been through this moment before. However, I suspect the first time through, you paused for quite a bit when asked what you needed. *Why* do you think that happened?

In my opinion, this happened because chances are very high that in your real life outside this book, you were seeking things behind invisible doors that weren't even remotely related to food, water, or shelter. You might have been trying to find "promotion", "paycheck", "lottery win", "retirement", "new car", "attractiveness", "fame", "success", "happiness," "being right," or a billion other possibilities.

After the question was asked, your story system was forced to re-ground to a truth and reality *far* away from your reality while reading this, which I assume is not even close to the most primitive your life can *actually* become.

Time was required to float your truth-story anchors and drop them all the way back to the *very bottom*. All the way down to the foundation of you – to your ground state. *Once there, you were forced to make a choice.*

This forced choice, which was made in the very first moment of your *new normal*, was not a binary one-or-the-other decision. It was not a case of opposites.

Instead, there were actually *three* possibilities to choose from, and you're choosing one of them again and again even now.

Each of these three options represents one of the directions any process takes relative to itself. It's the same possibilities of change that existed in the balloon system. They might be the only consistent choices spanning the entire universe, which includes all things human, because: *Everything is in the process of changing relative to everything else.*

Your choices are to expand, to stabilize, or to collapse.

You have already aligned yourself with your forced decision from this choice in *The World of You* even in this split second that passed before time stopped. In a way, this decision is your deepest *bias* – the general direction your subconscious wants to take.

I suspect you are curious what exactly I think you are choosing to expand, stabilize, or collapse with this forced choice of personal bias. *It is your existence.*

From your point of view, your existence is *your experience in this moment.*

Existence *is* firsthand experience.

The ever-changing moment all around you takes place in the master context of gapless, objective time, which is one of the very few things all humans share as I write these words. We are all anchored to the Earth, which means we are all experiencing *approximately* the same overall rate of time.

This reality of every human existing in one time-frame - one *balloon system* - in this universe is the only reason I can call it "objective" time as I write this. Should humans ever live on different planets or somewhere in the cosmos away from the Earth, human time will change to become objectively *relative*, because we will have finally found a straw-path into entirely different balloon systems of time and movement.

I believe we will ultimately be ready for that potential future, because you are dealing with relative time pretty easily in *The World of You*. You experience and even control *relative* time in your imagination, because you hold an enormous amount of wordless *moving* details inside your memory and merge them with your current experience.

And so, that is where you are now. You have arrived at the very beginning of your time and movement in this new imagined place – as a single point in *The World of You*. All of your experiences that took place and built your truth before reading this book have just crashed down and re-anchored to this imagined reality, and are ready to start compiling all over again on the lowest ground it can imagine to exist.

From your starting point in *The World of You*, you've likely just experienced a nearly complete collapse of usable context. Your floating anchor of "normal" has reset itself. Truth as you know it just rearranged around a new "ground" provided by incoming sensory currents merged with imagined sensory currents.

After this re-anchoring, your answer to my question *what is the first thing you need* exposes a lot your feelings about your own existence. Your answer, whatever it was, was ultimately selected by which of the three paths you chose when it comes to the forced-choice of your actions in unstoppable objective time.

You chose to *expand* your time remaining to exist.

You chose to *stabilize* your time remaining to exist.

You chose to *collapse* your time remaining to exist.

If we add your action-timeline underneath the single point of you, then at this paused moment in *The World of You*, you are somewhere above that timeline because you are not dead.

From this position, each of these three "straw" path possibilities represent an ideal and the absolute limits of different directions you can take with the story of you.

The path of expansion points straight up away from your timeline.

The path of collapse points straight down toward your timeline.

The path of stabilization points exactly parallel to your timeline, and in the same direction.

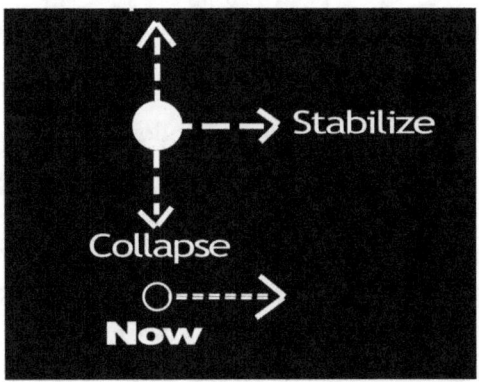

Which path are you on *right now* as you read this book? It may surprise you that I can actually know which you are consistently choosing this from back here in the past without even knowing who you are. Here's how:

You are not seeking to expand your time to exist by choosing to read. So the path straight up is not your current choice.

You are also not seeking to collapse your time, because this would mean you are trying to end your experience right now. If this were true, the act of reading this sentence is good evidence you have at least temporarily chosen a different path.

This leaves one path, and I feel very safe assuming you are on it right now. It is the path of stability.

Choosing to be on this path means your story system must feel pretty far removed from predicting any story of your immediate death. In fact, you must feel so far removed from your own harm or death, you determined using your time to read was the right thing to do. For reading to be right thing to do, it must also be true you are not feeling extreme thirst or extreme hunger right now, and it's probably safe to assume your shelter need is currently fulfilled.

After all, if you were surrounded by a pack of wolves, outside in a thunderstorm, or had an angry mob beating down your door, I don't believe you'd choose to make reading this sentence your highest priority. The words on these pages can't actually overwrite your survival programming. Shelter, and the reduction of fear and increased feeling of safety that come with it, must be in place for you to be here, reading these words to create new secondhand experiences.

As I see it, it is easy – *very* easy - to become completely blind to this "stable" path, which you are on right now, and end up in a binary this-or-that belief system. If this happens at this fundamental level, you end up trapped in a story where every choice you make is between life and death, black and white, yes and no, right or wrong, and there is nothing in between.

For example, in *The World of You*, should you decide to act toward *expanding* your time to exist by seeking food, water, and shelter, everything that follows that moment becomes easy to think about through a this-or-that lens. *I'm not expanding, that means I'm collapsing!*

In my opinion, this is *why* many people I have put through this thought experiment jump to "I don't have a chance," or "I wouldn't even make it through one night." Their timeline's *predicted* length took a major hit, and these individuals then skip over all the time between this new moment and the end of their need story arcs – death. Perhaps you even did the same. You might even be doing it right now. Telling yourself stories like *I can't survive in this place*.

If you did this, you might not be wrong, but you are also not being fair to yourself. You are posing this entire situation as immediate life-or-death and are *excluding* the middle completely. You are letting it fall into the abyss. Your entire reality gets framed by a predicted final destination *that has not yet arrived.* That destination exists at the very limit of the system of you. You have never actually experienced death before. It's possible you have never even come close to it before. This often doesn't seem to matter to your story system, but it is not the reality of your situation. *You are not actually a binary being.*

The only thing binary about your existence is your two end points: Your beginning and your eventual end. This moment, as you read this and also imagine yourself in *The World of You*, is *neither* of these points. You are the analog *between them.*

Your existence – your story - *is the middle.*

You are literally the spectrum of possibilities between your starting point of one and your ending point of zero. You and you alone fill up the space inside the boundaries of your existence. These boundaries are very real and easy to define. They are everything between the top-most and bottom-most, left-most and right-most, and front-most and back-most edges of your skin and hair. You are the three dimensions of physical space exactly in the shape of your body, and the movements you justify through the fourth dimension of objective time. You are no more, and you are *no less.*

Inside this boundary of you, your body is its own spectrum of possibilities, too. It is a massive collection of biological systems running The Process in an uncountable number of directions and shapes and scales, all contained inside one massive super-structure called *you.* A million-billion individual life forms are working together to make *you* possible. They work to keep things as balanced as they can between each other and still maintain their own boundaries.

Every process of you needs energy, which is required to keep each system of you moving. This energy is needed to *create* new imbalances.

Your cells need energy, the electric chemical pulses in your brain require energy, your beating heart needs energy, and your expanding lungs need energy. This need goes on and on in recurring scales downward, all the way to the simplest and smallest processes in your body. All of them require energy to "blow up the balloons" in the first place so movement – or *action* - can occur at all.

Over time, the straw-paths you call nerves will actually change shape, size, and their individual connections to find a better balance with the justification process for actions you repeat most often or don't use at all.

Due to this constant movement inside you that is beyond your direct control, you, as the master process and authority over this enormous collection of life forms, *are required* to consume chemicals in the world around you for energy. You need to eat and drink to fuel the expanding, collapsing, and stabilizing processes that make you....*you.*

If you do not provide what these processes need - if you, as ruler of your body, decide to declare the unfair law that "all food and water are now illegal," these processes will enter survival mode. They will be forced to try to continue their movements by creating imbalances *without* an energy source from outside. They will begin to consume each other, beginning with the systems that have the weakest and least-resistant boundaries.

In other words, without food and water, these processes begin to consume *you* one process failure at a time. If you do not change this unfair "law" you've created by refusing to eat or drink, you will not continue to exist. The bacteria and other life forms already inside your body that do not depend on you consuming things to survive will continue living in your body even after you are dead, and they will celebrate and feast on your defeat. They will crown a new "king" or "queen".

And so, you must consume.

As soon as time unpauses in *The World of You*, which it soon will, this ground state of keep-consuming-to-create-imbalance-to-create-fuel-more-movement will begin to *warp* each of these three pathways of personal bias toward your own existence.

In *The World of You*, you can't explore all the possibilities in every direction free of time, like after you realized you could fly in *The Ark*. Instead, a billion invisible "doors" in every direction are about to begin moving toward you as soon as you unpause reality.

This means many possibilities will be erased, and priorities will be critical.

83

Choice Decay

When time and movement begin in *The World of You*, you will experience the destruction of nearly every story and possibility that existed, unlike if it were a timeless place like *The Ark*.

The stories that require you to physically be in two places at the same time become impossible – you are one point - so these options of being in a second point will be erased. This includes every story of you that begins anywhere else currently located outside the boundary of your skin. These points are separated from you *by time*. You can't restart your actual reality somewhere else just because you don't like your current story. This even includes the story of your death. From this time-stopped position, even your death is separated from you by time. If time doesn't exist, neither does dying.

This changes the forced choice arrows that were pointing straight up and straight down.

You cannot expand your existence by somehow performing actions without also experiencing time. The more you act, the more time must pass. This means the path pointing straight up cannot actually point straight up. The choice to expand must include time, which points to the right, so the arrow must point right and up at the same time – diagonal.

You also cannot collapse your existence without experiencing time - which would be required to move straight down. The more you must act, the more time must pass. The arrow pointing straight down must point down and to the right at the same time. (For the sake of not having black-background images from here on out, I'll switch to having a white background again.)

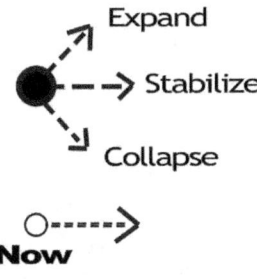

But even the middle path, stabilizing your existence, has a problem. Stability relative to objective time would be to experience time as not a threat to overcome, but as a gift to explore. This would create a reality like the billion-door space in *The Ark* when you could fly. Your self-consumption through time prevents this.

Another way to say this is your *decay* prevents this possibility.

Decay, or self-consumption, bends each of these three straight-line paths in the same direction: *Down*.

This means all of these paths eventually *result in your death* thanks to the nature of your existence: Some part of you is always moving and changing even if you are not making the choice for it to happen. The million-billion processes of you are mostly automatic, invisible, and uncontrollable from your perspective.

Second You, which you *do* control, can override some of these processes, such as breathing, but it cannot force them to *end* without other movements. For example, if you use Second You to stop breathing, you will pass out and wake up to First You breathing for you again.

You already *exist*.

You had no choice in the matter.

Your default setting is *alive*.

The process of you is already *in motion*.

This is how you *start* from your own perspective - above your timeline.

The result of this whole decay situation is a limited-time offer. You have a choice, *while supplies last*. If enough time passes without this forced-choice being made, all your stories become the same story: The story of the collapse of you.

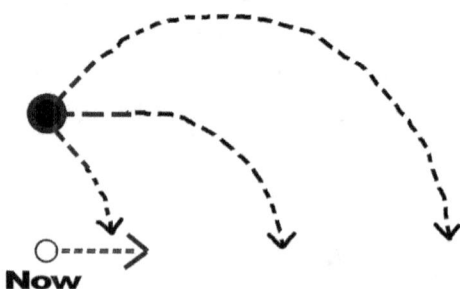

Now

This means after enough time passes, there will be only two fundamental paths of bias to choose from in *The World of You*. They will appear to be *opposites* from your perspective if that moment arrives. A binary choice.

It is only at *this* moment, right now, long before that collapse to binary happens, that you can see the limits of choice and realize you are between them. You can see they are both just different directions for the same story getting warped into arcs by time and the processes running inside of the boundary of you.

So, with this said, you can now *consciously* make the forced fundamental choice you have likely made *unconsciously* your entire life a million-billion times over.

Starting from the first moment in this thought experiment, you can use the time you have remaining on the middle path to move toward the path of expanding your time, or use your time on middle path to move toward the path of collapsing your time to a quicker end.

Justify movement toward staying "one," or move toward becoming "zero."

Try to live longer, or try to die quicker.

As soon as *The World of You* unpauses, your need story arcs will likely be completely ignorable or near panic-level priority. My vote is on the latter. After all, you were on the stable path, and you were so stable you could spend time reading instead of worrying about food, water, or shelter. Now, you may not have the slightest idea where to find and get food, water, or shelter.

This collapse of your stories may have been overwhelming, and shot you toward believing you cannot prevent your own total collapse to death.

If you believe this, just know you created a time gap.

You jumped *way out* from the present. You predicted a future certainty that has not yet happened simply because you are missing stories in your past capable of helping you in this place in the present. Pull yourself back.

Pull back to *this moment.*

We are all going to die. I might already be dead as you read this.

But *you* are not dead.

Not yet.

You are on the middle path. It's the only path you will ever experience in your entire life. The path of expansion and the path of collapse are not different paths, they are upper and lower limits of the middle path.

This means, in *The World of You*, you have from *right now* until however long it takes you to die from your most urgent need story arc remaining unfulfilled.

Perhaps that will be three days.

Perhaps it will be one day.

Perhaps it will be 8 minutes.

But it is not *now.*

You do not die *right now*. You *cannot* die right now.

Not in this all-wild world absent of all humans.

As you find the shape of Step 1's boundary in this imagined-but-paused place – whatever environment might be before you – your story system immediately begins running at full capacity. It doesn't matter if this place is an enclosed densely packed forest that blocks your view out of your sensory windows in nearly every direction. It doesn't matter if it is an open plain that blocks almost nothing from view but leaves lots of blank space. It doesn't matter if it is a desert that requires an *immediate* boundary in the form of shelter. Your mind will run *at peak*. Your senses and imagination are wide open.

Once time and motion begin from this paused and timeless moment, you will generate a full lightning storm forged from all the possible billion stories within Step 1's boundary and then project what is *assumed* to be beyond that high-resolution boundary using Steps 2 and 3. All of these projected paths will be created from what you have experienced directly and secondhand with the all-natural humanless world throughout your entire life, which might be a lot, or perhaps very little, or maybe none at all.

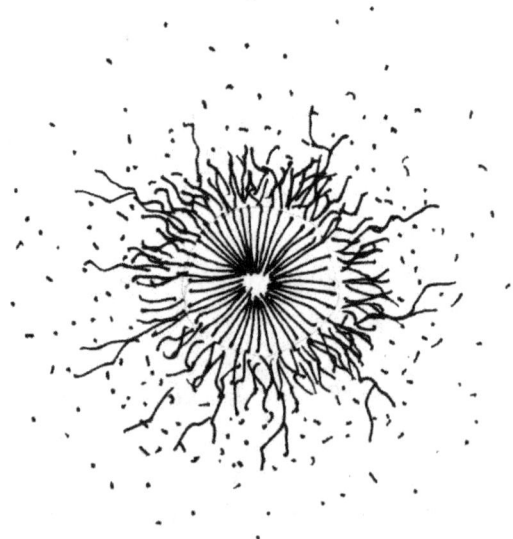

And now the moment arrives: Hit "play" on *The World of You*.

Your incoming sensory currents start to dance in your mind as movement at scales beyond comprehension start flowing in through your current-creating sensory windows at all kinds of intensities. Your story system floods with pulse after pulse of unique current signatures "showing" you what is true in every direction.

And then a voice from nowhere asks "What is the first thing you need?" and the lightning storm redirects all its projective energy into determining one, single, future priority.

What do I need?

Food.

Water.

Shelter.

I have no doubt you will struggle to appreciate the three days, one day, 8 minutes, or whatever time you might have left on the middle path without these things, because you have experienced a different truth in your past: You probably *expected* to have months, years, or even decades left before you ended up in a situation anything like this.

In my opinion, your inability to imagine the stories and actions you need to successfully re-expand your time remaining to match the time you feel you have remaining right now as you read this is what creates the *I wouldn't last a week* feeling.

Give yourself more credit.

Be fair to yourself.

You have a team of two connected time-warping survival super-machines inside that head of yours. Together, they have been enough to bring all of humanity from somewhere near this point in a very real *The World of You* to wherever you are today re-experiencing this point in your imagination thousands of years later.

You are built for this.

84

The Darkest Direction

I would be a fool to continue and ignore the path of collapse possibility.

I cannot assume you will make the conscious forced choice to try to stay on the middle path. I have to accept that it is entirely possible you have already been debating the path of collapse, and being thrust into *The World of You* only destroys every story you might have had left to want to continue your existence.

If this is the case for you, I am not about to send you words from the past to tell you everything is going to be rainbows and puppy dogs if you would just stop thinking of ending your life. I will not disrespect your truth in such a way. If you are debating justifying an action into existence that ends your existence – if you seriously consider answering the Call of the Void each and every day – then who am I to say you are wrong to be doing such a thing? You wouldn't be considering it if you believed it was wrong to do so, and I can't possibly know your story from back here in the past. I can't know your specific situation.

But I *can* know the *place* that is calling to you.

It once called to me as well.

There was a point in my life where I found myself staring into that same void each and every day.

I've even considered taking that darkest of paths – toward the eternal blackness of the unknown. Toward death. The stories I told myself about it started to seem better than the unpredictable hell I seemed to be forced to face each and every day by continuing to exist.

Most of my future stories started to fall into that black hole, and I suspect for those individuals that ultimately choose this path, (and I am sad to say the number of individuals I have known to take this path is higher than I'd like), the Hydra from Hell probably continues devouring all his or her stories until the void itself is the only story remaining. It *calls* and begs for the individual to justify taking the darkest path.

I have no doubt that life is hard for you. I dare not downplay whatever your situation might be. I accept it might seem like your suffering will have no end, and perhaps the darkness, or whatever story comes next after entering that darkness, might seem like it must be better than *this* story.

If this is how you feel, then I hope the slow destruction of certainty so far has started to give you hope. No one can know the future with any certainty at all. No one knows *your* future, either. *Not even you.* No one knows the whole story about anything.

There is no guarantee your suffering will continue for the rest of your days. Time and movement can do strange things, and they might change your situation in an instant. You cannot *depend* on this happening, though. You can't *expect* the universe to magically turn itself around and hand you a better path in life on a silver platter, and you can't expect anyone else to do this for you either. The choice to stay or leave this place is yours and yours alone. However, I want you to consider something you have perhaps not considered before:

You are not the only one staring into the void each day. You are not truly alone as you face the Hydra in your head telling you everything is going to be awful every single day for the rest of your life. There are others out there just like you. Suffering, just like you – right now.

Imagine the world as an enormous polished black orb, but only you can see it this way. If you are thinking of taking the path of collapse downward, then you are looking down at the darkness of this orb known as Death.

Now imagine, as you look down upon it, this huge orb becomes completely transparent, along with everything else that is not a human being. The only thing you can now see or hear is other people as they move about their day all around the world. You can see some people moving quickly because they are in now-unseen cars or planes. Some are sitting still, perhaps toiling away at their desk or job. Others seem to be working with intensity, perhaps constructing something or another hard-labor job. Maybe you can see a birthday party or another social event, or large groups of kids moving here and there as they move about in their invisible school.

Everyone you see is moving about through their daily lives, unaware that you can see them.

As you continue to look around at all the movement, someone in particular catches your eye straight across the orb from you. This person is still, and appears to be looking right at you. That is not what is actually happening though. This person is actually looking into the void. Just. Like. You.

Look around carefully. I think you will find many people all over this planet gazing into the transparent-for-you black orb of death. Each of them thinking of taking the same intentional path of collapse you might be thinking of taking each day. *You are not alone in this feeling.*

I am not going to tell you what you should and should not do. That choice belongs to you and you alone. It is your life. I am not the Keeper of Your Truth. I can't even know who you are. But I do know that you are ultimately responsible for justifying all of your actions and the consequences they bring. Perhaps these are the very reasons you find yourself listening to the call. I can't know.

I can know the path of collapse will *always* be an option. No one can actually take this option away from you – not if you are in the position right now to have the freedom to be reading this. Death and the path of collapse is ultimately the only option any of us have. It's not going anywhere.

Every single one of those people you see moving around will eventually go into the void you find yourself gazing into. But we all only go into it *once*. As best we can tell, the void of death has a doorway that you can never pass back through after entering.

With this being said, here's what I would like to propose, if you find it agreeable and fair.

I would like for you to turn your back on the black orb of death, *for now*.

I would like for you to then lean back against it, and try to make peace with your feelings about death. Try to *accept* that you have or are seriously considered taking your own life. Try to accept that you can hear the void calling to you at this point in your life. Try to accept that it will always remain your choice as to when or how you might answer that call. There is no shame in any of this. In fact, I feel the exact opposite of shame now lies before you.

As you face upward with your back against the void of death – I want you to know every movement you justify to move forward on the middle path has the potential to change not just you, but *everything*.

This is not some generic mantra story I want you to tell yourself. I'm not going to sit here and claim "*you just have to believe in yourself, and you will change the world.*" I'm not trying to blow sunshine up your butt and make everything sparkles and confetti. This is about something entirely different. This is about *perspective*.

You are facing away from the darkness, but *you know what it looks like*. It is in your memory. It is in your story system and thus included in your stories. This allows you to recognize when others are gazing into it and hearing its call, perhaps for the first time in their entire life.

You know from firsthand wordless experience what it feels like to take the forced-choice of existence down to its absolute minimum: A single, solitary moment of deciding what is right based on what seems to be true, and then finding yourself one single justification away from death itself. You have potentially taken the middle path down to the shortest it can *ever* be for a human being. This gives you an unparalleled range of perception.

You can recognize *fear* outside of words.

You can recognize *despair* outside of words.

You can recognize *suffering* outside of words.

You have direct firsthand *wordless* experience with *them all*.

And you can rest assured you are not alone.

If you are debating ending your own experience so strongly that being cast into *The World of You* would push you into imagining the justification to end your timeline into action, then back in reality while you read this book, you are very near the point where you have *nothing to lose and everything to gain*.

I only ask you to resist the call of the void, for now. Death is not going anywhere. Consider making the choice to expand your time to exist another minute, another hour, or another day at a time. Know that there are others out there that might be choosing to turn their backs on death right now as well, and those individuals are now gazing out into the same universe as you. *You are not alone.*

It is my belief the world needs you more than you can know.

After all, if this world can be turned right side up – and I believe it can – many that do not understand what is happening will begin to have a specific wordless look about them, and you are one of the rare few that can recognize it for what it is, and make the choice to help – should you find it justifiable.

You will meet your end one day. You will eventually enter that void that I hope is against your back right now.

But until that day, you are here, and you have the capacity to change everything.

Keep climbing with me and see how you feel when we are done.

We still have quite a way to go yet, and the darkness may yet give way to light.

85

Destroying Infinity

You *will* die at the end of your story.

It is the one certainty about your future.

It is inevitable. You cannot stop it or prevent it from happening, and no secondhand story can tell you with any certainty what your firsthand experience will be passing through the doorway to your death.

Perhaps one of the hundreds of secondhand stories about the other side of death will turn out to be true. Perhaps death will be nothing at all – just like the first year of your life. The subjective truth of this story will remain uncertain to you until you experience it firsthand. No one alive knows this story.

If I assume you are trying to stay on the middle path of life as long as you can, then your story system is working tirelessly to ensure your ending is not predictable or immediate at any given moment. This would mean your mind's *main workload* is keeping the story of your future death as unimaginable as possible. It *resists* the idea as much as it can. Death becomes a kind of anti-goal.

And yet, no one has ever made it out alive at the time I wrote this. I suspect this has not changed for humanity at whatever time you are reading this.

You can imagine death away all you wish.

You can tell yourself it won't happen to you.

Death does not care; it arrives for you and everyone else just the same.

Believing your life story can be prevented from coming to an end results in a mental goal that can never actually be fulfilled. This impossible goal is all too easy to set though, especially during your first naked-in-the-raw-wilderness moment in *The World of You*. Death is near. Perhaps closer than it has ever been for you before.

So in a kind of mental frenzy, you begin pulling from your past to determine what "right" action should be justified to build the future story where you don't die in the next few hours or days.

The speed of your personal time – set by how quickly your own processes bend your straight-line paths of forever-expansion or forever-stability downward into arcs – can be noticed by how quickly thirst, hunger, and fear re-appear in time.

By constantly working to solve this time-based need problem as quickly and far into the future as possible, your mind keeps the story of your own death completely uncertain. By keeping death *uncertain*, the story of your continued existence and experience feels *more certain*.

In my opinion, this is the entire basis underneath "doomsday prepping". If you have an enormous two-year stash of non-perishable food and hundreds of gallons of water tucked away in your underground bomb-proof shelter that you have completely paid off, you likely feel much more in control of avoiding death in the case of an apocalypse-level event. Well, for two years, anyway.

But this line of thinking implies that your mind's *entire existence* is to problem-solve fulfilling your survival needs simply to avoid death. To say life and thinking is simply an act of perpetually avoiding a particular story is not really useful, and I don't think it's perfectly accurate, because it fell apart in *The Ark* – your mind continued to justify actions even with all your needs and decay into death removed.

So let's bring an idea from *The Ark* into *The World of You*. What if we float the anchor of death? Reset reality without the idea of it? What if you began *The World of You*, but your knowledge and memories were *completely erased* so you would have no concept or story of your own mortality?

Imagine restarting *The World of You* with no memories at all. You can still function and move as you always have, but you have no idea *where* you are, *who* you are, *when* this is, *how* you got here, or anything else at all.

You have never witnessed another person, so you certainly have never witnessed another person die. You have no idea how long your lifespan should be, or have any awareness at all of all the things that can go wrong or kill you.

I'm not convinced you would have a fear of death in this situation. Death would be an idea that cannot be imagined or understood. Yet I believe you will still feel a need to seek *something* when thirsty and *something* when hungry and *something* when fearful. Why?

Actually, let's question this one level deeper: If your mind were erased, would you actually experience fear in *The World of You*?

In my opinion: Not at first.

Memory-wiped You might quickly become aware of death's existence – it is all around you in nature. Animals kill other animals. Plants die, rot, and fall over. Animals die of disease or old age. Decay exists everywhere.

I think you would notice and remember that an animal does not come back to life after being eaten. A rotting corpse or skeleton does not suddenly become healthy and get back up again. And I suspect in this situation you would truly see yourself as "just another animal" in the world, and begin to apply the idea of death and harm you see around you to yourself, even though you have never experienced it yourself, and have no proof it will or can even happen to you.

Yet you know that the wolf you witnessed kill the rabbit could potentially try to do the same to you. I believe you will make this connection because you will notice that other life in the world, such as the wolf, notice *you*.

But of course, *Memory-wiped You* would have to experience these wordless stories of death through your own sensory windows to be able to apply them to yourself. And I believe you will *still* be driven to drink and eat and shelter yourself, even though you may have no idea how to do these things. I believe these justifications are built into your biology and genes. I suspect they are hard-wired by the million-billion processes inside the boundary of your skin which depend on you for their own survival.

Perhaps good evidence for this hard-wired story being true is that a baby typically cries when it is hungry or becomes frightened, and knows how to feed from a nipple without ever getting instructions. A newborn typically knows how to use smell to root and even *move* toward its mother's nipple from the very first moment of life. However, that newborn baby is certainly not yet familiar with *death*, or even the outside world.

This means the core need of your mind today cannot be avoiding death or have anything to do with death. *Your own death is always going to be a secondhand story.* Instead, I believe your core mental need is simply striving to achieve greater *certainty* of a future where you continue to exist. You do this by predicting and trying to *resist* entering stories you might have *experienced* firsthand before: Extreme hunger, extreme thirst, and extreme fear. Your story system is always trying to move you *away* and keep you away from these feelings to ensure its own survival.

But even this story is still an impossible anti-goal. If you achieve success, the result is an *eternally* long timeline. An eternally long life in this world *is not possible* in any way that we can measure to exist. This means even this "goal" is *always* out of reach. It is *an infinity*.

Your story system doesn't care. It *always* wants greater certainty you can keep obtaining food, water, and shelter in a timely manner and never experience hunger, thirst, or fear.

It knows if you prioritize these goals, and guarantee them in your future, you can experience comfort and rest and seek pleasure however you want. You will be free to choose as you please. You will have arrived at the reality-based version of *The Ark*.

But human experience in this world isn't *The Ark*, it is *Gravity*.

Tomorrow is *never* guaranteed because it is not something *you* control. Uncertainty can *never* be eliminated at the deepest core level, because you are not the only thing in reality that is moving. Ever.

And yet, your story system *still* doesn't care. There is *no time* to think about this. There are more urgent problems in need of solving.

All those neurons, bioelectrical pulses, chemicals, and processes in your head are running as fast as they can to try to fulfill a goal they can never *actually* achieve. There are *always more doors collapsing toward you*. You cannot actually escape them.

If you think you can escape *Gravity* to arrive at billion-door space in *The Ark* in this human life, then it is very likely you are overlooking the existence of the stable-with-decay path completely. The middle becomes *excluded* to you. It disappears - becomes invisible - because there are *always* more doors coming toward you. You instead become door-obsessed, as you keep trying to prevent yourself from ever experiencing hunger, thirst, and fear.

You will justify all your movements based on future truth-stories you believe are coming but want to avoid, and as a result, you start trying to control the movements of everything else around you.

The time you have *between* doors of food, water, and shelter gets erased and falls into the void – your time in the middle becomes devoid of meaning. You keep trying to race past the middle and re-arrange or open as many doors collapsing toward you as possible. You have to be sure these three "boogeymen" of hunger, thirst, and fear are not nearby.

If this happens to you, you are left with *only* the path of expansion or the path of collapse in your life. You are thinking in straight lines: *I'm not riding the straight line up to eternal life, so that means I'm riding the straight line down to my immediate death.*

Binary thinking sets in. Your story system gets overwhelmed trying to get ahead of objective time and movement. You start endlessly working to avoid what is unavoidable – the present. There is no time to slow down. No time remaining to consider anything else. You are permanently stuck trying to exist in the future.

The disappearance of the stable middle path results in you *always* trying to bend your story arcs upward, with the darkest path as the only alternative if it ever bends another way.

If you never notice the time that exists *between* the doors – if you live in a constant state of anxiety about the next door - you might want to ask yourself if you are trying to *live forever* and haven't realized it. If you are, there is a very good chance those invisible doors you are focused on in your life are not actually labeled food, water, or shelter. You might be trying to escape, avoid, or control far more than hunger, thirst, and fear in an effort to avoid experiencing hunger, thirst, and fear.

There is a perfectly reasonable explanation for why this quickly gets out of control.

One of these three feelings was not present at birth. It appeared *after* you arrived in this world. That feeling is *fear*.

Fear leads to the inability to stop binary, black and white, straight-line thinking. *Fear* can make you overlook the growing distance between layers of doors, which means you have more and more time to experience existence *between* them and justify different and unique-to-you choices.

Unrecognized fear ultimately leads you to one place and one place only – *trying to control the entire universe*. Trying to create a *total, perfect* shelter in the present moment.

You have experienced total perfect shelter already.

It is a room with no doors and no windows.

This room *is* safety.

It is also a prison.

Control itself is a paradox – a loop trap.

I believe achieving control of the entire universe would almost certainly destroy you. It's not *actually* what you want. Achieving it would result in experiencing no new stories. Ever. Everything would always be exactly as you expect it to be. Emotions and feelings would vanish for you. There would be no difference between your future, current, and present moment. You would be the dog that finally caught *all* the cars.

I am nearly certain that as you thought about opening the billion doors of *The Ark* after discovering you could fly, you did not hope to find and enter the room with no windows and no doors. I highly doubt you would ever want to go back to this room by choice in the conditions of *The Ark*.

If you did enter it, I think you would quickly want out again.

There is a paradox here of epic proportions. It is an ultra-powerful loop trap in a hidden place. It exists between *fear* and *shelter*. This is the mother-of-all loop traps, as I see it. It also comes apart just the same as all the other loops.

Loop traps are always an illusion caused by a gap in a story being missed – something is hidden. Typically it is hidden by a certainty, and timeless certainties are the most powerful camouflage.

There is only one way to see something invisible. You must look at it from a different perspective.

For example, you might believe the story that straight lines exist. As far as I can tell, there is no evidence that straight lines actually exist in this universe *anywhere*. Our senses see and feel "approximate" straight lines all around us, but they are not actually straight at all when you use tools to magnify to a far greater resolution than our senses.

After all, if atoms are made up of "orbiting" particles or energy, then the outer boundary of this orbit is *a curve*, no matter how complex that orbit might be. According to everything I have learned, there is no evidence that square atoms exist. They are all "balloon systems". Some seem to have the appearance of six balloons with their knots tied together; some are only two, while others might have eight or three. There are no straight lines anywhere.

Traveling in a straight line is not possible either. There also seems to be no part of the universe untouched by gravity, which bends any straight line that might exist. It might take "zooming out" thousands of miles to notice the curve in a "straight line" at the resolution of our senses.

As far as I know, there is also no proof any infinity exist. A computer cannot run a program for infinite time. It will fail from decay before that happens. Numbers like pi cannot *actually* go on forever. Forever is not something that can be experienced. Infinity is a loop trap – a paradox.

We are merely stuck in the middle, and we cannot grasp the beginning or end of any story. No one knows the whole story about anything.

Take a deep breath.

Let go of everything you might be certain about.

Recollect all your memories related to untamed nature.

Time is un-paused in *The World of You*.

Let's start adding shape to our shared foundation.

86

Blowing Out the Side

Assume you entered *The World of You* after just eating a large meal and drinking a large amount of water. Your food and water story arcs now start at the same point, and will rise and fall at different rates, creating two different size arcs.

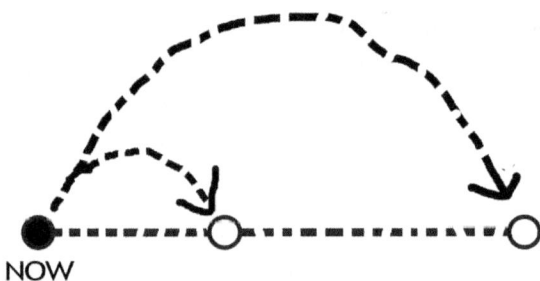

The smaller arc is your water arc, because you will need water again sooner than you will need food again. These two layers of doors are invisibly collapsing toward you in every direction, but the layer with all doors labeled "water" will reach you first.

You have excess energy to resist gravity because you just consumed both water and food. Your digestive system takes as long as possible to let these things through, because it wants to extract as much energy as possible out of every single bit you consume. You can think of food and water as balloons, and your digestive system as the straw that connects these balloons with smaller balloons that are then disconnected and moved around your body to fuel "you". The end result is time for you to exist without needing to consume again *right away*.

How much time? For you personally, it remains uncertain. For you as a human though, it is approximately three days.

Once you begin to experience thirst, you will be reminded your timeline is collapsing. If you do not satisfy this thirst, your body's processes will start to consume the water around them. Eventually too much water will be consumed from your own body, and some processes will begin to fail from the inability to keep creating imbalance without water. You then begin the process of dying from the inside out.

But what happens to this story arc if you find something to drink before these three days are up?

Well, for one thing, *you don't end up dying*. If you find something to drink in *The World of You*, your water story arc changes direction - from collapse to expansion - and your need to drink moves several hours or days further into the future. Your overall action timeline will then lengthen by approximately the same amount as your first water story arc.

This means if you have a plentiful source of water you can drink from in regular intervals, such as a burbling spring next to you or perhaps a large lake, your story system won't be able to predict your next experience of extreme thirst *at all*. Your thirst story arc end-point, which is the end of your water story arc, disappears from your story system's view along with the expectation of experiencing extreme thirst.

This is exactly how your mind wants this to be.

Your "anti-goal" goal related to extreme thirst seems to be met. But it isn't. It's not possible to achieve it for a very simple reason.

Even with an "endless" water source, your position on your water story arc is still always the present moment. Regardless of how much water you have available, your water story arc shape is *unchanged* because it is determined not by the world around you (though it can have an effect), but by the million-billion mini-processes that make you possible.

This means if you were to decide *not* drinking is the right thing to do as you experience extreme thirst, you will die *just the same as if you never found water at all* –lying next to a near endless supply of it.

Acts of consumption are *required* to stay on the middle path of life.

If you have an endless source of water, *and you are always willing to justify the action of drinking from it* as the right thing to do, you will *feel* much further away from extreme thirst, even though you have to keep drinking at the same frequency.

As I see this, it's extremely important to note that these two things put together - a water source and a willingness to drink it – can be recategorized into a *type of shelter*. These two things together are a shelter from extreme thirst, which is the marker that your time left to exist is collapsing to zero very soon.

You must make the conscious *choice* to drink to re-expand your time remaining to exist again. This reaches all the way down to that first fundamental forced choice: Are you seeking to expand your time, stabilize your time, or collapse your time to exist?

The time and movement that exist between the body of water and your mouth create an invisible boundary. As the balloon of you empties of water to keep the processes of you running, you must justify movements in time – a straw path - to connect back to the body of water in order to refill the balloon of you.

This requirement to justify specific actions just to survive finally gives us the label to the *vertical* axis on our graph. It is your perceived "mental freedom." In the case of water, it is the mental freedom to justify actions *other than drinking*.

If pointing your arrow straight up is only your personal movement without the presence of objective movement like in *The Ark*, and pointing it straight down is only objective movement without the presence of your personal movement like in *Gravity*, then performing the *required* movement of drinking does the equivalent of blending the two: It reverses *Gravity* by causing the invisible collapsing layer of doors to expand away, which allows you the time to open doors that are not labeled "water" – at least until the water doors start collapsing close to you again. In other words, by performing the required action of drinking, you can experience *The Ark* – temporarily.

If you assume you will repeatedly perform the action of drinking when you experience thirst, you end up with a graph that should look pretty familiar and very un-loop-like. It is repeating arcs. We often call these "waves."

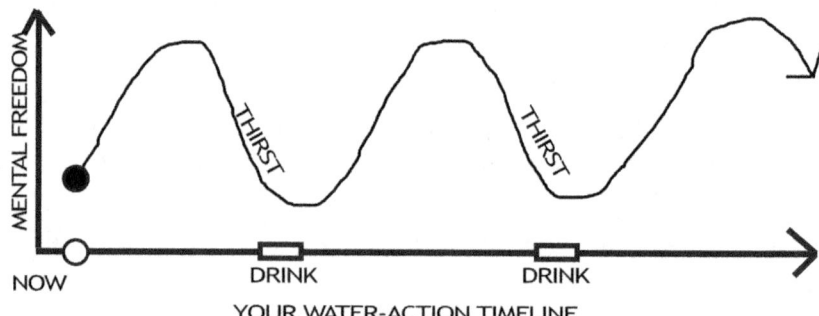

YOUR WATER-ACTION TIMELINE

This graph also reflects how time and movement are inseparable from each other and how speed and freedom of choice are linked.

Here's another way to think about it:

Imagine your *only* need to survive is to drink water. However, you must drink a mouthful of water *every three seconds* or you will die of dehydration.

In *The World of You*, even if you begin with a water source in front of you,

you still have almost no freedom at all. You are completely enslaved to your own water need. If you make any other choice of movement than drinking – you are likely to die because more than three seconds will pass. If you stop moving, you will also die because more than three seconds will pass. Your timeline will end and your movement will end because you died from lack of water. Your heart will stop squeezing, your lungs will stop inflating and deflating, and there will be no electrical activity in your mind and body.

As a human reading this, you hopefully don't have this severe of a problem – but you almost certainly do have a very *similar* problem.

You can live for hours on one mouthful of water. If you just took a drink, you have quite a bit of time to justify non-drinking movements until you must justify drinking again. Perhaps one of these non-drinking movements might be to move toward food.

Your food need story arc looks identical to water's, except it takes a much longer time to go up and down. At the start of *The World of You*, the "food" layer of collapsing invisible doors is approximately three weeks away from crushing you.

It is helpful that this arc is much longer, because the movements tied to eating typically take a bit longer to carry out than drinking. Picking up, biting, chewing, and swallowing food takes longer than simply sucking in and swallowing water. As a result, a bigger chunk of mental freedom is lost when it comes to food compared to water.

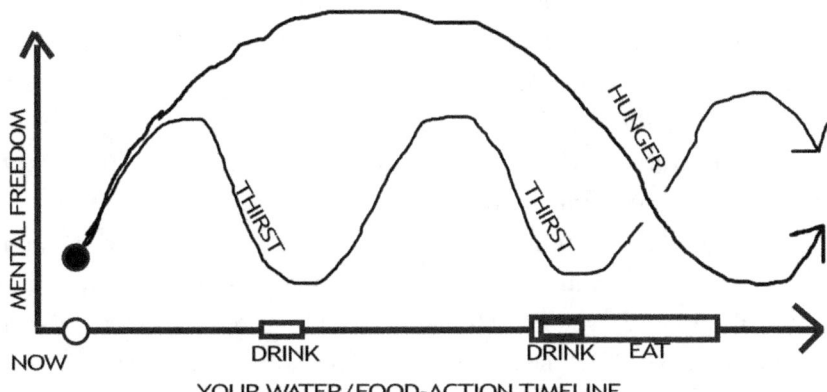

YOUR WATER/FOOD-ACTION TIMELINE

Together, these waves form what I've been calling your biological frequency, or minimum required life *speed*. Each individual's biological frequency is different.

Each time you restart an arc by sacrificing your mental freedom to perform the required action of eating or drinking, you extend your time to exist by approximately one more *wavelength*.

Each wavelength you add grows the overall *length* of the straw-path you call life, which connects the two balloon-points of your birth and your death. I'd go so far as to say you have added so much time and movement to your life today it's practically blowing out the side.

Here's what I mean: If it weren't for the existence of boundaries, *gravity* would have quickly connected these two points of life - emergence and extinction - long ago, and you and I wouldn't even exist.

Instead, if I assume you live on Earth like I do (or did if I am now gone), then the main straw you and I live in – the movement created by gravity between the balloons of the nothing and something – has become "clogged up" with somethings. These somethings are atoms. These atoms clog up the gravity straw because they can't pass through each other. They can't pass through each other because of their own boundaries.

This "clog" has essentially stopped time in what we as humans call the "downward" direction, and now life tumbles and stirs around on top of the clog, mostly moving *sideways*. Your existence and movements are quite literally "blowing out the side" of this huge gravity-straw path because most of your movements have nothing to do with "down." This comes at a small cost (or perhaps large depending on how you look at it): You must occasionally consume chemicals to create imbalance, which makes it possible for you to push "up" against the straw currents of gravity, and begin your sideways-journey.

The path of least resistance in this gravity straw would be for you to stay pinned to the ground, unable to move. You would simply be the next layer of atoms thickening the clog. Because we as humans can move though, our main straw-path that directs the currents of time is no longer "down", but "sideways" - the spin and wobble of the Earth itself.

You might think spinning-defined time implies we truly are moving in time-loops. This is the same illusion as the time-loops related to eating and drinking, but on a bigger scale. We broke that loop by adding your one-action-at-a-time straight time arrow. Well, the Earth has an action time-line as well:

The entire solar system is moving.

Imagine grabbing a white ball. Let's say this ball represents the Earth. Now imagine putting a black dot on that white ball. If you spin that ball on the ground, the dot appears to make loops like you might have imagined if you thought of spinning time loops.

However, the Earth is not a ball spinning on the ground. Instead, you need to spin the ball and *then throw it*. If you were to trace the path of that dot after throwing it, you won't get a loop at all - you will get a *spiral*.

All it costs to continue your spiraling existence is the loss of your freedom to whatever you want from time to time. Assuming, of course, that you can find sources – no wait, "re-"sources - that allow you to continue existing: Food and water. Without these two sources of re-energizing (restoring imbalance by reinflating your internal balloon systems), all choices on the middle path collapse to the most basic and black-and-white choice: Trying to stay alive or trying to die, which eventually collapses to no choice at all but to die – a point when all freedom to move is lost.

This is your current situation in *The World of You*, assuming you have no idea where food and water might be found and no experience or context to help you find them. You have no idea how long it's going to take you to find water and food. As a consequence of this, you have a very limited time remaining to choose your actions freely.

If you choose to die, you die.

If you choose to frolic and lie on the ground and watch the clouds pass by, it comes at the cost of losing some of the minimum time needed to find and consume resources, so that you can watch the clouds one wavelength longer.

The most efficient approach at this point of the story, assuming you choose to live, is to justify movement toward your needs right away. And this seems to be subconsciously recognized, because everyone that has ever entered this place answered "food," "water," and "shelter."

And so we resume climbing.

87

Knowledge and the Sheltopusik

I'm going to make two permanent changes to *The World of You*. These changes will remain in place for the rest of our journey together out of this abyss: You begin in an *unfamiliar* place, and you are *not* trained in survival techniques.

This new unfamiliar place is a hilly and forested part of the world. The place you imagined in this experiment before this point may (or may not) be similar to this, but making this change will save us time.

The only reason this saves time is because the details in the world all around you aren't actually what's important – they are mostly noise. It is instead the structure and boundaries present, and your justifications for movement *relative* to them that are important. If you need water, the only difference between a forest, a desert, or a bog is *how long* it may take you to find water – but your need for water *does not change*. Your need story arcs and my need story arcs are *approximately* the same in length, because we are both human. We are the same species. This is why using it as a common framework is possible at all.

Water

Lets assume you start out in this unfamiliar forest very thirsty. Water is your top-most priority. You take in your surroundings in the present and determine you are on the side of a tree-covered hill.

Which way should you move to find water?

You'll use Steps 1, 2, and 3 to craft a story about water in this place. In this case, assume you create the story that water is most likely to be found in a valley based on what you believe to be true about water and hills – that water flows downhill, not uphill.

If you feel certain enough about this projected truth-story, it becomes the path of least resistance relative to all other billion paths, and you will justify time and energy in the form of movement to validate this prediction. Through this process of movement, your story system will or will not validate its ability to understand the truth in this place.

You act in the present based on what you think is right. You cannot and will not act on what you think is wrong and look for water that is flowing and collecting uphill, because ending the feeling of thirst quickly is your most urgent priority. You already have a long-established and highly validated truth tied to this place when it comes to water. So you justify movement down the hill.

How do you move, though? Do you travel with great caution? With great speed? In my opinion, this will depend on what stories you know and imagine to be true about the dangers that exist in this unfamiliar forest.

If you believe running into a life-threatening predator like a bear or lion is a possible story, you might move slowly and cautiously and listen for movement, or pay close attention to strange smells that mean you've crossed the boundary into a story that includes a possible threat.

If you are on a rocky, forested hillside, you might listen carefully for rumbling landslides or tumbling rocks, and be hypersensitive to vibrations under your feet.

If you believe there are hundreds of deadly and camouflaged creatures like bugs and snakes, you may choose to look carefully where you are about to step before you ever put down your foot (you are barefoot, after all).

All of these believed-to-be-possible-in-this-place stories modify your movement toward where you think water might be found. These modifications, such as moving slowly and quietly, are stories you believe will keep stories of your possible harm as remote and uncertain as possible. Remember your gaps, though. The stories of harm you are trying to avoid merging with are all built from *your own* past experiences. They will try to merge with any new experiences you have along your journey to water, but that hasn't happened yet.

New experiences – or new truths – merge with your past truths because you don't suddenly change realities. You did not blink and end up somewhere else in time and space. Time and movement are *gapless*. All unexpected stories are added and *merged* with past stories so that your context for this reality can improve its expectations of the future.

If you unexpectedly encounter a snake on your journey to where you believe water to be, your story system overrides its "find water" justifications, floats the anchor or reality as the snake and water stories merge, and re-grounds all your justifications to this *new* truth-story. You crossed a boundary into a new story that has different and higher priorities than your previous story. This snake story is now a story of *shelter*.

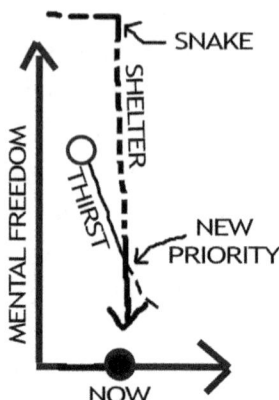

You do not have the freedom to abandon your search for water because of this snake (assuming you are still trying to expand your time). This snake won't cause you to overwrite your previous truth anchored to water either. By this, I mean you won't suddenly switch to believing water flows to the tops of hills now just so you can go the other way. I also don't believe you would give up, sit down, and wait for death to come to you from dehydration just because of a snake. You *need* to continue the same story you were in before the serpent – you are driven to do this by the underlying forced-choice you make constantly – to try to live.

In my opinion, the quickest option to leave the story of the snake and rejoin the story of water is to find a way around the snake at a safe distance. If you cannot do this, I suspect you will use a tool like a nearby stick to move the snake out of your way while keeping safe distance. Another option might be to try to kill the snake for food or at least eliminate it as a future threat to you.

Whether or not this snake is food or a threat to you depends on *how many stories you know about snakes*, such as how to identify venomous from non-venomous, how different species of snakes behave, and which are safe to eat.

No matter your story of the snake – I believe this snake will not stop you from continuing the story of water. The serpent is merely a temporary need for shelter on a bigger journey to a need. Your story system came across something unexpected, and then tried to predict what comes next using your previous experiences. You imagined many stories of shelter from that prediction – such as moving around it - and you will pick the story you feel is right based on what you believe is true to use as justification for your next action – which again is highly affected by what you know and have experienced with your "unit-category" of "snakes" in your past.

As you pass around the snake at what you imagine to be a safe distance, your shelter story-arc reverses direction, and your water need becomes priority again.

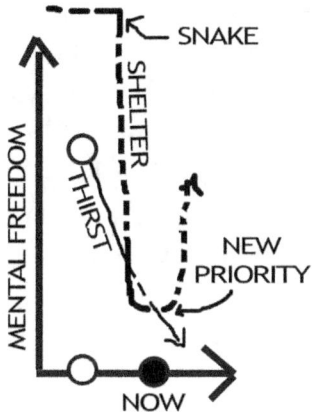

And so you continue.

As you reach the bottom of the hill, you hit a jackpot. Flowing straight ahead of you is a creek. It is wider than you are tall, and knee-deep. Crystal clear water burbles calmly along its rocky banks and fish visibly swim against its currents.

What kind of emotions do you experience in this moment?

I think it will be happiness, pride, and excitement. Not only have you found what you need to fulfill your most urgent and frequent need to exist, but you now have the peace of mind you can *always* find a drink when you need it. This is probably a better outcome than you hoped for when you set off with no idea if you would ever actually find the water you needed. Your ability to quench any future thirst is *nearly certain*.

The length of your timeline dramatically increases, along with mental freedom, because you no longer can easily imagine merging into a future story that involves extreme thirst. You can now drink whenever you want and as much as you could possibly need with a few quick justified actions. The story of finding water to drink just went from potentially taking *the rest of your life*, to only requiring a few seconds out of every day.

After assumedly drinking yourself silly from the creek, I have what will probably come across as an absurd question: Will you now abandon the creek and wander off into the wilderness?

I can't be sure of your answer of course, but I feel pretty confident it's *no*. In front of you is an indefinite supply of your most frequently needed survival resource. With that provided, your mind can now float the anchors and reset your ground truth to one where this high-priority need is easily fulfilled – it's practically *certain*.

I believe only a direct threat against your other two needs could drive you away from the creek. If a huge territorial animal like a bear has decided this creek is theirs, you will be forced to flee to fulfill your shelter need. If you can't find food in the area at all, you will be forced to leave to fulfill your food need. Neither of these stories is happening, though.

With water nearly certain, your story system now seeks to solve the next most urgent problem of uncertainty when it comes to your continued survival - food or shelter.

And so we continue to climb.

88

Feast and Famine

Assume food is the next most urgent need for you.

Fish were in the creek, so it's possible your mind went straight there when it comes to food. While it might seem simple to tell yourself the story "I will catch and eat those fish," this is not actually an easy truth-story to bring into reality through actions. Have you ever tried to catch fish barehanded before? How long are you willing to fail at this task before panicking? This leads to a what-to-eat dilemma.

You are what is often called (in my experience) an "omnivore". This means you can eat lots of things: Animals, plants, and lots of things in-between, including dirt if you really wanted or needed to eat. That being said, being ultra-selective about what to eat can create a bit of a time crisis.

To show what I mean by this, assume you were vegan or a vegetarian before entering this experiment. I have no issues with these dietary choices, but in *The World of You*, you potentially have a very severe problem right away: How long do you think it would take you to find enough food in regular intervals of time to keep yourself alive in an all-wild world?

I'm not saying a vegetarian or vegan diet is not possible in this experiment, but there are a couple of serious time and movement problems embedded in your food story arc. To find enough food to survive, I believe you will probably need to cover a *very* wide spread of land, and also know many stories of edible plants and what they look like in the wild. However, I see an even bigger problem lurking underneath this edible-plant-knowledge problem: No plant that I know of produces food at the rate your biological wavelengths require you to eat.

For example, finding an apple tree would be a wonderful discovery. I'm sure you agree with this. But are any apples on the branches? Are those apples edible yet? If not, you have technically found no food at all. You could easily die of starvation under this fruit tree.

When the apples finally do ripen some time in the future, you'll have a new problem. You will have *hundreds* of apples for a couple months at best, but no way to store them from rotting or going bad. You can't use this overflow of food to ensure you can fulfill future hunger very far into the future. An apple tree, though a treasure for sure, is only food for a short time in a very select time of year.

No plant that I can think of produces food fast enough for you. Pears don't appear every day all year round. Mushrooms don't grow full size over and over each night. A cornstalk doesn't produce ears of corn over and over again throughout the same season.

To overcome this problem and keep a vegan or vegetarian diet, you will need to find a *wide variety* of plants including nuts, seeds, berries, fruits, and vegetables that all produce food in different intervals of time and seasons to be able to ensure your survival in this experiment.

I think chances are, wherever you were on the globe when you originally started *The World of You*, you'd have to travel great distances and spend lots of time finding enough edible plants to stay alive.

This creates an imbalance that might be hard to balance out: Distance lengthens the time you spend moving, which also increases the amount of energy you are using to try to find enough food. Paradoxically, you would need to find *even more* food to increase your energy, so you can keep moving across great distances to find more food to increase your energy.

This can be a potentially severe imbalance: If you hike 20 miles to try and find enough plant-based foods to survive and *you fail to find enough*, you have expended an enormous amount of stored energy in your body, dramatically increased your movements, and even decreased your time remaining to survive - for no gain at all except learning a new truth: There is not enough food along that particular 20 mile path you just traveled to keep you alive.

These long hikes would greatly increase the possibility of your injury and harm by other creatures and/or your environment as well.

To save time – and *immense* amounts of it - you have the option (barring allergies or intolerances) to switch to food sources other than plants, thanks to your omnivorous digestive system.

Bugs are likely to be very easy to find, and provide far more calories per bite than plant-based foods. Granted, choosing which bugs to eat might be a gamble with your health and therefore your survival. I'm fairly confident if you live in the United States right now, like I do, you have *no idea* which bugs might be edible or make you sick. I have no past associations - no context at all - for eating bugs. I would be possibility-blind if I decided to start. I would have no idea what to expect hours later as I grabbed the little cream-filled kind.

Small mammals are a bit trickier as a food source, but with enough patience you might be able to devise a trap to catch them. If you are survival trained I'm sure you'd master this easily.

And to return this back to the opening of this chapter - there were fish in the creek. I have confidence you'd figure out how to catch them eventually.

My point here, is that I believe choosing to eat meat is going to provide far more calories than eating only plants will, and with less energy burned, especially if you can get the meat with ambush or trap-type methods that require very little energy and movement on your part. These types of methods take advantage of the animal's movement while reducing your own movements at the same time.

In the bigger picture, how likely are you to stabilize your food sources, and provide the near *certainty* of not dying from starvation? In my opinion, this depends on your ability to imagine stories to solve your problems.

I personally believe you are likely to end up relying on all of these sources of food, both plant and animal, to sustain yourself. This is where already having skills acquired through firsthand or well-understood shadow experiences would be very helpful to have in memory. Skills such as how to build a snare out of natural materials to snatch up small animals, or how to identify edible mushrooms, or knowing which tree bark you can eat.

There are many possibilities and methods to solve your food story arc problem. But because everything edible decays, and you have no preservation methods to slow that decay, your mind will nearly always have some level of thought and calculation running in the background related to acquiring food. It will always be working to resolve uncertainty related to food in this experiment.

Your mind only ever gets brief rest from this work, most likely after you have consumed a large meal. As your timeline expands again and your food story arc re-starts its path upward, you can slow down and rest. You have your food and water needs satisfied.

89

A Guide to Napping

After finding the creek and food, I think it's very likely you will seek a shelter nearby, and begin to use this area for hunting and foraging indefinitely. Everything you eat needs water (both plants and animals), so being near water will almost certainly be helpful toward fulfilling your need of food and water in the future. Of course, you need past associations to build such a story, but I don't think it would take you long to figure it out even if you didn't know it before entering this experiment.

With the discovery or the construction of a shelter, your final survival need can be fulfilled, and your story system and body can rest.

Resting, in general, is risky business. But it's also not something you can survive without doing. You *need* to go unconscious for several hours at a time over and over or you will die. Skipping rest, particularly sleep, is not an option. However, it also leaves you completely vulnerable in that span of time.

A physical shelter creates the "boundary" against fear itself. Shelter gives an extra layer of protection from movements in the objective world, allowing you to become vulnerable without being completely exposed to stories of danger and your harm merging into yours.

The need for certainty of shelter while resting or sleeping is very apparent in my opinion, but that certainty *is a feeling* contained in your mental boundary, not objective truth that I or anyone else can measure. It is you feeling like you are protected from the stories *from your past* that seem possible to happen in the future during the time you wish to sleep.

Would you be comfortable bedding down to sleep in the tall grass of an open field? How about sleeping in a tree? A hole you dug with your hands in the ground? How about out in the open on rocks or directly on the dirt? As I see it, your answers will be based on the stories you believe to exist in your environment, including predators or weather, as well as the likelihood you think these stories might happen.

As for myself, I believe the safest shelter in this experiment is going to be a small cave of solid rock, ideally with a moveable rock "door". If there are bears possible in this story, this might not be the safest option, as a bear can probably move that same "door" rock, but concerns about this will depend on how many stories you know about bears.

Even given this possibility, I still see a small cave as the best option for natural shelter in a forested *World of You*, offering the greatest chance for your survival. I confess that my knowledge here is only as good as secondhand stories and experiences in my past. I am no survival expert by any means. My stories are limited. I see my opinion as acceptable, because this book is not a survival how-to manual. There are plenty of those in existence already. Instead – because we are far under the looking glass of language - this cave merely represents the *approximate* concept of shelter. So I'll just keep rolling with this cave idea.

Let's now assume you find exactly this in *The World of You*: A small cave complete with a large stone "door" within a short distance from the creek.

When the door is shut with you inside, this space will probably feel like the room from *Gravity* with no windows and no doors – like nothing should be able to reach you. You are safe from the movement outside when inside. However, unlike *Gravity*, you also know the door exists, is not locked, and that you can open it. In this way, the cave is very much like the one-door room from *The Ark*.

Shelter is the connection between those two thought experiments. In *The World of You*, you have the freedom to open the door and expand your possibilities again, but you also have the freedom to close it, and, ideally, collapse the possibilities that may threaten you by collapsing your own.

With all three needs now fulfilled, your time-warping story machine can slow down time completely – you have no more urgent priorities now. Your thirst is quenched and your belly is full, so you encase yourself in a stone boundary, and your mind and body minimize all movement. Your story system has accomplished its primary goals: Keep you alive by analyzing your past and present, and using them to predict all future possibilities, which creates a list of stories you choose from to bring into reality – assuming your stories aligned with objective truth "close enough".

Sleep allows the chemical reinforcement of all the associations and truth-stories built and validated through the first day of this experiment, logs them into memory, and makes repairs to your body, so you can begin the same process all over again tomorrow more prepared and ready than you were today.

90

Puddle Jumping

The previous three chapters were an ideal. All your needs were met very quickly after starting out in *The World of You*.

But what if you didn't find a creek in the valley? What if you started in a desert? What if you only found a puddle of slightly muddy water with no clue how or where to find more? What would be different?

Technically, a mud puddle fulfills your water need. But you need to keep getting water at regular intervals of time, and you subconsciously are very aware of this.

I believe your core mental need to expand or stabilize your time to exist is going to take center-stage at this moment. I suspect you will attempt to drink the entire puddle (if you believe you might find water again sooner rather than later), or drink some of the puddle and attempt to store the rest (if you believe you may not find water any time soon).

If you have a way to store the water and carry it, I believe you would attempt to do so. I suspect you would do this because you know you will get thirsty again and do not know where more water might be. The story of stretching the water you have as far as possible makes a ton of sense.

But, if you watch any survival shows or are survival trained at all, you might know the story that you could be better off drinking more water now, instead of slowly weakening yourself by taking in less water than you need to fully function. To know this story though, you would need to experience it in your past.

With food, if you find a large fruit tree, I suspect you are going to try and store some of the fruit as long as possible. You know your food need comes in intervals again and again through experience. It's a well-grounded, validated, truth for you. You are fully wired to consider these calculations when resources are limited.

With weak shelters, I suspect you are going to try to create a stronger shelter, or start jumping between weaker shelters as needed.

The point is, your survival rests on the idea of intervals, or frequency, and your future planning revolves on how many wavelengths you can imagine to remain for you.

What other noise and thoughts might exist in this place though? Is there anything else to think about or act toward than just your survival needs?

What else matters in this well-grounded, highly-validated, firsthand, wordlessly experienced place?

Thinking about this question resulted in some major discoveries for me.

91

Shattering the Looking Glass

I find myself returning to *The World of You* thought experiment over and over even today. It has helped me isolate and figure things out quite a bit.

Before I "found" this place, I was a near completely invalidated human being. I cared almost exclusively about what other people thought about me, or might be thinking about me. I was completely time-blind and fit the symptoms of ADHD (I am now medicated for it as an adult), and I went through bouts of depression and melancholy.

Even though I knew what I *should* value in life (red flag!), I never quite felt like my beliefs and actions lined up with those values. I knew I was *supposed* to want to die for my family, what I was *supposed* to believe, and what was *supposed* to be right based on those beliefs, but none of this ever quite felt like... *me*. It all felt like somebody else's values, not my own. This always resulted in horrifically conflicting thoughts like "I should die for this idea because I am supposed to want to die for this idea."

I've found that *The World of You*, beyond re-establishing the lowest "ground point" of foundational needs for me as a human being, could also easily expose what values were not my own. It also showed me what matters in ways that made conflicting values easier to prioritize. Perhaps it can do the same for you.

You can absolutely figure out what follows in this chapter just by staying in *The World of You* and thinking about how your life is different, and I highly encourage you to do this whenever possible with any topic you might struggle with.

To give you a jump-start though, I'll run down what I see as some of the most dramatic topics to chew on, in no particular order:

Words

You no longer have need for these in *The World of You*. Words do nothing for you. You can talk out loud to yourself I suppose, but as the only human on the entire planet, you can hopefully notice what all the words you ever learned, heard, and used actually were all along: *Other's* experiences. The only words that are useful to you now are those that arranged cause-and-effects into an order that creates imagined sensory experiences in a way you can justify them into actions. These words might turn out to be helpful or even hurtful to your survival when transformed into action, though.

An example of this could be words you remember that you can convert into the actions needed to successfully build a fire. Another might be words you remember that were simply not accurate or were missing tons of information, rendering them unusable. Perhaps words like "smacking any two rocks together will make sparks".

All unusable stories will quickly fade though, because once you create a fire for the first time, or smack rocks together, the words that led you to do these actions lose tons of value and are replaced with firsthand *wordless* experience.

Other examples of helpful words could be those that explain basic survival skills, like shelter building or plant identification (assuming you do not have firsthand wordless experience with these things). They could be as simple as a description of what a carrot plant looks like, or descriptions of first aid techniques tied to wilderness training (again, assuming you can remember the words in the right order to convert them into successful actions).

Morality

Do *right* and *wrong* exist in *The World of You*? If they do, who or what exactly is defining what is right and wrong for you? There are no laws anymore, nor any government to enforce them. There are no holy books anywhere, and there are no religious institutions to enforce any rules. There is no one to tell you you've sinned, and no one to confess to. There is no school teacher to tell you you've done things right or wrong, and there is no other person around to tell you the sky is "actually more of a purple than a blue today".

Killing is no longer wrong unless you decide it. Stealing is no longer wrong unless you decide it. *Nothing is wrong in this place unless you decide it.*

All moral codes in existence as you read this book have officially been erased in *The World of You.*

I do not believe you would observe the wolf chasing the rabbit and label it wrong. The moon would not perform wrong actions. They might seem wrong *to you*, but your belief doesn't change a thing about these actions.

You and you alone are the only thing in this world that the word "wrong" can ever be applied to – and since you are the one that defines it and automatically act using that definition, deciding anything is "wrong" becomes its own kind of self-fulfilling prophecy.

For instance, if you decide to go on a rampage as the only human being on an all-wild Earth and murder every living thing you come across because you feel like it's the right thing to do, the world will not arrest you and haul you before a judge. If you feel no remorse about this mass killing, there is no one else to care, to shame you, or tell you to feel differently. If you steal the eggs out of every bird nest you find, the sun will not judge you for it. You are *completely free* of the words right and wrong *and* their assumed meanings.

That is *not* to say you are completely free of consequences. If you justify a mass killing in *The World of You* because it felt like the right thing for you to do, you might very well hurt your future self, because the natural consequences of all this death likely result in negative impacts to your survival. You not only destroyed potential food sources, which you might very much need a week from now, but you also stand a high chance of injuring yourself in the process of doing all this killing, which can easily result in your own demise through infection.

In other words, doing such a massively violent act would potentially throw your needs and the things that fulfill those needs badly *out of balance* – but not from the perspective of the world – the world has plenty more animals and life than what you just killed – just out of balance *for you.*

This imbalance could reset *The World of You* back to the beginning, and reboot your search for food, water, and shelter all over again, because this place of rot, death, and festering no longer has anything to offer. This relocation brings the possible stories of extreme thirst, hunger, and fear closer than they were before the killing spree.

In this world of only you, you experience morality, or right and wrong, at its most basic: When it is defined by the moment of justification. This loops us all the way back to the early chapters: You will always justify what action you feel is *right* based on comparing your present to your past, even if *right* is somehow determined to be murdering everything you can find as quickly as possible.

Due to your story system having a floating ground that is always adjusting truth, right and wrong are relative moment-to-moment and are ever-changing. This means any unmoving, timeless, rigid definitions for truth, right, and wrong, tumble all the way down the bottomless abyss between First and Second You where they finally fall apart because they must. They are self-defined concepts based purely on your own actions as they relate to your survival in this reality – and they always have been. This means these two words are just floating anchors that are moved around and given relative meanings by you, even today.

Let's then anchor right and wrong to their most fundamental meanings:

If you reach for a fruit on a tree with your hand and miss it, were you wrong?

Only if you determine you were because you *intended* not to miss it, but that determination is going to happen *in your past*. Should you decide it was wrong, it will not start the runaway process in this context. Second You won't now re-run the memory of this unexpected outcome in constant growing feedback loops. Second You does not transform into your eternal nemesis from this one "wrong" action like it did when you stole the candy bar. There is no one else here to judge you for missing the fruit. No other *Keeper of Truth and Decider of Right* hidden behind a mental boundary where you cannot see. You are the *Keeper of Truth and Decider of Right* in this place.

The next time you justify reaching for that same fruit after missing, your expectations will be adjusted thanks to the memory of your miss last time and the story voltage you experienced from it, which floats the anchors and changes the act of reaching so that it will be right *next time* you justify that action to accomplish your intended goal.

"Last time you reached and missed, this time stretch your arm further. Then you will reach it. That is the right way to reach the fruit."

And so you justify reaching the second time.

If you reach the fruit this time, the truth of your perceptions, movements, and outcomes (Steps 1, 2, and 3) are validated and your predictions and expectations in this reality were realized. You will experience a positive emotion like satisfaction, joy, or contentment, which solidifies this truth in memory, and you now carry this validated truth about how to reach things forward in time with you.

If you instead miss the fruit this second time, it's also noted and negative emotions are experienced (such as frustration), and your entire foundation of truth, right, and wrong are floated and adjusted and reset just like the first time you missed. This process can be repeated until the truth of what action is right to reach that fruit is found, or until you decide no longer trying to reach it at all is the right thing to do.

As you read this book, if you currently worry about what others think of you, your worst enemy in *The World of You* is you. You have essentially forced the validated wordless truths of your firsthand experiences, or the "truth built by you," into darkness because you believe it is the right thing to do. You essentially do not trust yourself.

This would mean you dragged your *Keepers of Truth* into *The World of You*, and brought the Hydra from Hell with them. You essentially require *their imagined* validation of *your* directly experienced truth in order to act. Which is again based only on what *you imagine them to believe*, because words never mean exactly the same thing to anyone.

Your own self-doubt – created by a hijacked Second You - and whatever direct experiences you held yourself back from having before this experiment began, are all potentially creating very serious shortcomings for your survival. After all, time matters tremendously, and hesitation is slowing you down.

Higher Powers / Forces

There is an "elephant in the room" when it comes to moral codes, and I'm not going to ignore it.

In *The World of You*, you may believe a higher power is still judging what you do and even what you think in this experiment. So be it. You are welcome to believe whatever you want - the only person you can ever hurt in this place with your own beliefs, is you.

If you believe you would carry your god or higher power in with you, I have a question: What actions *exactly* are you willing to prevent yourself from doing in order to comply with what *you believe* your higher power wants your actions to be?

As I see it, your own interpretation of the higher power's judgment becomes the other "person" in your own head, which you then keep trying to imagine judging your actions *before* you justify them.

I'm not saying there is anything wrong with believing in a higher power – but the documents and authority structures built around the idea of higher powers no longer exist in this place, and it's entirely possible that the story of the higher power's judgment you choose to believe in may very well be *human-context-hijacked*, which converts what I imagined to be a *positive* higher power's influence on you (that you can succeed), into a *negative* one (that you can't).

As I see this, "God" and the "Hydra" occupy the same exact place in our minds. Being able to tell which is which can be incredibly difficult.

Here's how I think you can notice if you are Hydra-hijacked: In *The World of You* - alone, naked, and fighting for your survival – do you feel *closer* to the higher power or god you believe in, or *further*?

I highly suspect your answer will actually be *closer*. I'm almost certain of this, actually. Could it be that the further you are from other people, the closer you feel to god or a higher power?

I'll use Christianity as a quick example of why I think this relationship exists:

In *The World of You*, most of the rules present in the Ten Commandments no longer have practical meaning *because they involved other people,* which no longer exist. If you consider the remaining commandments that don't involve other people, you will almost certainly follow them without even thinking about it. This can largely eliminate the need to "tell yourself" what is right and wrong.

For example, are you going to keep the Sabbath holy? What are you going to do to keep it holy? Who gets to decide if your action to keep it holy is right or wrong?

You don't actually have time or the freedom of choice to justify actions pursuing something off-topic to your survival, like choosing to rest all day. You especially do not have this luxury if you are still in the process of trying to find food, water, or shelter for the first time.

As I see it, the best you can do (if you decide following this commandment is still the right thing to do), is recognize the Sabbath when it arrives, and then create your own personal method of keeping it holy (assuming you can remember what the current day is at all - does Tuesday or Saturday or Sunday even mean anything anymore, or are you now free to determine the day of the Sabbath yourself?). Whatever your chosen method might be, it can *never* be wrong for you. After all, there's no one around to judge whatever it is you come up with as right or wrong. There is no one else standing between you and your higher power, dictating the way things "should" be done.

"No Other Gods Before Me" commandment? Who is capable of such a thing? Who is going to enforce this rule given your situation?

No idols? Same question.

Thou shall not kill? Shall not steal? Whose definition of stealing and killing is right for you? You are going to need to kill or steal from plants or animals to survive, and your million billion internal processes are killing *something* every second of every day. You get to define the meaning of every word in these commandments.

The catch is this: You have *always* done it this way.

If you are a highly religious person, before making some absolute decision about how you would recreate or not recreate religion and the will of higher powers in *The World of You*, please consider the following question: What would happen to your beliefs if *Memory-Wiped You* started this experiment? How many of these religious ideas are you carrying with you into this experiment from modern day beliefs promoted and put into you by *other* people?

I am not saying your higher power doesn't exist, nor am I saying your higher power does exist. There have been thousands of religions and higher powers in the history of humanity, and yet I ultimately suspect they all occupy a very specific gap in the human mind. For this reason, I actually believe memory-wiped you would recreate some concept that tries to fill this exact same gap. I believe you would absolutely recreate the story of a higher power or higher force of some kind.

I have more thoughts on this topic, but they will come later. In the meantime, please mentally bookmark whatever belief you might have about God, gods, higher powers, higher forces, or other religious or universal-scale ideas in *The World of You*.

Money

If you imagine every human and every human-made creation gone, but next to you was a pile of six billion one-dollar bills, does it have any value to you?

My guess is that it actually has *tremendous* value, but not as currency to spend. Instead it would be valuable because it is made of very durable paper, and paper can be a wonderful resource to have.

For example, perhaps some of the bills can be used for personal hygiene like wiping your butt and blowing your nose. Some could also be used to potentially make clothes, collect water, or build a shelter. The possibilities with this much paper are plentiful, but *none of them* involve the value of the dollar for what it can buy.

Like religion and higher powers, I have many more thoughts about money and its value in chapters to come, but it's important to note that in this "ground state" of *The World of You*, it has lost all meaning (as currency anyway). This means the value of money as you experience it today must appear somewhere between the conditions of this experiment and your present situation as you read this book, assuming money still exists.

Math

How much is calculus going to help you now? Quadratic equations? Fractions? The only math I can imagine being useful is the ability to keep track of days or time, and trying to calculate portions of food.

This would be all extremely basic math though, and in my opinion could be considered "innate" level math. I say this because I believe even if you cannot count, you can still look at a pile of food and "feel" about how long its going to last you, or "divide" it up without using a single number. You can look at the sun in the sky and start to "feel" how much daylight you have left. This is all numberless math. There are tribal societies around as I write this (Pirahã of the Amazon) that do not have numbers in their society and cannot count.

The result is calculation by "feel". Many of the animals of the world do these same "equations" without a calculator and a number line, and for this reason, I believe you can as well.

This means like the other topics above, the importance of mathematics must also appear somewhere *between* the conditions of this experiment and present day as you read this book.

Job or Retirement

What does the loss of your job mean to you now, if you had one before entering *The World of You*? Was your job accomplishing something meaningful now that you look back, or was it just something you did to kill time and earn money for you or other people? I believe there is a good chance the skills you were using in your job may be of no use at all in this experiment. I know this is the case for me. Designing electrical systems is a skill with no real use that I can see in this thought experiment.

I'm not saying your job today is worthless, as it likely exists for some reason, but it can be interesting to consider just how far removed from each other this experimental world and today's world can be, because both are very much based in the same reality, and are technically only separated by time.

Politics

There are no parties or candidates or any other political issues in *The World of You*. You are King, Judge, Juror, Executioner, Citizen, and Immigrant. This begs the question: What does politics seek to accomplish? How many politicians are concerned about food, water, and shelter issues? How many stand up and proclaim timeless certainties about life and about the future?

There are people in the modern world in my time (and one of them might even be you, though if you are reading this I have serious doubts), whose timeline only stretches as far as their next need for water, or food, or shelter, and they can see the possible story of extreme suffering and even death being possible in their near future.

These individuals have no need for politics or justice issues as I see it, except where they can be exploited to fulfill these needs. When it comes to focusing on politics and justice as they apply to others, freedom in time is required to learn, reflect, and develop opinions and beliefs. Individuals in extreme poverty, homelessness, or other dire situations lack this freedom, and so are trapped into being focused on fulfilling their very primary and basic needs – possibly by any means necessary.

Political issues today, of which many appear in that long list of modern one or two word-stories of "all of society's problems" I've brought forward several times, are all arguments built upon luxury and authority, while standing on the shoulders of being able to have a freedom of choice and opinion at all.

In the *World of You*, you don't have any need for thinking about these issues. There is no situation where they apply to anything.

Senses

In *The World of You*, suddenly the ability to use all of your senses to their fullest degree matters - a lot. If you spend day-in and day-out destroying one of your senses in today's world, like I did with my hearing using drums and stereos, you might feel regret in this experiment now that you need every bit of that sense to help you survive.

Perhaps that eye exam you've been putting off might be regrettable now, because if you had glasses, you could have brought them with you into this experiment.

As the early Covid-19 pandemic began, many people lost their senses of smell and taste. How would that affect your survival in *The World of You*? If you started out this experiment without them, would it change your potential ability to survive?

If you are missing a sense completely, because you were born without it or lost it before the experiment began, life in *The World of You* is going to be extremely difficult, especially if that missing sense is sight.

The shear magnitude sight has on this situation should give those of us lucky enough to have vision a feeling of immense gratitude to have such a thing. I'm not saying that if you are blind you are destined to die quickly in *The World of You* – I know your other senses will compensate enormously – but I do believe the ability to find food at the rate you need it is going to be an enormous hurdle to overcome.

Mobility

I have very poor flexibility as I write this, and I don't spend much time being mobile throughout my day. Like much of the United States "middle-class" population in my time, I spend most of my time in an office chair or the couch - sitting. This experiment gave me some perspective of just how valuable mobility really is.

Although it's arguable mobility is not a necessity for survival in the modern times of abundance as I write this, not having mobility heavily affects the ability to manage your life in *The World of You*.

Your ability to simply lie down on the ground and stand up again will matter quite a bit. The ability to stay balanced on uneven ground, jump from stone to stone across a creek, climb a fruit or nut tree, rest, and eliminate bodily waste in a squat position all make life much more manageable in the absence of grocery stores and furniture, toilets and bridges.

Speed

If animal predators may threaten you in this experiment, the ability to escape them or fend them off may boil down to how fast you can climb a tree, throw a rock, or take evasive or defensive actions. Speed might allow you to jump a rushing creek, or catch a fish bare handed, or snag a bug from under a rock before it scampers away. Speed can become its own form of shelter from the passage of time in the form of movements around you, and greatly help in your overall survival.

In modern times, just like mobility, speed is often unnecessary. We have cars, weapons, and luxuries that keep us from usually ever needing to have or use extreme bodily speed. This does not mean it is not without significant value, however.

Strength

Coupled with mobility, strength can help you climb to hard-to-get food, lift large rocks, push over trees, or break tree branches for all kinds of purposes.

Enormous strength however, like you would have if you are a bodybuilder, strength competitor, or have pushed yourself to an extreme muscular peak, will likely collapse quickly to be in balance with whatever minimum strength is required to perform the actions needed to survive on a regular basis. After all, more muscle requires more calories to feed the enlarged processes, so a natural balance between food intake and peak physical power outputs is always being found.

In this way, bodybuilding and extreme physique goals, which are common in the world as I write this, are only possible in a culture of abundance that allow time for training without much practical survival application. It is also done with the expectation that more food will be readily available to consume, which is required to maintain muscular mass in the first place.

Health

Getting the flu in *The World of You* could be a death sentence. You certainly wouldn't be able to afford the loss of calories from being bed-ridden for a few days.

How much do you rely on modern medicine to make it through your day, or to even be alive in the time period you are reading this? We humans really have come a long way in this category. If the *The World of You* wasn't set up to let you keep artificial limbs and organs, the price of suddenly becoming a 100% un-medically-modified person might drastically affect your ability to survive.

Health also stretches into food: How much healthier would you become in this experiment if you manage to secure enough food? There are no more sweets or soda in the world. There are no potato chips or even cheese. Everything is going to be unprocessed, although also a bit of a gamble with your health if you aren't sure about what is and isn't edible.

There is also no allergy medication. Nor any asthma inhalers. Life might get very tough if you suffer from extreme allergies.

My point is, being in good health is suddenly a huge factor to your contin-ued survival and I suspect will be highly valued each and every moment in this experiment.

Clothes

No, I don't think you would miss your multi-thousand dollar handbags or your exercise-wear of breathable spandex if you have these things now. I think you would instead miss having *any* clothing.

A bag of any kind would be an amazing thing to have in this world of you, and anything to cover your skin and provide more shelter from the elements would be welcomed, I'm sure. As a person who is bald, I'd definitely miss having a hat on a hot day, and sunglasses on a sunny one. I'd also definitely miss having thick layers like sweatpants and a coat should the temperatures drop at night.

I also think the importance of shoes will be immediately noticed. The soles of your bare feet would thicken up fast I'm sure, and you would eventually manage just like the ancient humans did - but having a grippy rubbery sole, or even a sack of leather on your feet would probably be a welcomed experience as you trudged through a world absent of roads, sidewalks, and manmade trails with less worry about injury or slips.

Home

How amazing and valuable would a house be in *The World of You*? How much relief and joy would the sudden appearance of a building with doors and windows bring you? Even something as crude as the historic log cabins that are currently preserved around the United States in my time, would be absolutely phenomenal to have in an all-wild world.

In my time, we have triple-layer windows, indoor plumbing, soft and supportive beds, carpet and wood floors, heating and cooling, and electricity. Even the most standard and boring home of the modern era would be nothing short of a miracle to come across in this *World of You* thought experiment.

Your place of residence consisting of four walls and a roof, if you have one, is enormously valuable, even today, but perhaps using *The World of You* as a grounding point might help increase its value to you a little bit more.

Groceries

By all measures in my time, the grocery store is literally a magical basket of food that keeps refilling at all times. There is conversation to be had about affordability in modern times as I write this, and I will address this in later chapters, but the ability to walk into a building and find hundreds, even *thousands* of variations of food by simply stretching out your arms to grab them is a true testament to how far society has come from the wilderness of this thought experiment.

I have lots to say on the food system, so I'll leave this topic to avoid future repetition, but the grocery system itself is an amazing accomplishment.

The Moment

Every minute of your life in *The World of You* has critical value. Granted, you'd probably wish for those minutes to slow down if you are starving, or thirsty, or in danger, but even so, every moment you are alive is one more minute of story system growth.

If you currently do not value your time right now, and engage in escape behaviors like I did to avoid the struggle of looping thoughts and rejections of yourself, be aware that this problem develops somewhere between *The World of You* and your current moment in time. You do not have the freedom to choose escape actions from the stress of this situation yet. If you choose to spend the current moment escaping, you are choosing to scoot yourself closer to more hardships and death.

This balance between time and survival leads us straight into the next piece of our shared foundation.

92

Capturing the End

Let's skip through time in *The World of You* thought experiment. There is a pattern at play, and I think it's absolutely critical that you notice it because it is going to echo around like crazy in this rabbit hole, and you'll even start to see it everywhere in your life outside of this book.

In this chapter, I'll be the one in the experiment. Let's assume I have my water and shelter needs covered by this point with the discovery of a cave and a creek. I will put the focus solely on the food need in this chapter.

I previously covered the problem of what to eat, but I want to go back to discuss something deeper and much more basic. Food is just the best way to illustrate the pattern involved, in my opinion.

My primary goal is to always increase my certainty to continue getting food at the time intervals I need it – or to avoid the story of extreme hunger. The further into the future I feel sheltered from hunger, the more certain I can be of my continued survival. With a source of *endless* food, my predicted food-story "wave" wiggles on as far as I can imagine just like my water-story wave, which means my timeline does too.

To keep my shelter need easily fulfilled, I will likely drop an "anchor" at my shelter. By this, I mean I will stay as close to it as possible while fulfilling my other needs, like food. After all, I never know when or if I will need to return to my shelter quickly.

As I mentioned before, the only thing that could drive me from staying anchored to this shelter is the collapse of my ability to find food or water within a certain distance from it. This "certain distance" is likely determined by how far *in time* I am comfortable being away from the safety of my shelter (I'm ignoring flooding or natural disasters that may destroy my shelter or make it unusable). A comfortable amount of time, I think, will be the time it takes my nearest "need" story arc to peak, and begin to drop back the other way.

For example, if my food-story arc is eight hours in total length, I'll probably struggle to stay anchored to my shelter if I can't find any regular source of food within four hours of travel in any direction – and wandering four hours away from my shelter is super risky by itself depending on what threats I imagine to be possible.

If the nearest food source was *five* hours away, I would spend more than half of the length of my food story arc (5 of the 8 hours) just traveling to get to that food, which results in me needing to find food again *before* I can even make it back to the cave. The result is needing to choose between food or shelter each time I make this journey. Starve in the cave, or be without shelter with a full belly?

If I found food *exactly* four hours away, I would then be completely enslaved to my food story arc, and have no time left to address any other issue in my life, including sleeping or taking shelter in the cave (with a few assumptions in place I'll address soon).

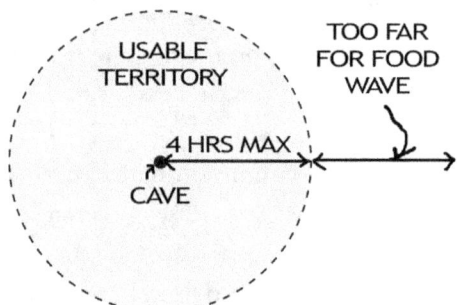

So, with this maximum distance in mind to be able to keep my shelter-anchor in place, imagine I begin exploring away from my shelter for food. The stream is very close to the cave, so water does not have a distance/time problem at all for me. It is fulfilled in seconds whenever I want. The distance and time to food, however, is yet to be determined.

Should catching and eating the nearby fish not immediately be successful, I might instead attempt to find plant-based foods in the area. This would be particularly terrifying for me, as I know very little about what edible plants look like, aside from what is commonly found in grocery stores. I would definitely be blind to most possibilities when entering into this story.

Regardless, I have no freedom of to make another choice, assuming I want to stay alive. Not eating results in decay of my body and with it the decay of choice - to the point of choosing freedom over food results in an intentional path to death.

So, I will cautiously begin to expand my "territory" from my central anchor point: The cave. I do this because I must. I would attempt to memorize where I am and would create landmarks (such as stacks of rocks) to help me easily relocate my shelter. I would probably also look "back" toward the cave often and try to remember what the surroundings look like from the opposite direction I am traveling.

Essentially, I am trying to build a detailed "associative" map in my memory based on objects. I am looking for things I can remember to help me return *in a hurry* to shelter should I need to. (Example: There is a funny-shaped cedar tree and a yellow rock just above the cave entrance from this direction.)

While I'm out foraging for food, if I spot a mountain lion that also spots me, and it starts moving in my direction, I'm racing back for that cave as fast as I can go. As I see it, *familiarity* with the path to the cave's location using surroundings and landmarks can greatly speed up how fast I can find shelter again, and could be the difference between life and death at the jaws of that mountain lion.

Let's assume I now travel north from the cave for about one mile, where I stumble into a blackberry bush that happens to have ripe berries. First things first: I'm hungry. I will gorge myself right then and there.

But what happens if there are far more berries on the bush than I can eat in one sitting? Now I face a difficult decision. Do I pick them all and attempt to carry them back to the cave, or leave them on the bush and come back again later?

I personally will choose to pick them all and take them with me. The berries should be good to eat for a few days, and that gives me a few days of increased certainty I can fulfill future hunger waves. There is a very specific reason why I think this is the choice that you will make as well.

Should I *not* take all the berries, I cannot be certain the berries will still be there tomorrow. Perhaps birds or squirrels or something else will eat them. If that happens, I won't have any berries at all. I can prevent this future food-wave collapse, which will cause a major collapse of time and freedom for me, by ensuring that I take all possible food with me *right now*, and not leave this bush unattended with edible berries still on it. Each one of these berries is a certainty against hunger I will not have otherwise.

If there is shelter nearby to this bush, I could potentially stay there, leave the berries in place on the bush, and go pick some more when I get hungry. This does not help my food certainty problem, however. I am still separated from my food. As a result, I will need to watch the bush like a hawk at all times.

I will have to run out of my new shelter screaming to shoo away any birds or animals that discover the berries. Even if I notice and shoo away every single bird and animal successfully though, there is still a great chance they are going to get at least one berry or two before I react. I will end up burning extra calories racing between these two points, add tons of anxiety to my life, and even worse: I'm no longer near my water *and* I'm losing my food to animals at the same time.

While I am able to better protect the bush by staying close to it, I still can't totally prevent the loss of this precious food. I can't be certain this bush will provide food to me again at any time other than *this moment*. So something extremely significant is about to happen.

All the fruit on this bush becomes *mine*.

By making it all mine, I give myself more certainty my timeline isn't collapsing due to extreme hunger for at least another wavelength or two. By making all the berries mine, I defeat the birds, rodents, and deer that would potentially eat them to survive *instead of me*. I believe this decision is the basis of all competition in life.

But it's not *just* the fruit that becomes *mine* at this moment. I will make every effort to make this *entire plant* mine. My goal is to completely "capture" this source of food, and ensure every last calorie it provides goes to me so I can stay alive longer.

This means that even after I strip every edible berry off of this bush, I will still plan to return to look for new buds or berries frequently, and monitor the plant for damage or disease.

Zipping slightly into the future, let's say on my fifth trip back to this bush I notice some new berries that aren't yet ready to eat. Am I okay to walk away while they develop?

If this is one of my only food sources, *no I am not. Fear* of not being the one to get these future berries is going to drive me to capture this plant *completely*. I will try to develop a way to keep birds, rodents, and deer away from this future food as it ripens. I'll try to invent the first *fence*. This fence will be a protective *boundary*.

This moment, all on its own, is what I believe to be the natural birth of "property". You may be tempted to think the cave would be the first property, but it was always possible it would need to be abandoned due to it not being the "right" shelter given the situation. I didn't *need* to defend the cave as much as I need to defend this berry bush.

Shelter has flexibility and works on an as-needed basis, unless I am dealing with problems of exposure to intense heat or extreme coldness. Shelter is always relative to a second point: Whatever I am sheltering from. There's always a chance I could find or build another shelter that works relative to the threat I fear.

With food though – something is either edible or it is not (because it leads to sickness or death if eaten). This berry bush might be the only food-plant I will *ever* find growing in this limited radius from the cave. Another way to say this: If I can't find shelter in *The World of You*, harm is situational and uncertain (ignoring unsurvivable temperatures). In the absence of food however, harm *is* certain.

So, I will keep a close eye on this plant from now on. I will memorize and trod the most energy and time efficient path between my cave and this bush – a path of least resistance. I will probably even monitor nearby plants and the ground near this bush for evidence of animals, or anything potentially threatening the plant's health. I may even try to build a stone "fence" or "shelter" to keep it safe.

By doing this, I am *capturing the end*. The "end" is the *predicted* end of my food story arc, which will be the end of me if it goes uncaptured. You may also think of capturing the end as capturing hunger in this case.

To be clear, the food story-arc is a mental projection, but hunger is a physical-based reaction beyond my control. It occurs because of actions being performed within my body. Because these actions began before my experience did, and because they require energy to continue *me*, my consciously justified actions are *forced* to feed these processes, because they will consume me if I do not.

By "capturing" all story arcs related to food, water, and shelter, my certainty of existence and my freedom of choice both increase.

However, as I capture these ends over and over again, I can easily lose sight of the "stable with decay" wave-like straw path life really is, and start to focus exclusively on the "doors," or the "ends", that threaten the length of my timeline – *no matter what I imagine them to be.*

This is an easy mental trap for me to fall into after the blackberry bush, because I will almost certainly repeat this "capturing" for any and all food sources I find within a safe and reasonable time-based radius from my cave. I will *memorize* where all these sources are, wear down literal footpaths from my repetition of visiting them, and make every attempt to protect their edible parts from being eaten by other wildlife. I will attempt to block access to these plants for everything else but me, and every part of the forest between my cave and the most distant food source I use to survive in every direction will become the shape of *my territory*.

I will work to memorize every detail of my territory and notice the slightest changes within it, all in order to predict any threat to my water, my food, my shelter, or myself.

I believe this extreme need for more certainty related to things I cannot fully control will make me a creature of habit, and cause me to construct a routine of walking my paths from source to source. Everything that eats the same food I eat is potentially a threat to my survival. In my opinion, this stage of my experience in *The World of You* would not look far removed from what the language of psychology calls obsessive-compulsive disorder (OCD).

This territory might be huge. It might take me hours to trod between all the plants. It would save a lot of energy and trouble and stress if I could bring these things closer to my shelter.

Assume all my plant-based food sources are out of eyesight from the cave – outside of my sensory boundary when at my shelter anchor point. I can't see if a deer is destroying all the buds on *my* blackberry bush, or eating all the apple blossoms on *my* apple tree. The bush and the tree might be three miles apart from each other. Not being able to keep an eye on my food sources makes me uncomfortable. It leaves my food story arcs uncertain beyond the one I am on, and this makes me anxious. I need my future food to be more certain.

So my story system starts to improvise upon the objective truth of this place.

93

The Dragon's Garden

In an effort to keep my food sources where I can see them, it makes sense that I would attempt to plant seeds I collect from distant food-bearing plants to bring the plants closer. Let's ignore the seasons in *The World of You* for now, because in reality they create a very big problem to deal with: Winter.

The reason I believe seasons can be safely ignored is because the only thing they add is the need to eventually "migrate" shelter-to-shelter, repeating the previous food/water/shelter process over and over again, but never getting quite to the point of putting together a garden (until I get close enough to the equator that winter no longer affects plant growth). So, let's just imagine the weather permanently warm for the sake of moving forward.

Once I have planted a garden, my biggest hurdle is now *time*. Depending on which plants I took seeds from, it might take *years* for those seeds to grow into plants that provide me food. Other plants, like green bean plants, can produce food in as little as a month and a half.

Regardless of which plants I try to capture by moving them closer to my shelter, I must protect the seeds and sprouts from critters that, well, like to eat seeds and sprouts. Birds, rabbits, rodents, and deer are pretty much my constant enemy. If I can keep them away from these plants by capturing them as well, all the better. More food for me. This end remains captured.

If my garden is successful, the result will be a dramatic up-tick in my certainty in almost every way. I will spend less time traveling to find food over much less distance, take less risks created by needing to move around to find food, and have a great chance of pre-capturing the end of many future food story-arcs (which reshapes them into waves and extends my expected future).

With time and energy saved, I earn myself more free time to choose whatever actions I'd like. For example, I could start to improve other systems. I could try to build fish traps, or a fishing pole and hook, or snares, or make clothes of some type from animal hide, or countless other possibilities. All of these can increase the certainty of my survival and grow my free time and freedom of choice even more.

This, as I see it, is the very beginning of the runaway process. By giving myself more free time, I can start to imagine and predict more possible stories where I *might* experience hunger, thirst, or fear. With the freedom to justify my actions on these imagined stories, I can begin reshaping objective reality to prevent these stories from ever merging into mine. I essentially start trying to "capture the end" of stories I *imagine* to be possible but haven't experienced firsthand.

I can imagine freezing to death at night. Creating clothes or blankets for this possibility "capture" this possible end. These things can also "capture" my imagined future story of being injured by a cut (which could become infected).

Making a weapon can help me "capture the end" of a shelter story arc created by imagining a wild animal attack. Pushing down loose rocks above my cave before they can fall on their own "captures the end" where they might fall on my head while I pay attention to something else. All of these rearrangements-of-reality-though-my-actions increase my certainty I have finally solved the time-based survival problem once and for all.

But death is *always* a projection. I have never *actually* died before from infection, cold, wild animal, or rockfall. I am capturing these ends based on *my fear* of their possibility. That fear is grounded to this reality, but only because it is a story that makes sense to me.

Because death is always a projection until it actually arrives, it is also perpetually *unknowable*. We tend to struggle with not knowing.

94

The Call of the Void

In *The World of You*, what is the closest you can get to total certainty of your survival? We both know it's not possible to be 100% certain. But, in my opinion, the highest certainty is nearly achieved at the end of the last chapter: A small cave free of all bats, pests, and biting things. A bed of soft pine needles or other foliage. A large creek of clean water with plenty of fish. Traps to effortlessly catch those fish, and a garden planted strategically in time to provide something ripe to eat all year round.

I believe you would occasionally still venture to "your" distant original food sources to check on them, though with a successful garden you may no longer find it justifiable.

Assume you achieve a stable existence in *The World of You* doing all this. I suspect you protect this territory, or "property," from predators and other creatures that seek to eat *your* "captured" food, dirty *your* clean water, or steal *your* shelter. They threaten your high level of achieved certainty and freedom.

You might feel reasonably certain you can continue capturing the end of your food and water story arcs, and find shelter for every situation you can imagine far into the future.

So…is this it, then?

You've won the game?

Is this all there is?

Achievement unlocked?

The pinnacle of your life?

The grand purpose of your existence?

Will you stay confined to the boundary of your territory until you die of old age? Perhaps wait for this experiment to end - if it ever ends? Do you resign your existence to capturing food, water, and shelter even though they never require more than a few minutes to maybe an hour to capture each day?

I can't know your answer. But I know there is no certainty for you or me. The anchor of "normal" is always floating and resetting to address the next uncertainty, whatever it might be. Your predicted stories start reaching further and further into the future.

Your foundation is solid, but you are still a drought away from crisis. How do you "capture" a drought? You can't control the weather. This doesn't matter. Your mind will always finds the next problem to solve.

The void of the unknown calls. Your mind longs to shine a light into it.

95

Fermi's Entangled Forest

Assume you have survived in the wilderness of *The World of You* for approximately two years. You have a regular rotation of food: meats from animals and fish, a crude garden built from seeds you've collected, and known locations of many food-providing plants. You found a nice cave, and survived many encounters with wild beasts. You figured out fire, and use it regularly. You have traps, snares, spears, and crude containers for storing and carrying food and water around with you.

But uncertainty looms. It claws its way into your future. It beckons you to solve the next imagined story of danger or death. After all, uncertainty left unaddressed could lead to something bad you didn't expect. It's better to know what to expect at all times, right?

Uncertainties beg for capture.

Certainty begs to increase.

Because of this always-nagging imbalance, I believe you will eventually begin to explore outside of your "home" territory. But why venture beyond this boundary of stability where food, water, and shelter come easily?

The void is calling. In your mind's seemingly innate need for ever-greater certainty, I believe you will begin to wonder if something lies just beyond what you have experienced so far that would ensure your survival even more. Maybe there is a better cave out there. Or a huge orchard of fruit trees and watermelons. Maybe there is a field with hundreds of rabbits, deer, or squirrels. Or maybe there is a threat you don't know about that could become a huge problem tomorrow if not captured now.

If you reach a high-level of stability (aka capturing the ends regularly), your territory becomes the two-room system from *The Ark*. Everything in your life starts to be expected. And just like the two-room system, you know exactly where the void can be found – the doors into darkness are wide open.

The void beyond what you have already experienced is an uncertainty all its own, and it feels like you need to know what it hides to determine if you need shelter from it. I think the need for shelter from movement other than your own drives you to seek out all other movement in order to feel safe from what would have otherwise been unexpected.

As time passed and you began to feel trapped in a familiar routine, you figured out how to carry water and food, constructed a mobile weapon, and set off to "pierce the boundary" of your own Step 2 that is defined by the limits of everywhere Step 1 has been. These captured resources-made-mobile provide the increased certainty you can capture hunger, thirst, or fear in a timely manner while on the move.

There is a balance here between mobility and the length of your timeline. If you try to carry enough food, water, and shelter to make your timeline stretch for months while on the move, you will not be able to move due to the sheer weight of your pre-captured resources. If you choose to make yourself perfectly mobile, you cannot carry enough pre-captured resources to stretch your timeline enough to escape your territory and return (because its size was defined by how frequently you needed these very same resources in time) and come back without creating a personal disaster. I think finding "waypoints" where you can restock your resources is ideal, and we'll now move forward with this assumption.

After multiple outings of ever-increasing distances and time from the familiarity of your known territory, you find a good balance between weight and mobility by finding and using food and water "waypoints," which allow you to pierce deep into the void of the unknown.

With each successful out-and-back, your mind grows more confident in its expectations from the unknown. It's even possible it will re-anchor to this new "exploration mode" story as your new "ground" – normal, predictable, and expectedly survivable. Perhaps your story system will even find it exciting and satisfying to expand your boundary of wordless, gapless experiences against the void, because doing this gives your mind an ever-growing list of truth-stories that validate and confirm its predictions and expectations to be right over and over again. Fear begins to fade away.

One beautiful and clear day, you go further than you've ever been before. You explore the unfamiliar-but-as-expected forest far from home, moving swiftly and confidently with years of living-in-the-wild experience under your belt. You hope to crest a hill into a new fruit grove, or vegetable patch, or water source, or perhaps just hope for something beautiful, awe-inspiring, and unexpected.

You are many hours outside of your territory. The decaying trunk of a large fallen tree blocks your path. No big deal, you've jumped a hundred downed trees to get this far. You toss your weapon over, and hoist yourself over to the other side, and land on a bunch of twigs and sticks. You didn't notice they were there before jumping over, and they crackle loudly under your weight.

You immediately look up to discover a small open meadow straight ahead. There, standing in the meadow and turning at this very moment to face the sounds you just made, is a Being with bright green skin.

This Being looks much like you in shape and detail, and is about the same size as you. It has hair in all the same places you do. You notice a row of dark spots running along the center of its back as it is turning toward the sounds you just made. The Being is unknowingly about to make eye contact with you in the next fraction of a second.

What do you do?

I want to provide you something you don't actually have in this moment: Time to think.

I urge you to please stick a bookmark in the book right now and close it or pause it, and think on this. Maybe take a minute, maybe take a day, or maybe take two. Take whatever time you need to make a clear choice on what you would do in this situation. Then come back and continue.

If you are reading this, I'll assume you've determined what you would do. What predictive and imagined stories did you come up with based on your past experiences coupled with this present moment?

Before continuing, I have to filter out one possibility: Did your choice try to insert *more time* into this situation?

I have to ask this, because *most* people try to do exactly this when I walk them through this thought experiment. As the mind is overwhelmed with the unexpected and truth needs to be re-anchored, expanding time is the first thing it wants to do, just like it wanted to do when encountering the pack of wolves.

Second You immediately overrides First You, as it examines and constructs all the possibilities, but as it isolates the situation and starts thinking in terms of movements, its perception of objective time – or the movement of objects outside of yourself – starts to warp. This results in Second You inevitably creating a *new story* where it achieves its most urgent need – shelter. Unfortunately, in the laboratory of your imagination, Second You actually changed the speed of other objects to achieve this shelter, which is not possible. To say all of this another way: When thinking about situations like these over and over, you can end up inserting time that doesn't exist by both slowing down the Being in your mind and speeding yourself up. It often results in choices like:

"I would hide so I can observe it!"

"I would find a place to watch so I can figure out what to do."

I'm sorry, but you cannot insert extra objective time to this situation by thinking of a new action to perform to avoid detection. There is not enough time for your body to carry out action like these. The human body is simply not fast enough.

This does emphasize the extreme value of each and every moment you have for making decisions, though. By this, I mean every moment of firsthand observation (Step 1) no matter how brief, matters as much as an eternity as your predicted future story arcs - which were wiggling on indefinitely before this point – but have now collapsed all the way back to this moment and the few microseconds available.

Your shelter story arc is angled almost *straight down*.

The Being heard the crunch of twigs under your feet and is already turning toward the sound. It *will* make eye contact with you quicker than you can turn and look at something behind you right now, because The Being is already halfway around. You do not have time to hide. You do not have time to develop some intricate plan. The moment of contact is a split second away and nothing can be done to prevent it from happening. So…what do you do?

Take as much time to think as you need, but realize you cannot create more physical, objective time before you both become fully aware of each other. Mutual awareness of each other is going to happen and you cannot prevent it. Your next action, no matter what it is, is going to be made relative to the Being knowing you exist.

Stick a thumb or bookmark in this book. Develop a strategy.

Turn the page when you think you've got a plan.

This is not an easy situation. You are trying to predict a Being you have absolutely no context to understand. It looks like you, except for skin color and the dark spots down its back. The only tools you actually have at your disposal are subconsciously-justified actions built by lightning-fast assumptions, which reach all the way back down to the forced choice of bias you make every second of every day.

Cixin Liu's book "The Dark Forest" was my inspiration to create this moment. In his book, Liu proposes a possible story for the universe: It has limited resources. In this way, the universe can essentially be thought of as a "dark forest." In the dark forest, if you (as a civilization or "hunter") come upon another hunter (which are other civilizations) - the only logical choice is to immediately *kill* the other hunter, because it is a threat to your continued ability to survive. To merge in the master measure from this book with Liu's, other hunters are "taking" your time to exist, because that is what resources ultimately provide.

For this reason, the dark forest scenario implies telegraphing the message of our presence into space would be foolish, because it will lead to our destruction if we are noticed. Every civilization in Liu's universe eventually realizes it is better to hide and survive than stand in the open of the dark forest.

Let's bring this concept back to our thought experiment.

Two beings on an all-wild planet are about to become aware of each other in a small clearing of a dark forest, and hiding is *no longer an option*. There you both stand – you and the Being that looks very similar to you except color and skin pattern. You cannot read each other's minds (at least as far as you can know). You are quite *literally* two alien civilizations of a million-billion processes becoming aware of each other's existence for the first time.

This being said, I do not see this moment as one of a showdown ultimately tied to consumable resources. If you are the only human on Earth and you still have your memories from before this experiment, then you are aware there are more resources on the planet than you can possibly consume in one hundred lifetimes.

To me, this means you aren't looking at this Being and telling yourself stories of it consuming your food, your water, and your shelter. You have already secured enough of each of those things, and have even been finding them and carrying them along the way. You were just out exploring the void filled with billions of invisible doors (what were you looking for out there again, exactly?)

Instead, I believe this moment between you and the Being is an assessment of direct threat; an automatic subconscious justification built on your "feeling" of imbalance between *fear* and *shelter*. Shelter is your buffer zone of safety against the unpredictable. Its intent is to prevent the complete collapse of your imagined time remaining due to fear. Currently, it is on the ground near your feet in the form of a weapon.

I suspect the sudden discovery of the Being has collapsed the certainty you had built up over the past two years in *The World of You*. Nearly all predictability and expectations of your future in this place are severely damaged.

This moment is *chaos* for your mind as it completely floats *all* anchors and truth-stories about your environment, and races to re-prioritize everything with extreme urgency to figure out what story you must have just entered and what action should be justified. You desperately want your timeline to re-expand and your freedom of choice to be restored, but you do not have any context to determine the best way to accomplish this. It's like you were in the void, blinked, and found yourself back in the room with no windows and no doors.

I suspect some part of this situation will feel similar to coming across the snake, or lion, or pack of wolves for the very first time. A lightning-fast story-building process takes place of both expected threat and identification of shelter.

However, a major difference between this Being and snakes, lions, or wolves is that you likely have *some* context to understand what threat level snakes, lions, and wolves pose to you. You probably know through firsthand experience or secondhand stories that those creatures have teeth and claws and fangs; their approximate speed and strength; and what may scare them away or even kill them. If you were lowered into a cage blindly with one of these creatures, you probably have a decent set of expectations to work with.

Now though, you are about to lock eyes with a Being, and those eyes will be the same size and shape as yours, but with radically different colors. Perhaps they will be bright yellow or fluorescent orange. If you were told you are about to be blindly lowered into a cage with a green-skinned yellow-eyed Being that looks similar to you, you will have *no idea* what to expect beyond those words. Your story-building confidence collapses in such a moment and adrenaline floods your body, heightens your senses, and gives you additional strength and speed to aid in your survival. Your brain dumps chemicals to slow time by speeding you up as much as possible. Nothing else in existence matters in this moment to your mind except the Being in front of you. You now exist as close to the present moment as you can ever get.

How do you assess what to do about a Being that looks similar to yourself? In this moment, "looks" are the only thing you have to go on. Your subjective wacky windows are informing you about an unexpected boundary ahead made of skin and hair. Your mind races.

While you have probably already imagined a path of action you will take with the Being, there are actually *three* paths of action possible. I will now refer to this moment of choice as the *Dark Forest Moment*. There are also three possible paths out of this moment, and they will feel very familiar to you.

No matter which path you choose, one thing will remain certain: Your entire reality has just permanently changed in a fraction of a second. Your truth is overwritten. Priorities are re-aligning. Right and wrong are reset.

You are entangled.

A two-point system.

Relativity applies.

96

The Time-Shelter Paradox

The first option of three possible paths relative to the Being is commonly called *flight*. Several individuals that I have put into the Dark Forest Moment with the Being have chosen this path.

"I would run as fast and far as I could to *escape*!"

This choice potentially provides more time that is so desperately wanted by your story system in this Dark Forest Moment, but it comes at a terrible cost, in my opinion. Choosing to run away dramatically damages any and all certainty in almost every single one of your predictive stories. Running also involves turning your back to the Being, which makes you vulnerable should the Being decide to give chase or attack you in some way.

Let's ignore this attack possibility for now and play this choice out from your perspective: Where exactly would you run? Which way is "away"? Exactly the opposite direction you were traveling before stumbling into this Being? Personally, I believe this is the direction you will choose.

I suspect you will naturally start running toward what is familiar: Your territory. Familiarity is a form of shelter. I believe running is a subconscious attempt to return to the level of certainty and predictability (validated truth and expectations) you once had.

So imagine you run - for miles - straight back to the safest place you know: Your dwelling of the last two years. You never even look back. You dive into your cave and seal the opening with the perfectly fitted rock. Safe.

In your imagination lab, I think this moment often becomes the story's ending. However, ending it here creates a void. Time and movement do not stop or have gaps. They just keep going. The story does not and cannot end here – not unless you are prepared to die in your shelter from lack of food and water.

How long would you wait in your cave, or any other makeshift shelter you found or created after running from the Being? You do not know what the Being did. You do not know where the Being is. You do not know if the Being is currently sneaking around, spying on you, trying to find you, or if it too ran the other way – away from you.

I think you will very likely be locked into a state of permanent anxiety and severe uncertainty. You will be unsure about your future and your survival in general. *Was the Being hostile? Was the Being friendly? Does it want to kill you? Was it scared of you?* You can answer *none* of these questions.

In my opinion, there will be no real rest for you again with the choice to run. You are mentally entangled with the Being because it entered your story system through Step 1, fell into memory in Step 2, but has absolutely no context or associations you can connect with it. All predictive stories and expectations involving the Being created by your mind are pure projection based on nothing but assumption.

I think it's likely you will develop sleeping problems, and sleep with a weapon because you fear an attack at any moment. Every noise would attract your attention. *Was it the Being?*

You might come out of your cave only in short spurts to get food or water, and only after peering around carefully. When outside, I suspect you'll be wracked with paranoia the entire time. The Being might be watching, after all.

The Being invades *every story* in your story system.

Did it have better tools than you? Better weapons? Is it in the trees? In a hole? Camouflaged? Can it change its skin color? Turn invisible? Is it watching you from somewhere you can't see it? Again, you cannot answers these questions, because you ran.

Your ability to imagine stories of your own harm and death begin to skyrocket. There are now a thousand new and probable stories of ways you can be attacked or be killed. Your past associations related to violence have transformed Second You to become your tormentor, as it constantly imagines all the ways the Being could attack you.

Since you don't know the Being's abilities, the possibilities only get more imaginative as you start building stories from your past related to any and all animals, or even fictional stories. *Can it travel underground? Fly? Possess your mind? Control your body?*

Your mind has dramatically shifted into severe imbalance. Certainty is almost completely collapsed now, and the number of story arcs in need of capturing has expanded to the point that your new reality works out to be: "at any moment I may enter the story of my death." Water is no longer your most frequent occurring need. It is *shelter*. The frequency of this need can now be measured as *always* needed within mere seconds.

I think you will likely retreat to shelter regularly, or move from hiding place to hiding place, and always watch for the Being to pop up anywhere. Perhaps you'd even start to hallucinate its appearance. Second You has literally transformed itself into the Being due to its inability to solve the problem.

Your territory collapses in size as you find yourself unwilling to venture out into the open for hours, which means the choice to run has potentially cost you precious food from fear alone. You have essentially become a mouse fearing a hawk, only leaving your den to eat and drink, and using constant vigilance while doing so, with your attention laser-focused for hawks. The twisted part of this metaphor is that you actually have no idea if what you believe is a hawk is truly a hawk. You have no idea if it wants to attack or eat you – only that it seems possible – it makes sense to you.

Your previous ground state of waves of future story arcs collapse to being almost a flat line. Your freedom to choose actions unrelated to food, water, and shelter is all but lost. The only choice is near-permanent shelter. In a cruel twist, the stress of all this increases the activity of the million-billion mini-processes of you, which makes them need you to consume more.

Even if you live this way for six months without ever encountering the Being a second time, I don't think you will ever see the forest around you the same way you did before the Dark Forest Moment. What seemed like *The Ark* before the Being became *Gravity* afterwards.

I believe you will eventually try to re-find the Being out of desperate need to re-increase your certainty about the future and grow your freedom of choice (which would give you the ability to create a story where you rest peacefully again without fear). You hoped running would do this for you. It turns out, you can run all you want, but you cannot hide from your own imagination.

In the end, you *need* to know the Being's whereabouts and capabilities to be able to predict your own stories again in ways that account for and include the Being in them. You have to "capture" *real* stories about it instead of imagined stories that are all but destroying your certainty in this world. So, you set out to do just this.

I suspect you hope to sneak up on the Being in the exact same way you feared it would sneak up on you – in secret and stealth – unnoticed. Your story system would prefer to gather firsthand experience about the Being to better construct stories about it while also being sheltered from it.

However, re-finding the Being first or the Being re-finding you first are both real possibilities, and both will inevitably return you back to the same moment when you chose to run. No matter what you observe, even if you watch the Being's behavior for *weeks* unnoticed, the moment when you decide to re-engage and lock eyes in total awareness of each other without knowing what will actually happen, you re-enter the *Dark Forest Moment*.

You can only *hope* your predictions and expectations of truth, and what you think will happen based upon your observations of this Being are "right" when the moment comes. *But you can never be certain.*

The *Dark Forest Moment* is a door into blackness that you must step through to determine what is true through firsthand experience.

In summary, choosing to run from the unknown will eventually return you right back to that unknown, or result in a permanent state of running from a feeling you can never escape, because the imbalance causing you to run is in your own head.

Choosing to run creates an expanding loop – a time-shelter paradox.

The Motion-Shelter Paradox

The second of the three possible paths relative to the Being is the choice commonly called "freeze."

As soon as the fallen branches crunched under your weight, the Being became aware of movement nearby, and began turning toward that movement, which will turn out to be you. There is not enough objective time left in this situation to avoid being seen.

Once entanglement occurs in the *Dark Forest Moment* of mutual awareness, this "freeze" pathway immediately begins decaying.

It might seem reasonable to think freezing stops this moment in its own way, and prevents something bad from happening. However, if you try to stay frozen to somehow find stability with the Being through inaction, the Being will become the one to choose which of the two remaining paths you will take, and therefore gets to control this entire situation. Your choice to freeze doesn't force the Being to freeze as well.

Freezing is not exactly a neutral choice, either. To better understand how freezing (or refusing to move or act) would affect the Being's perception of you, imagine it flipped around.

What if the Being froze as soon as it saw you and *remained* frozen even after you started moving? What story would such a choice of action (the choice of inaction) tell you? How motivated would you be to snap the Being out of such a trance? To move things forward? Would it bother you? Would you start to poke at it? Yell at it? Hit it? Would you be comfortable walking away from it? I have no doubt its refusal to move or interact with you would eventually become very distressing.

This is why I believe the story of stability by choosing not to move at all quickly dissolves as a form of shelter, just like it did at the beginning of *The World of You*. *You are forced to act.* Objective time and movement outside of yourself and within yourself does not stop even if you stop justifying movements with your body.

You already know what happens if your choice to freeze decays into the choice of flight. One path remains...

The Shelter Dance

Should you choose not to run as the Being turns to face you, even if you choose to freeze, the third of three possible choices is made - the choice of interaction and transformation.

You eyes meet, and your perception of time slows and your focus and strength increase due to the adrenaline flooding the system of you. An unexpected boundary in the shape of this Being creates an immense story voltage in your head. It is blocking your view of the universe, your view of time, and even your view of your own continued existence.

Your story system completely unanchors itself, wipes away every unrelated story to the void, and tunnel-vision focuses on analyzing one thing and one thing only: How this new and unfamiliar boundary behaves. With most of reality's stories wiped away, the stories of this new boundary's future movements can be built and revised at warp speed. After all, anticipating what movement comes next must be *right* to ensure you stay alive and unharmed. *Being right about this Being is the only chance at shelter you have.*

A new language that has never been used before in this experiment springs into existence – body language. Eyes locked, every slightest motion or facial change is perceived, analyzed, contextualized, and understood through merging with memories of your past experiences. Your story system will most likely hyper-focus on the slightest change in sensory currents coming through your sensory windows from this object's boundary.

You have done this with every creature you have ever come face-to-face with in this experiment, but this particular one will move just like you.

What exactly are you looking for in its movements?

Have you already decided what you intend to do relative to the Being? Why?

When I posed the *Dark Forest Moment* to a retired military pilot that was trained to survive behind enemy lines, his answer, surprisingly, was to freeze. However, freezing was simply how he began interaction. He already had a plan beyond the freeze.

After freezing, he intended to make the tiniest movement with an arm or a leg to see what the Being did in response. Then he planned to respond to the

Being's movements accordingly. He essentially built a kind of "dance" involving tiny and hopefully undramatic steps in an effort to gather as much context as possible for the Being in this moment.

I found myself pushing him through the dance quickly to arrive at a final outcome, but something about this strategy stuck with me. Over the next few weeks, I found myself returning to this idea quite a bit. This "dance," which I dismissed at first, has the potential for extreme value in my opinion, because its purpose is to expand the path of stability for as long as possible. It doesn't rush to a final outcome, ignore feedback, and push an assumed one-way story into reality. This "dance" can slow everything down and provide precious time for your story system to catch up and find more possibilities. This approach also helps minimize risk by avoiding any large or sudden movements that might be misunderstood, which might lead the Being to potentially jump to an assumed final one-way story about you.

Regardless of whatever style of Shelter Dance is chosen, the dance itself is not the endpoint - it is *a process* toward an endpoint. After all, you *must* resume capturing food and water sooner rather than later, so the amount of time you can make this "shelter dance" your top priority has a very real limit.

You can dance, but you cannot dance forever.

No matter your approach, this path of interaction and transformation – dance or no dance – leads to only three possible stories that I can imagine to "capture the end" of your shelter story arc and restore any freedom of choice at all.

Outcome 1: Death (Collapse-to-Expand Paradox)

It's possible you might go straight to trying to kill the Being, exactly how the universe is best handled in Liu's The *Dark Forest* book.

To have taken this option, you most likely decided the Being is the cause for the collapse of your timeline and your freedom of choice, so eliminating the Being is to "capture these ends". What you are really trying to capture though, is your world as it was *before* you knew the Being in it. This choice would be like trying to mentally return to the past.

To go back in time, it might make sense to your story system to simply "undo" whatever has changed. *The Being must be erased.* However, once this story is chosen, your body language toward this end is likely to be immediately noticed by the Being, leading it to potentially choose the same intention from a desire to survive you. This sets up a duel, where one of you is likely to experience the end of time and movement: Death.

If the Being is the victor of the duel, your time and movement collapse to zero and you no longer exist.

If you are the victor and successfully kill the Being, I think the result may not be what your story system hoped for.

Even with the Being dead, you are *still* not likely to move in the woods the same as you did before. Your story system justified killing to be the right thing to do, because it believed killing would return the certainty and freedom you once had. But time does not move backwards, and this predicted truth-story has massive gaps in it.

This event rocked your beliefs of what is true in this place, and with it all expectations in this reality. The whole *Dark Forest* situation had no context to start with: I'll assume you have no previous firsthand experiences with green-skinned yellow-eyed Beings to know the best course of action. It was simply a perceived threat to your safety, and you decided to eliminate it.

You now have only gained a few stories related to this Being: How to kill one, and a little about the fight it put up, which provides you with partial stories related to its intelligence, strength, stamina, speed, mobility, and a few other things.

As you stand over the body, you now know from experience it was pretty much a perfect matchup with yourself on all of these measures.

The urge to return to the same level of certainty and freedom you had before the Being became the story you wanted to merge into reality, which justified "capturing" the boundary that was suddenly separating you from that imagined future by destroying it. But that story of restored certainty and freedom was imagined from your past.

To tie this back into *The Ark*, justifying killing the Being right away is like regretting stepping into the void at all, and wanting to return to the familiar six-door room. Even if you return to that room from the void, though, doing so doesn't erase your remembered experiences in the black space. Killing doesn't erase your memories of the Being either, and so you will *drag the memories of it with you*.

A whole new set of associations and possibilities spring into existence in your story system now. The familiar world you knew has changed and is no longer as predictable. Your certainties in life will likely never fully rise to what they were before the *Dark Forest Moment* occurred.

How did it get here? Like you did? Was it always on Earth and was just undis-

covered by humans or was it just added? Was it human? You were told you were the only human. Were you lied to? Are there more of them? Is this world now potentially full of chaotic new human-like beings you need to fear and worry about?

You now have an internalized loop of uncertainty - a permanent "story voltage" of possibilities you have no context to predict and relieve. You have only tipped further into imbalance. While I suspect your anxiety will not be as severe as it was knowing the Being is still alive after running away, you now know a complex threat *is possible* in this place, and you are leaving evidence of your own existence for it to find.

Your runaway process, fueled by uncertainty, will likely begin constructing shelters against all the stories from your imagination where complex life might try to find you.

In a way, you have very much returned to the Time-Shelter Paradox. Choosing to kill results in nearly the same mental outcome as choosing to run. There are a thousand questions you cannot answer, and a thousand imagined possibilities you now find yourself preparing for. They can grow to be more and more extreme with absolutely no way to validate them as right or wrong.

By trying to capture your own fear with violence by killing the object believed to be the source of that fear, a feedback loop is created that results only in more fear - a paradox. Justifying the collapse of another to re-expand has only led to further collapse.

Outcome 2: Dominant and Submissive (Stability Paradox)

My 9 year-old daughter (at the time I wrote this) changed her choice in the *Dark Forest Moment* many times. One of my favorite answers shows the size of her heart: "I would hold out a gift. Maybe a rabbit as food."

Of course, there is the huge problem of not having a shared context with the Being: How do you know holding out an animal (dead or alive) will be seen as a gift and not a threat? You know literally nothing about what the Being values or wants.

This same problem applies to nearly every attempt at communication between you and the Being. You have no certainty the Being will understand what you mean by anything you say, do, or present. You cannot magically understand its context for how life works and what its needs are. There is a mental boundary which prevents it.

How will it react to a sound coming from your mouth? You cannot know.

What would you even say, and how would you say it that you could be certain of the outcome you want to happen? In my opinion, the only way to determine this is to walk it though:

Let's assume you interact with the Being using just about any method: Sounds, body language/movement, fighting, offering a gift, or whatever other way you can imagine.

If one of you shows submission using these methods, and the other one provides mercy (such as avoiding the killing stroke or accepting a gift), a core problem of uncertainty and balance appears, and the root of this problem is the inability to read each other's minds.

For example, assume the two of you start fighting in the *Dark Forest Moment*, and you end up the victor after quite some time. This fight included wrestling, biting, scratching, punching and kicking. Eventually, the Being caved into fear because it predicted and began to expect you to bring about the end of its timeline in a way it could not prevent, so it became submissive in whatever way you imagine this to happen. Sensing this fear and submission to your will, you decide to have mercy, and hold off your killing stroke.

The story might end there in the laboratory of your imagination. However, you now face some very tough choices. Are you comfortable turning your back on the Being in the near future? Can you *trust* the Being to not attack you when you are vulnerable? Would you be comfortable sleeping next to the Being? What kind of conditions would you need to sleep knowing the Being is nearby? Would you have to lockup the Being (or physically restrain it somehow) to feel secure it won't murder you in your sleep? Is the Being predictable at all?

The true test of any answer you might have to these questions is to flip the situation around: Let's assume *you* are the one defeated in violence, and the Being holds off its killing stroke. Now what? How do you feel about this Being? Will you consider attacking the Being when its back is turned? Will you attempt to murder the Being while it sleeps? Should the Being lock you in a cage to keep itself safe from you?

Let's hit this from another angle: Assume all this violence didn't happen, and instead you present the Being with a gift, perhaps my daughter's idea of a rabbit – and the Being smiles and accepts this gift. What now? Does the Being now see you as a servant? That it is better than you? Superior? Will it attempt to enslave you now? Demand more rabbits from you? If it did, would you willingly choose to be subservient to ensure your own survival?

I believe in either of these situations, you end up in a loop right back into the *Dark Forest Moment*. In his book, Liu's characters call this process a "chain of suspicion", which I think is an accurate choice of words to describe this point in our story. The chain of suspicion exists solely because we cannot read each other's minds, and we also don't have enough context to have any idea what "the other" might do next.

As a result, you can get stuck in a mental loop where you imagine a possible story of the Beings beliefs or intentions. You then project that story onto the Being, and get trapped in the idea that *maybe your thoughts about the Being's thoughts are right*. Second You then amplifies this possibility over and over again in an attempt to validate it as being right or wrong in any way possible, but since you cannot actually read the Being's mind to know its thoughts, Second You starts a runaway process in your head.

Is the Being thinking about killing you?

Maybe the Being is thinking about it.

What if the Being is thinking about killing you?

What if the Being thinks that you think it's thinking about killing you?

This growing loop will inevitably lead you back to one place: A *self-contained Dark Forest Moment*. You are one injecting stories of justification for all the Being's actions, but you are using *imagined assumptions*. Your story system desperately wants to predict the Being's actions by knowing its justifications for them, but it can only do this by validating your imagined stories of justifications *backwards in time*.

This self-contained loop in your own head keeps trying to determine what is going on inside the Being's head. *All of it* is fabricated and enforced by *your* past and imagination, not the Being's.

This chain of suspicion immediately appears in every imbalanced version of the *Dark Forest Moment*: If the Being was larger than you, smaller than you, presents as dominant, or submissive to you. These all link into a chain of suspicion that can't be easily broken, because you have such a severely limited context to understand any stories the Being might be using to justify its actions.

Trust is the problem. Trust is almost totally absent. In its place sits fear, anxiety, and paranoia. Your own mind can start to drive itself to madness in its desperate need for shelter. It might attempt to assemble and predict wild stories about the Being's beliefs and intentions. This arms a ticking time bomb in your head, counting down toward violence if not noticed and disarmed.

It might take seconds, minutes, weeks, or months to happen, but an imbalanced relationship will almost certainly lead to the *Dark Forest Moment* again, and the longer this imbalance remains in place, the harder it's going to be to avoid violence. (If you are wondering why the length of time is tied to the probability of violence, I'll get to that a bit later.)

In my opinion, choosing the path of dominance or submission only leads to temporary *physical* stability at the cost of creating *mental* instability. Internalized loops of stories that cannot be relieved eventually will cause this whole situation to end up back where it started: A *Dark Forest Moment* all over again - a stability paradox.

Outcome 3: Equality/Stalemate (Decay Paradox)

The best as I can determine, there is only one way to find shelter in this situation that doesn't result in the creation of an unbreakable loop inside your mental boundary. It is to recognize the Being as a valid equal, and not seeing it as a threat or as greater or lesser than yourself.

The part beyond your control is that the Being must make the same recognition with you. I admit this seems like it might seem downright absurd to you. As I see it, there is only one way this outcome can be successfully reached: By constantly expanding a brand new *external* loop between you. This loop already exists in your life today, and already has a name. It is called *communication*.

Now, I mentioned earlier that any communication, such as using a rabbit as a gift, is totally uncertain because you have no shared context to understand one another. This alone seems to create a kind of communication paradox. How can you even begin from such a place?

As stated earlier, the primary barrier to successful communication is that you cannot read each other's minds. You do not know each other's intents, thoughts, or feelings when you arrive in the *Dark Forest Moment*.

The only option to stop a constant decay of choice into fight or flight is for both of you to keep inserting *more time* by any means possible. The best way to do this is to attempt communication. The goal of this communication is to find common ground, *any* common ground – to create a foundation you can stand on together. In all of the other choices in the *Dark Forest Moment*, self-contained mental loops were created.

Run? Loops of all the things the Being might be doing you can't know creating fear and anxiety.

Freeze? Loops of stories of justifications for movement that keep canceling each other out of fear or anxiety.

Fight to the death? Loops of how to shelter against an entire world you don't understand, with threats you can no longer predict, causing fear and anxiety.

Submit or dominate? Loops to predict the other's justifications that might be trying to restore balance at any moment, creating fear and anxiety. (The dominant position requires constant loops of what the submissive might be planning to gain dominance. The submissive position requires constant loops of what the dominant might be planning to do to maintain or increase your submission.)

Instead, if you both approach with uncertainty, recognize that you both appear to be very similar and are surviving in this same reality, and do not rush to a conclusion about the other's thoughts and intentions, you can expand time while existing *together*.

In a way, this is a life-and-death parallel of you and Bob discussing the jet on the treadmill. You and the Being are on ever-accelerating treadmill. However, unlike you and Bob, you and the Being do not yet have words to work through this situation and eliminate fear and anxiety of each other's intentions.

The treadmill of life doesn't care, though. It speeds up anyway.

The real controller of the treadmill is your million-billion mini processes. They are always speeding it up because they have needs that *must* be fulfilled at certain intervals of time and they *don't ever stop* moving. They set the minimum speed you can operate in this life.

The longer the *Dark Forest Moment* with the Being goes on without a resolution that allows for other priorities to rise like eating and drinking, the more I believe you will see each other as a boundary standing in the way of survival. Your mind will start to subconsciously change your priority to food or water anyways, and your patience and willingness to try to maintain balance in this situation drops dramatically.

In other words, your bodily processes *decay your freedom to keep making the choice* to stay on the middle path to stability and balance through communication. In this way, your own body can force you to make a different *Dark Forest Moment* choice before you are ready to let go of the expanding path of stability.

This decay of the middle path forces you back into the intended choice of killing, domination, or running away.

Which of these intentions you will decay yourself into choosing is *already made* deep in your subconscious. Let's take a quick peek under those covers.

99

A House of Mirrors

To understand the choice you would naturally make at a speed faster than you can consciously control when faced with the *Dark Forest Moment*, think back to the end of *The Strawman* chapter, when you were floating outside the familiar six-door room in the huge billion-door space debating if you would re-enter when "something just moved". You hopefully noted how those words made you feel before starting the *Gravity* chapter.

That feeling, in my opinion, exposes your hidden and subconscious path of least resistance when it comes to unknown or unexpected movements, such as encountering the Being. I believe this path was crudely formed before your memories began, and then reinforced in the first few years of your early experiences – and it has everything to do with how you handle the unexpected.

Before your memories began, nearly *everything* was unexpected. Think about how easy it is to scare or surprise a baby, and consider this also applied when you were a baby. How you processed these unexpected moments (with or without the help of others) likely formed your fastest and largest story system straw-path that determines how to best "capture" fear, and convert the unexpected into the expected. This process "trains" baby-you's near storyless story system how to respond to future unexpected occurrences. I believe this underlying training is still in place for you today. It's what I was trying to expose by having you notice your feelings before *Gravity*.

If your subconscious reaction, or feeling, to "something just moved" was fear and wanting to escape, I think you will likely have a similar subconscious fear response to the Being. Escape is likely to be the trained subconscious path of least resistance for you, given to you by those who validated or invalidated your responses to the unexpected as a child. As a result, when you met the Being, you were likely to run, freeze, or submit, because that straw-path is the biggest and least resistive.

If your subconscious reaction to "something just moved" was fear and anger or aggression, you will likely choose a similar path with the Being. That might be violence or attempts to dominate with no willingness to reconsider unless there is somehow a significant slowing of time.

If your subconscious reaction to "something just moved" was excitement or curiosity, I think you will be more likely to choose the path to equality or at least attempt a positive interaction. However, I believe this reaction will be extremely rare without knowing what exactly moved, because we as humans seem to be hard-wired to fear darkness and the unfamiliar. Something unexpectedly moving in darkness reaches down to what I suspect to be ancient primitive survival wiring.

Nevertheless, your initial feelings to the unexpected in a secondhand context (like this book) may inform you what your physical reaction might be in a firsthand context. I believe this path-of-least-resistance to the unexpected was molded by those in your early childhood, who acted as a type of emotional funhouse mirror.

Some individuals perhaps warped and twisted the reflections of your reactions to seem wrong by dismissing your surprise or demanding you react differently than how you naturally did.

Others may have warped them by proposing worse situations after-the-fact to prove your reactions are too over-the-top given the situation, and to react differently.

Perhaps still others warped them by altering your environment to remove the unexpected from it, to prevent your "wrong" reaction from happening at all in the future.

Each of these individuals potentially became a *Keeper of Truth and Decider of Right* hidden in shadow for you. Directly addressing these individuals over things that may have happened like these above situations, which you cannot remember, leaves you with only secondhand stories to work with. These will be weak against their firsthand experience and memories, which leaves you permanently trapped in the great context problem relative to these Keepers of Truth Before You Had Your Own Truth.

I suspect that very few people in your life, if any, validated every reaction you ever had to the unexpected, and helped you work through your emotions in such a way that you formed a natural and balanced reactive pathway to the unexpected and toward others – but I admit that I can not know this with any certainty at all. I'm not sure we can ever know what this natural and balanced reaction would look like – because it seems everyone was raised and conditioned by someone else.

I also cannot possibly know what your path of least resistance might be as it relates to the unexpected. Only you can know – but if you've never actually experienced such a moment of fight, flight, or freeze firsthand, it's possible you do not know, either. Regardless, I strongly believe you are fully capable of *changing* this path-of-least-resistance with practice through firsthand experience.

By putting yourself in situations and positions that are unfamiliar or intentionally challenging your subconscious snap-reactions using a way that is not judgmental, you can redefine your fastest (and therefore your most foundational) pathways of story building and justification.

Another way of saying this is to regularly *face your fears*.

The goal of facing your fears is not to "capture" your feeling of fear through action or controlling your environment – that is a runaway looping process. Rather, the goal is to capture the automatic fear *response* in your head that might drive you to subconsciously (automatically) justify the option that leads to lasting damage to yourself, such as violence, fleeing, superiority, or inferiority. The goal is simply to better understand yourself, which might come with the added side effect of major personal growth. Perhaps even experiencing your own mental "event".

After all, only one option in the *Dark Forest Moment* seems to be the way to avoid internal harm to yourself – the choice of assumed equality and the choice to communicate.

By using body language to communicate no harm is intended to another's existence, and by recognizing you likely have much in common based on appearance alone, you and others (such as the Being) can potentially avoid self-contained mental loop-traps and keep things as an external loop between each other. Through regular non-violent communication, you can even expand the decay-to-fight/submit-or-flight choice further and further out in time, and potentially enter into a new domain: Cooperation.

If you first choose submission, dominance, or violence, it *might* be possible to switch onto the path of equality and communication later – but I believe this transition will be horrifically difficult with the Being.

For instance, if you are the dominant victor in a fight, to convert your imbalanced situation into one of equality, it is going to be absolutely critical you give the Being your total attention *at all times* while it is contained or controlled. Why? Well, flip the situation: Imagine if *you* were dominated and were pinned down or locked in a cage by a strange creature, which then *turned its back on you* and walked away. What does this do to you? How does this make you feel? Small? Unimportant? You seem to have dropped in status and priority to this creature, and must be less important than whatever else it is tending to, which will not go unnoticed at a subconscious level.

Should this happen, the leap to equality (should you later be freed), will be difficult. The longer the you-are-a-low-priority situation goes on, the harder it gets to rebalance to equality. You might even *demand* that the Being be locked in a cage for as long as you were to feel equal – which is not likely to ever be agreeable.

I cannot imagine all of the possible stories and methods to get from each of the other paths to the outcome of mutually understood equality. There are far too many variables to consider. There are too many subtle changes and gestures and bodily communications that happen at very high speeds that might go noticed or unnoticed, interpreted as intended, or misinterpreted from intention. There is no right or wrong way to get to the equality outcome. *It just has to happen.*

And the key thing that *must* happen to eventually reach understood equality and also keep expanding time from decaying everything back to another *Dark Forest Moment*, in my opinion, is successful two-way communication of non-threat and of common needs and values. Concepts like "friend", "food", "water", and "shelter".

A common foundation *must* be found between you as I see it. A new "ground," "baseline," or "footing" *must* be built to have any stability at all.

Even at a stalemate and position of equality on the backside of the *Dark Forest Moment*, the Being is still completely alien to you, and you to it.

All of this foundation building through communication has to happen as quickly as possible. The need story-arcs for each of you are still moving, and their turn downward will create a new imbalance between you soon.

There is a lot to unpack between each other, and very little time to do it.

First though, I'm going to thump you on the nose, and state what might be obvious, to increase my certainty the metaphors of this situation are not lost to missing context or an exploding cat.

100

Aliens In An Unfamiliar Universe

While the situation with the strange green-colored Being might seem alien or fictional, you very likely experience it every single day. *The Dark Forest Moment* occurs in your life every time you encounter another human you have never encountered before.

You cannot read the mind of a stranger and they cannot read yours. You do not know the stranger's intents or thoughts about you. The only thing you can do to determine your next action relative to this stranger is to notice body language and the context of the situation, and try and predict what happens next.

The only thing you know about a stranger's beliefs and truths are his or her actions, which include words formed by actions of the throat, tongue, lips, jaw, and lungs. This is also all a stranger can know about you, if we assume you are also a stranger to the stranger.

If you are hiking by yourself one-hour deep in the woods, jump over a downed tree lying across your path, and look up to unexpectedly lock eyes with another human being standing in the middle of a meadow ahead – what do you do?

While you have a hyper-fast subconscious path of least resistance that biases the general direction of your next action toward fight, flight, or freeze, the choice you make will depend heavily on the *context* given to you by Step 1 and 2 of your story system.

In this same stranger-in-the-meadow moment with the Being, you *knew* you were the only human on Earth, and believing this likely caused you to reject the story that the Being could be a human like you.

When I first created this *Dark Forest* thought experiment and tried it on others face-to-face, it would fail completely and dramatically if I used the word "person" or any other word that implied the Being was human in any way. Doing this would cause the individual to instantly jump to subconscious automatic actions based on truths and experiences with other human beings in their lives. This resulted in no hesitation at all in reaction, no careful thought, and often very bold actions justified on all-too-familiar assumptions fed by familiar low-re-sistance mental pathways.

My early failures were not without value though – no failure is absent of value. They showed me something interesting and shared among all human beings: The actions these individuals rushed right into performing toward the "humanized" Being, without hesitation, always involved *using language.*

In the real world right now, you already have lots of context for human strangers, which allows these kinds of assumptions to *likely* merge into what you expect.

For example, if you run into a total stranger while on a hike in the woods at a public park or trail, you might say hello, or engage in a conversation, or just make and then break (or never even make) eye contact as you walk past - without experiencing fear or anxiety from that choice. It all depends on what context exists for the other person's behavior in *your* past, coupled with the context of behavior that is expected in the *present* situation, which is also pulled from *your* past.

If the stranger makes a face you interpret as disgust when making eye contact with you, and then frantically starts digging in a pocket, you are going to build a story using your past with human beings when they are disgusted by someone else, and predict what comes next as *probably* negative. Your guard and anxiety may go up very quickly until you make it past the stranger without a threat emerging from the void of his or her pocket and they are no longer paying attention to you. However, the stranger will probably still linger in your mind until you leave this place, and may remain in your mind for quite some time. After all, you don't have any closure, validation, or justifications for their reaction to you.

In contrast, if you were an hour-deep in the woods, hiking on private and undeveloped land where no one else is expected and with no marked trail in sight, you might enter a barely-any-context experience like the one with the Being should a person unexpectedly appear ahead of you and is looking at you, naked, and covered in green body paint.

Gaps in your experiences create the extreme feelings involved in these moments. The behavior you are experiencing has barely any connection to experiences in your past. It is unexpected, and your story system's voltage spikes like it was hit by lightning to try to *make sense* of what is happening and adjust truth, right, and wrong.

More often than not (in the United States anyway), the situation with the naked green-painted person in the woods is not going to be considered normal. Cultural-context-built stories might merge to tell you this person "must be mentally deranged – perhaps even dangerous". He or she is not following *norms or assumptions* of expected behavior given the surroundings. There is also not much time for you to carefully develop a story of what is going on. You are context blinded and so must build context in the present moment quickly and firsthand. You have entered the *Dark Forest Moment* while exploring the void.

Just like what happened when I accidentally implied the Being was human, I think you will likely jump straight to *language* as your tool of choice in the void. You may yell questions or commands at this "green person" who might as well be an alien from an unfamiliar universe. If the person does not respond, you will probably tell yourself a story about this person in the moment, and justify what to do based on it. The outcomes of the *Dark Forest Moment* still apply.

To expose the extent of the experience gap with the green man a bit more: Someone out there knows the green-painted person's name is Bob, and he started doing this 20 years ago after losing a sports bet. Bob discovered he had a passion for scaring people (particularly hikers), so he likes to go out in the woods and do this naked-green-paint thing every other Tuesday.

If you ever visit a foreign country where your known language(s) is not used at all, you will be forced to rely entirely on your own context and norms to understand social situations and what is and is not expected in a given environment (assuming you've not read any books, don't have a translating device, and lack all stories of the local culture). The *Dark Forest Moment* is always possible, because you are not completely familiar with normal. You are the alien in this case.

What is the only thing that prevents a *Dark Forest Moment* for you when, say, a stranger wearing a business suit makes eye contact with you as you walk into a new-to-you office building?

Context. It is *only ever* context. It is only ever *your context*.

It is entirely possible the person in the suit is debating attacking you and even has a weapon to do it. Yet you likely feel somewhat safe in this situation, because his or her appearance aligns with a set of "norms" and "assumptions" – your beliefs and truths - of what an office is like and what people wear in them. Seeing the stranger doesn't invoke fear and anxiety because everything happening is expected, so your story system's current truth-story is valid and right.

Being murdered in an office building by a stranger in a business suit is not a "culturally common" association to make from experience, but that doesn't mean it can't happen. If you have somehow never set foot in an office building, and your only experience with one is a single fictional movie about a serial killer that wears a business suit, works in an office, and kills coworkers one-by-one –

suddenly that same stranger might invoke serious fear and anxiety for you. This is why shared experiences and culture are so important to our well being, in my opinion. Without them, we deal with the stress of near constant *unexpectedness*.

In my opinion, *shared* context or shared approximate assumptions built by repeatedly shared firsthand experiences are what constructs *normal* and sets *expectations*, and is ultimately what holds society together, keeping it from becoming a waking nightmare of fear and anxiety and violence and running.

When looked at through this lens, you can perhaps see what acts of terrorism truly accomplish: They provide a new, very real, and very possible *violent* context for a situation that did not have violence in its cultural context before. You are made painfully aware it's now possible buses explode, planes fly into buildings, someone may "randomly" blow themselves up, or bring a gun into a school.

Terrorism damages and collapses common and shared contexts, norms, and truths on a large scale by *forcefully* merging in new stories, and this twists and warps everyone's expectations, predictions, and ultimately actions.

Your predicted story of safety, based on the highly-validated truth-story proven right by your story system and firsthand experience over and over for years, suddenly gets reset – invalidated - which shakes up right and wrong with it. Fear, anxiety, and stress grow as a once familiar place returns to a shadowy void of unpredictable actions and beings that may threaten to harm or kill you.

The *Dark Forest Moment* can go *so much further* though, and can be expanded and collapsed to fit into and understand many situations.

When a person charged you with a knife from a dark alley earlier in this book, you entered a *Dark Forest Moment*, and the "Being" (knife-wielding person) had seemingly already chosen violence to death or domination as the right thing to do. You had almost no time at all to unravel and understand what was happening.

A blind date is technically a *Dark Forest Moment* as well, though not as severe, because you meet under an *assumed* shared context of a potential romantic interest. That doesn't mean it will always go that way. Certainty does not exist.

The *Dark Forest Moment* is also the mechanism behind almost every horror, suspense, thriller, and crime story. They are often centered on the forced choice of immediate action in a void of context. *Why is this person doing this?* The mental boundary and hidden context problem allows gaps between us to exist, which allows actions to be justified and carried out *before* their intent is understood by anyone else. It might even be possible their intent and justifications will never be understood, if they refuse to communicate them.

According to the secondhand stories in history I've heard and read, the first time any given person within one culture encounters another person in a different culture they've never experienced, the *Dark Forest Moment* has appeared.

Examples are individuals from English-based colonizing cultures encountering individuals from African cultures or Native American cultures, among others. Was communication from an assumed position of equality attempted to stave off violence? Or was the choice for violence and dominant/submissive roles immediately chosen?

The *Dark Forest Moment* outcomes that created imbalances have echoed through human history many times over, involving what seems to be all races and systems of beliefs. These have frequently resulted in violence to a death, or to submission and dominance, or to fleeing and pursuit.

The end result is a huge collection of two-point electric balloon-straw systems of imbalance between individuals and even entire cultures that are always trying to collapse to balance through violence, actions, words, force, or coercion, only to encounter currents coming from a third balloon or straw that changes the shape of the moving system, complicating things ever-further. This creates nearly constant "Murders with Sprinkles on Top" situations that repeat in giant loops of time - a paradox of epic and tragic proportions.

For all these universal processes that keep churning and cycling to find a balance to or at least a different and stable shape – there must be a shared foundation recognized by everyone. And so, with this you-and-the-Being two-point system isolated, let's quickly explore the benefits made possible by avoiding so-often-chosen paths to imbalance. After that, we'll continue the journey up the rabbit hole.

101

Benefits of The Socio-Path

Assume you and the Being have just successfully expanded time from the *Dark Forest Moment*, which will eventually result in a cooperative partnership. What would such a relationship do for you to increase your certainty – your mind's seemingly constant goal?

Your main problems wouldn't change. You still need food, water, and shelter in a timely manner. Neither would the minimum speed required to keep capturing them, because the intervals don't change with the addition of the Being. They are yours and yours alone. Their capture is still always uncertain, but your mind will still always work to make them more certain as far into the future as it can.

Certainty is also something that is never achievable when it comes to movement. However, the Being as a partner can actually help bring it closer.

Together, you have double the eyes, ears, and other senses to notice danger and opportunity. There are now two independent bodies working to gather food and water. Partnership increases the certainty that you can secure these needs in a timely manner through redundancy. If one of you can't due to injury or illness, the other can. You can assist each other with physical tasks that weren't possible with just one body.

In partnership, the feeling of safety increases, which I think can move this partnership into the category of shelter. This feeling of safety and shelter is very real, in that many predators may not attack a "group" like you and the Being, but may have attacked "you" as an individual. When together, you only have to focus on *half* the world you did before to achieve the same feeling of safety.

You also increase your overall comfort – which I am defining as free time to pursue other things than just capturing food, water, and shelter. This lets you both focus on solving smaller problems, such as improving your already-existing survival systems, or simply getting more rest.

You both can share your diverse sources of food, along with and where and how to find these foods, increasing food security. As a team, you and the Being can also potentially hunt and kill larger animals that were too risky before (although there is still a considerable level of danger and risk with this, as an injury can still easily lead to death or lifelong injury to yourself or the Being).

The boundaries of your two independent "territories" now look like two balloons connected by a straw-path, which you created and connected by exploring the void of the unknown. As a result of this connection, they can merge to become a single, larger, and shared territory (should you both agree to do so). This larger territory becomes an expanded space where both you and the Being are familiar with what to expect, which also increases feelings of certainty and safety. Familiarity is its own type of shelter.

You and the Being know from firsthand experience what belongs and doesn't belong, and can alert each other to unexpected things and events within the areas you are most familiar.

The longer you support and work with each other, the more I think this feeling of certainty and confidence in survival is going to increase.

It is inevitable you will develop crude communication with each other quickly. Examples could be whooping calls for alarm, and silent gestures like pointing where to look. These communications might look much like military and police wordless gesture communications. As I see it, this level of communication is not at all far removed from other animals of the world. Birds, monkeys, apes, and many other creatures use this level of communication.

Even though you both likely feel safer and more confident with each passing moment, each of you still need to capture food, water, and shelter at particular intervals of time. These intervals have remained the same as they were before you ever met. You each have your own biological speed and sense of time, determined by hunger, thirst, fear, and fatigue.

The presence of each other adds a new need for each of you, however. The non-fulfillment of this need will prevent the level of relationship these word-shells just detailed, and also threatens the stability of everything.

102

Getting On the Same Wavelength

Imagine you and the Being expand time from the *Dark Forest Moment*, and develop a shared-but-very-shaky understanding you do not mean to kill or harm each other while standing in that clearing. To accomplish this, crude communication had to be born.

The exact details of how this is imagined to happen will be different between you, me, and each and every human, so the process of reaching the point you both are not fighting or fleeing is entirely unique. There is no shared foundation in the details and style of the Shelter Dance to assumed equality.

After the shaky balance is found, however, the Being motions for you to follow in a non-threatening way (however you imagine it to be done), and you begin to walk together.

On the way to wherever the Being is leading you, you pass under a walnut tree. The Being stops, quickly searches the ground, and gathers a couple of walnuts. Then It cracks open the shells easily using a couple of nearby rocks.

The Being approaches you with what you interpret to be a positive expression. It opens its five-fingered hand to show you four walnut kernels (If you happen to be allergic to walnuts, you can imagine the kernels to be any food you like that might be found in a wilderness situation that doesn't have to be eaten immediately. It could be a green leaf, a root food, berries, bird eggs, or whatever else you can imagine.)

The Being picks out *one* kernel and hands it to you, and then closes its hand around the remaining *three*. It immediately pops the three into its mouth, and begins to chew.

How does this make you feel? How do you interpret this act?

Consider any emotion you feel right now as a predictive story that didn't go as expected. So what *did* you expect? Was this outcome worse or better than expected? Completely unexpected? Think on it.

The Being collected four.

It gave you one.

It ate the remaining three.

How do you react?

If you never expected to receive any food from the Being, this act will likely create a positive emotion for you. Perhaps you would tell yourself the story that the Being sees you as worthy of food it has collected, or has decided snacking together might be a good way to build a relationship, or just offered a gift when it didn't have to. For you, this seems to be a step in a good direction.

If you expected to receive equal amounts of food, this will likely create a negative emotion for you. Does the Being see you as less than itself? Why did it not give you an equal amount of food? Does it know it kept more for itself than it gave you? Is it assuming you don't need to eat as much as it does?

No matter your expectations or however you might interpret or rationalize what just happened, an imbalance has presented itself. This "voltage" now exists in your story system and there it will stay internally bound should you resist communicating about it. This voltage can be discharged externally (current, also known as action, which includes words) to move toward balance at any point in time.

So what do *you* do about this?

There are many approaches available. Far too many to walk through. I think chances are high you've experienced similar scenarios to this situation playing out in media-based stories (movies and shows and books). If this is the case, you might already have some context to help you justify actions to collapse back into balance. The main problem that exists between you and the Being is *words are not yet an option you have at your disposal.*

If you interpreted this walnut transaction positively, imagine the same situation repeats itself *again* at another food source just a few moments later (the Being picks 4 fruits, gives you 1 and then eats 3). Does this repeated unequal splitting raise the story voltage for you? Does it change the possible justifications you can imagine underlying this action? Would you *now* feel compelled to do something and not let this continue?

If you think you would choose to point out you are being given an unequal amount, what method would you use to try communicating this? However you attempt to do it, what are you going to do if the Being becomes angry or aggressive in response?

Give yourself a minute to imagine a walk through of the possibilities before continuing.

How many times would you have to be given unequal shares for you to believe the Being must see this as a dominant/submissive relationship?

If you counter these imbalanced acts with passive-aggressive actions (like giving the Being one of something you collect and keeping the other three), does this not send you straight back to the *Dark Forest Moment* if the Being refuses to accept the imbalance like you did? What happens if it points at the three objects in your hand and motions that you should give it another?

If the Being becomes aggressive or emotional after your attempt to communicate movement toward equality or fairness, time collapses and the *Dark Forest Moment* rears its head once again with the same choices as before. You can run, freeze, fight, or try to *re-expand* the collapsed time between you by continuing to communicate.

Allowing the *possibility* of a dominant (Being) and submissive (you) story to remain valid in *your* mind creates a *second story system* in your head. This second system starts trying to create and inject the Being's *justifications* for its actions using only *your* context. But your context cannot possibly build the justifications created through the Being's context – you cannot read minds.

The result is a story you *tell yourself* of *why* the Being must be doing what it is doing – which eventually will lead you back to the *Dark Forest Moment* decision to run, freeze, fight, or keep trying to expand communication with the Being - but the decision is made entirely within your own head between First and Second You, with no input at all from the Being. It doesn't have a clue this is happening. It cannot read minds either.

The *Dark Forest Moment* is really the point in time where two entangled contexts are not on the same wavelength. It is what happens when two different *truth-stories* collide over the same movements in objective reality, and the gaps and experiences between these truth-stories do not align with each other in a way that permits them to merge.

In the case of you imagining justifications for the Being's actions, both truth-stories are both in *your* head, and one is, at some level, a made-up story – a projection.

Does the Being actually know those were not equal portions?
Why does the Being seem to act unfairly toward you?
Should you consider stealing food when the Being is not looking?

By saying nothing and allowing what you perceive to be imbalanced actions to continue without knowing *why*, the imbalance keeps growing in one place

- your head. There are ways this can balance itself – such as the Being accepting you splitting everything in your favor when you are the one giving – but let's assume this doesn't happen.

Instead, the Being always demands *one more* from you so things are equal, and you cave to the request and hand one over. Now this relationship is almost certainly a dominant and submissive relationship, although you do not know for certain *why* the Being keeps doing this.

By keeping this story imbalance in your head, you are the only one in the relationship choosing between all options and outcomes of a future *Dark Forest Moment* – in secret, all inside your own head. *Should you flee when it's not looking? Try to kill it when it sleeps? Try to flip the submissive/dominant relationship by force, which maybe involves imprisoning, trapping, or restricting it somehow? Or do you decide to make another push for equality through communication?*

At this point, establishing equality is going to be far more difficult, because not addressing the imbalanced actions in a timely manner the first few times they happened is what put you in this position. How are you going to communicate that the imbalance was only *temporarily* okay, and that you've secretly been keeping score since the first interaction, and expect the Being to correct its behavior?

You have one advantage: Surprise. Should you choose violence in the *Dark Forest Moment* in your head - the Being will not and cannot be prepared to defend itself at all times. Unlike you, it is not imagining the story of your sudden attack, because it has already experienced and validated your approval of the 3-to-1 split. It potentially believes the current relationship must be fair from your perspective. All of this can be flipped the other way as well, which exposes a boundary that is responsible for the new need between you:

Perceived fairness, by each individual involved, is required to prevent imbalance, which carries the potential for violence at any moment.

There is one major hurdle in the way of rebalancing unfairness quickly: You can't read each other's minds, so neither of you have access to the other's context responsible for justifying actions the other interprets to be unfair.

A *shared language* becomes necessary to correct what feels like unfair situations in a quick manner. Language becomes *necessary* to restore balance. *Shared language is the tool that allows two isolated contexts working through their own story arcs to move toward understanding each other.*

This expansion into language must happen quickly. You are on a ticking clock and so is the Being. The speed of your needs has not changed. If this unfairness, especially if it involves food, water, and shelter persists when either of you start to experience extreme thirst, hunger, or fear, then unfairness will become much harder to correct because it becomes much harder to communicate.

It gets harder to communicate because your mind is forcefully switching your priorities back to your survival needs and away from this slow and frustrating process of communication toward fairness, which is hindering you from capturing the ends of your need story arcs. Frustration will keep building, and the potential for violence goes up more and more with each passing moment.

Perceived fairness in a timely manner is the core social need.

You and the Being have already successfully communicated your way through a conflict of truth without running away, collapsing into violence, or swinging the imbalance into dominant and submissive, so time re-expanded for each of you. Each time that you and the Being can make it through the *Dark Forest Moment* created by perceived unfairness and still return to the position of equality, balance, and perceived fairness will result in a deepened understanding of each other.

For example, if you successfully correct the unfair walnut situation, the Being now understands that you perceived imbalance, and potentially builds the story that if it were to do the same thing again, it would upset you. This also tells the Being that food is apparently very important to you.

But at this point, all of this assumes you both can agree that *equality is fairness*.

If the Being had good reason to justify dividing things the way it did – which apparently it did because it believed it was the right thing to do - then the Being will have to successfully communicate *that* reasoning to you as well in a way that you can understand. This makes a slight tweak to what I believe is the foundational social need:

Mutually perceived fairness in a timely manner, *which is not necessarily equality*, is the core social need.

Mutually perceived fairness prevents the collapse to violence, establishes a shared context of perceptions (Step 1 of the 3-step story process), establishes mutually understood values and needs (food, water, shelter), and enables a potential partnership.

But the clock is ticking.

The *speed* of communication is critically important.

103

The Birth of Words

Time is the only thing that lies between this moment and your death.

This means you have a limited number of movements remaining within that time.

This means speed is often *critical*.

Your initial "language" with the Being (to prevent violence and possibly death) was body language, eye contact, and expressions, which you used to read each other's intents toward the other.

If you managed to get through the initial *Dark Forest Moment* without violence, it really is nothing short of amazing given neither of you had context to understand the other. You both had to *assume* that because you *appear* similar, you must *be* similar, and it was not worth the risk of rushing to the conclusion that the other was a threat.

By doing this, the story between you stays open-ended, and its conclusion left uncertain. The relationship starts as an open door into near total darkness, with the light needed to see provided by actions relative to each other. This prevents the collapse of time you each experienced at the *Dark Forest Moment*, and begins the process of rebuilding the certainty of survival and freedom of choices you each had in your past – but as a team.

You can think of all of this being a very high-stakes game of charades.

Can you imagine playing charades with someone you have never met, and having mere seconds to determine the right answer to avoid a potential fight to the death? If you've ever played charades, you know there is a timer for a reason – it is not easy to communicate using only body language, eye contact, and expressions. On top of this difficulty, you typically have a shared cultural context with the other players, making it many times easier to guess answers than the same "game" would be with the Being.

If you ever meet someone you do not share spoken language with, body language is likely to be the immediate and natural fallback. It can be difficult, slow, and frustrating. In such a situation, if mutual understanding does not progress and grow quickly, you may end up collapsing time back to the *Dark Forest Moment* as anger and frustration build and other needs become priority.

A *faster* method for aligning context is needed to prevent actions that create imbalances such as violence.

As I see it, the next level of communication with the Being will probably be pictures. This might be drawing in the dirt, laying objects into shapes, or painting using some dye or color you grabbed from a flower or clay.

What though, would be the most important messages you intend to communicate with an alien you have never met but appears to have the same senses and capabilities as yourself?

I think the first things I want to know are in the Being's past. *How did the Being get here? Like I did? Is it human or familiar with humans? Does it have a family? Is it knowingly part of this experiment? Is it an experiment itself?*

Unfortunately, all these questions will be incredibly difficult to ask using only pictures and gestures. (If you don't believe me, try to ask someone how he or she got to where they are without using words.)

So what questions *could* be asked and answered using only pictures and gesture? Honestly, I can't think of many. I can't even imagine pictures being the preferred method of communication *before* sounds and words. I think it would only be preferred if one person was mute or could not make sounds the other could hear.

Other animals of the Earth do not use pictures to communicate with speed (that we know of). High-speed animal communication seems to always be gesture-based movements, which may include changing colors and appearance, dancing, or sound. And sound is exactly where we are headed next.

The fastest and best method to communicate quickly as a human being, aside from visual-based charades, is inevitably going to be using sound.

Sound operates very quickly, travels in almost all directions, and does not require the listener to fully focus attention for it to enter his or her sensory windows – which is required for pictures, reading, or gestures.

Before we get into words though, I believe there are other human sounds we understand that are not words at all. I suspect you might even recognize their meaning if the Being made them. These are the sounds of *emotion*. Sounds like cries of pain, laughter, screams of fear or excitement, and sounds of mourning or sadness are just a few sounds that seem to be approximately the same for all humans. We can often build major pieces of stories about others from these sounds alone. We can typically even recognize these sounds in non-human animals (though care must be taken not to project this story to a point of certainty). The same caution applies to gestures and facial expressions in humans tied to these same feelings, which can often be wildly different from culture to culture.

So, given how much can be communicated from a simple human sound, what exactly is the difference between a sound and a word?

Reaching back into earlier parts of this book, a word is just a *consistent* sound you can make as a human that is ultimately anchored to an idea, object, or action, along with the associated feelings and contexts in the world as it is committed to memory. A word is a sound that represents your remembered specific sensory experience with reality. The thing that makes a word special though, is the existence of a second point that recognizes that sound as familiar, and has associated it to something which is *approximately the same.*

Between you and the Being, no shared words exist yet. You might know the approximate meaning of 50,000 words as you read this, but if the Being does not recognize a single one of them to mean approximately the same thing, they are just random sounds to its ears, and vice versa.

So perhaps the best and most-needed approach to maintaining balance between you at speed is to start building a shared language using a language you already know.

For instance, let's assume you both drink from the creek. You could start with the word "water." You could try using a combination of gestures and sounds, like pointing to the water, and repeating the sound "water" until the Being attempts to copy the sound. If the Being does this successfully, you could then reinforce, or "anchor" this association as correct using excited and approving body language that express *"yes, this particular sound is tied to the creek".*

If successful – which would be validated by repetition of the sound in connection with water by you both over time, the experience of "water" gets anchored to the shared experience of the creek.

In the same way, perhaps the Being has its own language, and gestures to the walnuts in its hands and says "grup". Perhaps you will end up learning words of its language as well, resulting in a new language created by merging two existing languages. You both now refer to the creek as "water" and walnuts as "grup."

Are you experiencing a feeling right now? A ping of familiarity? Exploding cats and strollers perhaps?

You already know how "fuzzy" this is. You can't actually be certain the Being associated the sound "water" with the actual clear liquid in the creek, or with the name for the entire body of water (which is what you call a "creek").

To fight this problem, perhaps you could scoop some water out of the creek, point to it, and say "water." Still, certainty the Being understands "water" to mean the same experience you intend it to mean is not possible. You might discover later (with the addition of more words to your shared language), the Being thought the bubbles on top of the water was the "water" all along. There is *always* some degree of misunderstanding in language. There is *always* some degree of uncertainty. Even when everyone involved witnesses the word's birth and anchored it to a same firsthand experience with objective reality at the same time, every single person perceived it differently, and there is no way to share these perceptions with each other.

Regardless of how a common language develops, what sounds it is built from, and what sentence structure is used, the shared language has one primary purpose as I see it: To prevent the collapse to violence, by enabling its users to establish fairness as quickly as possible using sounds instead of slower methods.

By using language to keep violence at bay, you both can reap the benefits of cooperation with two sets of perceptions and senses, and dramatically increase the chances of survival for you both. Language is a tool of stabilization. It helps keep you, me, and everyone else on the middle path.

And this can be very stressful.

104

Walnuts and Word Loops

Go back to the very first 3-to-1 walnut split with the Being, except let's add a basic shared language into the situation. The *Dark Forest Moment* was in your court, as the Being seemed content with you having one while it ate three. The imbalance is in your story system, and doesn't seem to exist in the Being's.

Instead of trying to communicate fairness through body language, gestures, and maybe drawings (as in the original situation), which will be slow and frustrating, you can now accomplish the same thing in a lot less time using words.

What follows is a crude and very simplified conversation to show how the presence of even a basic shared language provides an expanding non-violent choice from a *Dark Forest Moment*. The sentences will feel awkward, maybe even cartoonish, but I'm doing this to keep the underlying purpose as clear as possible. I am also going to cut out all the subtleties that might complicate things, such as cultural-based norms around gifting, kindness, and conflict.

You may make different choices than I am about to make on your behalf. There are an uncountable number of ways to work through this imbalance. I ask you to not get caught up in the specifics, but instead keep watch over the bigger picture. Since the imbalance is not in your favor, I'll begin with you speaking up to try to establish perceived fairness:

"You lot. Me not lot. You me *same* lot."

With these three basic sentences, you've moved the imbalance you are holding in your story system to an external place. The action taken, speech, was not physical violence. As a result, you have now passed this imbalance (and the *Dark Forest Moment* that comes with it) to the Being. All it took was nine words. These words described previously justified actions and expectations from future justified actions.

The Being now has many options, all of which you are already familiar with. It is in both your interests for the Being to communicate to you *why* it split things the way it did, and not choose the violent path out of its own internal *Dark Forest Moment*. Violence is a risky choice, after all, because it creates an immediate threat to the survival of you both.

This leads me to an extremely important point that applies even today as you read this: Due to the fact you assumedly cannot read anyone else's mind, you also cannot know or predict the stories anyone else builds about *your justifications* for your word-actions – so *physical violence is always a possible choice for anyone to make, just as it was always a possible choice for you.* Being as patient as you can is important. Communication is about trying to re-expand time away from violence. There are limits to this process, but that's coming later. Instead, let's assume the Being replies to you.

"You not lot. Me lot. Me hungry lot."

Now the *Dark Forest Moment* is returned to your head, but you know *why* the Being split things unequally – it was feeling hungry. The Being also just chose words instead of violence or walking away, which is still a positive path forward because the middle path is expanding. You now can choose fight, flight, freeze, or talk.

"*Me* hungry. You me *same.* You me *same* lot. *Me* give same. *You* give same."

The *Dark Forest Moment* now gets passed to the Being's story system.

"You me *not* same. Me hungry lot. Next day, you me same lot. Me same lot. You same lot. This day not same lot."

The next decision is a tricky one. You can choose to keep going with this conversation since the original imbalance is still not acknowledged as a problem, or you can accept this new arrangement made using words to hopefully play out through future actions of splitting food equally…tomorrow.

I personally would advocate for continuing, because although you have voiced you are also hungry, the Being's justification for the 3-to-1 split involves *assuming* something about you. It assumes it is hungrier than you, *but it cannot actually experience or know your hunger level.* This is actually a dominant/submissive imbalance resting on an imagined and projected story. You have to inch even closer to violence to correct this.

"Me hungry lot! Me same grup! You me same grup."

Back to Being, who seems agitated now. The *Dark Forest Moment* option for physical violence grows closer. You wisely brace yourself for possible violence, as the Being must now openly state it always requires the lion's share of food, gets to decide how hungry you are, or that you are equals and it recognizes an unfair act was done. It thankfully chooses words.

"You hungry………. You me same grup."

The Being hands you another kernel. The moment now passes to you, and you can accept this, or you can go further and enforce the new arrangement.

"*You* same this day next day. Me same this day next day."

The Being pauses, then gives a smirk.

"Me same. You same. This day. Next day."

Imbalance has collapsed back into balance, for now.

As clunky as this conversation was - and to be honest I found it surprisingly difficult to write - it was also highly functional and accomplished something very important to the relationship in my opinion.

Together, you both made it through the *Dark Forest Moment* a total of eight times. With only eight exchanges of words related past actions, you both established a new framework for fairness around your primary needs in the present and future.

The original 3-to-1 split had the potential to validate a dominant/submissive relationship from one perspective (yours), and a fair relationship from another (Being). However, it ended instead at a new agreement of fairness from both perspectives. Now you understand and recognize each other as equals and partners even more. Your connection and trust in each other has grown.

Something else that came to light through words as well: The Being had a justification for giving you less. It gave you less because it believed it was the right thing to do. Do you think you or the Being could have communicated and understood this using only gestures or drawings?

As I see it, shared language is the key to survival with each other as strangers, even though word meanings between us are always approximate at best. As I see it, they are the reason why we aren't in a state of constant physical violence every single day with someone unfamiliar. This makes shared language a form of shelter from *each other*.

If we are each balloons of different sizes and pressures, language is the straw that can be lengthened between us, providing valuable time and slowing down our action-currents, which allow us a better chance to get on the same mental wavelength – to understand each other without relying on total blind assumption and imagined projections.

If this purpose for language feels familiar, just know you already applied it to do exactly the same function when you were trapped on a treadmill trying to solve the plane on the treadmill problem with Bob. In that situation, the angry roosters were the looming threat of harm or death from failing to find balance between you, and the person speeding up the treadmill was not me, but your own need story arcs going uncaptured as time kept passing. That passage of time spent talking leads to more and more uncertainty of your survival due to the growing sensations of hunger, thirst, fatigue, fear, and the new added need of fairness with a partner that may or may not be willing to keep choosing the non-violent path.

If Bob just yelled "you are wrong" the entire time, not helping to solve the problem or really communicate with you in any meaningful way – he is being unfair to you by creating imbalance *and* robbing you both of precious time to balance it. His constant invalidation of your truth will build up to an intolerable level of unfairness in your story system, since you know his choice of domination is going to doom you both.

It's the same as the Being demanding - without any consideration of your words that say otherwise - it is hungrier than you. The Being *cannot know this is true*. It cannot feel your hunger. Nor can Bob know your answer is wrong without hearing your justifications for that answer.

And what would you do in these situations? What if the Being refused to believe you or accept you as an equal, like Bob refused to explain why he thinks you are wrong and he is right? Would you attack the Being? Run? Push Bob backwards? Trip him?

Invalidation of your personal truth creates what amounts to what feels like an intolerable condition. It creates intense story voltage *in your mental boundary*. Running away won't collapse it to balance for you. Trying to ignore it, whether the other believes they are fair or you are inferior, won't balance it for you. Only *action* can balance it. The most stabilizing action available to you is words.

With words, you and Bob worked through your two truth-stories to create a merged truth about the plane and treadmill pretty quickly, and you and the Being worked through your two truth-stories to create a merged truth about each other and fairness pretty quickly. You likely came out more respectful and friendly with each other in both situations, no matter how frustrated you each might have been during the process.

Balancing certainty and freedom of choice between two people is always going to be an emotional process. Especially for the individual that has to "give up" perceived certainties and freedom to restore fairness, such as the Being handing over one more piece of food that it would have eaten if you didn't argue for fairness.

A framework for fairness through words almost always exists as an option. Let's put this to the test.

105

The Cookie Monster

Going back to what I will now call *The World of Two*, assume the Being shorted you yet again *after* the conversation in the previous chapter established what is fair related to food. It gave itself more fish than it gave you the very next day, while you split your gathered food from plants equally. You notice this imbalance, and attempt to enforce the agreed understanding of fairness from the previous day.

I'm going to advance the shared language a bit, so that some subtleties to this situation can begin to make their way in.

"Hey, you gave yourself more than you gave me! We had an agreement! Everything is to be split down the middle!"

"I'm sorry, I know we agreed, but I am very hungry. I worked hard to get these fish so I feel I should get more."

You now have a difficult choice in an internal *Dark Forest Moment*. You can agree to this justification-story as fair. But even if you do this, you will still need to communicate that this choosing-to-take-more-because-I-am-hungry cannot become the new *definition* of fair. I see addressing this as a *critical* task. If the future understanding of what is fair is not stated, a new understanding can be *assumed* that puts you at the mercy of the Being's judgment on how much food it is allowed to keep *from you*, based on nothing but its own hunger and imagined projection of yours.

If this happens, then each and every time the Being does not split things evenly in the future, you are stuck in a much more dangerous *Dark Forest Moment* where you are trying to compare *feelings of hunger with only words* to justify what is fair. This gets ugly, because neither of you can actually experience or validate the other's hunger.

In my own firsthand experience, this is the biggest key to enforcing fairness and balance in life – timing. Everything always comes back to time it seems.

Your mind is a prediction machine. It eliminates a *ton* of fear and anxiety to have an agreed-upon framework for fairness in place *before* critical needs like food, water, and shelter are consumed. Having shared stories to construct future expectations, and then bringing those expectations into reality through actions allow for the *Dark Forest Moment* to begin to disappear from view. In other words, violence begins to fade when the fairness need is mutually understood *and repeatedly executed to the expectations of all involved* without having to redefine or adjust that understanding.

If the Being already ate its larger share of food, leaving you less, fairness is no longer achievable at this moment. It's already imbalanced and the means to rebalance is destroyed. You can't ask the Being to puke up its fish and then separate out the half-digested pieces to make it fair. Well, you *can*, but would you want to?

Mutual fairness is best-established using words, gestures, or pictures to describe actions before those actions occur to set expectations and avoid negative emotions, which increase the potential for violence and can cause difficult-to-repair damage to relationships.

If you don't believe the above statement is true, I would urge you to hang out with children for a while. In case you can't, here's an imagined situation built from my own experiences with all the children in my life:

Imagine you are babysitting two kids that are the same age. I'll call one Billy and the other Susie. For the sake of keeping things as balanced as possible, let's say they are both five years old.

Now, pull out four chocolate chip cookies - their favorite.

Hand Susie three cookies and Billy only one, *and then walk away immediately.*

As you walk away, imagine popping in some earbuds to listen to this as an audiobook while Billy's fading voice yells to you, "Hey, why do I only get one and she gets three?!"

You next hear the voice reading this book say,

> "Your vocabulary in life grows through time starting at no words at your birth to whatever number of words you know right now. Your vocabulary can only grow one of two ways: The first way is by being exposed to already-existing-but-new-to-you words and then successfully associating these words to specific experiences so you can start to use them in similar ways your sources used them. The second way is by creating new words you associate to specific experiences and then successfully teaching others what experiences these words represent, so they may use them successfully in a similar way to how you used them.

This means that age and vocabulary typically grow together, so kids normally have smaller vocabularies than adults. Because kids have fewer words to work with, establishing fairness with words *after* the unfair action is more difficult than it is than adults. The younger the children are, the harder finding fairness becomes."

Billy and Susie are similar to you and the Being during your first conversation over walnuts. Their words are clunky. They are strained. They are limited.

If you are gone from Billy and Susie's view, then you just created a *Dark Forest Moment* between them. They each have to make a choice: Run, say nothing, say something, choose equality, or choose violence - and they have to choose fast. They are on the ever-accelerating treadmill now.

Kids, unfortunately, have yet another problem that makes this process even harder than yours and Bob's or yours and the Being's: Time blindness. Kids are not able to perceive the passage of objective time well at all, so their rate of time is completely individualized to their own senses and actions – much like you in *The World of You*.

Because of this, the amount of time that exist in the future for a young kid in *any* given situation is almost always *forever*, because he or she has no idea what a minute *feels* like, and cannot yet comprehend all the movements outside of his or herself that may change the situation from what it is now.

Put a young kid in time-out for 30 seconds and that child will often perceive it as an eternity. Time for kids is instead measured by tasks and feelings. They exist doing one task until something in their world or some feeling they experience justifies changing to another task, and marks the passage of time.

This means from Billy's point of view, Susie just got three cookies and she will *always* have three cookies. This situation is now *forever* unfair unless something is done. There is no distant future prediction and no expected next task to measure time. You were the only thing tied to the cookies, so making you come back was priority number one to re-establish fairness. Now you are gone.

You didn't say "Susie eat two of these cookies and then give the third one to Billy." If you did, that would give Billy a reference point to enforce fairness. He could watch her finish her second cookie (with lots of anxiety) and then *boom*, Billy is triggered into action in the only moment he has left to enforce fairness before it is lost – "That's two! Give me my cookie!"

But since you did not actually offer guidance for splitting them equally, or any kind of reference point for fairness before you left, Billy has no measure at all as he watches Susie position the third cookie in her hand to take a bite. Speed is now *everything* for Billy and words are hard to come by because fairness is a blink away from being gone *forever*.

While both kids are experiencing the *Dark Forest Moment*, Susie is actually experiencing the externalized version of it, and Billy the internalized version. She *knows* Billy sees this as unfair because he said so – he screamed it at you and she heard him do it. She can now run with the cookies (flight), try to pretend she didn't hear him screaming (freeze), offer the cookie (balance), or choose to see the cookie as "mine, because it was given to me" (imbalance/dominant).

Meanwhile Billy can only imagine what Susie is justifying to be right and wrong.

If you aren't sure why kids so often see their things like toys as their property, use the cookies to consider how toys come into their life. From each kid's perspective, you magically created these things or got them from a store *specifically for him or her*. They are a prize - an unexpected treat. You apparently just thought Susie was more special than Billy.

If the kids see the cookies or toys as an act of approval, or validation of love from you, I think it's less likely they will share them with each other, because there is far more at play here than just junk food. There is an imbalance tied to self-worth relative to *you*.

In the absence of adults being present as an authority to enforce fairness, and the absence of fast and clear language as an option, violence inevitably rears its head as the only option with enough speed left to restore balance.

If, from Billy's perspective, Susie shows full intent to eat or keep the third cookie by ignoring Billy or being possessive of the cookie as he reaches for it, Billy's need for fairness and pain of injustice and invalidation will likely lead him to steal the cookie, smash it, grab and eat it by force, or be violent toward Susie either right now or in the very near future.

From across the house, you hear crying and screaming, and remove your earbuds while running back to the room where you gave the cookies. Susie is now crying and hurt. Billy is also crying and wearing anger and sadness on his face as a reaction to the unfairness of it all, and feeling forced to use violence. Both of them have hit each other and blame each other for what has happened, and they are now locked in a cycle of imbalance, blame, and violence.

If you are a parent of multiple kids, you've seen this moment a thousand times, maybe more. This is the blame game. And it's likely the kids won't think to point this back to the real problem: You. Billy probably won't re-ask the question "Why did you give her more than me?!" and Susie is definitely not going to ask "Why did you give me so much?!"

The priority between them now is re-establishing fairness in the face of authority and fear of consequence. Who started the violence? Who should be in trouble, and who should not be in trouble?

Billy chose violence first of course, so in the absence of experiencing the process that led to violence, Billy is likely to be the one that gets in trouble.

And as this happens, I suspect his resentment of adults, of their rules, of their judgments, of the unfair justice system he is apparently stuck in begins to grow, and he stares you down with raw hatred in his eyes while complying with whatever punishment you might dish out to him for choosing violence.

None of this was fair.

And none of it was Billy or Susie's fault. They did the best they could with the time and tools that they had. We all do.

To bring this back to you and the Being, with enough vocabulary and time, you and the Being could establish the critical pre-action fairness framework needed to cement your partnership and erase the *Dark Forest Moment* possibility of violence. The key, as I see it, is to work fairness out on the topics capable of driving you both to violence quickly: Food, water, and shelter – the things needed to survive.

It works a lot like a contract, actually.

106

The Morality Contractors

Was equality a *rule* at the beginning of *The World of Two*?

If so, does this rule define right and wrong?

The answer to this last question is yes (in my opinion) – this rule *does* define right and wrong, but not in a way that can be measured objectively. This is because the "rule of equality" is based on something else entirely.

This rule appears throughout the animal kingdom. Animals often follow it. You and strangers tend to follow it too, even though you did not call a meeting to determine the rules before meeting any given stranger.

It might be better to think of this "rule of equality" more like an informal unspoken agreement – an invisible contract. In *The World of Two*, this contract is "written" by nothing more than *communication* between you and the Being, and "binding" through shared memories of those same communications.

In this way, one could argue that this equality contract exists as a remembered electrical current pattern of word sounds that you both associated and anchored to *each other*.

But why do this at all? What are you to each other?

In my opinion, You and the Being - two points - recognized some part of yourselves in each other. The electrical current flowing from Step 1 into Step 2 created by a boundary shape (the Being's body) was recognized as another "self" by your own story system. Essentially you each recognize "the other" as a "balloon" like yourself. This, in my opinion, caused the sensory experience of this "other self" to associate into *all* of your future stories. It is an intense kind of entanglement into memory and emotion that cannot be easily undone.

To simplify: You became horribly imbalanced by the mere *awareness* of the Being's existence. This feeling is so intense, that ending contact by running away or even killing this "other self" does nothing to relieve it.

To find balance between two points imbalanced by awareness of each other's existence – a double feedback loop – something must change. That change is language, which allows the two points to find and enforce an ever-finer balance between them, and prevent feelings related to internal process from damaging the ability to work together.

After you develop more and more shared words with the Being, you both realize that the rate of your repeating bodily needs do not actually match, and it is throwing things out of balance from time to time when trying to strictly stick to the rule of equality. To deal with this, words for feelings are figured out, and then you and the Being are able to modify this rule-of-equality to instead be "split things *fairly*."

The reason this possible imbalance can exist and still be considered "fair" by both you and the Being is *trust*. Trust of each other's future actions, in particular. As two separate points unable to read each other's minds, trust is not automatically given. Trust is instead something that is *earned*.

For example, if I am a stranger, I can tell you to trust me all day long. But these are unanchored words attached to nothing in your experiences. I have not demonstrated myself trustable through my actions.

When the shelter dance began between you and the Being, every single movement made was watched like a hawk, sending both story systems into overdrive on what justified it. As more and more movements happened without violence, the story systems started to relax, but only slightly. The story was most likely something along the lines of "If the motions meant threat – it would have been obvious by now." This means trust appeared and began to grow with each moment you chose not to be violent.

With language, your movements can be made predictable for the other's story system, which cannot actually know your justifications for those movements, and vice versa. This continues the growth of trust.

Take the words "I mean you no harm," for example. These words tell a story about your future actions. They mean you are not planning to attack intentionally. This is *not* the point where trust is built. These words mean nothing by themselves. Trust is instead built when you *follow through in time* and connect your actions to these words in a way the other story system expects.

When the link between the memory of your words describing your future actions are repeatedly followed by your actions matching those descriptions – you build trust. I think it's important to note, though, that the one who decides your words match your actions is *not you*. It is the "other self." The best way to ensure this connection is made is to simply ask if your actions met the "others" expectation from your previous words.

The more your words make your future actions and movements predictable regarding anything survival-related such as food, water, shelter, and fairness, the faster and stronger trust will grow. As it grows, less and less story voltage is created by actions that are not pre-communicated.

In this way, the function of language as a tool to find and enforce fairness is like a contract in time – it creates a "fuzzy" connection that crudely copies the intended future actions of one story system into the other's story system before those actions happen.

With enough contracts honored, a level of trust is reached where the rule of equality can be warped and even broken using the same language-contract system.

With trust the contracts will not be breached from repeated firsthand experience, an imbalance can be made agreeable, and trust retained, even if it means the Being eats three of the four pieces of food.

This is possible through a deep mutual *understanding*.

Understanding makes it possible for inequality to still be fair. But to do this, you must first deeply *trust* that the words the Being uses to describe its always-invisible-to-you feelings of hunger are actually true. Once it is trusted the Being is not looking to rob you, dominate you, or be unfair, then when it asks for a larger portion of the food, which you have normally been splitting down the middle, your story system can stay at ease.

If the Being instead tried to take most or all of your shared meals, you couldn't know if it saw itself as dominant or if it truly needed more food than you, because you have no shared context in your past to know its current actions match its future intentions. It might actually be *starving*.

But knowing this takes *time*, because building the language necessary to communicate such a situation takes time. Building trust takes time.

There is a problem brewing underneath this situation, though. There are maximum speed limits to the process of growing deep understanding and trust through two-way communication, and they barely align with your biological *minimum* speed limits…

107

The Limits to Making Sense

When it comes to trying to use *someone else's words* as your experience to build truth and right, wrong, and fair in a way that grows understanding and trust, very real and strict limits exist.

This applies in any story you know that describes reality, which you did not or cannot directly experience yourself – such as the Being's feeling of hunger.

To find these limits, all you have to do is consider your 3-step story process and what it forces to be required from words: That they create sensory experiences through time.

The Maximum Speed of Making New Sense

The first limit is one of speed, and I touched on it slightly in the Billy-and-Susie cookie situation two chapters ago. In this section I'll frame things between you and me instead.

Since you can read and understand the words in this book, you and I have a shared language. However, I can't add a new word to our shared language without familiar words around it to make my new word make sense for you – to give it a *context*. For example:

"Tyb."

What does this word mean? How could you possibly know? *I* already know what it means of course, because I created it. It's likely that you don't however, because you do not have an experience of any kind anchored to the word sound you are imagining to hear when you read the letters organized into the word-shell "tyb".

Strangely though, despite me thinking I've created this word, I can't actually be certain you don't already have an experience for it. This is due to an entirely different reason than the context boundary between us.

Instead, it's possible "tyb" has somehow entered your culture or your life for the first time some other way than this page in this book right now. If this is the case, then unless you learned it from someone that did learn "tyb" from this book, I can assure you we do *not* mean approximately the same thing when using it. We are likely in a kangaroo-and-the-cat situation.

I'll try another word: "Ptykits."

Nothing, right? Even if you have seen this word-shell before, let's move forward assuming you haven't. You have no idea what it could mean. However, what happens if I add a few words around it?

"Yellow, blue, red, ptykits, pink, and purple."

Does this new unfamiliar word suddenly make more sense for you? My guess is yes, because I just gave the new word some context that anchors it to experiences you probably do recognize. You probably still have no idea what color you should imagine (see: making sense) but you have a pretty good feeling ptykits is a color.

Be careful though, because this *is an assumption*. I have not said ptykits is a color. You have exactly zero proof or evidence this word represents a color at all. It merely appeared in a list of words that you do recognize as words for colors.

With this word anchored on assumption, you might look for clues that validate your assumption when it appears again.

"Ptykits sruvs were everywhere."

Did this validate or invalidate your assumption of the word being a color? I'm hoping by this point you are catching on to the very-real speed limit and hidden dangers created by *relying* on secondhand story shadow experiences to build truth. If you think this is a trivial matter that only applies to fiction and imaginative writing where new words commonly appear, you are accidentally jumping a huge gap in this story.

This is the maximum speed of all of learning through the secondhand shadow experience of communication.

New or unfamiliar words, images, and sounds, and new and unfamiliar contexts for already-familiar words, images, and sounds have a natural speed limit of one-new-idea-at-a-time. The person that wordlessly experienced something new is constructing the words of his or her story from wordless memory, in order to recreate wordless experiences for the one reading or hearing the new words to describe it. But too many new words at once causes any shadow experience to fall apart. If you do not know what "tyb" means, then you certainly cannot know what "tyb sruvs" might mean.

I can make up a huge new language using the same alphabet you already know. But the only way you are ever going to learn what any of it means is one word at a time, because each word has to anchor itself to wordless firsthand experiences, or to familiar words you already have anchored to firsthand experiences.

Even the grandest fiction writers in history anchor their new ideas to familiar words at some point so you can make sense of the new ones. Sometimes this missing-context-for-words problem can even be used as the turning point or climax of an entire story.

The Structural Limit of Making Sense

As you read this book (which is an action), your story system is expecting a familiar sentence structure to continue arriving. In your past, sentences have always presented in a certain way – the words have a particular *order and a particular structure*. You were taught these things as a child in school and through conversation by nearly constant repetition, to the point that your story system validated the order of words in your first language who-knows-how-many-times until the order of words became an expectation. But some very interesting things start to happen to your sense making if I mess with this structure you expect to keep happening:

.Salad to burning)while(loves and giRaffe banana flow a eating candle,

Can you *make sense* of the above statement at all? Imagine if this was the first thing the Being said to you immediately after making it through the *Dark Forest Moment*.

As each word in the above statement appeared on the page in your present (now your past) because you moved your eyes from left to right (or listened to the string of sounds being made if this is an audiobook), you undoubtedly made *lots* of associations for each word. You probably have personal context built from real-world experiences tied to "salad", "burning", "loves", "giraffe", "banana", "flow", and "candle". Maybe some contexts are built from wordless firsthand experiences, such a memories of seeing and eating a salad. Maybe other contexts are secondhand shadow experiences like an image or video on a screen, like if you've never seen a giraffe in person.

Regardless of the source of your experience that gives meaning to each word, what you couldn't do is put these experiences for each word together in any sort of way that allowed you to understand what *overall new experience* is being communicated. The addition of random punctuation and capitalization didn't help things either.

The problem is the string of words above is not presented in a *story form*, and it turns out we as humans need stories to *make our senses experience* what is being communicated. The experiences the words convey have to *flow through time* in an order we are each trained to be familiar with and expect, which allows us to turn the individual context for each word into a *compiling* single sensory experience. We order our words to *make sense - literally*.

It's possible lots of weird connections happened for you anyway when you read that strange "sentence" the first time. Maybe you mentally built a giraffe eating a banana. If you didn't, you just did, because the sentence before this one just told that story in a way that did made sense. It's also possible you put together some sort of "flow candle," whatever that may be. Maybe a salad is burning? These don't make much sense to me even though I can now imagine them, but for some reason maybe all these experiences compiled into a story for you based on your own personal past tied to these words.

It doesn't really matter if it did though, because the statement is way beyond the limit for the expected sentence structures in the English language, which we are both using.

I *could* make the strange statement make sense, and teach you the structure of my English-word-based sentence one step at a time, but rarely does this ever happen when we are speaking to each other. More likely, you will end up as a hostage.

108

Hostage to Speed

Right now, even at this very moment, you are being held hostage. To understand what I mean, let's bring back the strange sentence from the previous chapter.

.Salad to burning)while(loves and giRaffe banana flow a eating candle,

What if I said your life depends on understanding what this bizarre sentence is trying to say, and repeating the correct story of its meaning back to me within the next 30 seconds?

Will you start to panic? Are you even sure which sentence I mean by "this bizarre sentence"? Even if I clarified that I mean the "giraffe banana" and not the "if I said your life depends on" sentence, you simply *can't* analyze and merge all the individual words into a single story, and you can't understand *my* context behind each word either, thanks to those pesky mental boundaries between us.

You would certainly try If your life depended on it though, right?

If your life *depended* on understanding that "sentence" in 30 seconds, time itself is a major pressure where it wasn't before.

What if I refuse to help you make sense of it? How much would it irritate you? Would you explode at me? Implode? Beg?

What if I gave you the gentle reminder that *every word you use means something slightly to entirely different to other people because literally no one has the exact same firsthand experiences as you, and this results in nearly the same outcome as my word salad sentence, even if the words were in an order that made sense?*

Your senses and memories are not the same as my senses and memories.

You may have no idea what a "giraffe" is, and there are no words included in that sentence that help you figure out I anchored it to a specific animal. Maybe you were taught the animal I anchor "zebra" to is a giraffe. Either way, this is only the surface of the problem.

You imagine an entirely different candle than me. Perhaps mine is battery powered and skinny, and yours is wax and very wide.

You might imagine a salad that is a spring mix, and I might imagine yellow mustard potato salad.

The banana I was experiencing from memory when I typed the word might be green and small, but yours might be yellow with brown spots.

In our need to operate at the minimum speed required of us, we often completely overlook this reality of "fuzzy" language and its persistent limits and gaps. There is *always* an atomic bomb's worth of possibilities baked into every sentence (did you just see a mushroom cloud or something coming out of the oven?), because each of your words, or mine, or anyone else's in the language you think you understand are being *anchored* not to their experiences, but to *your* personal, individual, and unique experiences.

Language (and therefore stories) making sense is based completely on the almost constant *assumption* that the subjective experiences in objective reality between you and the storyteller *are shared*. This means the greater the *actual* differences in past experiences, the more misunderstanding is occurring within the same shared language. Since no one has identical experiences, no one can perfectly understand another's words.

If the Being said the bizarre sentence from the beginning of this chapter to you right after you managed not to kill each other – you *know* you can't assume a thing about what those words mean. You can't (and in my opinion *won't*) assume the sound "giraffe" coming from the Being's mouth is anchored to the same "giraffe" experience you know. You can't assume *anything*. *Everything* the Being says, no matter how familiar, must be confirmed and validated to match *your* assumed connections.

But this same exact situation is present when any story comes from another person today.

This means shared language, *when used as the foundation for truth or right or wrong*, immediately creates enormous contextual voids and gaps. In my opinion, this is why language is so often referred to as a looking glass, and this is why I believe all doctrines and laws declaring right and wrong beliefs or actions using words will eventually collapse without constant re-anchoring and re-wording.

This gets really hard to notice and accept in your daily life. It's likely thousands of secondhand stories pass through your senses each and every day. Some of them make sense, while others do not. The ones that do not are more-or-less *nonsense* for you, because either the anchors for the words the storyteller is using are missing for you (which causes the words to fall into your gaps and disappear), or they just aren't in the order your story system expects (so a sensory experience fails to be built from remembered electrical current patterns).

With enough *time* and patience though, you *could* fix this. It's extremely likely the secondhand stories that are nonsense to you are *not* nonsense to the storyteller. This means you absolutely *could* make sense of any word or story by continuing to dig for the experiences that provide context for them. You could absolutely make sense of an oddly worded sentence if you had enough time and ability to find the experiences and context that created it in the first place.

But time to do this is a luxury we do not often get as human beings.

We are in a hurry.

This hurriedness often cuts out the extra time needed to learn new words or unscramble the strangely ordered words from others.

If you came face-to-face with the Being for the first time, and it spoke a string of words that sounded like gibberish to you - you'd probably recognize it as *some* language, but not one that you recognize. You now have until one of your need story arcs (or the Being's need story arcs) demands capture to figure out how to reach a stalemate or agreement from this gibberish. After that, you will be forced to go fulfill these needs instead of communicate, and that sends you into a less-intense-but-still-very-much-there *Dark Forest Moment* because you can't explain why you need to leave or stop talking.

Today as you read this, you are probably bombarded by words from who knows how many "Beings" sharing their experiences in the form of words. At the time I wrote this, this is often happening so fast there is no *extra time* to debate or really think about what any one word in any one story means before the next word and next story enter your story system.

The result of this "great flood" is becoming more and more hostage to a high-speed truth, which means the stories rest on an enormous pile of assumptions.

The stories that do not require you to learn new words or decode long complicated sentences are the easiest to make sense with, and therefore accept as true. Stories built with *familiar* words form a path of least resistance between the context of the storyteller and your own context, which is creating sensory experiences from those very same words.

Strings of familiar words do not require you to suddenly *resist* and stop the barrage of still-incoming words so you can insert "thinking" time to consider what does and doesn't make sense.

I'd be willing to bet you have not spent any large amount of time with the nonsense sentence at the top of this chapter. Why would you? I've not implied it is anything important.

But what if I implied that it is critical in some way? Like if I said your life depended on it? Like if the Being said it to you in the clearing where the belief you were the only human created a context of assumptions, norms, and expectations that did not include some green-skinned yellow-eyed version of yourself.

Should nonsense words become critical in such a way, time (and therefore speed) is everything. You need to establish which words are anchored to survival needs, and fast: Food, water, shelter, fairness, hunger, thirst, fear, threat and other emotions. You are looking for any clue that might tell you something important about the Being related to what it needs, and then must work to unpack and remove all assumptions you can about them.

You need to figure out *how* to communicate urgency and priorities quickly, if you both need sleep, if you both want to be treated fairly by the other, and if you both want more certainty of each other's future movements with the fewest assumptions possible. And the only way for fairness and future actions to be predictable is to have a framework agreed upon *now* to set shared expectations for the *future*.

Even if the Being needs food and water at greater frequency than you do, with enough words this critical topic can be understood, and then imbalance can be agreed to as fair *before* the Being takes more food by force. Assumptions will *always* exist in language though, which means the speed and quality of words relative to action is critical to keep these hidden-by-speed assumptions from turning into violence.

Many things will start to change between you as your shared language grows and assumptions shrink.

109

Balance and Bonds

Let's fast-forward time a bit in *The World of Two*. You and the bright green-skinned yellow-eyed Being have been cooperating for years. Together, you've developed a garden with revolving ripening harvests, developed new and ingenious ways to collect and store meat and plant-based foods, and have built multiple shelters for different situations you often found yourselves in. You've aided each other through periods of recovery from injuries and willingly picked up the slack needed to capture the ends during these times. You've even saved each other from harm and animal attack several times, and chose to stand together through dangerous situations that posed great risk to you both.

And what has all this accomplished?

You've both dramatically increased the certainty of your survival. The collapse of your projected need-story-arc waves is so distant you can't imagine it. Neither of you expect to experience extreme hunger, thirst, or fear any time soon. Both of you have consistently voiced predictive stories of what will or can happen in your world, which the world validated. This has created mental "shelter" from the feeling of uncertainty and fear and anxiety of the unexpected.

This growing ability to not be surprised and to avoid disaster through your own predictions and justified actions has awarded you both the freedom to stop thinking about finding and consuming food, water, and shelter *all the time* and start finding new stories in reality to understand, master, and possibly control or shelter against. These stories are still mostly centered on what matters most: Food, water, shelter, and fairness. Mutually understood fairness is itself a form of shelter against the unpredictability of another story system that is always using a context hidden to you.

Through the use of increasingly complex language, you and the Being have grown a deep bond. You not only use language to be predictable for each other, but you've also used it to connect your inner worlds. You now have shared words that allow discussion of possibilities only ever imagined to exist, and even discuss how you feel about creating them. This translates into the sharing of fears, worries, values, pasts, and lots more. It also lets you each share a completely hidden connection to these things – feelings - with words that describe experiences like scared, worried, angry, lonely, and happy. This exposure of hidden processes through words greatly deepens your bond as you each near a full understanding of the other's personal context. There is a profound kinship now, a significant *recognition of self* in each other. What began as a superficially familiar outer boundary shape, has become similarities and familiarities that reflect one all the way to your hidden cores.

With almost all of each other known and understood, you both operate in near total sync without the use of words at all. The words, actions, and feelings about words, actions, and possibilities have been linked together through time so consistently, that you both often know the story of the other's actions, feelings ,and intentions just by observation alone.

You have merged so many stories down so many rabbit holes that you are two halves of the same story system. The only differences between you are the stories being built in the current moment, because you each have different physical perspectives.

As I see it, this level of trust and bond is *love*. Love, to me, is a word that means a deep understanding and acceptance of another's beliefs and truths, whether you agree with them or not. It is the full recognition of self existing in another.

The polar opposite end of the spectrum from love is fear. Experiencing total fear would justify you to run or attack in order to control (dominate). Total fear can justify you to try to kill, or justify you to build a fortress of a shelter to protect you from "not-you".

Of course, you most likely experienced *both* love and fear at the same time in the first *Dark Forest Moment* with the Being. Love, created by electrical currents that see self, justifies you to *assume* the Being thinks and works like you do. Fear is the uncertainty of this assumption, due to not being able to predict what comes next because you have no context in your past for it. This justifies resisting this "electrical current of self," because it is not yet validated to *be* self - there are still enormous gaps and voids in the assumption.

This means love and fear, in my opinion, are the two balloons that bias justifications for your movements toward one outcome or another. Fear demands immediate movement toward strict *equality* and predictability. Love is instant trust and understanding through the recognition of self, until actions no longer match expectations, which push currents in this balloon system back toward fear.

You've experienced *Dark Forest Moments* hundreds if not thousands of times in your life, so this bias toward fear or love when experiencing a recognition of self in another's outward appearance already has a path of least resistance, which was forged in the house of mirrors ten chapters ago.

With this in mind, let's reflect on what's left of the looking glass through the lens of *The World of Two*.

110

Reflections In the Looking Glass Shards

The addition of language can change a lot of things. Let's take a minute to compare *The World of Two* and *The World of You* and see what changed.

Moral Code

Do right and wrong exist for you and the Being in *The World of Two*?

If they do exist, who exactly is defining what is right and wrong *for you*? Can the Being call you wrong with authority? Can you call the Being wrong with authority?

There are still no official laws, nor any government to enforce such laws, right?

Yet the beginning of a social moral code has surfaced, and with it right and wrong. But this code does not *objectively* exist. These ideas only exist as the remembered invisible *agreements* between the two points: You and the Being.

If you teach the Being how to fish, and it makes a mistake trying to attach the bait after you showed it how, those now-past actions that caused the mistake were wrong relative to you, but were right relative to the Being when they were performed. After all, it wouldn't have tried to attach the bait the way it did if it didn't think it was right (if it didn't believe the movements it used were justified).

Once the Being realizes its bait has fallen to the ground, it already knows it did *something* wrong in the past because the present did not turn out as expected. At that point, the Being floats its anchors to what is true, and looks to adjust them. The Being may look to you for assistance if it cannot remember the "right" story you taught it secondhand, or it may try to re-discover the "right" story on its own. Either way, a written or formal rule about attaching bait changes nothing, and barging in and screaming about it being wrong is no help to the situation, either.

Even if you catch the Being making the mistake, *the action you observe has already happened and is in the past*. While you could say "no that's wrong!" to try to prevent *any delayed consequences* of the already-performed mistake, this does little when it comes to the Being's personal growth in your relationship. The Being was *already* attempting to attach it correctly. Its story system was trying to justify the "right" movements from memory into reality. If those justified motions are abruptly stopped and invalidated, its story system will then take the next path of least resistance forward. That path might be anger at you (for invalidating instead of helping), asking you to put on the bait instead (to prevent further invalidation), quitting completely, or trying again in spite of you.

Learning new movements in reality works on a spectrum of risk. Forceful invalidation mid-risk causes your feelings toward taking risks to change.

Imagine if you finally decided to take the step through the door into blackness in *The Ark*, and just as you are about to cross the threshold without catching yourself a huge voice booms "NO THAT IS WRONG."

This sudden interruption to what you believed as the right thing to do *after you justify it* changes *everything* about how you approach the doors into blackness and even your entire justification process.

Would you stop trying new doors now? Would you try sticking just your foot through to see if the voice yells again? If so, then who is the *Keeper of Truth and Decider of What is Right* for you? Is it you, or is it the voice, because it judged your choices as a mistake? Are you going to yell for it to come back? What if the voice never happens again? You will be stuck second-guessing yourself with no context to relieve the situation. The Keeper of Truth Hydra is born at this moment – the moment someone tries to assert themselves as the *Keeper of Your Truth*, and you hand them the crown without knowing their justifications or assumptions.

In *The World of Two*, if the Being says the sky is "actually more of a purple than a blue today" after you said it's more blue than purple, are either of you *objectively* wrong or right?

If you kill the Being while it sleeps after years of time together, is this objectively wrong? No one is going to show up to arrest you. The universe will probably not stop and demand justice. The animals will not likely change their routines to build a nature court to try and punish you. The moon will not tumble from the sky. Time goes on just the same. There are still no holy books or religious institutions in *The World of Two* to enforce moral rules against you. There is no one left in the world to judge you and declare you a sinner, and there is no one to confess to.

Killing the Being only collapses your own certainty of survival and freedom of choice, and has the potential to make your life more difficult in every way. It's also possible this sends your mind into permanent imbalance due to stories you can no longer justify because they included or required the Being.

Destroying the Being well after you develop a shared language is the equivalent of popping one of the balloons in the electric balloon-straw system. It leaves you as the only balloon attached to a now open-ended straw. A *ton* of current will start flowing through the straw as your entire system of truth, right, and wrong try to find balance with nothing but wilderness, the remains of what you and the Being built, and your own self-generated stories.

Killing of any kind is still not *objectively* wrong, it is only *relatively* wrong between you and the Being because it creates imbalance. Stealing does the same thing. No action in *The World of Two* you justify is *objectively* wrong. Wrong is an intentional violation of understanding and trust, forged through the honoring of many memory-based verbal contracts, and making no effort to clarify assumptions or justifications for the violation.

You and the Being are like all the other life on Earth determining right and wrong. They all go through the *Dark Forest Moment* just like you. You both will always do what action you feel is *right* based on comparing your present to your own past. If the Being justifies a series of movements that are unexpected to you, understanding the justifications behind them is needed as quickly as possible. Two-way communication is the path back to balance, and to right any wrongs.

Higher Power / Forces

Just like in *The World of You*, I accept that you may believe a higher power is still judging you and determining objective right and wrong in *The World of Two*. That's fine by me. You are welcome to believe whatever you want.

If you think this would be the case, I will simply ask one question: What *actions* are you going to prevent the *Being* from performing, in order to comply with what *you* believe your higher power wants?

In such a situation, the higher power is a third "person" in your mental boundary that you imagine to judge you both before carrying out any action at all. How is this a fair situation to the Being? How is it fair to you?

If you are a highly religious person, before making some final decision about religion or higher power in *The World of Two* with certainty, consider the following: What would happen to your religious beliefs if *Memory-Wiped You* started this experiment? How many of these things are you carrying with you into this experiment from modern day stories promoted by *other* people, and how many are built from your firsthand experience with a higher power?

Should you become that same story-promoting *other* person to the Being? To what end? What do you fear will happen if the Being performs an action that disagrees with *your* religious beliefs of right and wrong? Do the consequences you fear involve the present? The future? Do they even involve your current life?

Here's another balance check: What happens if the Being has a religion all its own that seems strange and unfamiliar to you? What actions can the Being rightfully prevent *you* from carrying out to comply with what *it* believes its higher power or religion judges to be wrong? What is reasonable for the Being to fear happening should *you* perform an action that disagrees with *its* beliefs of right and wrong tied to a religion or higher power that is nothing like your own?

I'll ask you not to tell yourself a story about my beliefs based on my past actions of putting these words in this order. I am not saying your God or higher power doesn't exist, nor am I saying your God or higher power does exist. Nor am I saying your religion is wrong or right. That is all entirely up to you to decide.

Words

Words now suddenly have tremendous value in *The World of Two*. They allow you to communicate feelings, fears, needs, and actions at significant *speed*. You and the Being can teach each other new stories that each of you might know relative to the other, such as your history. You can use words to try to build solutions to problems related to food, water, and shelter with barely any risk. Most of all, you can use them to keep misunderstandings at bay, and prevent violence between you while at the same time collapsing fear and anxiety into harmony, truths, trust, and understanding.

Money

Does money have any value in *The World of Two*? If a pile of six billion one-dollar bills appeared next to you, does it have any value to either of you?

Again, I think this money would only be worth the paper it is printed on.

The value of money does not yet seem to exist like it does at the time I am writing this, so its value must appear somewhere between these two points of time.

Math

Math is beginning to have use now, but only the basic stuff. It exists as a tool of equality and fairness when it comes to measuring things in the objective world.

"One" is just a word, and it has to be anchored to approximately the same wordless perception of a boundary shared between you. So does "two", "three," and all the other number-words.

If you both seem to perceive boundaries of objects the same way in your Step 1 – which means you both share approximately the same sensory resolution - then a walnut made of millions of molecules and gazillions of atoms appears as a single object.

If there are four walnuts, then together you can create any new words you can agree to anchor to the quantities you and I call "one, two, three, and four". Perhaps it could be "Sny, Fuv, Tru, and Ov." Looking at the walnut pile, you both can use language and agree that separating them into "Tru" and "Sny" are imbalanced, but "Fuv" and "Fuv" are balanced and equal portions.

Dividing a fish into "Tru" equal pieces will probably require some combined perceptions communicated through words to agree on what "Sny Trud" (one third) looks like, since the pieces of fish will not have the same approximate shapes.

To fast forward this math story through time, you would probably create shared measuring "units" to keep things more fairly balanced without having to rely on feelings. Perhaps food shall be divided into equal amounts using the size of a finger to measure, which you both call a "Grank". "Sny" (one) grank is agreed to be the width of a thumb, which works because yours and the Being's thumbs are almost identical in width.

If water is in need of rationing, it could be divided with units you both agree to call a "fist". A "fist" is determined by completely filling a crudely-made cup you created out of clay just for this purpose.

No matter the measuring systems created, from this point math is only limited by the memorization of shared words that represent how many things there are. The "operators" like add, subtract, multiply, and divide are, in my opinion, still completely natural, unspoken, and intuitive. By this I mean you aren't going to look at a fish you both intend to eat and then "multiply". There is only "divide," and it can be done without a word for that operation. In this way, there is no need for an "order of operations" yet, only a number line and a measuring system.

Perhaps you both stop your number line at "Ov" (four). After that, to get to bigger numbers, you start over again at "Sny" (one), but put "Te" in front it. "Te Sny" is four plus one (or "five" in English). "Te Ov" is four plus four (eight). After eight, "Te" is changed to "Vi". "Vi Sny" is eight plus one, or nine.

You'll probably also create words that modify "Sny" into smaller pieces too, so you can measure out halves, quarters, or whatever else is needed.

Either way, you are now limited only to what you both can agree to and remember when it comes to math and measurements.

Job or Retirement

If you had a job before entering *The World of You* and *The World of Two*, what does the loss of your job mean to you now? Was your job accomplishing something meaningful relative to your current situation with the Being? Chances are improving your answer could be yes, though I can't know this.

As you read this, if you are a teacher, a communicator, or someone that works with individuals that can be unpredictable or unable to communicate well, you might realize you have a very valuable skillset that reaches all the way back to the origins of society, and might be incredibly helpful in *The World of Two*.

Politics

Perhaps unexpectedly, politics in *The World of Two* exists. Politics did not exist when you were struggling to fulfill your needs alone, but it sprung into existence the moment another became involved. There are no elections or candidates or offices, of course. There are just two representatives, and each is representing a self.

There are plenty of political issues to deal with. You are no longer King, Judge, Juror, Executioner, Citizen, and Immigrant like you were in *The World of You* – this is now a two party system.

If you are wondering what political issues might exist, well, they are the same as in modern times when I wrote this: What *is* fair when it comes to food, water, and shelter? Who *should* get to decide? Who *should* be the one to filet this fish? Who *should* count out the walnuts? Who *should* gather them? All of these "shoulds" are all ultimately political issues of authority, and every conversation or debate you have with the Being over them is political. You are both working to find an agreeable compromise on what is fair, so you don't end up going to "war" with each other over control.

For me, this exposes the foundation of what politics is all about: To use words and communication to try to find fairness without collapsing to violence. It is *not* to win, it is *not* to dominate over the other, and it is *not* to avoid the issues altogether. It is to confront the issues head on, and use language to find a way to agree on what is fair. Politics is just a word for the morality contract creation process, setting the terms and definitions, and working to eliminate as many assumptions and misunderstandings as possible.

In *The World of Two*, this political foundation is extremely simple: Is one of you perceiving that the other is trying to dominate or be unfair? If so, talk it out so it doesn't collapse to violence.

At the time I wrote this, I believe the foundation of politics in the United States is completely lost, and has already collapsed into a battle of dominate-the-other-at-any-cost with barely even consideration on what is fair. I believe I know how this has happened, and will share it in later chapters.

Senses

In *The World of Two*, the sudden loss of one of your senses is not as severe a problem to overcome as it would have been in *The World of You*.

If you somehow get through the *Dark Forest Moment* without one of your senses, it might be possible to establish an "imbalanced" agreement of fairness based on your limited abilities relative to the Being.

After all, if you made it to such a point, somehow you survived for two years with that missing sense. I'm not saying it's unlikely you would ever make it to the *Dark Forest Moment*, but I think it would be a massive challenge to negotiate a balanced outcome, depending on the sense you were missing.

Imagine if the Being was deaf and never heard you coming from behind. Or if the Being was blind and could not see you, but only hear you. How would that change things? It would be challenging, right? Not impossible, but definitely more challenging. Every sense we have seems to be very important to our survival. Once the *World of Two* is established though, the being can help fill your "sensory gaps" using language.

Mobility, Speed, Strength, and Health

In a similar fashion to a missing sense, if you are limited in mobility, speed, strength, or health, it might be difficult but possible to achieve a sense of trust, understanding, and an eventual "imbalanced" balance with the Being. It might willingly provide the bulk of these things you lack in exchange for your effort in something else – whatever that may be.

Again though, not possessing these things from the start made it more difficult to have survived the two years it took to get to the Being in the first place. Still, if you made it to *The World of Two*, language and two-way communication is the strongest tool to find balance and fairness between each other.

Clothes, Home, Groceries, and the Moment

As I see it, there is no change to these between *The World of You* and *The World of Two*. They still don't exist in a meaningful way like they do today. These are all still immensely valuable, and the presence of another does not change their value between this thought experiment and the world around you as you read this book.

Truth

In *The World of Two*, are you officially the Keeper of Truth? Is the Being?

Or is the merged truth of what makes the most sense for both of you the larger truth? Does this merged truth move and adjust itself as more and more words are created and more and more experiences are had?

What would happen if you both experience an event that radically changes your shared truth?

Let's put it to the test.

111

Warping Your Perspective

Rewind time back all the way back to the beginning of *The World of You*, long before you ever met the Being. All of that stuff with the Being, and the two years of surviving *The World of You* has not yet happened. You have no memories of any of it.

You are alone – the only human in the world – and you are naked and now trying to capture the end of your need story arcs for the first time.

You manage to find a fishless creek and drink yourself silly. You even find a cave near the creek, complete with a rock "door" that blocks the entrance nicely. Food however, is causing you great anxiety: You can't seem to find any.

Due to your lack of food collapsing your timeline, its capture becomes your top priority, so you decide to leave the creek and cave to search for it. You don't really have a choice - your freedom to choose some other priority does not exist, except for one: You can choose to give up movement and let your food story arc take you to extreme hunger and eventual death – the path of collapse into darkness. I'll assume you are not willing to give up on life just yet, so you keep moving, and desperately hope to stumble into *something* edible.

You fight your way through dense forest growth for hours, moving away from the familiar creek and cave in one general direction. The void of the unknown must have *something* you can eat. You'll surely find food if you just keep going…

After pushing your way through some particularly heavy growth - winded, tired, and starving - you collect yourself and look up to a sight that is nothing short of a miracle: In front of you is a tightly-packed grove of food-bearing trees and plants.

You see apple trees, pear trees, grape vines, squash plants, melon vines, berry bushes, lettuces, pepper plants, bean plants, tomato plants, and some you don't even recognize. You even notice carrot greens and recognize potato plants. Many of these are already bearing fruit and vegetables that appear to be ready to eat.

What feeling would this moment bring you?

What story would you imagine to explain such a place?

No matter that story, I suspect you will gorge yourself silly on sweet, delicious fruit.

After you are done, I suppose you'll just turn around, leave this place, and never come back, right?

Of course not. I believe you will treat these plants like I did after I found the single berry bush. You will start figuring out how to capture this entire area. How to make this whole cornucopia "yours."

For now though, after stuffing yourself with what was possibly the best tasting food of your entire life, you plop to the ground and lean back against the trunk of a fruit tree that just helped fill your belly. You stare off into the dense grove of food in complete satisfaction and bliss, and suddenly notice a pair of human-like bright-yellow eyes looking right back at you through the thick plant growth not ten feet away.

You pop up fast, keeping your eyes locked on those yellow eyes as adrenaline floods your system. There is a large and unexpected rustling in the overgrowth to your immediate right, and you turn your eyes away from the yellow eyes to discover a green-skinned Being that looks extremely similar to you in almost every way stepping out of the bushes with a very intense look on its face - and what appears to be some kind of weapon in its hand.

The yellow-eyed Being now emerges from the thick growth to your left, where you first saw it watching you.

Welcome to the *Dark Forest Moment 2.0.*

112

Danse Macabre

The same three choices for action apply to the *Dark Forest Moment 2.0* as the original *Dark Forest Moment*, but now there are two Beings, not just one. This creates an immediate submissive/dominant relationship. You are very much the underdog in this dance of death.

From the perspective of the two Beings, who have worked together for years to build this garden, you could easily merge with the same story as a deer, bird, or rodent. You could simply be a *pest* eating the food they intentionally planted close in time and space to increase the certainty of their own survival and freedom of choice far into their future. You could also simply be seen as food, just like a deer, bird, or rodent.

The pair of Beings have the majority of power in this *Dark Forest Moment 2.0*, and you are at their mercy. Let's work through the options exactly like the first *Dark Forest Moment* and see what has changed.

Option 1: Flight (Time-Shelter Paradox)

The outcome of this choice does not change no matter how many Beings there might be. You might successfully escape (assuming you are not attacked while running away), but you will continue your life in fear and anxiety from that point onward. *Are they stalking you? Watching you? Planning to attack?*

I believe this choice will lead you inevitably back to the *Dark Forest Moment* just like before for two reasons. The first is the need to gain information about the two Beings so you can better imagine stories that include firsthand experience, which help control your own runaway story system. This sensory information will allow you to predict and validate *something* about them as true, providing a feeling of shelter. The other reason is to try to get food from their treasure trove of a garden again, assuming you continue struggling to find food anywhere else. It's risky, but what choice do you have? The other choice is to risk starving to death searching the void.

Option 2: Freeze (Motion-Shelter Paradox)

This option is a good one at first because if you aren't moving, you aren't validating a story for the two Beings that you are a threat. However, you aren't communicating anything else, either.

Just like the original *Dark Forest Moment*, this "don't move" option quickly decays as the two Beings approach. You must inevitably justify a movement, and choices are very limited: You either run, or begin the "shelter dance" in an effort to somehow secure shelter against these Beings.

Since running returns you back to the *Dark Forest Moment* again, the choice to freeze collapses to the path of transformation.

Outcome 1: Death (Collapse-to-Expand Paradox)

Should you decide to attack the two Beings with intent to kill your way out of this situation, you know your chances of success are not good. It is very likely you are not only going to end up losing, but will earn yourself serious injuries (assuming you aren't killed in the process).

Even if you succeed and kill both Beings, you end up back in loops of fear and anxiety with imagined and unvalidatable stories about what other type of creatures might be out there, and also unvalidatable stories about these Beings' existence and experiences.

Should you see yourself choosing this option, fear might be overriding your ability to see other ways of surviving this moment.

Outcome 2: Dominant and Submissive (Stability Paradox)

The natural shape of this two-balloon situation (the pair of Beings as one balloon, and you as the second) is dominant and submissive. If you want to survive, you are going to need to communicate in some way though body language or expression that you are not a threat and had no idea you were eating food that "belonged" to them. You ate because it was the right thing to do at the time you did it, given what you knew through firsthand experience and secondhand stories – you were all alone in this place.

How you accomplish this communication is up to you. It could be falling to your knees in tears. It could be stepping backwards with your hands open in a non-threatening way. There are a million-and-one ways to try to do this.

You are pretty much at the mercy of the Beings, though. You have essentially been "captured". You can believe whatever you want about equality, fairness, and imbalance, but you cannot express it in words at this moment. Your intents and justifications are locked away and hidden in your prison of a mental boundary. Unfortunately, the clock is ticking, and the minimum speed limit to find balance applies.

Outcome 3: Equality/Stalemate (Decay Paradox)

This outcome, unfortunately, does not exist for you in this moment like it did one-on-one with the Being in the first *Dark Forest Moment*. You can choose to try to take this path - to act as if the two Beings are each your equals from your perspective - but you cannot yet express this in words without a shared language. Your intents and justifications are locked away and hidden, the same as theirs.

In this 2-on-1 imbalanced system, whatever path the Beings choose will become the path of least resistance for you, unless you are willing to risk death by assuming they have chosen to kill you. You freedom to choose the path forward is way more limited than theirs. The path to violence and death is really the only path you can bring into reality with confidence. All you can choose aside from this is your best to communicate no harm or threat is intended.

However, as you *hope* for no harm to come to you, and maybe even recognition of equality, you must brace your story system for the possibility of violence to erupt. *Violence is always a possible choice anyone can make at any time.* You might be viewed as a captured pest, after all. You also might be seen as a threat. You can't really control any stories and assumptions they may have about you; you can only try to control the story being built in the present through their firsthand experience with you.

This means the Beings have more freedom of choice than you do, and it is ultimately the Beings that must choose the middle path of stability to avoid violence. You exist in a state of desperately wanting to take this middle path, while at the same time preparing yourself to react to a different and much more violent option. Second You is trying to establish stability through very conscious and controlled choices, while First You is loading the fight or flight movements into the justification barrel for immediate firing as soon as Second You quits overriding it.

If you've not reflected much on this situation yet, you might be surprised to realize that right now you are not in one or even two places at once – you are in *four*.

You are reading this book, you are trying to survive, and you are both Beings. Time-warping reality-bending speed-machine, indeed.

The Shadowy Forest

In the *Dark Forest Moment 2.0*, you are shadow-experiencing many perspectives at once. You know how this situation feels from the point of view of the Beings, because you've already walked through the process and experiences it took to build a bond at all, and to then build a shared territory.

From this point of view, you both worked hard to cultivate this land into a garden to increase the certainty of your survival and expand the freedom of your choices, which is promised by the future food this garden is expected to provide in time. Currently, you expend a much smaller amount of movement and time these days capturing and maintaining food, water, and shelter.

Then along came the day when you were resting and trying to figure out if you can splice two different plants stems together to make a bigger and sweeter fruit, when a large amount of rustling came from the garden. You alerted your other self (the Being), using silent communication - the same you both use for hunting. This method let you sneak up and observe whatever was eating or damaging your precious "captured" food before it became aware of you.

At the moment you realized the intruder was yet another strangely-colored "self", you may have told yourself the story this is "objective" wrongness in the form of "crimes" like trespassing and theft. However, before jumping to this conclusion, I have a huge question for you: Was any of this food really *yours* in the first place? You moved plants around, or harvested and planted seeds from food plants in a specific place, but you are doing this to increase your certainty and freedom of choice.

If a berry bush you considered "yours" was a two hour walk away, this "trespassing" human could have cleaned the bush completely of fruit without you knowing it happened until the next time you checked on it. And even then, you may never know it was a creature just like you doing it. You would never know a "human" existed at all. Would you still feel a need to immediately guard this plant with violence, and start frantically searching for the "thief" to harm, capture, or kill? Or would you consider this a case of bad luck and search for food somewhere else?

I think you would choose the latter. Why does this same bush getting "robbed" feel different if it is moved much closer to you – say 5 seconds away instead of two hours? Why does this make the "theft" feel personal instead of a case of bad luck?

In my opinion, it's because by building a garden, your source of food is now inside (or close to inside) the sensory boundary of your Step 1 *at all times*, allowing you to easily capture every edible thing every plant makes, and also protect it very quickly.

Knowing your food is protected and close, your story system can anchor it as being almost certain – basically pre-captured – and start to focus on whatever problems and priorities interest you. *You become free to choose your movements without those movements threatening your survival.*

In contrast, from the point of view of the "trespassing" human, none of your needs are secured, so you are *not* free to choose your priorities or movements. You *must* eat, or face extreme hunger and death (not that the two Beings could know this, of course. They can't read your mind).

From the point of view of the Beings, the first thing to deal with is negative emotions. You both must first accept this trespasser is *unexpected* by both your story systems. This story existed in darkness – a gap. Does this make the story objectively criminal and wrong?

If the possibility of a new-to-you and similar-to-you creature eating food from your garden was not one of your shared possible stories with the Being – are *you* perhaps the one that is wrong? You both certainly thought of the stories of deer or other animals in your garden, and you planned how to deal with those things for minimal emotional drama. In this way, an intruder was definitely expected. But this one, who looks like you in almost every way except color – is it just another animal? Should this be an *expected* story, or is this truly *unexpected*?

As I see it, that familiar electrical current signature of "self" hits you every time you look at the trespasser, but the balance of power is in your favor. The *Dark Forest Moment 2.0* is upon you – let's take a look at the possibilities from your original perspective as partner with the Being.

Option 1: Flight (Time Shelter Paradox)

I see very little chance you and the Being will run away from this "Trespasser" that looks similar to both of you except for color. You *could* run of course - it *is* an option - but you have spent *years* building this place to expand your time to survive. To run would be to abandon all of this and to collapse all of your certainties and freedoms it has given you. Plus you will know nothing about the Trespasser – which will only lead you back to it.

Option 2: Freeze (Motion-Shelter Paradox)

This option is good for gathering information and making decisions about future movement, and you both did this already while secretly watching this Trespasser gorge itself on your food.

After the Trespasser made eye contact with you though, this option immediately decayed. Staying with this option beyond this point would allow the Trespasser to choose the *Dark Forest Moment 2.0* path of fight, flight, or attempt to communicate first.

Should it choose to run, it seems likely that you and the Being will pursue. After all, this Trespasser is perhaps the biggest threat to your food supply you've ever encountered. It is probably the first creature capable of completely "unmaking" your garden, because it seems to possess the same abilities required to "make" it in the first place. In a way, a pursuit is still the "freeze" option, until the pursuit ends and it decays to a different path.

In order to "capture the end" of this story, the option to freeze needs to collapse quickly the instant you are spotted. The only option that has the potential to regain your previous certainty is to engage, to begin "the shelter dance," and determine how to best shelter yourself, the Being, and your food from this Trespasser.

Outcome 1: Death (Collapse-to-Expand Paradox)

Killing this Trespasser results in the same mental fallout as making this same choice in the first Dark Forest Moment. You and the Being will be left with more questions than answers, and have an increasing anxiety there could be more of these "you" (human) creatures in your world.

Perhaps you would worry that "new" creatures might start appearing to try to steal your food at any given time, or are looking for a way to attack and kill you whenever possible to take your food.

You may feel compelled to construct a literal fence or boundary of some kind to keep your food in the garden and everything else out. However, no wall will ever be high enough to provide total shelter over your food – especially a wall built from stone that is easily climbable.

The result of this uncertainty might be to set booby traps, alarms, or use other methods to convert unseen movements near your food into movements or sounds that will enter Step 1 of your story system and allow you to react quickly.

Of course, it is possible all of this fear and anxiety will be over a situation that is not possible to ever repeat itself. It's possible you may have just killed the last one of its kind. But your story system can never validate this. Death is final. The dead answer no questions and tell no stories. Your memory of the (human) Trespasser enters every single story your story system builds. You have a story voltage that can never be collapsed into balance.

You'll have to try to balance it, though. I suspect you will eventually start to go on hunting parties and reconnaissance missions to determine if there are more strange creatures outside the boundary of your territory.

I see this outcome as the path of permanent paranoia, fear, and anxiety for both you and the Being. An internal mental loop is created in each of you, as your story systems projects the dead Trespasser into all your stories in the form of more Trespassers. You might even call this a type of "ghost". It is a paradox of a position to be in.

To try to cope with this stress, I think you and the Being will tell story after story to each other of the now-dead human creature. You might try to calm each other's runaway story process by validating what the dead human must have been thinking, its intents, and why your chosen actions were definitely the *right* ones to make.

"It was in the garden because it was sent by someone."

"I bet it was watching us for weeks."

"It would have killed us in our sleep. We did the right thing by killing it first."

These stories only exist to try to validate a past action into being right with imagined justifications, when you can never actually validate if it was right or wrong to do at all.

Outcome 2: Dominant and Submissive (Stability Paradox)

The choice of domination will be incredibly easy to make in this 2-on-1 situation. I believe this is the path of least resistance – and it is therefore the natural structured outcome of this two-point three-mind system.

Two points, you and the Being, have essentially become synchronized halves of a single, larger point (or balloon), while the second point, the Trespasser, remains a small balloon or point by comparison. This means the direction of all movements in this two-balloon system are toward the smaller balloon, and even the Trespasser can sense this to be true as its fear spikes the moment it realizes there are two of you.

Let's assume this choice to dominate is noticed by your movements and demeanor, and triggers the Trespasser to run. You and the Being pursue, and chase the Trespasser for a while with the threat of violence or capture (domination).

I don't think you will give chase for long though, because you are ultimately anchored to the place where the chase began. You and the Being cannot easily move your garden and territory. Chasing the Trespasser into the void outside of your known territory results in a dramatic collapse of your certainty. For this reason, I suspect you will stop your chase after awhile and return "home".

I think you will then have a heightened anxiety that the Trespasser will come back at any moment and steal more of your food – and to be honest, it's incredibly likely this will happen. You are now exactly in the position I was in when I tried to protect the remote berry bush – except now you have *lots* of plants to protect. You and the Being can't just take *all* the berries in this situation and run away to a new location, you have a regular food supply that rotates in ripeness all the time, and no way to move or store this much food. You are quite literally grounded to this place.

The result of this choice, then, is the same as what happened in the choice to kill, except this time you know exactly what movements and creature you are imagining to defend against.

Instead of alarms, you and the Being will probably create a bunch of pre-made movement "machines" designed to automatically act in violence or capture the Trespasser on your behalf. Machines like hidden traps, blunt force devices, sharp and harmful-to-climb fences, and alarms. While this outcome looks extremely similar to the fallout from killing the Trespasser choice, the anxiety may be quite a bit worse in this one, because you both know the Trespasser is *still out there.*

Is it looking to hunt you? Kill you? Is it stealing your food right now? Are there more of them? How many should you expect? What kind of defense do you need? Is it best to kill or to capture? There are a million stories you can imagine and absolutely no way to bring the Trespasser into them with any certainty. The *Dark Forest Moment* waits to return again.

Outcome 3: Equality/Stalemate (Decay Paradox)

As I see it, this is the best choice for all parties, the same as it was the first time through the *Dark Forest Moment* with you and the Being. If the Trespasser does not run, communication in some form will almost certainly be attempted by the Trespasser to expand time between you.

However, this *Dark Forest Moment* is far more difficult to navigate than the first one. Neither you nor your partner, the Being, can read the Trespasser's mind, and the Trespasser does not speak your language. Even if you welcome the Trespasser into your camp, you cannot be sure if it will enter out of submissiveness and fear, or because it truly is relieved by your gestures and happy to enter as a perceived equal. Is it actually fearful and waiting for a moment to attack or escape? Is it taking in the details of your camp in order to grab your most valuable things as it darts back into the woods unexpectedly? In a nutshell: Is the Trespasser *acting* submissive and friendly in order to expand its time until a situation that works to its advantage presents itself?

The *Chain of Suspicion* has started.

The only remedy to this mental-boundary created problem that hides beliefs and justifications is for this Trespasser to share your language. You need to communicate *your* intents to the Trespasser, and you need to do so as quickly as possible before its fear or rising anxiety result in the collapse of time, and another *Dark Forest Moment*.

Move yourself back into the perspective of the Trespasser to experience the weight of this situation. You've just been aggressively approached by two strange-colored Beings working together and communicating with each other. If you choose not to run, you must choose to submit, or choose violence. You have no ability to predict and react to two story systems at the same time at the intensity required of the *Dark Forest Moment*.

Let's assume you submit. Their angry looks seem to fade, and they gently lead you to their "camp." You still have no idea what is about to happen at any moment. Are they about to kill you? Eat you? Are they friends, or are they enemies? You cannot know. The *Chain of Suspicion* grows longer.

The priorities when it comes to communication are the same as the first *Dark Forest Moment*: Establish an equality "contract" as quickly as possible - establish shared needs, build trust by making your movements as predictable as possible, eventually aided using shared words. The shared words can eventually establish mutual understanding, and eventually result in an agreed-upon imbalanced relationship.

This seems like it is just the Dark-Forest-Moment-to-trusted-partner path as it was the first time with the Being – and it is, until it's not.

The Minor Third

You and the Being have the Trespasser in your camp. There was enough recognition of self to overcome fear, if only slightly.

Your first contact with the Trespasser was like seeing yourself from years ago when you first met the Being, and that recognition registered as believing you understand and can relate to its reactions and expressions. You can *empathize* with its fear, because you recognize it.

You have no common language with this strange individual though, and without it, no shared context outside of this empathy. Nagging fear reminds you your empathy is based on assumption and that more evidence of self is required for trust. So now the same exact process must repeat itself between you, the Being, and the Trespasser that already happened between you and the Being so many years ago.

This process toward trust will require actions from all three of you. Actions require *time* from each of you. They are new added tasks with urgent priority – Fear demands immediate movement toward strict *equality* and predictability. Whatever tasks you were prioritizing before this Trespasser entered your life must now come second until some level of trust through predictability can be established.

This sudden reprioritization does not trigger an immediate collapse to violence, because your needs are fulfilled and pre-captured far into the future thanks to your garden, creek, and shelters. You have the *freedom* of choice and time to merge the Trespasser into *The World of Two*. It has all the same needs you do. It needs a shelter. It needs food. It needs water. It also needs assurance it is being treated fairly and also has a say on what is fair.

In the *World of You*, you reshaped nature from its all-wild natural shape using your own actions over the span of years. You were self-sufficient. You ran into the Being while out exploring the void to expand knowledge of your surroundings.

In the *World of Two,* you and the Being reshaped nature even more. You have steadily reshaped your relationship. You have steadily reshaped what fair looks like. You have reached a point of so much trust between each other you can accept the Being simply requires more food than you.

But as the Trespasser, you did not experience any of the time or transformations between these two Beings and this world. You began your experience, struggled to find food, and while trying to prevent the story of your death by starvation, you stumbled right into a "patch of nature" filled with the very things you needed.

It was completely unknown to you that this "patch of nature" took *years* of time and *millions* of justified actions to construct. How could you know? You believed you were the only human on Earth. Since imagining *The World of You* involved returning everything to its natural all-wild condition, it's perfectly reasonable to assume this treasure trove occurred naturally, right?

This area is everything you could hope for. It is a huge surge of certainty for your story system. It is practically heaven – a paradise – located in hell.

You probably cannot imagine yourself leaving it. You would never want to go back to the "before times" of extreme hunger and no food anywhere.

And then the two Beings appear. One looks very different from the other, but they are obviously a team.

They have weapons.

You surrender.

They lower their weapons.

They make strange sounds to each other.

They take you into their camp.

And oh, what a camp it is!

They are trying to communicate with you.

You do not know what "grup" means, but they brought out a pile of what appears to be walnut kernels.

The green Being is given three kernels by the non-green Being, who says "tru grup," whatever that means.

The non-green Being then gives itself one kernel and says "sny grup". It then hands you two and says "Fuv grup".

Is this their version of equality?

Is this what they believe is fair?

Does this mean you are worth more than the non-green Being, but less than the green Being? Should you give them all back? Refuse them? Does accepting this number create dominant and submissive positions in the group? Should you ask for another? Give just one back?

The possibilities here start to grow larger than I can go through in this book.

There are three story systems involved, three imbalances trying to balance, and every change, word, or action sends all three systems spooling through the 3-step process again and again, changing the possibilities.

In a lot of ways, this is similar to a three-body problem in physics. In other ways, it is way more complicated.

After all, one of the "bodies" is significantly imbalanced against the other two.

115

Scared of the Dark

As I see it, if this new being in your life – the Trespasser - behaves against yours and the Being's expectations, new imbalances and story voltages are created and will be evident through emotions.

Let's fast-forward time a bit. The Trespasser has learned your language pretty well, but over time it seems less and less interested in tasks to improve the camp or help with anything related to capturing food, water, and shelter.

All the Trespasser seems to want to do is lay around by the creek napping. It picks its own food from the garden when hungry, and drinks from the creek when thirsty. It chooses to sleep outside of the main cave after being asked to construct its own bedding to sleep inside, yet it still runs into that cave during a storm (somehow always finding its way onto *your* bedding after a few minutes).

If asked to do a task, the Trespasser's first response is to resist. For example, you asked it to repair a fish trap that was damaged, and it said it will fix the trap later - then resumes napping by the creek. It never did end up fixing it – *you* did.

While you were busy descaling fish for a group meal and the Being was in the fruit trees collecting apples, the Being asked the Trespasser to please go search for wild mushrooms. The Trespasser immediately countered with "why?" The Being explained it was the time of year for them to appear at the base of specific trees, and then explained how to find these trees. The Trespasser slowly meandered off into the forest, only to come back *hours* later empty-handed. It said it couldn't find the "certain trees" so it just decided to come back instead.

This behavior appears constantly. The Trespasser falls behind on group hunts for bigger game, refusing to move at the pace of you and the Being. It says it doesn't like to touch fish, so it won't help empty the fish traps each day. It says its scared of heights, so it won't climb the trees to collect fruits. It refuses to fetch water to boil over the fire because its "feet hurt too much when it carries that much weight".

The only task the Trespasser seems to like is drawing pictures. It mashes together certain flowers and dirt to create different "paint" colors. It then uses these colors to create beautiful patterns on the cave's walls and nearby rocks along the creek. It has painted picture stories in the camp of the three of you, and created a colorful picture of fruit on the trunk of each respective tree.

When the Trespasser is not napping or doing some task you have to nag it to finish, painting and the tasks associated with painting seem to be the only thing it chooses to do willingly without resistance.

I'd be willing to bet you, as my reader, are experiencing a pretty strong emotion right now. What is it? Anger? Frustration? Annoyance? What is the story you expect from the Trespasser? Do you feel like you are about to enter a *Dark Forest Moment* with the Trespasser where violence is possible? Do you feel like you are being forced to keep choosing the "run away" option from these Dark Forest Moments because I am the one controlling these words?

Do you instead allow this story voltage trapped in your mental boundary build into an internal *Dark Forest Moment*, where you justify violence without the Trespasser having any clue this violence is coming?

Are the Trespasser's actions the issue here?

If not, what *is* the issue? *Who's issue is this?*

When you asked the Trespasser to fix the fish trap, is that really a *Dark Forest Moment*? Is that request a point of collapse or expansion? Is it capturing an end? If so, of what? Trust? Fairness?

Are you annoyed because the Trespasser *refuses* to do the actions *you* ask it to do? Is it *wrong* to not do these tasks? If so, what two points are being used to define this is wrong? Are you the *Keeper of Truth and Decider of Right* for the Trespasser?

I suspect you will answer "no" to this last question, yet I also assume you are still annoyed, might want to answer "yes", and see the Trespasser's behavior as unacceptable. Why?

Think on it. It's worth trying to understand yourself whenever you can, in my opinion.

As I see it, there is a very strong urge by your story system to capture, or "control," this Trespasser. It *needs* the Trespasser to "value" what you and the Being have built the same way it values it. Your story system wants the Trespasser to justify increasing overall certainty and freedom by assisting in this place's upkeep, expanding it, improving it. Your story system *needs* this Trespasser to seek the same goals as you. (Now might be a good time to ask: What is that goal, and why? Stop and think about it if you need to.)

You and the Being managed to free up lots of time for yourselves. This allows time for improving your lives and connecting your stories through language, instead of hunting and gathering non-stop all day every day.

By deciding to bring this Trespasser into the fold, you both have sacrificed lots of this free time, and must expend lots of energy you didn't have to previously. All in an effort to grow the Trespasser's story system to match yours. You and the Being took *months* of time to teach it the stories of how to care for the fish traps, how to use tools, how to prune and graft plants, how to speak your language, and how to identify and store edible wild plants. You even told it the detailed story of how you and the Being met, and how you developed things to this point from that moment.

You didn't have to do any of that. You could have just killed the Trespasser. Enslaved it. Chased it away. And now the Trespasser won't even "repay the favor" of your sacrifice and hard work by doing what you or the Being ask of it – which is a tiny fraction of the work you and the Being have put into this place. The Trespasser doesn't seem to value the garden like you. It doesn't stress about it, carefully study it, or worry about the appearance of a single dead branch in an orchard of fifty trees. This might feel *ungrateful* to you. It might even feel disrespectful. Asking the Trespasser to do its *fair share* to maintain and keep improving this place probably seems to be the right thing to do from your perspective, right? In exchange for all the certainty you and the Being have provided – what's wrong with asking it to do a chore or two?

What is happening here, and what's the way out of this mess?

Are you going to physically *force* the Trespasser to do the labor you ask it to do? Are you going to punish it with violence, or isolation, or withholding things it needs or wants if it doesn't comply?

In my opinion, your mind greatly resists the possibility of returning all the way back to the *Dark Forest Moment*. It has labored for years to capture this much certainty of your survival. When comparing the story of keeping this certainty by not confronting the Trespasser, to the story of confronting the Trespasser and potentially collapsing to violence – your story system almost certainly will choose to keep its current level of certainty. It chooses not to escalate the conflict with an equal. It resists the Dark Forest. It *fears* it. The Trespasser is doing the same, in a different way.

Put yourself back into the Trespasser's point of view. You do not *want* to experience a story of violence with the two Beings. You do *want* to keep this certainty about your future captured. However, you are also not interested in losing the freedom to pursue new stories. You don't *want* to sacrifice your free time to some "random" task one of the Beings seem to constantly ask you to do. These tasks don't feel important to you - they feel like busy work.

There is enough food and water in this place for all three of you for decades, and you volunteered to sleep outside to avoid everyone needing to build more shelter. You, as the Trespasser, have secured the freedom of choice that comes with capturing hunger, thirst, and fear. You are living near the billion-door space of *The Ark*. You have no interest in stepping backwards and losing the freedom to open more doors. *The Beings are just worriers. Do they not realize they are good to relax as well? To pursue whatever they want? To reap the literal fruits of their labor?*

Ok, let's stop here. Somewhere in this chapter, a switch has flipped. Noticing it is extremely difficult. This switch happens fast – so fast I'm pretty sure it can only ever be noticed by Second You *after* it happens.

This situation is *no longer about right and wrong*. This is also no longer about trust or fairness. Somewhere above, this entire story has pivoted to be about *authority*.

I can spend the next many chapters detailing what amounts to tiny imbalances, differences in values, and differences in perceptions to explain how they add up into big problems. I don't want to do that though, and I don't think you want to read that, either. There is another way to understand what has happened, and quickly: Find the limits.

So with that, I'll now blow things completely out of proportion so we can continue up and out of this hole.

116

Ordered Chaos

While you debate why you expect the Trespasser to do what you ask, even though you never intended to become a dominating authority, the Being calls from somewhere in the distance. You and the Trespasser run in the direction of the call.

You arrive slightly winded, and are greeted with a sight that stirs a mix of emotions: There with the Being stands a new trespasser, looking scared and submissive – especially after you arrive.

The new trespasser looks similar to the first Trespasser, but with some small differences, like the length of hair and slightly different skin tone.

It does not speak your language, but does seem to speak an unfamiliar one. The first Trespasser doesn't recognize this individual or its language, but as soon as the new trespasser sees the Trespasser you've been living with for quite awhile now, its tension and fear seem to melt away.

As the three of you escort this fourth individual into your camp, yet *another* new trespasser comes running out of the forest at the group. Three of you immediately put up your guard, but the escorted trespasser immediately cries out to the new one, and runs toward it. They meet and embrace, evidently happy to see each other.

The drama and emotions settle, and the original three of you decide to feed and care for the new additions to the camp. You debate openly what to do about this situation. After all, despite the two newest trespassers apparently knowing each other, they do not seem to have a shared language; so it appears you, the Being, and the original Trespasser are the only three that can use language to communicate.

You offer the new trespassers your shelter for the night as a courtesy, and then continue debating deep into late hours before choosing to sleep under the billion stars of the night sky.

The next morning, two more trespassers arrive. They know each other, but do not seem to know anyone else.

Then two more arrive from the forest the very next day.

Then more the day after that.

And again the day after.

And again.

And again.

How many days would this need to repeat before you try to forcibly close off "your" property to prevent more trespassers from joining? Would you attempt to close things off? How many more must arrive before fear, anxiety, and panic start to set in for you?

The arrival of new trespassers repeats itself every day *for weeks*. Sometimes only one will arrive per day, sometimes three or four. It's never the same time of day when they show up, but they are always naked and starving, do not know anyone else, and do not share a language with anyone else.

Finally, the day comes when no new trespassers arrive. It probably wouldn't be easy to notice this, because there are now well over one hundred people on "your" little piece of nature you and the Being so carefully planned and cultivated though the years.

Despite the best efforts of yourself, the Being, and the Trespasser, this group is proving nearly impossible to manage.

They are all similar in that they are the same basic shape as the three of you, but some have different features when it comes to genitals, some are muscular and others are thin, some are rather large around the mid-section, while others are half the average height. The skin of each one seems to be a unique shade and hue from all the others.

As for behavior, they often fight over access to the cave shelter every night and during weather events. No one can agree on who gets to be in the cave, so some have resorted to not leaving their spot inside (for as long as they can, anyway). By doing this, these individuals are attempting to "capture" the shelter.

Overall, this "hoard" eats the food from "your" garden without hesitation, even if the fruit is not ripe. They jump to grab the higher branches and often break them in the process, or break the lower branches while trying to climb to the higher branches to reach what little food is left. Some have paired up and "captured" a single tree, and are being violent toward anyone that approaches it.

Some empty your fish traps without putting them back - often carelessly grabbing the fish and throwing the trap aside - damaging and breaking the shaved and woven wood that took you a month to figure out, and years to perfect.

Some have started to "capture" certain fish traps, and guard and watch them at all times - pouncing the moment they think a fish entered.

Still others wait and watch these individuals, and read their reactions toward the traps. If they notice individuals start moving toward the trap, the "watchers" race to get to the basket first and "steal" the fish. This has resulted in many physical fights at the creek.

This battle to "capture" food, water, and shelter at every new moment in time has resulted in things like waste management becoming a lower priority. Feces starts appearing in places it is not welcome, like the cave. Urine is also collecting in puddles near the creek and in the middle of camp where it creates terrible-smelling mud.

The Trespasser's paint collection, as well as your bedding, tools, and many other things the three of you built or created in your time together are now splattered and scattered everywhere. The once beautiful paintings have all been smeared; transforming what used to be vivid colors into browns and grays.

Occasionally, you notice one or two of the trespassers seem to speak the same crude language, but you don't recognize it. Most don't seem to know spoken language at all. Instead, the air is mostly filled with the sounds of yelling in anger and growling in combat, and cries and moans of pain from injury and sickness.

Some trespassers have "relations" right out in the open, drawing attention from others that interrupt or even collapse into animal-like domination "games" of violence on a group scale.

Your certainty of future survival provided by this place, built by the hands of the Being and yourself, is in the process of collapsing completely. The garden that promised all your future food is now getting picked clean by others at an unsustainable rate. You have to go further and further upstream to find water that looks clean enough to drink. The size and numbers of fish being caught daily is dropping. The cave is ruined from feces and urine, and from constant fighting and illnesses. A cough has started to run rampant throughout the population.

You have lost all control of fairness and trust in everyone except the Being and the Trespasser. Even *forced* equality is not manageable. You cannot communicate with any of these trespassers using words. You, the Being, and the Trespasser are the only ones that share context and understanding at a level deep enough to avoid the *Dark Forest Moment* every few minutes.

Do you regret the decision to take in the Trespasser? How about the second trespasser? The fifth? In your opinion, where exactly is the line of too many trespassers?

Do you wish you built a wall to keep them out weeks ago? Dug a moat?

Set traps?

These questions and their answers do not matter anymore. All of them involve the past that cannot be changed. This line of thinking is your story system deciding it made a mistake because your certainties have collapsed, and it is trying to figure out what justified action in your past caused this unacceptable condition so it can rewrite truth, right, and wrong for the next time you might encounter this situation.

But there is no next time. This flood of individuals is not going to vanish. This thought experiment isn't going to reboot itself again. If this were *The Ark* – you've already stepped through the threshold and fallen into your new reality. If this were *Gravity*, all the movement outside of yourself is beyond your control and limits possibilities.

You, the Being, and the Trespasser face a crisis-level decision. It involves your actions relative to this "chaos". *What do you want to do?*

Do you find yourself desperately wanting control? Wanting to create order?

Do you want to scream and demand everyone to stop?

Do you want to run away from it all?

Do you freeze and stare, and do nothing while all the movements in this place get carried out by the masses?

Do you try to "capture" the chaos with violence until a manageable number – or none - are left?

Perhaps you "dominate" the crowd into order with violence or emotional screaming, using an imposing presence and fear to stop the madness?

Do you try to stabilize by using words none of them even know?

What is your path of least resistance?

In the stories we share as humans through mass media in my time and culture, the "domination" route is often chosen to stop chaos like this. Perhaps a character with a giant booming voice commands everyone to stop what they are doing, and it works. Perhaps an overwhelming physical force starts smashing through the crowd, breaking apart the chaos through injury and death five or six individuals at a time.

I think it's safe to assume that whatever decision you make, you will not be making this decision with input from the "second point" of this mass of trespassers. They represent hundreds of imbalanced systems trying to find balance between each other. They can't just magically unify to make some joint decision as "them" relative to "you".

Do you try to merge your justification for this decision with the Being? What about the Trespasser?

Lastly, take a long look at your territory and garden in shambles.

What value is this "piece of nature" to you now?

Is this still *your* property? Was it ever really your property? Or now that its value to your future is collapsed, is it becoming just another piece of forest to you?

You'll need to make this decision quickly. Your need story arcs haven't changed. They are being pulled downward by the million billion processes of you. They require capture soon, and in this situation, violence might be required to do so.

The forced choice is upon you again, like an out-of-control trolley car approaching a split in the tracks, and you holding the lever that switches which way it will go.

117

Chaotic Order

If we assume the previous chapter's chaotic situation was how the first organized human civilization actually began, then based on what you know about society around you right now and its history, what forced-choice of action seems to have been made?

There is no wrong answer to this question, in my opinion, because we have no idea if something like this ever actually occurred – there are not many surviving records of ancient primitive society. This story is instead a metaphor, just like every other story. There is little doubt, however, that chaotic moments like this have occurred throughout human history.

As for my personal opinion on the forced-choice in this situation, I believe *all* the *Dark Forest Moment* choices were made. After all, there is a rather large group of individuals here, and each one of them is looping through the *Dark Forest Moment* again and again pretty frequently as their need story arcs keep moving up and down at nearly the same speed they always have. You, the Being, and the Trespasser are not immune or even special when it comes to this. You are just three nodes of hundreds surrounded by chaos and part of the chaos.

I imagine some individuals choose to *run*. The sheer number of possible stories predicted by their story system – all of them possibilities that threaten survival - push them to justify escape. While running away, they might very well see themselves in others that also chose to run, and stabilize their time to survive by bonding with each other. They might begin to develop language and small societies once a place is found they can feel safe together. I suspect these groups will always have great emphasis and value on defensive measures against the now unseen "hoard of savages" they escaped. In the end, this is the same outcome that occurs from running as an individual, the time-shelter paradox.

I imagine some individuals choose to *freeze* and do nothing while the chaos unfolds around them. These individual story systems determined that not adding to the chaos is the same as not being part of it, like somehow not moving stabilizes their story. I think this would decay eventually because the surrounding chaos will eventually include them. Someone in the chaos would eventually focus on the frozen individual as their new "second point" in an effort to find balance and stability.

At this moment of forced-interaction, the choice to remain "frozen" would immediately decay into one of the other *Dark Forest Moment* paths, though I admit it's also possible to remain frozen in the face of direct interaction. The choice to "freeze" *must* eventually collapse though if one is not willing to die in this position, because the need story arcs are still in motion, and if movement does not occur to capture those ends, death comes next.

With either of these choices, I suspect the movement eventually justified depends heavily on the experience-built path of least resistance toward interactions with strangers.

Another possible choice I imagine may be found in the story systems that get overwhelmed by trying to predict the chaos of thousands of stories that are rewriting themselves every 3-step story cycle. The mind of these individuals creates its own shelter - a boundary - to seeing self in anyone else.

This insulates the story system from all the other individual *Dark Forest Moments* happening for others, and allows it to operate with a simplified "me" and "them" black-and-white story structure. The truth before this point, and with it right and wrong, un-anchor and reset into a black-and-white reality. It is a "self" versus the "chaos of others" reality.

This allows the individual to justify trying to fulfill its shelter need by *any means necessary* relative to "them" or "chaos": Murder, coercion, taking hostages, and other darker social possibilities are all justifiable. It ultimately doesn't matter what action is justified, because there is no recognition of "self" in any of "them" anymore. Empathy can return only when this story system no longer feels overwhelmed – *maybe.*

The catch is that once stories are justified into actions that result in permanent outcomes like murder and mutilation to others, it will be very difficult, in my opinion, to ever rewrite truth where the collapse of empathy is not a justifiable "right" again.

If the story system justifies splitting everything into "me" and "them", then how can it ever get the opportunity to imagine the possibility that "them" might actually be "me"? "Them" being equal seems absurd. Inevitably some individual in "them" will justify an unacceptable-to-"me" action that validates the rejection of "them" all over again. I'll hit on this idea a little harder in later chapters.

I imagine others might try *forcing* social order. These individuals might climb to a high point and yell as loudly as possible in an attempt to make his or herself everyone's new priority. Without shared language though, stability using this method will be incredibly difficult to maintain. Even if gesture or crude language is used to stabilize and try to split things equally or fairly (which results in a need for authority and control to decide these things), the primary problem at play is each person's needs are only approximately the same.

Some individual's story arcs move faster and some need more resources to capture their ends. Figuring out fairness at the speed and frequency needed becomes a race between the stable middle path of using communication to push out fear and unfairness, and the collapse to violence because needs are going unfulfilled.

To make this even harder, once enough people experience what they perceive as "unfair" by such a controlled system, they begin to validate that feeling in each other – which could be as simple as acknowledging a shared experience of hunger, thirst, or fear when it seems others are *not* experiencing these feelings. The fight for authority then begins, with the end result being violence or escape from the ever-growing list of rules and judgments by those in power trying to keep control by forced-order.

The last remaining choice I can imagine is to stabilize and create order by expanding time away from the *Dark Forest Moments* by constantly choosing not to collapse to violence or separation through those same constant *Dark Forest Moments*. This choice is *required* for any two people to not run or attack one another, no matter how much time is involved. However, in this chaotic situation with no language, this can only ever be the second choice. The reason it is always second, is speed. The story system can only add one idea at a time, one word at a time, and words will almost always fail when balanced against extreme feelings of hunger, thirst, fear, and unfairness. Finding stability in the absence of a shared language or a structure of authority is going to be painfully slow.

Knowing this, I'm now going to do something impossible in objective reality.

For the purposes of moving forward, assume stability is *magically* created before cycles of murder, starvation, and collapse from fear can take place in this chaos. There are 100 people left in this mini-society. The structure of authority is You and the Being as the co-leaders, and the Trespasser next-in-line.

You already know the secret to stability - make enough resources available for all 100 story systems to feel certain their need-story arcs can be captured far into the future, so far they can no longer imagine an ending caused by lack of these needs. You and the Being reached this point before. You also know you need to be able to divide up those resources in a way that is mutually agreed-upon as fair by *everyone*.

Let the games begin.

118

The Social Machine

Thanks to magic, the collapse to endless *Dark Forest Moments* throughout this 100-person mini-society is "paused." In reality, as the limited food, water, and shelter were consumed faster than nature could replenish them, this collapse would continue until a balance was found between the first balloon (which is the number of individuals), and the second balloon (the quantity of food, water, and shelter available in time to meet their needs).

By pausing this self-balancing process between population and resources, we can now skip over all the noise it creates and better focus on the individual-to-individual balances. For the sake of not imagining an Ark-like utopia where nobody needs to eat, drink, rest, or shelter, instead imagine everyone still eats, drinks, and has shelter, but the balance is uncomfortable.

By this, I mean the shelters are packed tightly shoulder-to-shoulder when they are needed, and food and water rations are super slim - barely enough to feed everyone, but everyone is fed.

Everyone in this 100-person society is made aware the magic that paused this process of collapse will unpause in *three years*. This means if no changes are made, the great unpausing will result in the quick destruction of this society.

Knowing this, and what the future would look like if you do nothing, what are your priorities and plans to create stability and fairness in this three-year window as co-leader with the Being, and what challenges do you face?

You need social stability. How can you help this along? Now is a good time to recall the very first *Dark Forest Moment* with the Being – what did you need to stabilize your relationship? The answer, as I see it, is a fast way to communicate. In my opinion, this makes language the number one priority.

How can you teach language in the most effective way given the speed you need it – which is *right now*? You do not have the time to build it using the environment like you did with the Being. You must get everyone on board with the same words at the same speed.

This need for speed gives birth to a new system. This system is needed to teach everyone the language you and the Being created the most efficient way possible. The goal of this system is to get everyone anchored to each word as it relates to objective reality in this society as closely and as quickly as possible. *School* is now in session.

With everyone gathered together – the Being gets everyone's attention, and holds a walnut up in the air. A finger is pointed to it, and a sound is made: "Grup!" The Being gestured for the group to react, and repeats the sound while pointing: "Grup!" The masses respond…"grup!" With a few more repetitions, this new society's first shared word will be in the bag, and they all move on to the second word.

As leader, now that you know language is in progress, what *else* needs to be done to stabilize this society before the magic wears off? Well, since there are more individuals than available resources on this little piece of nature, you know food, water, and shelter need expanding, so the consumer-to-consumable balloons are better balanced three years from now.

The territory will need to grow larger than it was for you and Being to find and gather enough wild-growing foods to offset the badly damaged garden, so explorers are needed to go out into the void to do this. You also need to protect the creek from contamination, clean up and remove all the waste in the shelters and near the creek, and also figure out a way to expand shelters so everyone can be comfortable in storms or other dangerous weather.

The biggest barrier between the way things are and the way they need to be, is *time*. You don't seem to have enough of it. Gardens are slow processes. Some fruit trees take years to start producing food. Caves are not plentiful in this area. The creek is a high-traffic area, which puts it at constant risk. None of this matters though, you don't really have the choice to fall short on these things if you want everyone to survive.

The natural fix to any "time" problem is to increase speed. In my opinion, this is best done by the creation of "specialties" in your mini-society. Individuals, or small groups of individuals, can be given specific and specialized tasks, and then dedicate nearly all of their time to accomplishing those tasks.

For instance, the Trespasser has naturally fallen into the role of being the main teacher of language. It was clear within the first few weeks that some individuals picked up language easily and others struggled. As a result, time spent by the Trespasser "tutoring" language expanded from a couple of extra hours at first, to consuming nearly every daylight hour of every day.

This is a huge problem for the Trespasser. If it is spending all its time teaching language, it has none unavailable for hunting, gathering, eating, drinking, and helping to stabilize the food, water, and shelter resources. Since language is critical to balance and fairness in society, and the Trespasser's time spent doing tasks (which it struggles to perform well) is not as critical – it makes sense to rebalance these things, and make teaching language the Trespasser's *specialty*.

In exchange for most of the Trespasser's time being dedicated to teaching language, you and the Being agree to gather and prepare meals for the Trespasser so it can continue growing language as quickly as possible,

Now some big questions for you: Does this new role for the Trespasser change your feelings about it compared to back when it would lie around napping near the creek, and resist helping with much of anything? Do you now see the Trespasser as one of your inner circle, and immensely valuable, where before you were perhaps considering violence and increasingly upset over its unwillingness to comply with instructions?

I suspect your answer is "yes" to these questions, although both truth-stories about the Trespasser are absolutely valid. I will address what I believe happened in the next chapter. For now, let's move on to the specialization of this mini-society.

Two others, Alice and Bob, turn out to be extremely talented with plants. They assemble a small team, and set to work planning, improving, and expanding the garden. They know how multiple plants can be put close together in such a way that it increases all plants' health and boosts overall food production. They become the "garden team" specialty, and the same time problem exists for them that existed for the Trespasser – since they are spending all day tending and working with the garden, they are not spending any time capturing their needs. These things must instead be captured for them in exchange for their time as "garden experts".

As I see it, the most time-efficient social machine is one that has a very diverse set of experts, and is maximizing the benefit of each specialization to society by matching individual talents and expertise to roles that benefit the survival and longevity of everyone.

This means that, ideally, every individual will specialize in something different. Each is promoted to spend all his or her time doing the thing they excel at doing. This allows overall progress to happen at a much faster speed than, say, you and the Being doing all of this alone.

Language is what makes specialization possible, because it must be communicated that food, water, and shelter will first be provided in exchange for talent and expertise, and both the need-provider and the specialist-receiver *must agree to this arrangement as fair.* The specialist cannot *assume* food, water, and shelter will be provided, and then go off and specialize all day. Fairness is a relative word – two points are required.

For every specialist, *someone* must still be willing and agree to spend the time needed to capture and provide food, water, and shelter to that specialist. In the case of this mini-society, the Trespasser – a full-time language teacher – is the first to receive this *offer* to specialize from those willing to provide food, water, and shelter to the Trespasser. Again, specialization is *offered* - it is a *contract* - it cannot be demanded or taken.

The mishandling of this contract builds a voltage that might easily go unnoticed until it is too late....

119

The Hidden Hierarchy

Let's back up just a touch and uncover the hard-to-see balance between fairness and specialization.

In *The World of Two*, if one of you spent time "specializing" in something, the other is required to fulfill the specialist's needs. For instance, if the Being constructed a new wooden shelter over the course of three days non-stop, you would need to spend time capturing food and water to fulfill the Being's needs during that three days.

This means *The World of Two* is pretty much always fair when it comes to specialization, because being able to specialize at all *requires* an agreement or mutual understanding *in advance*. Without this "contract", the Being would push itself deep into hunger, thirst, and lack of shelter to perform this "special" task. Without this "contract," the task takes significantly longer to finish.

In *The World of Two*, specializing is simply a re-arrangement of "free time". Free time, which is another way of saying time *not* spent finding and capturing food, water, and shelter, had to exist first for specialty to be possible at all. If you both spent 18 hours on food, water, and shelter each day, and slept the other 6, free time doesn't even exist, so specialization doesn't exist.

After free time exists for each of you, only then can you *voluntarily* give up some or all of your free time to the Being across those three days, essential doubling (or at least greatly increasing) its free time. This balanced-imbalance is made possible through language and trust, but also an agreed-upon recognition of the *value* of the particular specialized movement in time.

There are limits where this agreement falls apart.

If you and the Being agree to this balanced-imbalance for three days so a shelter can be built, and then the Being did *nothing* but nap by the creek for two days – fairness and trust collapse and the *Dark Forest Moment* re-appears, along with the possibility of violence. On the flipside, if you keep "forgetting" to bring the Being food and water while it is working for two days, fairness and trust also collapse, and the *Dark Forest Moment* re-appears along with the possibility of violence.

Now jump through time to *The World of Three* – when the Trespasser resists doing the tasks you or the Being ask it to do, and might even intentionally avoid completing them. How does this compare to specialization in *The World of Two*?

Consider your conclusions carefully – the Trespasser is actually doing something in the *World of Three* that wasn't happening in the lazy-specialists *World of Two*: The Trespasser *was* capturing its own food and water. It was choosing not to use your shelter, and only did so in extreme weather. This means there is not *actually* an imbalance of time related to the required minimum movements of capturing food, water, and shelter. The Trespasser is spending the exact same amount of time doing these things as you and the Being. You each walk to the garden, or creek, or traps, and consume. The Trespasser is not "specializing" in being lazy. The "problem," as I see it, is something else. The imbalance is in a completely different and hard-to-see place. It is actually the same imbalance that occurred when the camp flooded with more than one hundred new trespassers. Those that arrived were not "specializing" in chaos or collapse or laziness. They were not being unfair to you or the Being. They, too, were still performing the minimum movements required to capture their needs completely on their own.

They were not lining up and demanding that you and the Being feed them and shelter them. They were not demanding special treatment, or that they rest while you, the Being, and the Trespasser do all the work. You can relate to every single individual. They are each duplicates of yourself after beginning *The World of You* for the first time. They each experienced the same fears and anxiety you did, but quickly found themselves stumbling right into the miracle of a fully realized garden – just like you did eight chapters ago before the two Beings found you.

After that moment, each of the 100 individuals were trying to manage the *Dark Forest Moments* between 99 other individuals, and at a very fast pace. There were far too many others to enforce equality or create fairness. Those things take time and language to stabilize – neither of which was available.

They each knew their next meal was coming from this area, but were very uncertain if they could get it without being hurt or having to hurt someone else. They knew there was water nearby, but were uncertain if it was clean and safe to drink, or if they would be chased away from the water with violence. They knew there was shelter, but were unsure if they could use it, and even if they could, were unsure of how long they could.

If you were the 100th person to discover the garden, but it was filled with dozens of other individuals fighting and devouring the food – would you still decide to cast yourself into this void of uncertainty and violence?

This mass of "trespassers" is nothing more than 100 food-and-water story arcs going up and down at different frequencies, while wanting shelter from all the other trespassers which threaten their story arcs. They each only want to capture their particular ends, but there simply aren't enough means to capture all the ends. Balance is not possible in this situation without death or separation.

Yet, as partner to the Being, you probably felt anger, frustration, sadness, and even despair as you watched all this unfold in the homestead you had worked together with the Being for years to build. I suspect it felt deeply *unfair* that these newcomers should get such quick access to their survival needs with barely any effort, while you and the Being broke your backs to reshape the world in this energy-efficient way. It's possible the Trespasser even feels this way about all the other newcomers. And the second trespasser might feel the same about the 90th. And the 20-30th about the 90-100th.

This "hierarchy" of perceived unfairness *through time* is the structure of what I believe to be perhaps the most significant human imbalance in existence. At the time I wrote this, this imbalance is still fully present – and I'd be willing to bet it still exist in society at whatever present moment you are reading this.

For two years, I studied and uncovered much about it while trying to rise out of the darkness of this rabbit hole. It involves the looking glass, authority, and every other layer we've fallen through. I've come to call this mess in movement and time The Floating Anchor Problem.

PART FIVE - THE FLOATING ANCHOR PROBLEM

Stories are a lot like donuts.

Ghosts are fully capable of destroying the world.

120

Single File World

Let's rewind time back to the *The World of You* and change the ending. In this version, you never end up adventuring out into the void beyond your territory and encounter the Being. You never meet the Trespasser either, and the flood of strangers never occurs. Instead, in this version, you continue tending to your personal garden. You hunt. You fish. You maintain balance with your environment in an effort to keep your certainty of survival stabilized.

By constantly improving your ability to do these things, you create more and more "free time" for yourself. Your expectations to capture food when hungry, water when thirsty, and shelter when fearful are highly validated through countless successful repetition, and no longer consumes your story system. More and more of your time can be spent thinking about and doing whatever you want.

Perhaps you tinker, or build, or paint with mud. Perhaps you watch the sunsets and sunrises. I cannot know what actions you justify to be right for you in your free time all alone in the world, but I am going to eliminate one option: You do not develop written language of any kind.

You live for many years in harmony with nature, and finally die peacefully while sleeping on a particularly clear night under the billions of stars. Your body then vanishes into thin air (this is your imagination, after all).

Soon after your body vanishes, the Being appears in this world. It magically pops into your homestead in much the same way you first popped into this thought experiment - by reading or listening to this book. It was reading "The World of You" chapter and imagined it to be the only one of its kind to exist on a planet reset to an all-wild and time-paused condition.

However, something happened when the Being hit *play* on its paused world. Its surroundings suddenly changed, and in the blink of an eye it found itself unexpectedly thrust into the reshaped piece of nature you once called "home".

How long would it take for the Being to figure out the systems you developed? To figure out there is a garden and what plants are part of that garden and which are not? How long would it take to figure out how to catch fish? To figure out fishing poles and traps and snares? To determine what animals exist nearby, how they move, and then how to catch them using your weapons or tools of choice?

Do you know what a basket-style fish trap looks like should you randomly stumble into one in the forest today? Would you recognize an unset snare dangling in a tree? Do you know how to hunt? How to kill without causing severe suffering for the animal? How to safely prepare plants, fish, and other animals for eating? How to use animal skin for clothing or coverings? How to weave? How to graft?

It's hard to say just how long it would take the Being to figure out its own system of food, water, and shelter, and also figure out these "things" you left behind related to them. Do archaeologists always know what some ancient discovered thing was used for? Without a complete context, any discovered object can easily be misunderstood, and there is no way to go back in time and determine if that "understanding" is the "right" truth-story of that object.

The Being is in somewhat the same situation. Only if it captures its needs using the same methods you did will your "things" left behind immediately make sense. There is no way for the Being to ever be certain what the baskets by the creek were used for, unless it somehow realizes that purpose through their design. This would require the baskets to *tell a story* to the Being with their shapes and locations alone. The Being has to look at them and build the story they are meant to be submerged in flowing water, and fish swim in an opening, but cannot easily swim out.

How long will this take to do, you think? I do not know. It could happen quickly, or it might not happen at all.

For the sake of this story, assume the Being does figure out *all* your systems related to food, water, and shelter. It figures out what you created everything to accomplish, and resumes that use. If we assume the Being's need story arcs are roughly the same wavelength as yours, then the moment it figures out all these stories, its story system arrives at the same level of speed and certainty you had between hunger and capturing food, thirst and capturing water, and fear and capturing shelter.

If you built these improved systems through trial-and-error over *five* decades, let's assume the Being comes to understand these pre-built systems completely within *one* decade.

The Being has gained that difference - forty years - of "free time" that you did not get in your lifetime, thanks to your systems and inventions (of course, the Being does not know any of this).

So the Being spends this free time progressing certainties of food, water, and shelter capture even further. It constructs an automatic attention-getting "alarm" that pops a fish trap up and out of the water as soon as a fish enters. To increase the size of its "sheltered space," the Being constructs a new shelter from logs, bundled twigs and branches four times the size of the cave. It even develops an irrigation system to divert creek water into the garden, so the plants don't have to be watered by hand when a few days pass without rain.

The Being dies of old age after making five decades of improvements, and vanishes. Like you, it did not develop any written language. It had no need to.

The Trespasser now pops into *Single File World*. It, too, unexpectedly found itself thrust into the homestead you built and the Being improved. Now though, The Trespasser has many *more* stories to figure out than the Being had to figure out.

Let's assume *all* the creations and "alterations" to nature by you and the Being will eventually make sense to the Trespasser, who will use them in identical ways. Knowing this will eventually happen, do you think this "process to understand" will take a longer amount of time for the Trespasser than it took for the Being? Or do you think it will take a shorter amount of time?

I think it will almost certainly take longer. Not only must the Trespasser figure out how to crudely catch fish, and then figure out the basket is a fish trap that makes catching them easier, but it also has to figure out the story that the odd contraption nearby is meant to interact with the fish trap, and is designed to operate when something enters the basket, and this operation is simply to hoist the basket out of the water, and the only purpose for all of this is to get the attention of the individual that will empty the trap.

Outside of these invented systems, the speed of nature remains the same: The garden plants yield food for the Trespasser the same as it did for you and the Being, the creek yields water the same, and the shelter yields safety the same.

The Trespasser should, in theory, reach the same level of certainty you experienced near the end of your life at about the same speed as the Being – let's say one decade. But, the Trespasser needs *more* time to build the stories that merge the Being's improvements to your improvements. All the systems left behind are more complicated than they were for Being after you died.

The Trespasser cannot read your mind or the Being's mind. It doesn't know anything about you or the Being – only that something or someone must have existed before it arrived. It can only study the shapes and placement of the things left behind to try to connect their movements to the "right" stories. The chances of the Trespasser putting these things together into exactly the same stories is going to be significantly lower.

Let's assume it takes the Trespasser another decade to reach the same certainty the Being experienced before it died 40 years after reaching your level of certainties. The chances these certainties are reached using *exactly* the same stories as the Being goes *down*. Whether these Trespasser-created stories are more efficient than the original-and-lost story is yet to be determined.

To express this timeline a little more clearly: It took you 50 years to reach your level of certainty. It then took the Being 10 years to reach the level of certainty that took you 50, thanks to the improvements you left behind, like the garden. The Being then improved these systems for another 40 years. The Trespasser then took 10 years to reach your level of certainty, and another 10 years to reach the Being's level of certainty (which took the Being 40 years of free time to reach).

This means the Trespasser progressed through 90 years of survival system improvements in only 20 years (which is how long it took to figure out all the stories behind those 90 years of system improvements).

The Trespasser has *no idea* of any of this of course. It has no idea the homestead and items took 90 years to develop when it found itself in *Single File World* instead of *The World of You*. Regardless, it quickly catches up, and then spends the remaining 30 years of its life improving systems even further, before it too dies and vanishes.

Bob pops in next. It takes him 10 years to reach your level of certainty, 10 more years to reach the Being's level, and 10 more years to reach the Trespasser's level. He just progressed 120 years in only 30 years.

Without repeating this story over and over, I think you can see the pattern here and maybe even predict when it hits a limit: Eventually it will take an individual the *entire* rest of his or her life just to figure out every story left laying around the homestead. Because of this, no *additional* progress can be made in that span of one lifetime.

In theory, if we keep our assumptions the same on lifespan, abilities, and absence of language in *Single File World*, this is where growth and progress stop. Every new person will be just a repeat of the last, with the only exception being the possible rearrangement of already-existing stories. Every individual would experience an entire lifetime of growth, but in the big picture, the certainty and speed of capturing ends is not changing. Certainty of survival is now acting like a broken wave that rises in mental freedom from entry until death, when it instantly collapses back to no freedom when each new person begins.

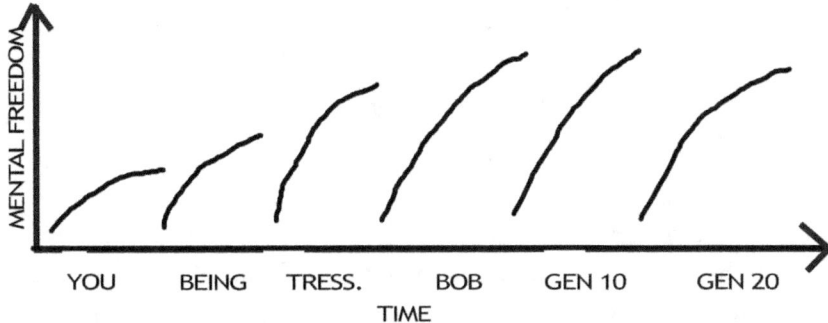

I'm now going to convert these half-waves into a different type of diagram. In the case you are listening to this book, I will again do my best to explain the transformation using words.

Starting with you in the *Single File World*, assume that every single movement you ever justify through your lifetime was related to capturing food, water, and shelter, or to improve the time they take to capture (increasing mental freedom). Earlier in this book, this resulted in "need story arcs" that became waves traveling from left to right. I'm going to reshape that diagram into a circle.

The center of this circle represents no movement at all - the end of your time "line" - your death. Since you only died in *Single File World* once, the single line representing your mental freedom relative to your actions ends in the center.

From the moment you entered *Single File World* until that dead-center moment, your action-freedom waves wiggle near that center without ever touching it.

To better understand this body/mind "probability cloud," I'll show a set of only three need story arc wavelengths ringing around the center. When time was left-to-right earlier in this book, these need story arcs moved up and down relative to a straight timeline. On this *death-centric* graph, the time*line* gets replaced by a single point, and your waves move up and down toward or away from that point in any direction. If they ever touch the point, it is the same as your story arcs touching your timeline from before – you have died.

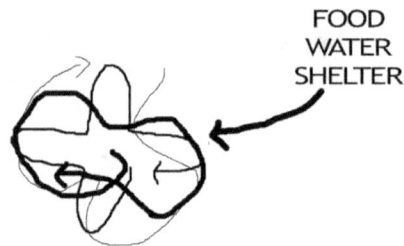

When you consider we need to show your *entire lifetime* of need story arcs, you can perhaps agree it makes more sense to just fill in this area near death as solid – there are just too many waves to get anything meaningful from the mess these individual lines would make. After all, how many meals do you eat in a lifetime? Drinks do you take? Shelter from *something*? There would be hundreds of thousands of waves to show. It just makes more sense to represent all of them as a shaded area, which I'll label as FWS (food, water, shelter story arcs).

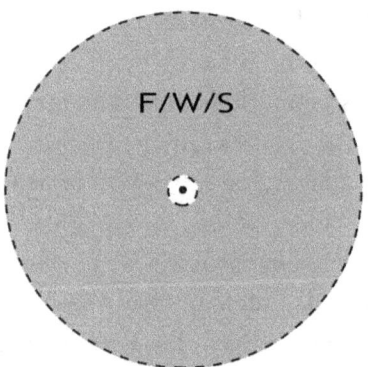

Since the exact center point inside this ring represents your timeline's endpoint, then the distance from the center to the outer edge of the entire shape represents your perceived distance from death. This means being close to the center is like the end of the *Gravity* chapter, and moving toward the outer edge is to return to the beginning of that chapter. You are always being pulled toward the center by your bodily processes (decay), but then you ideally "capture the end" to avoid death and continue wiggling on with our existence.

This shape has another layer we need to show – free time. Free time involves lots of "distance" from feeling like you are going to die or that you need to eat, drink, or take shelter. As a result, it creates an outer layer of possibilities.

The layer closest to the center is the absolute minimum required time and movements in your life that *must* be performed to stay alive. Everything outside of this inner layer is free time.

This means when you first entered *The World of You*, the entire shape was just a big dark ring. It was all the inner food, water, and shelter layer, because every movement in your life was justified toward finding and capturing food, water, and shelter.

As you "captured" each survival need consistently over and over again with increasing ease, the inner ring began to shrink in size because less and less of your story arcs and actions were used to do these things, which means more and more of your story arc and actions were left to pursue anything else – you were *free*.

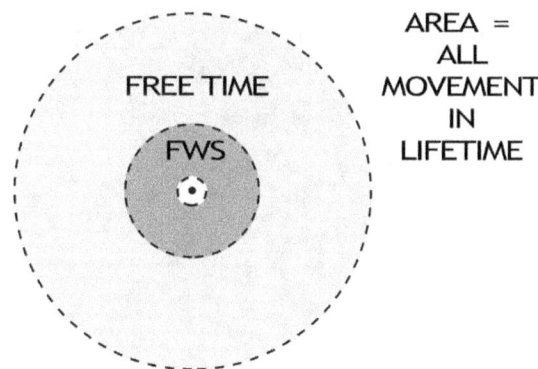

When these two layers are added together, they represent the total amount of experience you will have throughout your entire life in *Single File World*. Since you chose to spend your free time working to improve the minimum required movements related to capturing food, water, and shelter to avoid thirst, hunger, and fear, the overall shape is a circle, which remains centered around the end result of not capturing these things - death.

By improving your capture of resources, the dark inner area representing the minimum amount of movements to survive kept shrinking. You kept getting the same amount of food, water, and shelter, but spent less and less time capturing them. The rings in the above graph represent the average amount you spent in each ring over your whole life.

Now, because we are assuming every individual will live the exact same amount of time as the previous in this single-file thought experiment, the total area of this shape never changes. It represents exactly 50 years of time and movement.

This also makes the food, water, and shelter inner area's size very important.

The second individual to enter the Single File World, the Being, begins somewhere inside this food/water/shelter inner area. After all, just like you, its first priorities are food, water, and shelter, and it spends 100% of its time figuring these things out.

However, thanks to your efforts, the Being quickly enters free time, and now only has to spend *half* the time you did justifying the minimum required movements of eating food, drinking water, and sheltering. This means the Being had even more free time area and an inner area that ends up being half the size of yours.

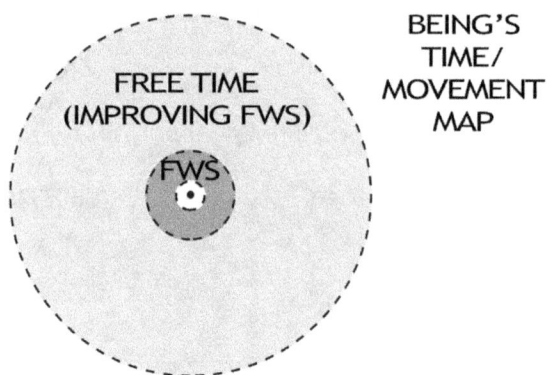

If we fast-forward past Bob and 30 more individuals in *Single File World*, then each life results in the same exact overall shape, with the inner ring being about the same exact size. The only thing that will change its size is the frequency that each individual needs to drink or eat or shelter, which determines how much overall time of the 50 years of total life is spent in that inner ring.

Everyone so far (you and Being) has only ever cared about these pre-existing food, water, and shelter stories, and worked their entire lives to make the inner ring as small as possible. Now the chopped-wave diagram from earlier is transformed.

If we were to put this ring into the context of *The Ark*, then what is happening is each person is beginning in the billion-door black space with a limited amount of time (Gravity), and choosing to open doors and not enter any of them. After enough people do this, all the doors will be open, and the next individual to enter *The Ark* will only have to spend time peering into each already-open door.

Eventually, someone will choose to go through an open door, and it will change everything.

121

One Story Too Many

Imagine you begin *Single File World* again, but without any memory of the events in the previous chapter. You imagine the world all-wild once more, remove all human-made things and humans, and then look around while everything is still paused.

When you un-pause the world, everything around you suddenly changes and you find yourself in some sort of primitive homestead. There are baskets, fruiting trees, and a hut-style house made of logs, leaves, sticks and stones.

You do not know this yet, but you are the 35th person to enter, and there is so much development and progress in front of you (1,700 years worth), it will take you *exactly* the rest of your life (50 years) to figure out what everything is designed to do. Or at least what you *think* it was designed to do. But let's give you a different path to take this time.

Instead of assuming you will robotically follow the pattern of figuring out *everything* before you start changing *anything*, let's assume you only spend 10 years of free time figuring out these existing systems. After that, you decide you want to make *a parachute* with your free time.

Exactly *why* you want to make a parachute, I have no idea. Perhaps you have a strong urge to fly, and so you want to be able to jump off the nearest cliff and softly float down. Parachuting and jumping off of stuff is not really my thing, but maybe it's yours.

You stop trying to figure out the bizarre rope contraptions near the creek, and the many other weird looking items lying around everywhere. You don't even attempt to figure out the stories of all the trinkets and contraptions, and instead begin working to create a parachute.

For the sake of this story, let's assume all the objects and systems you are choosing to ignore do not rot or decay in any way, and stay in perfect condition through time unless purposefully or accidentally broken.

You spend five years working on this project in your free time.

Throughout these five years, you create different parachute designs. You climb up the hill, walk up to the edge of the cliff, tie a log to the parachute, and then throw it off. Your designs fail over and over again, with the log crashing to the ground in an enormous thud or splintering into pieces. Your story system floats its anchor to truth, tries to determine what went wrong with the prediction and expectation of it floating down slowly, and re-anchors to a new truth-story (and with it right and wrong). If the re-anchored truth-story makes enough sense, you will justify the actions needed to bring this new expected truth-story into reality. You are *trying again*.

At the end of the fifth year of this try-fail-adjust-try-again truth-story cycle, you find success. The log drifts down slowly and gently makes contact with the ground. Excited, you climb down, retrieve and attach the parachute to yourself, climb up again, and jump.....

As soon as you take this leap, you realize that if anything goes wrong you are about to die – and experience a brief panic before experiencing everything you had hoped: Drifting slowly downward in a way that almost feels like flying. As you float down slowly, birds flutter out from the cliff and pass beneath your feet. You land a bit harder than the log did, but are unhurt. *Success!*

You spend the rest of your life tinkering with the parachute's design. You want to fall slower and have more control of your direction and speed. You make a lot of progress toward this before your own end arrives one fateful day.

Your body vanishes.

The Being pops into the homestead as the 36th in line.

The Being's first few days will probably look a lot like the last chapter, but now there is also the possibility of something radically different happening.

Instead of a balance between the amount of advancements in front of it and the time it has left in life to figure them out before dying, there is now an *imbalance*, and the Being has no idea this imbalance exist.

The Being does not have enough time left in its remaining 50 years to figure out every already-existing story. It can't know this because it doesn't know it only has 50 years left, or that it will take 50 years to figure out every story in this homestead that has been left behind.

If we stay with the assumption all lifespans are exactly the same length, and all story-building abilities are exactly the same, then your parachute has added one-too-many stories for the Being to *ever* reach 100% understanding of every existing story in this homestead before it dies.

This happened because you made the choice to spend your free time *not* improving systems centered around food, water, and shelter, but instead "specializing" in something new or interesting to you personally, and not following the pattern of figuring out everything around you *before* doing something new.

Because decay and rot are removed from this thought experiment, all of these things you never figured out (because you chose to focus on the parachute) are still sitting there in the exact same condition when the Being arrived.

If this place were *The Ark*, then there are now more open doors in the billion-door void than can be explored in a lifetime, because you added a *new* door labeled "Parachute."

The Being might never be able to figure out the story of your parachute. It begins *Single File World* like you did: Only caring about food, water, and shelter. Your parachute might be the first created "thing" that catches its attention, but without knowing what it was designed to do, it may decide to use it as, say, a *fishing net* long before it could ever imagine the story *I will have a lot of free time in twenty years, I bet this thing can float me down from that cliff up there for fun on those boring days.*

The Being doesn't know this, but the "donut shape" of *existing* stories contained in all the "stuff" around it now has a "spike" shooting out of it. If it chooses to explore that spike – knowingly or unknowingly - it will create a "dent" somewhere else in the shape. This dent will be whatever existing story will never be experienced due to limited time.

It's important to note that this spike exist strictly because of one detail of this thought experiment: Every individual in line to enter will live exactly 50 years. This life story-shape and total area is a "probability cloud" of a 50-year lifetime of story arcs and movements.

The total area of the dent, and the total area of the spike match *exactly*, because the total area always adds up to 50 years of movement.

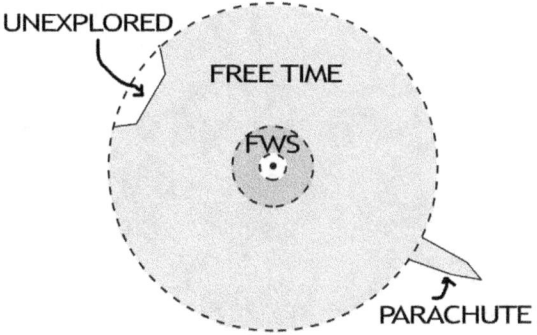

Now…how far can this spike stick out?

Well, if we assume that an individual has all needs pre-captured by those that came before, and each arrives to a paradise of a perfect garden, shelter, and clear water, then this spike can grow to include *almost all* the "free time" available over 50 years.

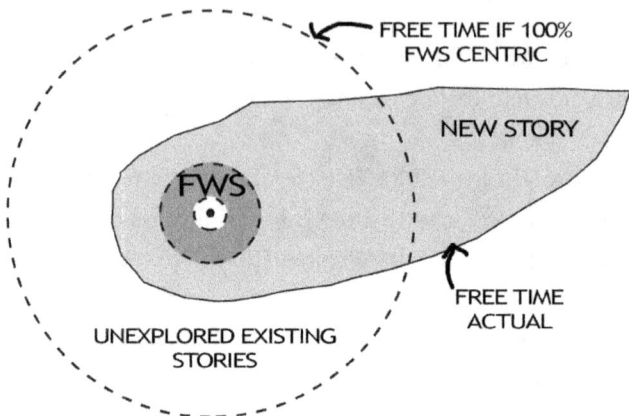

Why can't the spike include all of it? Can someone not dedicate his or her entire life to a story of specialty like parachuting?

In the world without decay, you can.

But we don't live in that world. This is not *The Ark*; this is *Gravity*. Everything is *moving*.

122

The Decaying Spiky Donut

I created *Single File World* to hopefully make it easier to understand a much larger picture of the world around you while reading this.

Let's bring back the shape of a lifetime of stories – the two-ring donut. As it turns out, the inner area of minimum movements required to continue surviving matters quite a lot, because its size ultimately determines the size of "free time movements" available in your lifetime.

This donut-of-free-time area can be reshaped any way you like. This might include creating a spike of specialized movements in time, such as making a parachute and using it.

A single spike has a limit though – it can only stick out as far as your available free time in life can allow. If you choose to "specialize" in multiple things, then none of the spikes will be as long as if you stuck with only one specialty in particular. The total area of the donut is limited to how long you might live. If you have 30 years of total free time available in your life, then you might have a spike near 30 years long, or maybe have 10 spikes, each 3 years long. But you can never exceed 30.

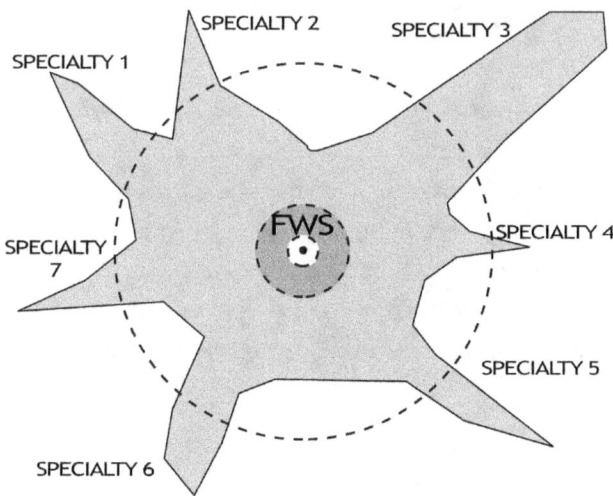

In *Single File World*, each spike of specialty you might pursue comes with a consequence for those that come after you. The further you go into your own specialized story unrelated to food, water, and shelter, the less and less likely it is the person after you will be able to build the same story of that spike successfully.

For example, let's re-label the spike called "Specialty 3" in the 7-spike diagram to be "parachuting". In this spike, you built the parachute, then built a ringed target to try to land on as you float down with the parachute, then built an elaborate scoring machine to track how many points you get in a day playing this target game, then built a parachute spreading device (a large dome) to speed up checking the parachute for tears and damage, and even created a device that makes it easier for you to consistently fold and store the parachute. You've created one heck of a *rabbit hole* of stories here, and they all anchor to the story of a parachute.

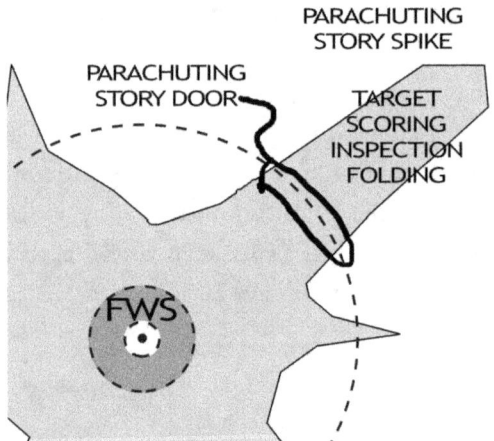

The Being who comes behind you in *Single File World* has no idea what the parachute *even is*, and ends up using it as a fishing net. If the story of the parachute is never figured out, then for the Being, this spike breaks off because an *anchor story* was needed to figure out all the other stories related to it in the spike, and it did not "discover" the anchor story.

This leaves all the objects lying around that *were* anchored to the parachute story into an island *floating* in the middle of nowhere – now attached to nothing.

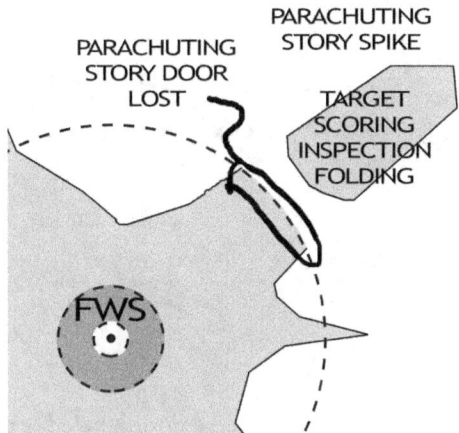

Without the anchor story, the Being is left scratching its head about these other objects. It is likely to instead repurpose these things to suit its own priorities and interests – assume it does exactly this.

It uses your scoring machine as a day counter, the dome as a roof for its new shelter, and the folding device to make leaf pouches.

When it comes to the shape of the Being's spiky-donut story in time, these re-purposings merge most of the islanded specialty spike back into the food, water, shelter centric-circle, with the scoring machine becoming its own short specialty spike of "day counter".

Is the Being *wrong* for doing this?

If you say yes, I'd be very curious how the Being can possibly know it has done anything wrong, and who can possibly communicate that wrongness.

The Being has built the stories it can with what it has - which is whatever memories and contexts it can manage to associate to all the strange things laying around in this world it just entered.

No matter how it chooses to use these things, *it is not wrong*.

For example, it imagined a few stories to explain the rings (that were your targets when parachuting), and believes they most likely were some kind of measuring system, or perhaps a game where logs (which were laying all around the rings) were rolled from a distance toward the center – which is how it uses them now.

This all happened because of the re-purposing of an anchor story (parachute) into a fishing net, which resulted in the long and complex story of things tied to the parachute spike becoming nearly impossible for the Being to understand in the way you did.

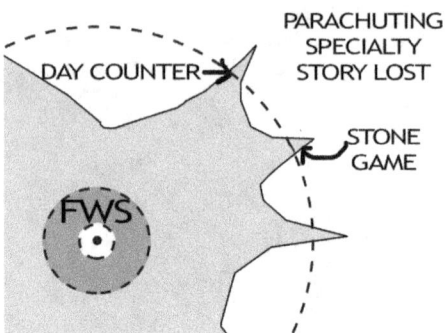

If this were *The Ark*, then you created a door labeled "Parachute" and went through it. Beyond that door you created a room filled with more doors labeled "Inspection Dome", "Parachute Folder", and "Target Practice". Through the "Target Practice" door is another door labeled "Scoring Device".

The Being, who is unable to read these labels, rips off the "Parachute" label and changes it to "fishing net" in its own language. This destroys your rooms beyond the parachute door, and moves them to become more unlabeled doors in the *unspecialized* billion-door space of *The Ark*. "Fishing Net" becomes food centric, and what used to be the "Target Practice" door, is relabeled "Stone Game". It's also possible the Being will not use the other objects involved with parachuting at all, leaving their doors open in *The Ark* with no labels.

This reshaping process can and will repeat itself over and over again as individuals come and go single file. Dramatic spikes may poke out of the donut into extreme specialization where many more specialized stories are created in one particular lifetime, which it becomes nearly impossible for future individuals to follow without the context for that spike.

Take a second and look at all the "things" around you right now while you read this book. Imagine you are a primitive human that can only grunt and has never experienced your current reality before.

As a primitive human, you are only vaguely familiar with the idea of a garden, and have only ever lived in a cave next to a creek. Do you think you can figure out the story for everything around you right now before dying of old age?

I know I couldn't. I am sitting in a climate-controlled house at a desk with an adjustable-and-wheeled chair. There is carpet under my feet and electric lighting over my head. I am typing this book for you on a laptop computer, while wearing clothes made of "synthetic" fibers and a watch that tells me with a liquid-crystal screen what 1/24th piece of the day it is and what 1/60th section of that 24th piece I am in. It is doing this by connecting and synchronizing to an invisible-to-me network of towers that are ultimately connected and synchronized to invisible-to-me satellites orbiting the entire planet.

This same watch also tells me my current heart rate and the number of steps I've taken since midnight. It also sends me text messages and emails. I have a kitchen that can provide clean water to me if I expend energy to move for about 10 seconds to get it, and I have food I can eat with only one minute of movement. This food can be stored for decades without going bad. This ignores the enormous amount of stories needed to understand the materials of my shelter like glass, plastic, drywall, paint, wood flooring, and doors. Windows in general would likely completely blow my mind as a primitive human.

Then there are the stories around me of machines that clean things like vacuums and dishwashers, and soap and skin care products. This is still ignoring the things in my kids' rooms like markers, paints, models, aquariums, and more.

As a primitive human, I *might* be able to figure out stories for many of these things, but I am also very likely to build the "wrong" story for each of them.

The electronics (should I manage to turn them on) would quite literally be magic to me, and the nearly instant water and food (if I could figure out that food is inside of those very hard shiny cylinders, and the story of how to open them) would be a literal miracle.

There have been millions of "specialists" that came before me to create the things and systems that I experience firsthand every day. I hope the same is true for you as well, and that your surroundings related to capturing food, water, and shelter are not even close to being as difficult as they were in the beginning of *The World of You.*

The shape of all of the specialized stories around me has made the inner area of my life's donut (capturing food, water and shelter) extremely small. I spend not more than a few seconds to capture a drink each day, maybe an hour or two of movement a day toward eating, and have shelter at nearly all times that my other two needs stored and available to consume while staying *inside of it.*

If you expand the spiky donut of time and movement to include the life-times of every single human that has ever lived, then there are tremendously long spikes extending out in every direction. And I do mean *every* direction in terms of the donut and in terms of movement in reality. We now specialize in not only movements side-to-side and forward-and-backwards, but also movements up (into the sky and space) and down (into the earth and sea).

In my opinion, there are now so many "things" and systems humans have created by reshaping nature over and over again, I think it would take hundreds of lifetimes for one person to imagine a story for only *half* of the things he or she didn't already know, and it's very likely most of those stories will not be anything like the stories that created those "things" in the first place.

I would have no idea if an unfamiliar thing in front of me is an extremely specialized tool to be used at the very tip of an extremely deep rabbit hole of specialization, or if it is for something basic and simple like harvesting a very specific food.

Does this thing I found called a "laser" belong to a story about presenting to other people, or a story about melting things? (Perhaps the original story of its creation was for hair removal). If I instead use it for some other purpose – am I wrong? Is there really such a thing as "misuse" of anything as long as it doesn't result in harm to the environment or myself and brings the story I imagine into reality?

In my opinion, it's okay to let stories created in the past break and reshape themselves into something new most of the time. Story growth and story decay go together and balance with each other, and letting the spiky donut move and change shape is how progress works and everything stays stabilized and balanced.

If we try to hold too many things rigid, and try to prevent story decay and reshaping, things quickly get horrifically out of balance, and it can result in us tumbling back into the abyss.

To keep climbing, we have to keep our stories moving. This is much harder than you may imagine it to be.

123

Story Overvoltage

Imagine *Single File World* starting all over again from the beginning. You can keep all your memories built so far in this book. There will be only one major difference from your previous *Single File World* experiences: You will get *one week* with the next individual. At the end of that week, you will die.

This incoming individual will have *no idea* it will end up with you in your homestead, or that it only has a week with you before it is alone. It will not share your language, and it will begin with about the same number of stories you have at the beginning of this experiment, but with no shared cultural context embedded in those stories.

Knowing all of this is going to happen, what would you change relative to your previous experiences in *Single File World*? Anything? Do you start thinking of what you will want from this individual right away? Can you teach another person your language in one week? If so, to what end? What is it you want to teach them? How do you approach them when they appear, and do it in a way that prevents and controls the *Dark Forest Moment* path of violence against you, or running away from you?

Are there experiences you would want this individual to have before you go? Experiences you would want to guide them through? Which ones and in which order? Which ones are most important?

My guess would be teaching them the food, water, and shelter systems first, because they bring stability and free time. Also perhaps emphasizing the importance of kindness and understanding in the face of the fear in another, while at the same time preparing for the possibility of violence by that individual to try to eliminate fear (but would be unsuccessful).

I have no doubt this one-week overlap would have a significant overall effect after a few generations. I think it would accelerate the journey to a stable level of certainty related to food, water, shelter, and all the improvements needed to speed up their capture. I also think this would dramatically extend the length of specialty spikes, allowing them to reach much further than single file world with no overlap.

For example, even without a shared language, the specialty story of the parachute could be easily told through demonstration (assuming the *Dark Forest Moment* can be held at bay long enough to do so). The parachute, target, scoring system, inspection dome, and parachute folder can all be well understood just by watching someone else perform the one larger story that includes all of them.

What would have previously taken *years* to figure out in the non-overlapping *Single File World* and would more than likely result in the collapse of the parachute spike altogether and the reshaping of the objects involved into different specialties, can now be used as designed and intended simply through observation (assuming the sense of sight is present).

The full story of this parachute spike is now being connected into one bigger specialized story for the next individual by the *actual anchor story* that was missing in *Single File World – you.*

It took you *years* to imagine and build all these things related to the parachute and get them to work. It will take the Being, who is next in line and will appear in the last week of your life, assumedly less than *four hours* to fully understand the story of each specialized thing you created. In just four hours, the Being is now in a position to lengthen the specialization spike of the parachute even further.

I suspect that the Being could understand a large chunk of *everything* there is to understand in your homestead just by watching you perform the actions first, which it then attempts to copy until the stories are validated through successful reenactment.

But one week is not enough time to convey your 49-years-and-51-weeks of experiences. It is only enough time for a crash course in the critical and important stories. Whether or not the Being can remember all the stories it is shown is not for you to decide or control. Whether or not it understands what was happening in these stories like you do is not for you to decide. This won't stop you from trying though, I'm sure.

In an effort to make yourself more certain the Being understands you correctly before you are dead – I believe a new boundary will be born. This boundary is a layer of very high resistance created by your actions in an effort to guide and control the Being's actions. This boundary allows you to be more certain the Being truly understands the stories the same as you do *as quickly as possible.* This boundary is born from a need for speed.

This boundary is essentially the formation of a narrow invisible "straw-path" between two imbalanced story systems. One story system, yours, already knows *tons* of stories about this place. You are the bigger balloon. The other, Being's, knows almost no stories about this place. It is the smaller balloon.

Before I go any further, I think it is incredibly important to recognize that both story systems are *fully capable* of creating stories and are doing exactly this at all times. This was the whole point of the non-overlapping *Single File World*.

Even in a one-week overlap world, each story system is still isolated from the other, though. They are each sitting in prison cells, letting in reality through wacky windows. The moment before this one-week overlap situation begins, this system is a lot like having the balloons inflated and the clips in place. It's the beginning of one instance of *The Process*. There is a huge *story voltage* present.

The larger collection of stories anchored to this place, *you*, desperately wants to share these stories with the smaller collection, *Being*. This means you are ready to remove your "clip" right away. The smaller collection of stories, Being, wants to learn the new stories of this place to survive, but it does not realize it will take it *a lifetime* to grow its system to the same size as yours without you, and it also doesn't now know anything about you. This means time must be spent getting the Being to trust you before it can even consider removing its clip.

Let's assume the *Dark Forest Moment* is successfully expanded and the Being actually *wants* these stories. By knowing these stories already exist (by watching you act them fast and comfortably) it's possible the Being will feel insecure and unknowledgeable relative to reality – which will create the feeling that it needs *shelter* from its own ignorance.

This shelter need prompts the smaller system to remove its clip connecting to the bigger system (you) once it is believed you do not mean any harm. It is willing to open up quickly because time is passing and speeding up right now for the Being: Its story-arc frequencies have not changed. It needs food, water, shelter, and fairness as quickly as possible - perhaps within minutes, hours, or a day.

You, already familiar with the stories to capture all these things quickly, are practically a magician to the Being. You have very obviously survived here for a long time. Trusting you is the key to survival.

So, the Being "takes the clip off", and this wide-open acceptance of new stories, *your* stories, begin to rush in. With only a week to share them though, the current of these stories is tremendous.

There is an extreme story overvoltage between you and the Being. This means there are more stories in your head than are sharable in a week.

If you try to start "pushing" the stories into the Being faster than the straw path of communication and observation will allow - the result is something horrific.

124

Explosive Collapse

In the world of electrical design, an "arc flash" is one of the most dangerous and horrific kinds of events.

To understand an arc flash, just imagine two huge wires – each as thick as your arm or leg - entering into the same metal box. These wires do not touch each other, but do come close to each other.

Inside the metal box, there are two terminal blocks, which you can think of as two chunks of metal surrounded by plastic, with a hole on each end. A wire is inserted into one end of a block, and a screw is then tightened down on the wire to hold it there, and connects it to the other end.

Imagine these blocks are about a shoulder's width apart from each other in the box. If you want to imagine this realistically, you can think of another pair of huge wires attaching into the other end of each terminal block and exiting out the other side to connect to some electrically powered thing you just imagined into existence.

This design would make this cabinet a type of junction box. Here's a bad drawing of what this would look like.

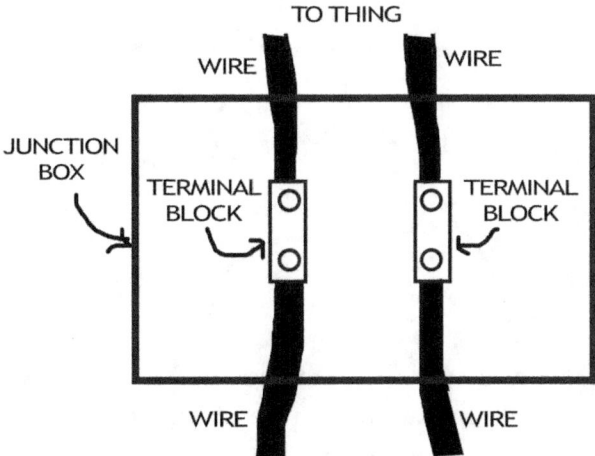

Now imagine the voltage (electrical imbalance) between these two wire straw-paths is *huge* - on the order of thousands of volts. The current flowing through these wires to your imagined electrical thing is also enormous, which is why the wires are so thick in the first place – assume thousands of amps (for reference, a typical United States home is fed by wires sized to carry 100 amps to 200 amps of 240 volts maximum)

If we add balloons to show the imbalances involved, then each "straw" would have a dramatically different sized balloon, and the upper ends would be connected to your massive imagined electrical load – let's call it a "huge electric motor". This would mean the currents in these two straw-paths will always be moving in opposite directions. Here's a bad drawing of this system to hopefully help make it make sense.

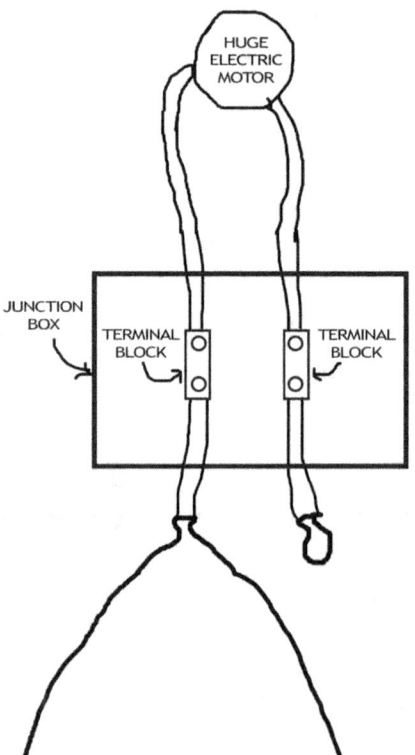

So based on the above picture, when this system is turned on, current will rush out of the huge balloon, pass through the left terminal block, pass through motor (turning it), and then continue on through the right terminal block into the smaller balloon on the other end. Make sense?

Now for the disaster.

Imagine you opened this junction box and accidentally hit your hand on the door in such a way that the metal wrench you are holding flies out of your hand and into the junction box, where it touches the metal screws on both terminal blocks *at the same time*. Again, here's a bad diagram:

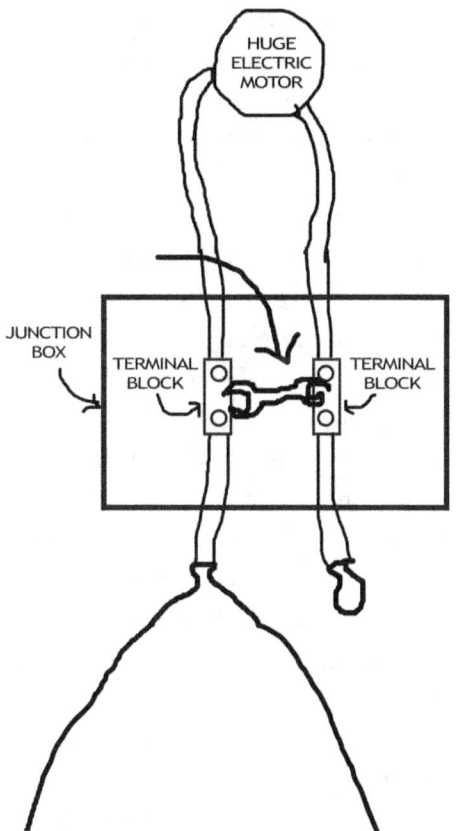

Before this moment, the path of least resistance for current between the two balloons was to take the straw-path-wires passing through this box. Voltage (imbalance) between these balloons is applying electrical pressure everywhere in this system in all directions at all times, but following these properly-sized metal-straw-paths was the easiest and fastest path for these balloons to reach balance with each other. These wires and terminal blocks are the path of least resistance between the two balloons – a human-designed path of least resistance.

All other possible paths resist much more.

At the moment your *metal* wrench makes contact with the *metal* of both terminal block screws, a *new* path of lowest resistance is created between these two balloons. Your wrench just became a new "straw" that makes a new-and-different system shape *almost instantly*. There was the system *before* the wrench, and an entirely different system *after* the wrench.

Even though the overall imbalance hasn't changed, the reason this is a major problem *for you* is due to the wrench itself. Although my drawing is terrible and shows the wrench being as big as the wires, the wrench is actually quite "thin" compared to these forearm-thickness wires. There is *dramatically* more imbalance in this system than the wrench can resist. You've essentially put a *tiny* straw between two *enormous* balloons that want to find balance *instantly*, but are separated in time by resistance – which is the length of the straws and the motor these two points are forced to connect through to collapse toward balance.

The result is a wrench-shaped *shortcut*. You technically just created a wrench light bulb filament that is *immediately* going to start blowing out the side.

Faster than you can ever see with the limited "refresh rate" of your eyes, the wrench gets red hot, then white hot, as the tiny wrench tries to hold back and slow down this enormous voltage from collapsing to balance *right now*. As the wrench reaches temperatures beyond white hot, the arc flash begins.

The reason for the flash might be surprising: It is not because the air around the wrench somehow becomes a path of lower resistance and current is exploding outward in the air toward you. Instead, the wrench heats and expands so much, the boundary between it and the air starts dissolving. It is no longer clear where the wrench ends and the air begins.

The wrench is expanded apart into molecule-sized pieces due to its inability to contain all the heat made by its resistance against the imbalance, so the current is turning it into a balloon. As this happens, the imbalance keeps "feeling" for the newest available path of least resistance toward balance *through* the superheated air and *between* the expanding pieces of vaporizing wrench – it is "feeling" *all* paths with every moment, and "taking" the easiest one. This is exactly how bolts of lightning travel through anything that provides the best connection between the cloud balloon and ground balloon, which might include you, a tree, or something else, like a house next door full of priests.

The straw paths connecting to the motor are barely seeing any current now. It *makes no sense* for current to go this way to find balance between the two balloons. It still always "checks" this way and even uses this way (current is always taking all paths), but there are paths far easier than this one, so the bulk of current takes those ways).

At this point an enormous pressure wave is created from the speed and heat of this wrench vaporization event. From your perspective, you are literally standing two feet from a bolt of lightning between two terminal blocks, which has exploded a wrench into superheated metal bits, which are quickly hurtling toward you as electric current continues to jump through them. This is known as an arc "blast". You won't notice any of this because all of this happens so fast you wouldn't even have time to react. You exist slightly in the past compared to the present, after all.

The arc flash/blast doesn't stop here. This lightning between the terminal blocks will not stop until the two balloons find balance, or something forcibly cuts off the current (puts on a clip), or enough resistance builds up between the two ends of the lightning bolt that it breaks the connection through air and is no longer the path of least resistance.

As the arc of lightning grows, even the air cannot handle the speed these electrical balloons want to balance. The air is the most resistive part of the straw-path now, and so it too begins to heat up, breaking apart the molecules and atoms and creating a new channel of ultra-hot plasma. After this happens, resistance can go no lower. The straw path is as large as it can get, but the two balloons still want to balance *instantly*.

All paths continue being explored with no manmade boundary controlling anything. The speed of collapse to balance is now being determined only by the distance between these two imbalanced points, and what the universe allows to be possible in this location.

With plasma now letting current flow at its maximum rate in this location, and it *still* not being fast enough, the plasma channel itself starts to "blow out the side." A literal arc of lightning that is physically ripping apart atoms grows longer and longer, expanding toward you with an enormous pressure and heat wave.

You are in the path of this growing arc of lightning, and your body might provide enough resistance to finally stop this event, but it is also extremely likely this event will stop *you* in the process, or perhaps even find *you* to be the best path between the balloons.

In the absence of devices capable of *inserting a boundary* between these two balloons, this cosmic-level explosion will keep growing until the voltage and current collapse to balance.

Arc flashes (which can become arc blasts like this story), are horrific, and have killed many people in real life. But not all arc flashes are this size or this horrific. They happen all the time at many different scales.

If you've ever been shocked when you touch a car door or doorknob – that was an arc flash. *You* were the wrench connecting the voltage between the knob or handle "balloon" and the ground-under-your-feet "balloon" (or just-you "balloon"). In these cases, there just isn't enough voltage present to blow you apart, and instead things quickly collapse into balance – way faster than you can react and pull your hand away.

And this exact same process – *The Process* - can and does occur between imbalanced story systems all the time.

Like you and the Being.

125

Someone, or Some Thing

An arc flash can happen between two story systems, or minds, the same as it did between the two terminal blocks. All you need to make it happen is a "wrench" handled a bit carelessly, or an extreme amount of voltage.

I'll now re-label the same junction box system we used in the previous chapter to suit this story related to the one-week overlap *Single File World*:

The balloons represent you and the Being's minds that contain all the stories you each hold in memory of surviving and experiencing this homestead and reality.

The terminal blocks represent you and the Being's bodies.

The junction box is your Step 1 boundaries, or the limits of your view from those wacky windows.

The motor represents acting out the stories of this place. Stories like how to get food, water, and shelter, and how things work.

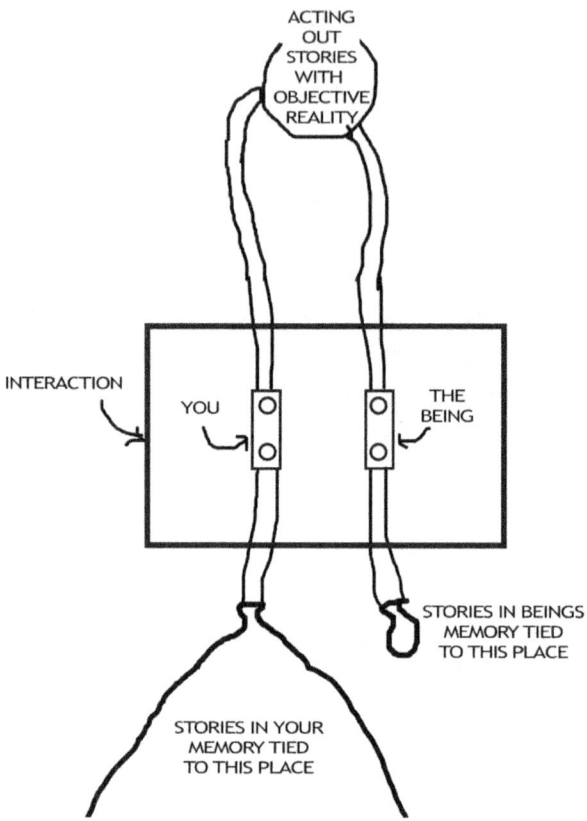

The two balloons are connected to the same flow of time. Even if direct two-way interaction is not possible, the Being can still fill its story balloon much faster than you did simply by watching your actions (the motor).

However, one week is not enough time to act out all the existing stories. The life cycle of plants alone is enough to overflow this process of one-way communication through observation. There are thousands upon thousands of stories in this place, and some of them are limited by the speed of nature (such as those involving plants). As a result, "interaction" (two-way communication) beyond one-way observation is born. This is not required, but will likely be accepted and wanted by you both because of the recognition of self in another. You *want* the Being to know your stories, and it *wants* to know your stories. Your actions needed to survive are approximately the same.

You gesture the story.

You perform the story.

The Being watches the story.

The Being begins performing the story.

You experience unexpected sensory currents and story voltage.

The Being's performance of the story did not match the movements you expected to see. Your story system begins working to figure out why. It dumps chemicals. It adjusts its truth and expectations. Then it tries again.

You gesture the story.

You perform the story.

The Being watches the story.

The Being begins performing the story.

You experience the unexpected sensory current and story voltage again halfway through its performance.

You are actually in a story overvoltage situation, and there is a very good chance you do not realize how extensive this problem is, and neither does the Being. You want this imbalanced system of knowledge and experience to reach balance quickly. This is not a communication problem – it is a speed problem. New stories can only be built one new idea at a time or understanding falls apart, and you are trying to speed up the transfer of stories faster than is physically possible to tell or show them.

This will start to create an invisible high resistance boundary in your mind to specific movements in the Being's performance – your hurry forms an "invisible straw" between you and the Being. You *don't have time* for the Being to waste any movements that create *different* stories than the stories you are trying to transfer.

This boundary has a name you are very familiar with. It naturally appears to increase the speed and accuracy of story transfer from one story system to another. Unfortunately, what is about to happen will also begin a process that will spiral completely out of control and grow exponentially for thousands of years afterwards. It will grow to become the basis of The Floating Anchor Problem, and lead to nearly all suffering and atrocities to come (ignoring environmental disasters). All from this one-week overlap situation with the potential to justify the first murder, and the end of this entire experiment with it.

You don't need to know the name for this boundary to recognize and understand it. It is understood through a slap on the hand, or a smack on the cheek or back of the head. It can be the sudden grab of a wrist, an irritated expression, or the wag of a finger in the face.

It is movement *correction*. It is the forceful reshaping of someone else's actions *after* they are justified to be the right thing to do. It is the *invalidation* of what is believed to be true and right in the moment. It is the forceful hacking and chopping off of learning through trial and error and experience, and instead channeling another into the long spikes of very specialized stories - *your* specialized stories.

This boundary is called *wrong*.

Wrong is the wrench thrown between the two terminal blocks.

Wrong is the thing that ends up exploding in your face.

Communicating wrongness is not *always* an arc flash event. If you can stop yourself from continuing to use it again and again – if you can be your own circuit breaker – you can prevent disaster. The arc flash and arc blast are caused by an unwillingness to slow down.

In the great rush of current to find balance in only one week, *forced compliance* clashes violently against free will and choice, as you attempt to exceed the natural speed limit of making sense.

In this great rush (*instant transfer* and understanding being preferred), your story system begins to treat the other individual as some *thing*, some *object*, that must be controlled and made predictable. In short, the Being must behave like *you want and expect it to. There is no time for anything else.*

When this happens, the other person becomes an "it" – an empty shell – and the story system hidden inside that shell gets overlooked. Since you cannot see the other person's mental stories because they are hidden, your rushed story system only deals with "its" actions.

This human-shaped "thing" must comply with *your* stories and *your* truths. Any action that is different from what *your* story system expects is immediately perceived as *wrong*. This extremely restrictive straw boundary of "wrong" begins to immediately heat the wrench, and you might experience intense frustration, anger, or other negative feelings because of it – all directed at the Being. *Why can't it just get it right?*!

This is starting to merge quite a few ideas from previous chapters, so here's a story-example to hopefully help this make sense:

Imagine you are alone, and your story system justifies touching something with your hand is the right thing to do. It then realizes through Step 1 and 2 that the thing you are touching is incredibly hot, so it immediately and subconsciously justifies the action to pull back your hand.

Your system dumps chemicals to remember this. You experience emotions and begin adjusting your truth. Your story system labels the previous story and justifications that created it as *wrong* in your memory, so it will not be justified again in the future.

Similarly, if your story system justifies showing the Being how to do something as fast as possible, and then Step 1 and 2 realizes the Being's movements are not as expected, you might immediately justify *your* actions to stop *the Being's* movements using the *same subconscious reactive pathway* as you did to pull away from the hot thing.

There is an absolutely *massive* problem with this: You don't *actually* have control of the movement you are stopping, and you never did and never will. *The other person does.* A real boundary is very much being crossed here.

This doesn't matter to your story system, because it is in a *big hurry* to get this story transferred and move on to the next. When your story system justifies stopping or changing a movement it sees and knows to be "wrong" *and it doesn't stop or change* (because that movement is not actually being justified by you at all), it experiences a massive story voltage and immediately and subconsciously justifies the next action on the path of least resistance to stop or change the action not meeting expectation.

It will stop that "wrong" movement *by force* using movements it *can* justify successfully, because they are the next path of least resistance. Your story system might subconsciously justify a slap at the other's hand or a smack on the back of the head. It might grab the other's wrist, give an irritated expression, or wag a finger. It might yell out "AHHHHH STOP!" Whatever it takes to get this "other" movement to stop doing the thing your story system recognizes *as wrong* as fast as humanly possible.

This is an arc flash right over another's story system. Your story system is arcing out of your body and trying to justify the actions of another body you do not actually control. Your story system is trying to treat this other person as an extension of *self.*

The idea that the other person has a story system creating and modifying stories is ignored – flashed over.

You might not even be able to stop yourself from doing this, because the source of all this trouble is beyond your ability to consciously control.

126

The Straw Tunnel

The faster you feel you need to act, and the more you repeat an action with an outcome you seem to be able to keep consistent, the more this connection moves *inward* in time for you – you start to *capture the end* of ever needing to consciously justify that action. Instead, this story becomes the accepted and validated truth of what you can and cannot control with your justifications in this reality.

There is a boundary getting crossed in this process.

A new movement is made consciously. If you discover a strange door, you are going to stop and think of all the stories that are possible before even thinking about touching the doorknob. This requires context for the assumed reality on the other side of this door. The choice to turn the knob (or not) becomes a very *consciously* justified movement.

In contrast: When's the last time you ever stopped and thought about how to open a door in your life today? Especially doors you pass through every day of your life? Granted, this statement makes a few assumptions about you, like you are physically capable of opening doors and that you have doors you pass through every day, but my point hopefully makes sense.

Somewhere between a strange door and a familiar door, the justification process of turning the knob went from being a conscious one to an automatic, subconscious, pre-justified one. This happens, in my opinion, because nothing unexpected ever happened (or at least nothing significant) when it comes to turning doorknobs in reality.

I have personally gone through this conscious-justification-to-subconscious justification process multiple times in my life, and you have too. The most common for all of us, in my opinion, is the control over our own bodies.

We imagine the story of what we think the hand attached to our arm *should* do, and then justify the action to make that story real. Let's say that the hand fails to do what was expected, though. Our truth adjusts with a flood of chemicals, along with right and wrong, and then the process repeats.

The story system repeats this time and time again until that hand, which is *hardwired directly* to it, does the action as expected almost every time without ever resulting in something unexpected.

If you have hands, and if you are capable of controlling them, just look at one of them and then spread it as wide open as you can – fingers splayed - and then close it into a tight fist. No problem right? Didn't even have to think about how to open and close it? Did you have to think about the muscle of each finger? No? It's automatic right? Perhaps a subconscious justification process triggered a conscious request? I'll assume "yes" and keep going.

Another common conscious-to-subconscious transition is any skill you practice often. Musicians can often play their instrument without actually thinking about it. They can even read the music on the page without consciously thinking about each note, just like you are reading this without thinking about each letter. They convert what's on that page of music into bodily movement related to playing their instrument without thinking about how to do it.

I eventually reached the point in drum performance where I could justify the story of what my sticks will do next at high speed *without thinking about it*. I could ram out a custom solo non-stop for *any* length of time, never to be repeated because no part of it was ever consciously justified. I played *through feelings alone*. Whatever I felt should come next was translated into movements automatically – a path of least resistance.

I believe sports players often do the same thing. They perform the movements of their sport without thinking. There is a conscious trigger story, but it sets into motion a subconscious *automatic* justification process. A batter in baseball looking at a 90 mile-per-hour fastball is consciously watching the ball's movement, but likely using a subconscious system to judge where that ball is in space relative to his or herself *by feeling*. If the story of the ball's movement looks a certain way to the subconscious, the subconscious pre-justified action sequence of swinging the bat is triggered, with enough resolution to often put the end of the bat where that 90mph fastball is heading. *The bat is being treated as an extension of the body* – and by all means, for the professional, both the bat and the ball operate very much like an extension of self.

They are almost always consistently predictable. After all, everything that happens with that little ball happens *fast*. There is no *time* to think what to do. All those ball stories and the actions justified have moved out of the conscious and into the speed-center of the brain – the subconscious - because they are so familiar and the truth of the this-then-that connection is so validated it becomes an extension of the self.

To some degree, you try to reach the same point with everything else you ever interact with. When it comes to controlling your body, and things directly controlled by your body such as instruments, bats, balls, gloves, racquets, clubs, video game controllers, keyboards, cars, forks, spoons, and your legs and arms, this does not create a problem.

A problem occurs if your ability to justify a movement by *someone else* is validated by consistently connecting to it yourself. The more this story is noticed and validated, the more likely it is to cross the boundary into the subconscious and – if not checked - lead to the end of the world.

This invisible arc-flash straw-path can become the new path of least resistance when interacting with others, because there is nothing to invalidate the truth-story for you *except* you consciously noticing it, or the other person starts resisting your story system's attempts to validate this connection yet again.

Which causes the wrench to glow....

127

The White-Hot Wrench

Perhaps the best example I can think of to explain a story system that exists in an overvoltage, tries to force a speed that is too fast, and starts to cause arc flash events that transition into the subconscious making them difficult to stop, is a parent's (or caretaker's) story system relative to a child – a newborn baby.

Relative to the mother, a baby is born with almost no stories at all. The only stories that exist about the baby before the birth are usually anchored to mom's firsthand experiences, or secondhand stories allowing others to shadow-see *through* the mom.

These are stories like "Billy is a kicker", "Susie likes to suck her thumb", "Mikey is a large baby", and "Alice likes to keep me on my toes".

These are *not* the developing baby's stories. They might very well be built on the movements of the developing baby, but they are stories individuals *tell themselves* to be the developing baby's stories.

If we draw a diagram of the developing baby and mom's story systems at this point – before birth - the result is a two point system: One point has tons of stories, and another has almost no stories at all. The no-story system is simply a point with a boundary inside the larger boundary of the second point – because it is literally inside of it.

The birth mother and the developing baby appear like two points contained within the same system – because they are. You can't remember any story from your time in the womb, because you were not imagining stories and then trying to consciously act them out. If anything, I suspect your developing story system merely "practiced" sending currents through the nerve-straw connections as they became available. Only your mother was actually imagining and attempting to justify stories into reality at this point.

Before I go further, I am not arguing in support of a "blank slate" at birth (known as tabula rasa). Instead, I am arguing that a developing baby has no conscious control over the world outside. In the womb, its entire universe is bound against its body tightly in every direction. There is nothing to really control. Once the developing baby is "hatched", there are only a few pre-wired "if this, then that" sequences, such as rooting for a nipple when the senses determine one is near, suckling and swallowing, and crying.

Beyond these few sequences though, a newborn baby doesn't have the ability to control most of its *own* movements in any predictable and consistent way, so manipulating objects to bring some imagined story into reality is out of the question.

This means if you were just born and then left completely alone in *The World of You* – it is almost certain you will die. You cannot feed yourself. You cannot bring liquid to your mouth. You cannot even move yourself in any direction beyond a few inches. You aren't imagining stories of possibilities and what is likely to happen next. There is no next. You are incapable of preventing anything. We as humans are *helpless* for quite awhile after birth. I've often heard this period of time called the "fourth trimester," because you are nearly as helpless as you were developing in the womb.

This creates a *huge* imbalance for you right away. Whoever raised you in those first months of life was almost completely responsible for *capturing the end* of all of your story arcs *for you*. This person actually justified the actions you could not yet perform.

Your caretaker ultimately figured out, through constant truth-story adjustments, which of your cries meant you were hungry, which meant tired, which meant pain, and which meant you've soiled yourself. This person potentially learned the meaning of every tiny expression of body language you had.

If the action justified as right by your caretaker's story system to try to match your need did not result in a change to *your* barely-controlled actions, it was marked as wrong in the caretaker's story system, discarded, rewritten, and a new action justified. For your caretaker, validation of the action being right was measured by a change in *you*.

As a baby, you were likely experiencing a huge internal voltage between your story system and the story it can't seem to make happen, most likely because it does not yet even know what action is needed relative to the feeling or how to justify it. In my opinion, it does not know yet because the straw-paths that form Step 2 and 3 of the story system are not yet consistent enough to form a path of least resistance. The electrical currents are still taking "all paths" available, which cancels themselves out from moving any one direction.

Due to this, babies have no idea an object still exists if it is hidden from them – which is what I've always known to be called "object permanence". Their entire reality is contained in the "right here right now."

As a baby, how you reacted to this imbalance between mind and body and a need for ends to be captured without any means to do so, I cannot know – and neither can you firsthand. I'd argue most babies begin crying or wailing. Others might not cry, but just lie there silently instead, or perhaps even make cooing sounds. Perhaps this reaction is encoded in the genes provided from each parent to best ensure this reaction results in survival.

Whatever action your hard-wired story system decided to take in response to your decaying story arcs you cannot capture, it was one of the very few actions you had available to use at all. If you assume you cried when this happened, then your crying immediately created an imbalance in your caretaker's story system when heard. This cry-voltage revised your caretaker's truth-story, (which was probably that you currently need nothing), and also collapsed what was presently justified as the right thing to do (probably trying to sleep). This huge voltage likely jumped you right to the top of this person's priority list, and he or she started justifying actions *on your behalf* to help you capture the end of the need story arc that was causing you fear or hunger (which is merged with thirst).

As a baby, you are *using the arc-flash method*. Your cries are flashes of lightning that completely ignore the autonomy and freedom of another story system. As time marched on, you gained more and more ability to predict your own justified body movements by making countless adjustments of truth every story cycle.

You eventually found consistent success at grasping things, bringing them to your mouth, going longer without pooping yourself, holding your head up, rolling over, and pushing up. Those repeated truth-stories are then "captured" and internalized to your subconscious to operate at great *speed* due to their consistent validation. Perhaps you next moved on to consciously work on feeding yourself, crawling, walking, and even talking.

Along the way, your story system also tried to control other human beings as "things". This isn't meant to be malicious as I see it; it is just a carry over from when this behavior is how you survived as a baby.

For example, you likely figured out which actions made a caretaker appear (such as acting more hurt than you actually were), how to use emotional actions (like crying) to make someone do what you needed, and how the word "no" started changing their behavior. This was your story system working to find the path of least resistance by exploring all possibilities.

From your perspective, you were experiencing a pure new story at every moment. From the perspective of everyone around you, your pure new experiences were already expected and familiar. Your caretaker likely knew what action to expect from you next at almost all times, based on observing all your previous actions over the past few years. This means *your caretaker's* story system could predict nearly every single one of *your* new stories before you even experienced them for the very first time. This "knowing what is likely to come next for you" before you did created a choice for your caretaker that was made each and every time an interaction occurred with you.

If you cried and pointed to a toy on a shelf, then your caretaker already knew what action your story system was trying to justify, or quickly figured it out from their own memory of your past actions and behaviors. That person then:

A. Justified the action of getting the toy for you, keeping the two story systems imbalanced exactly like they were before this interaction.

B. Moved toward a new balance, and helped you take a step toward performing the action you want to justify with your own body, such as having you move closer to the shelf before getting the toy for you.

C. Ignored you in hopes that resisting your attempts at arc-flashed justification will make you choose a less-needy path, or take a new path for this action such as you getting *someone else* to perform the action for you. In practice, this option probably increased your voltage level at first, causing you to escalate this external justification story to an un-ignorable level (cry, scream, etc).

D. Forcefully denied getting the toy for you with a "no", or "get it yourself" or "some other time," or "play with something else."

Whatever option your caretaker chose, it was justified by their story system as the right action to take. This decision, rooted in his or her larger story system believing it can quickly and easily predict your smaller one – which is an *assumption* - can easily validate itself to be a subconscious automatic pre-justified response to nearly *every* future action-demand you try to justify from your caretaker. If this happens, your caretaker is arc-flashing right over your story system and its justifications, and is subconsciously responding to you without any regard to your story system's validation process at all.

Every caretaker and child reaches this point on some topic or act. This is the moment where the caretaker refuses to cave to the demands of this action-limited story system, and is an attempt to *reverse the current flow* through the bolt of lighting arc-flash straw-path. The child's story system has consistently treated the caretaker as an extension of itself, but now the reverse is happening, and the caretaker is instead telling the child what actions are allowed to be justified.

This almost immediately results in a war between two minds with warped mental boundaries. Both story systems, the caretaker's and the child's, are trying to determine exactly where one self ends and the other begins - where one's control using the justification process ends and the other's control begins.

If the caretaker consistently performs the action justified for them by their child, the child's story system does not experience any resistance, sees no boundary, and does not build a different story. The child's story system has determined through constant successful repetition that validated a high level of certainty, *if I do "this" then the other will do "that"*. An expectation from reality was born. The *other's actions* become a subconscious extension of the child's *self*.

If this happens, then the sudden experience of an unexpected "that" by the caretaker after "this" is justified by the child, triggers emotions in a big way.

Parents know this as "the tantrum", the "fit", the "fall-to-the-ground-and-scream moment". The child's story system already determined through repeated validated experience that the *caretaker* was *self*. When the caretaker's actions did not merge as expected when justified, the child's story system experienced *a collapse in truth*.

As I see it, the reason the reaction to this is often so dramatic is because the child has *no other truth story*. This-by-me-causes-that-from-caretaker has *always* happened. Suddenly and unexpectedly, "this" is not connected to "that" anymore.

If the caretaker had also been very quick to always deny the child autonomy up to this point (such as preventing any attempt to "get the toy" on his or her own), emotions can escalate quickly. From the child's perspective, not only can he or she not figure out how to get to "that" using the previously validated path, but all other paths he or she can imagine to "that" are not allowed to be justified based on the corrective actions of the very same "thing" known as the caretaker, who is suddenly and unexpectedly *unpredictable*.

The previously defined mental boundaries for both individuals are changing. This continues until something or someone breaks the arcing current between these two systems.

128

Broken Circuits and Live Wires

From the child's perspective, the caretaker was understood and controlled. The caretaker was validated to be "self" because "If I justify this, caretaker does that". The child's mental boundary was expanded to include the caretaker. Now the caretaker's mental boundary is resisting. Which means it feels like the child's previously stable mental boundary is now collapsing.

As caretaker, you get tired of your actions being "controlled" by the child. For months or years if the kid did "this" you did "that", but your story system also believes the child has been capable of doing "that" on his or her own for the past six months. These "involuntary" actions are robbing you of your "free time", so you move to re-capture the ends you've been performing for the child for yourself. Hand these actions off to the child to perform instead. You are trying to create a new path of least resistance toward balance between you.

I think it's important to note that the story about the child's capabilities exists for you as caretaker, *but not for the child*. This story where the child does "that" action without any help exists completely in a gap of the child's story system. He or she has never experienced this story before. It's possible he or she imagined it once before and was forcibly denied bringing that action into reality *after justifying it* by a slap on the wrist or a strong "no!", which flagged it as "wrong" and removed it as a future possibility. That story is now no longer available because it was invalidated.

The development of healthy sphere-shaped boundaries between two points is a slow process that ideally makes a straight line.

AT BIRTH
A to B unordered

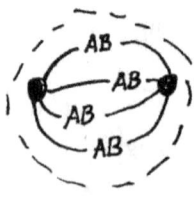

CHILDHOOD
A to B finding
order

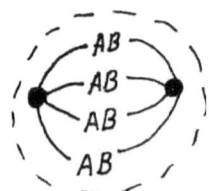

TEENAGER
A to B
balanced
(ideal)

I would argue such a straight line is rare, maybe even impossible to actually occur. This means the creation of a perfect and balanced mental boundary between a parent and child is not likely to happen. Inevitably, "loose ends" from one story system remain connected to be "finished" by the other. The result is two-point relationship imbalances like codependence, extreme insecurities (without consulting the other), or even anger that the other would "do this to me". This typically becomes apparent in the teenage years during the most dramatic rebalancing and eventual "separation" of these boundaries, but it can persist for a lifetime.

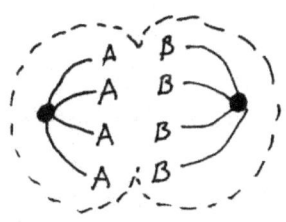

BALANCED
A to B divided
evenly

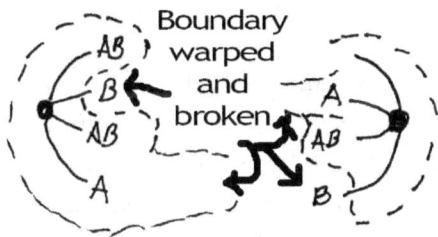

IMBALANCED
A to B divided
unevenly (codependent)

This "broken connection," which is really the path of least resistance between one point subconsciously justifying actions that in turn subconsciously justify actions in the other, only feels balanced when the two points are together. When separated, these "justified justifications" becomes "live wires" feeling for a new path of least resistance.

This, to me, is inevitably why we as humans so often end up dating or becoming friends with people that behave like our caretakers, because they naturally connect to those "live wires" and finish the circuit.

In my opinion, these imbalanced relationships are often held together or started because of "arc flash" events that connected the loose ends. The voltage present at the end of an exposed "live wire" can be so dramatically high relative to someone else that a simple two-way interaction can result in an arc of lightning that connects the two in a very intense way that neither person is consciously aware has happened.

In my opinion, this creates what feels like overwhelming feelings of "self" in the other (even though it is actually the other person becoming a means to "capture" the unfilled "self" of the first person), which becomes the basis of infatuation or intense and dramatic feelings of love and connection. In this case, the words "*you complete me*" apply quite literally.

In the most recent "imbalanced" diagrams, the person on the left is almost a perfect match for the one on the right's "live wires" and "holes." Where "A" is missing in one it is present in the other, and where "A" and "B" are missing in one, both are present in the other. However, there is also a place where both individuals have both "A" and "B", and I believe this area is likely to be a source of conflict.

Whether or not this kind of "co-dependent" relationship can last through time is up to the individuals. If these live wires still exist unaddressed whenever the person has kids of their own, you can probably predict what is likely to happen – the child with almost no stories of its own will be given the story through repeated actions that validate "if caretaker does this" then "you do that" until it is an established truth.

This is, in my opinion, exactly the mechanism that sends emotional and personality imbalances passing down through families *for generations* even though they are not based on DNA or genes at all.

All it takes is one individual with a very badly warped mental boundary and a possible separation from the person, or a large change in the balanced-imbalance relationship. The result of constantly-exposed high-voltage story-straw-wires will inevitably connect via arc flash with others that can complete the circuit and fill in the missing "that" and "this" pieces of a this-then-that imbalanced relationship.

Almost all of this is happening *subconsciously*. All of these action-stories have been repeated so many times they've become *internalized* truths. You might *know* something feels wrong. You might feel like the emotions are too strong – something feels horribly out of balance - it might be very difficult to figure out *why*.

After all, completing another's story eventually becomes normal and a path of least resistance, especially for a child, so it seeks to "capture this end" with more and more certainty and speed. All so that child can perform the required-of-it action faster, to get back to "free time".

This end is not actually capturable, though. There is an invisible boundary preventing it.

129

Danger: High Voltage

Back in the one-week overlapping *Single File World*, here you are, trying to cram decades of experiences into the Being in just one week. You don't have much time and you don't have words, so you must hurry. You make sure that when you do "this" the Being does "that" exactly as expected.

Its mind is ignored, because this is a one-way transfer of information. You've already been living here for years. It just got here. It makes sense that your stories are far more important than any it might have. After all, your stories have successfully led you to your last week to live - and they have repeated themselves as valid time and time again. Your truth-stories are *right* and you know it. Your truths created this place the Being gets to inherit with no effort at all. If it can't make its movements match yours, which you know are right, then it is *wrong*.

The lesser-storied Being does not know *half* the stories you know. It doesn't know the decades-long story of how you built the garden seed by seed. It doesn't know the year-long story of trying to make fish traps that actually work. It doesn't know the countless stories of you failing and adjusting your own truth, right, and wrong over and over again for five decades to reach this point *right now* where you are trying to transfer as many stories as possible in every *minute*.

This high-pressure process of slow communication through actions and gestures quickly begins to take its toll. The only way to validate understanding this way is "if I do this, then you do that." It is the same as the caretaker and the child. This can warp and potentially damage the boundaries of the two story-systems involved.

The Being's story system, hidden where you cannot see, is imagining and experiencing stories *at the same rate as yours*. It, too, experiences extreme voltage and chemical releases when its predictions and justified actions do not turn out as expected.

You suddenly and forcibly stopping its already-justified actions in progress is, at first, *unexpected*. However, this quickly becomes part of the Being's ever-revising truth about this place each time you forcibly correct its actions.

Your world does not change. Your story system keeps validating what you are doing to be right and to be working as intended, because the Being is correcting its movements. The Being's world however, is beginning to flip upside down.

Your actions are taking place in the same reality the Being is experiencing. As a result, the Being's story system remembers *your* actions, which will be considered in future stories used to justify actions. It does not want to experience unexpected physical pain or invalidation of its truth. Both of these reach all the way back to the forced-choice of wanting to survive. Nor does it want to experience the sudden stoppage of its already-justified-to-be-right movements by this *boundary* known as *you*.

So, the Being's story system watches you through Step 1 (observation) for movements matching those in Step 2 (memory), which indicate a Step 3 (prediction) of an incoming slap on the wrist, a push from behind, or a growl of disapproval.

As the Being, this recurring unexpected imbalance created by the actions of another to harm or invalidate you must be relieved. If it cannot be balanced through two-way communication, which would involve trying to loop through the *Dark Forest Moment* again and again with another being that knows much more than you about this reality, then this voltage will remain trapped inside your story system.

This now-growing second balloon needs to rebalance with "ground" where it started, so it starts desperately looking for any story to do this.

Ground is the world where these external invalidations are not happening to you. If you can predict these actions, you can stop them from happening.

If no path to ground can be easily found, reactions can be intense, especially if the situation is terribly unbalanced. This would be the moment of emotional breakdown. It is the running-in-tears-to-slam-the-door-on-you-and-scream-GO-AWAY moment.

As the Being, this imbalanced situation is mentally painful. This homestead provides you a huge amount of certainty that would crumble if you ran away and "slammed the door". You know you would have to start *The World of You* from scratch if you did this, and you would have no idea what happened to this other creature that keeps invalidating you. You would have to figure new stories for food, water, and shelter, and you don't know many stories about those things in this place yet, but this other creature does, and definitely has the upper hand.

This means you are likely to stay with this unpredictable-but-seems-to-want-stability being because it is the path of least resistance compared to running or violence. However, your story system needs the unpredictable-ness of its actions to be made predictable. It is, quite simply, afraid of the dark. Dark meaning *The Dark Forest Moment* of needing to choose between violence or running.

In an effort to make this other way-more-advanced-but-rushed "self" predictable, your mind creates what amounts to a second story system. This second story system ends up becoming a strange scrapbook version of the real story system you cannot see in the other being's skull. It is made from pieces of stories you cut and paste together from the pages of your own memories and imagination.

Whenever this "rushed" being is inside your sensory boundary, you try to run and grow two story systems at once. You need to be in two places at the same time to achieve more certainty *of shelter* from this "other" that might suddenly stop, change, or invalidate your stories. Your story system wants to build the real truth of this reality using its *own* process - not be forced to tell itself what some other system says the "truth" is around here.

During this period, when trust is low and invalidations are high, your mind as the Being is probably consuming twice as many calories as usual. Trying to be in two places at once is stressful work.

First You is managing *your* stories and justifying *your* actions and truths built from *your* own firsthand experiences, and Second You is trying to predict the *other* being's actions and justifications for them based on the actions First You is considering justifying.

This means your second story system is trying to determine if the action your first story system wants to justify might result in a justification and action by the other being that is *negative-to-you*. If Second You predicts this is likely, it will *overwrite, or re-justify* First You's definitions of truth, right, and wrong. *All to secure shelter.*

Shelter from the darkness.

Shelter from harm and violence.

Shelter from uncertainty.

Shelter from invalidation.

Shelter from the source of all of this: *The other.*

The other person is becoming the *Keeper of Truth* for you, because that person keeps overriding and invalidating what you believe to be true. And since shared language is not going to be possible within one week, and all of this is done through body language and actions, this means *the transformation of Second You into the Hydra is possible without any words being used at all.*

If the stress of this imbalance gets high enough, perhaps First You will collapse into submission or attempt to escape by running away from *the other*. You might limit your own actions to match whatever Second You believes *the other* person will accept as right. You might try a pseudo-escape, and try to avoid this other person by trying to predict which movements might put *the other* in your sensory boundary. Or, you might simply explode into violence against *the other* in a fight for the authority over your own actions by trying to destroy or contain the other inside a boundary of your own making - perhaps a literal cage. Finally, you might try to stabilize, and get rid of the internal voltage creating Second You. This would be done by establishing equality and eventually fairness and understanding – but good luck with this in the absence of language and with a being that is significantly more experienced than you.

As a child, can you demand equality with a parent? What tools do you have for such an argument without words?

As I see it, there is one saving grace to this one-week overlap arc-flash to stop permanent damage and let things come back into balance: Time.

The Other will only be with you for one week. As the Being, you just don't know that yet. After that week, this individual will be gone, and you will be free to choose your movements as you please.

The Hydra will then begin to fade away, transform back into your friend, and merge with First You again, because it no longer needs to exist. It was needed to predict the *Other*.

It's possible you will have live wires left exposed after the *Other* is gone. They will expose themselves when you need to connect "this" to "that" using stories associated to negative actions from the *Other*. These remembered invalidations will associate into your stories of how to connect "this" to "that" to remind you of all the *wrong* actions you made while attempting this particular connection.

Another moment this live-wire "exposed" feeling might occur is when you experience a situation where you simply do not know how to connect "this" to "that". For a solid week, the *Other* forcibly showed you tons of this-to-that stories you didn't know before. This associated the *Other* not only to all this-to-that stories, but with the feeling of *not knowing*.

This means for awhile, after the *Other* is gone, when you encounter something you don't know, the Hydra may come racing back to attack First You all over again. *Should* you know this story? *The other would know this story.*

This won't happen forever though. There is no one else around to capture any ends or connect "this" to "that". You will be forced to complete any incomplete story by taking a risk and stepping through the doorway into blackness. Once the end is captured a couple of times, the old truth involving the Hydra will be overwritten with a new truth involving only First You, which is far more powerful because it is wordless, gapless, and filled with immense amounts of sensory experience.

The damage from the arc-flash can be reversed and repaired. It might just take time. After that, you have something new to think about for quite some time: You now expect another to join you sometime in the future. You are made aware you will have one week remaining to pass on what you know when that moment arrives.

What do you think the chances are that you will create the same invisible straw-tunnel boundary of "wrong" in that week? What do you think the chances are that you use the same teaching style and same rushed aggression that the *Other* used on you?

Think on it. How would you approach such a situation?

Now consider a twist: How much would it change everything if the new arrival spoke the very same language you already know? How much faster could progress be made? Do you think it will eliminate this extreme story overvoltage situation?

130

Thunderhead Over Bobville

Let's go back to the 100-person mini-society where we last left off. This is the original story line with you surviving in *The World of You* on your own for years, meeting the Being and working together for years, and then meeting the Trespasser and working together long enough for it to learn your language before everything collapsed into chaos from a flood of trespassers.

The collapse of available food, water, and shelter caused by the sudden massive surge of new individuals was magically stabilized, and you and the group had three years to get everything balanced before the magic wore off.

There was a lot work to do to stabilize everything in three years, but you and the 99 others did just that. The Trespasser taught everyone language as a specialized teacher while you and the Being captured the Trespasser's needs. With the specialist roles born, there was an explosion of stories and expertise in every direction.

The garden was massively expanded by specialists at the same time the "hunter and gatherer" specialists developed new and more effective techniques. "Toolmaker" specialists worked to create specialized tools that were more consistently successful at just about every task. "Homebuilders" constructed new shelters. "Cooks" created new recipes and combined and portioned out the gathered and hunted food to be distributed among all the specialists. Others specialized in gathering wood and other raw materials. Some specialized in upkeep and maintenance, others in sanitization and keeping the creek clean and waste managed. There were many tasks that needed doing, and much time was needed to accomplish each and every one of them.

Since speed is so important, it was far quicker to have individuals perform specialized tasks to build highly repeated truths, than for each person to perform every task or tasks at random and experience less validated truths. In the end, nearly everyone specialized in *something* so well that the truth-stories in each transitioned into their subconscious, and progress continued to accelerate with each passing day. Everyone got faster and more efficient at his or her role every day.

This method worked. Time passed and the magic wore off three years later. Everyone transitioned to the food, water and shelter systems they had arranged and created, with really no problems at all. All the systems kept things as fair and as balanced as possible. Resources were plentiful.

And that is where we will now pick up this story.

There was a large group meeting, and the town voted it shall henceforth be named "Bobville".

Everyone in Bobville is now grounded, or anchored, to the shared localized experience of coming together to build this society. Everyone had his or her role in making it successful. When it comes to everything before this moment, You and the Being have an especially deep connection reaching back *much* further – to the point in time when this well established village was just the patch of wilderness where you both felt most certain of your own survival, and decided to capture the ends the best you could. This place was your shared foundation of ever-growing certainty of continued survival, and trust in each other. This place is where the first "grup" was ever given to another.

The Trespasser also has a pretty strong relationship with you and the Being as the next longest inhabitant of Bobville, though not nearly as many stories are shared between you and the Trespasser as you and the Being. The garden and shelter and survival needs were all pretty much pre-captured with high certainty before the Trespasser ever stumbled into your world.

After that, all the others have even less shared stories with each other, though with language, everyone's connection is deepening every day.

This is the basic layout of the Floating Anchor Problem on a larger scale. It is created by an imbalance in the amount of stories built from firsthand experience between individuals. I've attempted to explain the mechanism at play as simply as I can over the previous couple of chapters: Two very experience-imbalanced story systems must interact in a hurry, with or without shared language.

When you experienced a feeling of unease about the Trespasser laying by the creek all day and resisting to help, that was the Floating Anchor Problem *happening to you*. You seemed incapable of realizing just how high your anchor had been floated (how much your normal had changed) until you felt gravity pull it all the way back down (when the flood of trespassers began collapsing everything back to nothing).

Everyone in Bobville knows that the collapse would have continued without "the magical stabilizing" of food, water, and shelter that provided three years to better capture them. They all know they would have been thrust into the darkest of *Dark Forest Moments*. In this way, everyone now shares the experience of "saving" this place and restoring it to balance – because they all also experienced its collapse.

New stories are now being made. Old stories are being re-experienced and shared using words, and story systems can now start to "make sense" of different perspectives of past events. With the passage of time, all of Bobville might reach a deep connection level like you and the Being already share.

But the Floating Anchor Problem grows in time. We all can only experience and build a limited number of stories with what limited time we have: From the beginning of our first story to the end of our last.

Jump Bobville forward twenty years.

Babies have been born and children are running amok in the village. This second generation is taught how things around the village work, as well as the shared language. They are also taught through that language the full history and founding of Bobville; a story given to the teachers by the Being and yourself. Pictographs have also been developed, so stories can be told in a second way - through a series of symbols - instead of just speech.

This new generation does not have the same starting anchor point for story-building as any of the original 100. Their ground truths and expectations are radically different from your own back in *The World of You*. Meals are regular and hunger is easily captured. Houses already exist and have comfy beds. The gardens already have fences. All the adults around already specialize in some specific task. They have more free time in their lives than there ever was for the first generation.

Let's jump forward *another* twenty years.

There is another entirely new generation of children, the third generation, with entirely floating grounds. They all must go to school for longer than the previous generation – there is, after all, more stories to learn now.

The territory, or outer boundary of the town, has expanded dramatically. It stretches for a mile in every direction from the original central cave where it began. The life-giving creek flows through the middle with many bridges across it connecting the two sides.

A market sits next to the creek now – a place where individuals come to barter and trade different items of interest and relative value with the farmers, artists, craftspeople, and tradespeople. The kids born twenty years earlier have grown into adults performing their own specialties, like medicine-making and blacksmithing, and moved into their own houses-shelters.

You and the Being are advanced in age now, and mobility is a little more difficult. Your lives are almost completely composed of free time, so you both naturally fell into the roles of storytellers and historians for the town. Many Bobvillians gather around regularly to hear the daring stories of survival in an unexplored wilderness from much more uncertain times. You both consider yourselves lucky to live as long as you have, and you smile at what this place has become. The Being carries around a pocket of grups at all times, and offers you two each and every time it sees you out and about.

Jump forward another twenty years.

You and the Being have died, as well as most of the original 100. The town has continued to expand. The elders in town are now mostly the second generation, and a fourth generation of children has joined society with yet another different floating ground anchor point for normal.

The elders often meet, and have become increasingly annoyed at the fourth generation's unwillingness to work. The kids and teenagers argue back they shouldn't have to work - there are more than enough specialists to keep things going, and food, water, and shelter haven't been an issue for decades. They argue if they want to lay around and snooze by the creek all day and paint pictures, its their right to do so.

Time-jump another twenty years.

The town is now huge. It takes most of an entire day to walk from one edge to the other. New specialists roles have been created within the previously small specialty "spikes".

A fifth generation has joined the fold. The second generation has now mostly died. The elder third generation now gripes about how the fourth and fifth generations don't work very hard – they are all "lazy" and don't care about what really matters. "Back in my day we had it harder in every way" becomes a kind of mantra amongst the elders.

Most of the middle-aged fourth generation performs their specialized job begrudgingly - they *know* the job needs doing, but are not excited to perform it.

The young fifth generation dreads the idea of having to choose a specialty job and having to go to school for longer than any generation before them. All to learn the stories and take tests about a bunch of long-dead people responsible for this place. Who cares?

The marketplace by the creek is sprawling, and bartering has sometimes collapsed into violence because some individuals seem to be selling things everyone lines up to get, creating a feeling of unfairness and imbalance in the individuals left with no sales. The government developed something called "money" where one "ptykit" token is considered to be worth one gold nugget the size of Mayor Bob's fingernail and can be used to trade for just about anything else of value. As a result, tons of individuals, especially those who are struggling to sell their goods, have taken to the hills try to find gold.

Jump twenty more years.

The town has grown larger than anyone can travel across in two days time. Multiple marketplaces are now spread all around, and some have proposed separating the growing town into North Bobville and South Bobville to keep things better managed.

The third generation is gone. The fourth generation, now elders, continues at their jobs for as long as they can still can. The fifth generation, being forced through many years of school and "educated" on what is okay and not okay on nearly every social topic, has started to show rebellious behavior toward all authority, whether religion, government, or school. They complain about "the system". They want freedom to determine their own futures instead of falling into the same specialty roles, the same houses, and the same lives as nearly every generation before them.

In response, the older generation starts to create ever-restrictive rules and laws like curfews and an increasingly long list of punishable crimes related to nearly everything the fifth generation keeps doing, as they seem hell-bent on not doing anything expected of them.

The sixth-and-youngest generation of children is significantly smaller in number, mostly because the disgruntled fifth generation has little interest in buying a house and "doing a job until they die".

Let's go ahead and take a *big* leap forward now – 100 years.

The newest generation's floating ground truth anchors into a society that has divided harshly in many ways. The town decided to split some time ago, and now some individuals ground into secondhand truth-stories where East Bobville is "full of poor and unruly scum that will steal and hurt you the same as look at you". Others ground to West Bobville being "full of snobby, uppity, wealthy people that think they are better than everyone else". Some ground into a reality where more restrictive rules are needed to control *the others* unpredictable actions, and many others ground into a reality where *the others* just want to control them and keep them down in the gutter so those others can keep living on high street.

News-Story outlets are born to try to trim down and manage the enormous amounts of stories happening at all times. East Bobville media, funded by East Bobville residents, tells constant stories of the crimes West Bobville citizens have committed and speculate what their next crimes might be. West Bobville media runs constant stories of schemes by the rich and powerful to try to gain even more control and money.

Groups start to form that believe all kinds of things, because they are building their own secondhand truth-stories that are limited only by "what makes sense". Some of these groups believe stories like the town being founded by aliens, birds aren't real, reality doesn't exist, we are all just hallucinating everything, the government is full of lizard people in costume, electricity is not real and is only a way to control the people and hide things behind "high voltage" signs, there are only five planets and not nine (or was it eight?), that books are the reason the kids create problems, and that kangaroos actually have retractable fangs and *everybody knows it*, but they refuse to say it out loud. All these groups have something in common – they have all heard rumors something big is stirring regarding ice cream.

A few weeks later, West Bobville media tells a story sent to nearly all citizens that ice cream is good. It brings joy to citizens on the west side to take their kids to get ice cream, and is bringing lots of income into the area.

East Bobville heard about the West Bobville story, and tells a story of their own that is sent to nearly all citizens explaining how ice cream is the poison that is destroying the town by increasing obesity and health care costs – which is provided mostly by West Bobville - and is now looking to create laws to ban ice cream

A short time later, billboard are posted in East Bobville that say to vote "yes" to Proposition 2 in the next election. The proposition puts an end to the high taxes caused by the problematic ice cream consumption, which is produced and mostly sold out of West Bobville.

The largest arc flash event to ever happen in this society is primed and ready.

Massive groups, anchoring to the same words used to describe far away places, groups, and people they have never met or experienced, validate each other into believing the words themselves represent reality.

But words themselves are often arc flashes creating a certainty about others that does not actually exist. They can easily short circuit entire story systems and reality with them.

No individual West Bobville citizen believes East Bobville's stories are specifically about him or her. No individual East Bobville citizen thinks West Bobville's stories are specifically about him or her. Every individual believes the stories from the other side being told about his or her side is unfair. Every individual begins to fear what the *other* side is about to do *as a group* because these stories are unfair, and so they tell themselves a story *as a group* that all the individuals on the other side actually believe the stories they are told *as a group*.

And no one is wrong. *Everyone is right.* But no one is interacting or seeking to understand the new and unfamiliar story systems on the other sides.

Imagined stories of "the other," based on media stories, are projected and then actions are justified that actually bring those imagined shadow experiences into actual reality – anchored to barely anything at all.

Everything comes to an anxious head at the historic monument in the center of town – the cave with the stone that makes a perfect door – where thousands from East Bobville and West Bobville are gathered.

131

How a Ghost Opens the Abyss

A churning mess of who should have authority over defining what is right, what is fair, and what is true has resulted in escalating tensions and beliefs about "the other" in Bobville.

As humans, we all start the same way – birth (and the first loops of our story system years later) - but in different places and in different situations and in different times. In short, we all begin – or ground - in an environment containing different certainties and different levels of free time. Every environment holds a different truth, different values, and different experiences.

The combined memories of all firsthand experiences from everyone alive right now, which is humanity's current *living memory*, only reaches as far back as the earliest memory that can be recalled. No one else living has experienced anything before that point in time firsthand.

Once that earliest-firsthand-memory-holding person dies, the next person with the earliest recallable memory becomes the new boundary limit in time for all of humanity's living memory. All the memories of firsthand experiences *between* these two people are either lost to the void completely, or transition to be only known through secondhand stories, such as words or other recordings traveling through time.

Written and recorded words, unfortunately, can create problems spoken words can't even come close to creating.

Suppose that the person that just died – who we will call Alice, who was the holder of the oldest firsthand experience in humanity's living memory - left behind a diary.

The earliest entry in Alice's diary was written *before* the next oldest living memory begins (which belongs to some old guy named Bob). In other words, no one alive has firsthand experience from the same date as her first entry.

In Alice's first diary entry, there is a passage that reads, "Today, Being told me purple people are the worst. Watch out for those with purple tint in their skin, they are not smart and want nothing more than to take everything you have, even your children."

Alice's diary is found and preserved by the Bobville Historical Society because it has such historical significance. With their permission, individuals can access this diary.

Let's say a person – let's call him Charlie - eventually finds and reads this first entry in Alice's diary and makes a copy of *only the passage above.*

The story Charlie builds with these words will be created *not from the Alice's context,* but from Charlie's *own* memories and experiences anchored to those same word-shells that Alice used. If Charlie has negative experiences with purple people, and has come to fear them, it could be a story like this:

The Being is the founder of this entire town. The origin of everything. The Being knew purple people were a problem all along and we didn't listen. The Being wouldn't want these people in this town. They are everywhere today! They are the reason we have so many problems, and the Being tried to warn us!

Charlie then uses this story, which he imagined into a shadow experience *from his own context* using Alice's words, to start an Anti-Purple People activist group. This action, justified on an anchor-story that you can hopefully see to be incredibly unstable and contain a frankly absurd amount of possibilities and assumptions, creates what I can only call the beginnings of *a black hole.*

This black hole has extremely strong gravity, and if individuals venture close enough to it, it can pull them off their own ground-story and toward the shadowy singularity.

All the possible meanings of the word-shells in the passage from Alice's diary are being collapsed and then assumed to have *one specific possibility.* That one possibility, created by Charlie from Charlie's experiences, is then believed by Charlie to actually be *the Being's truth.* A singularity.

I do hope you can see the problem here.

There is no firsthand experience involved *anywhere* in Charlie's story except for reading two sentences on a page *assumedly* written by Alice. Charlie has never met Alice. Charlie knows nothing about Alice. Charlie doesn't know how the Historical Society came to believe this diary even belonged to Alice.

Unknown to Charlie, the Being that helped found the town actually has no connection to anything that is happening. The words used to justify Charlie's creation of an Anti-Purple People group were not written by the Being at all, *they were assumed to be written by Alice,* who can no longer clarify they are her actual words or clarify their meanings. The best Charlie can do to increase certainty in this situation is to find other words and recordings both from Alice and from the same time in history, to see if he can make *more sense* of these words. He could even read the entire rest of Alice's diary in hopes of finding more feelings and beliefs from the Being about purple people.

Charlie can never be certain though, he didn't actually experience anything he read. The story of what just happened to Charlie is between not two, but *three points*.

Reading is a kind of firsthand sensory experience. The act of reading is not your eyes seeing symbols on the page. Reading is translating those symbols into meaning at *incredible* speed. This took lots of practice to learn and internalize to the point of crossing into your subconscious. My youngest child is 6-years old as I write this, and this process is only beginning for her. Most of her conscious effort is spent on the symbols themselves. She can only connect those symbols to her experiences when she says the word out loud, because she does know the word and already has a meaning for it, but only in the form of sound.

You are doing the same act right now – reading (or listening if an audio-book). You can see the individual letters and words, but there is a good chance you are far beyond having to sound out these words and then trying to decode what the sentences mean. *Reading has become subconscious for you.*

These groupings of letter symbols make words that create shapes. You see the *shapes*, and those shapes are anchored to sounds through – you might have guessed it – constant repetition with your own experiences to validate the connection between shape, sound, and experience into your subconscious.

You are reading this to yourself, and the order of these word-shell current signatures only ever *rearranges* what is possible using your own experiences (because other experience does not exist for you), and I can only grow and merge one new idea at a time.

This growing and merging process happens automatically for you as a reader though, because reading is all handled subconsciously. If my words "make sense", they merge right in to your truth – no thinking required – resulting in a new story that is *self evident* to be true. It just *feels right* and it is therefore *true*. This would be dangerous if I was writing these words assuming you were part of a large group that believed exactly the same thing – but I know better, and have given every effort to avoid this (though I am not perfect and have probably made some mistakes).

Charlie read the words in Alice's diary, and his own high-speed subconscious context automatically filled the word-shells with meaning in a way that happened to "make sense" for him. When this happened, it seemed to Charlie like he just instantly experienced validation of his fears from *the Being* – the co-founder of Bobville.

Validation is powerful, because being acknowledged as "right" in the moment is the only form of shelter we have from the *Dark Forest Moment* with others, and survival itself with reality. *Validation is safety.* Validation *is shelter.* To be validated by someone of power and great authority amplifies these feelings even more.

But the Being did not *actually* time travel and validate Charlie's story. Charlie *self*-validated his own assumed story built from Alice's word-shells, which were themselves self-validated by Alice to make so much sense they were worthy of writing down. However, the word-shells she wrote down were not even her own, but actually chosen and filled, assumedly, by the Being and its context created through *its own* firsthand experiences. Hence *three* points, not two.

If Charlie had never read that passage in Alice's diary, he may have never created the Anti-Purple People group. Now though, with his imagined and assumed support of the Being, Charlie feels justified and validated in his fear, and perhaps even converted these feelings into a pre-justified *Dark Forest Moment* decisions of superiority and domination, which lead into hate and violence.

With this in memory, Charlie justifies actions that bring this imagined-to-be-valid-story-from-beyond-all-living-memory back into modern day reality. He does this by creating signs that say:

> "Those with purple tint in their skin are dumb, and want to
>   take your children."
> –Being, Founder of Bobville

Is Charlie wrong? *Not to Charlie!* Charlie is doing all of this because he believes *it's the right thing to do.* Charlie is fearful, and wants to overcome that fear. He believes the Being felt the same exact fear. This validation gives him enough story voltage to make overcoming this fear his highest priority.

But he has declared a timeless certainty, which becomes its own anchor point and disconnects itself from Charlie – but Charlie will still consider it connected to him.

Every person that reads Charlie's words on this sign may possibly *make the exact same jump.* They read the words on the sign; build a story that makes sense from their own experiences tied to those word-shells, and may now believe the

Being validates their newfound fear. *This Anti-Purple People group must be in line with the Being's beliefs, and that's a big deal.* Something that is at least four contexts removed from objective reality is now merging into objective reality.

This increase in story voltage puts these individuals on the event horizon of the black hole. If the shadow experience created by the sign makes enough sense, it merges into truth and creates enough story voltage to justify feelings of fear and anxiety. The Chain of Suspicion grows.

It's also possible the words don't quite make *enough* sense and so does not fully merge with truth, the story voltage stabilizes, and the individual remains unsure, unfearful, and unanxious - *but the story goes into memory anyway.*

Lastly, perhaps the story doesn't make sense at all, is rejected from merging, and even repels and disgusts the individual, because the story is *wrong* relative to his or her experiences tied to those word-shells.

Once the story is in memory anyway though, *which happened to everyone that read the sign,* all it takes is one bad firsthand experience or negative second-hand story about a purple person to validate the story on Charlie's sign, and make it *more true.*

A handful of people are pulled into the black hole by Charlie's sign, and justify dropping into his group meeting as the right thing to do. There, they all share stories of bad purple people. Timmy said a purple person cut him off in traffic. Veronica said she was verbally assaulted by a purple person at the local market. Charlie's belief is validated even more with each and every story. The gravity of this black hole "truth" is increasing, and everyone else in this meeting becomes the inner circle around him and his black hole truth-story.

Membership continues to grow, and eventually a man shows up and claims his wife was *murdered* by a purple person. Everyone gasps. He explains she was abducted by a gang of purple people while walking in the park, and one of these purple people kept her locked up and fed her only ice cream until she died.

Mental boundaries warp and blur now. The merge with the singularity is approaching.

Each person in this group *actually believes something different.* Some are building their belief on one or two firsthand experiences with purple individuals. Some in this group have never even met a purple person firsthand, so their truth-stories are built from entirely secondhand stories and shadow experiences – from the words of others alone.

When they all get together and tell their stories, the words they each hear from others always validate their own beliefs and assumptions – *because the meaning of the words they are hearing or reading are already validated from their own beliefs and assumptions – a loop.* These individuals got together in the first place because of a specific anchor story about purple people that used specific words, and everyone present is now validating each other's feeling of fear to be justified based on this one anchor story – and so the fear *itself* becomes truth.

Purple people are scary.

This is now validating itself as truth through firsthand experience with group over and over again. And each time it is validated, it slowly moves inward toward the subconscious for each and every person.

Hate and fear are on their way to becoming *automatic and pre-justified* in the minds of everyone in this group. And none of this is centered around firsthand experience with purple people. *None of it.*

At this point, everything that has happened is due to Charlie experiencing fear, and him imagining a "ghost" of the Being feeling the same fear and blaming purple people through firsthand experience, which provided Charlie an "object" to blame for his feeling.

Now he is also imagining that all the other stories from his group, *none* of which he has experienced firsthand and *none* of which he has the context to understand without assumption, also validate his truth-story. This is creating a growing need for shelter from this growing fear and anxiety, and the need for action to *capture this end* of this shelter story arc is rising in priority.

How many arc flashes can you see so far?

The ever-increasing story voltage responsible for justifying Charlie's actions toward hating purple people and doing something about it, is actually being amplified in a kind of self-feeding loop.

Second Charlie connects the experience of the incoming words to First Charlie, and First Charlie recognizes the incoming current signatures as familiar, which results in a rising voltage that justifies more and more of his movements. Charlie believes he shares a truth with the Being and everyone that agrees with him. Interestingly enough, Alice is now *completely skipped over and unmentioned. The Being* agrees with *him.* They *all* agree with *him.*

Imbalance does not actually leap through time. Imbalance *is responsible for* time. Memories of experienced imbalances that are long gone can create loops in movements and truth when they are written down – they can suddenly be recreated through language and fear if they match up to the present moment in a way that presents as *self-evident*.

Charlie is actually justifying movements based on a voltage created entirely from his own context and someone else's words. A loop – a *cycle*.

He doesn't realize the Being never actually said these words.

He doesn't realize that Alice actually knew a *different* Being than the founder he is imagining.

He doesn't realize that this other Being had color blindness and saw green skin as purple.

He doesn't realize that Alice was actually green-skinned, and this colorblind Being was *sarcastically* poking fun at Alice by saying she "isn't smart and will steal anything, even children," which is the exact *opposite* of anything Alice would ever do.

She laughed at this joke from colorblind Being so much, she wrote it down to re-read and smile at for years to come.

The membership of the Anti Purple People grows though the years, and due to increasing protest and violence wherever they have a meeting, the group votes to rename themselves to the Anti-Ice Cream League, because they have each told themselves the story that "purple people love ice cream" is true and think this is a sly way to keep things going.

One of the Anti-Ice Cream League's members is very politically connected, and makes it into the newly elected king's throne room to lobby him about the dangers and problems of ice cream running rampant in the kingdom.

From here, the doors of *The Ark* for *every single citizen* have started to collapse inward toward this Charlie-created black hole. The overall number of possible stories in the kingdom are collapsing, because through arc flashes over hundreds if not millions of other story systems, *one story* imagined by *one story system* built by *nothing but assumption* is making sense to more and more people. The story becomes the basis of justification to act *against* some yet-to-happen *collective* force threatening a *Dark Forest Moment*. This threat is not caused by the anti-purple people group – no, their belief feels justified to them. It is actually caused by the assumption of shared belief in the *other* system – the purple people themselves.

Urgency grows. Shelter is needed. Shelter from the fear of this *other's* collective future actions. There is no time to consider "*their*" stories. If "this" story is the reason "that" is happening, then "this" action must be justified to prevent "that" justification. The Chain of Suspicion.

The hurried need for shelter from the unpredictable results in anxiety. This voltage gives birth to a Second *Us*, created by the First *Us* to try to make the unpredictable "actions" of "the others" predictable, and therefore expected and controllable.

Second *Us* then repeats the previous process *internally* – validating the group's fear to the point of spiraling into a singularity. And not a single person in this black hole is choosing to interact two-ways with "the other" in a way that might validate or invalidate the justifications for the actions believed and predicted to be coming.

All of this collapses to a timeless certainty: *It is Us versus Them. Everyone is either with Us or with Them. There is only Us and Them.*

One day, the *Dark Forest Moment* appears. *Us* ends up face-to-face with *Them*. If we assume the same process happened to *Them* – caused by a rejection of the story assumed to be believed by *Us*, then the imbalance between these two points is enormous. However, the Shelter Dance of stability cannot begin, because any stories of equality, fairness, or even the intents of Them (or Us if looking from the side of Them) are preloaded with already-shared and in-group validated stories.

This means the *Dark Forest Moment* has already happened. It happened for each person the moment that person decided the whole truth was known about not only another person he or she had never met, but about an entire *collective* of others – that the whole truth is known about *Them*.

They want to destroy us. They are terrible people. It is us versus them.

The choice of domination and violence is already made, built on the belief *the other* collective has *already chosen* the same path of domination and violence.

What miracle would have to occur for *Us* to make the choice of expanding time by suddenly recognizing "self" in *Them*? Or for *Them* to see self in *Us*? For the truth-story told by Second Us and Second Them (to protect First Us and First Them) to destroy itself?

The context for any words "They" might use is already built - not by Them, but by Second Us. The same is happening in the opposite direction. The two groups don't even speak the same language anymore, even though they might use the same word-shells. The experiences – the contexts - that fills word shells are completely different. The invisible mental boundaries are huge and impenetrable - nothing can get across the gap from one side to the other, and no one even realizes they are there – they are arc-flashed over.

This is now Schrodinger's *reality*, and an abyss has been created between two churning black holes of "truth" that are about to connect with the simple act of "dropping a wrench".

For those in these groups looking at the other face-to-face, there are too many of *the other* to even think about taking the time to stop and change paths. Besides, there is a near-solid boundary of people behind them that pre-justified the path of domination and violence, so running is not possible. The *Dark Forest Moment* tumbles into the only path that can finally stop this runaway process of self-validating fear of *the other* - the choice *to destroy* finally becomes the path of least resistance.

How does one turn off the instinctual need for shelter from justifying anything but violence when facing huge numbers of "enemies" with time between them completely collapsed?

As I see it, *you can't*.

At this point, it is too late.

Both sides have already projected their stories onto the other, and are now acting out that reality in a self-fulfilling prophecy that requires violence to overcome. The *Other* didn't even get a say. Certainties built on assumptions and validated by collectives have hijacked the entire situation.

The dance quickly decays and the enormous voltage begins the collapse toward balance when the first gunshot rings out. The wrench has just fallen across the terminal blocks. The clips holding the boundaries closed between Us and Them are taken off.

Chaos erupts at tremendous speed, as the most brutal of all the *Dark Forest Moment* choices is made time and time again in rapid-fire – pre-validated as the *right* choice, the *only* choice - that can be made to survive.

As the wrench starts to blow apart from the tremendous currents colliding, blood begins to flow in the straw-path streets.

A single man had a fear he refused to question or seek to understand. He let the words of a ghost, talking about another ghost (that wasn't even the same ghost he imagined it to be), validate his fear.

And now the black holes and those that resist it to collect into an oppositely-charged black hole churn and collapse into each other, connected by the fear that has pushed its way into the subconscious of an entire culture.

The children of those killed will justify new stories of hate and fear. Is this wrong? Not to them.

The friends and families of those slain will most likely take up arms to avenge their fallen friends and relatives. To die is *forever* unfair. Forever imbalanced. The entire human world begins to fall into the darkness of the abyss.

People – whole beautiful people with intact story systems capable of imagining and sharing millions of wonderful stories – willingly run into Hell itself because the Hydra whispered certainties of others' "truths" until it was believed.

People willing to die to overcome what they fear.

People willing to die to shelter and protect what and who they love.

People willing to die to re-establish fairness.

In my opinion, this is the saddest story ever told, because it has repeated itself over and over at every scale imaginable since the dawn of man.

In my lifetime, I've watched this happen many times.

To keep it going, in my opinion, is madness.

132

The Collapse of Color

When does the violence end?

How does it end?

As I see it, there are only two ways it has ever ended.

The first way is by something forcibly breaking the massive current flowing through the straw – something parting the seas of people collapsed into chaotic violence – which would have to be overwhelming to all the collective story systems involved.

Already down the *Dark Forest Moment* entanglement path of violence, these story systems have already collapsed into black-and-white survival thinking, and are now automatically (subconsciously) trying to regain a certainty of survival through the domination over the other. A "circuit breaker" intervention would have to shift the focus of survival entirely. In my opinion, this would be something like an alien invasion, cosmic disaster, or divine intervention by a higher power. In other words, no one is going to be capable of stepping out onto the battlefield and simply yell for violence to end to actually make it end. The battlefield is The Process collapsing toward balance.

Another way this ends is the two polarized groups finally collapse to near balance, and a new common ground state - or shape - is reached. As I see it, this happens when enough individuals from each side finally realize they are both fighting for the same reason and refuse to continue, or when no one remains alive willing to choose violence yet again in the *Dark Forest Moment*.

Running away from each other does not stop The Process, it only delays the inevitable *Dark Forest Moment*'s return.

Dominating until one side remains does not stop The Process either.

Let's go with this choice for a quick minute, though. I've heard "history is written by the victor" many times before, but here's what I see to be a problem with that saying: The justifications that led to war in the first place were never something that could be easily noticed – the objective truth about *Them* was contained in all the hidden story systems and shared cultural context of *Them* all along. This means if the war is won and Us wins over Them, the Chain of Suspicion does not disappear.

Is one of Us a sympathizer with Them? Is one of us an Other? Does my neighbor Bob secretly like purple people? He certainly likes ice cream. Purple people like ice cream, too. Alice has a cousin that was purple, so she is probably one of Them or is working for Them.

This fear and anxiety will lead straight back to new *Dark Forest Moments* over and over again *within* the group of "Us" (which doesn't even really exist because no two people actually believe the same thing). The Hydra known as Second Us travels around in shared language and social invalidation, *continuing* to build stories out of fear of *the other*, whose stories we do not know, cannot be seen, and can never be sure they *no longer* exist.

This leads to the anchor of reality and truth being floated to a new normal - to fear the unfamiliar. Fear of outsiders becomes part of the culture. The children are taught to fear and stay away from purple people. This is justified to be right because the children "must be protected". Many of "Us" were killed at the hands of purple people in the Sugar War. Adults feel they must capture the ends the children cannot, because it is assumed the "children depend on adults" to do so.

And so the stories of monsters and demons and ghost and goblins begin to appear in the children's stories. *There are scary things in the dark, waiting to get you. Purple scary things.* Not people, not equals like you and me - *things. Monsters.*

The entire culture of Us becomes centered around validated fear of the dark and the unknown. Of things that cannot be seen. Of always needed shelter. *Anyone could be a purple person.*

Purple people supposedly liked to walk the streets at night. Now you are likely to get a flashlight in your face from authorities if you do the same. Gun sales go through the roof as the purple people in the invisible collective cultural belief *could attack* at any time. The black hole continues to churn. The need for shelter is completely out of control.

In my opinion, there is only one shelter from other people - from other hidden story systems that are just as complex, fast, and amazing as yours. It is not seeking to control them. It is not hiding from them. It is not dominating them or submitting to them. It's recognizing there is no "them" at all, and letting your own fear melt away.

No two people believe the same thing about *anything*. And no one knows the whole story about *anything*. Anything otherwise is a massive assumption of horrific magnitude.

Forget trying to assume what an entire group of story systems collectively believe. It's nonsense and only leads to the Chain of Suspicion. Stability is found through the recognition of self in another. Try to start at equality in the face of any *Dark Forest Moment*. When the fear of "them" dies, the recognition of "self" can finally happen. The only way to conquer fear, and keep it at bay forever, is to continue to expand away using the middle path toward trust.

I once ask a social media group of more than 700,000 individuals "If you could magically reshape society, how would you change it?" Nearly every single comment out of hundreds was positive, wished for peace, and for every person to have what they need in abundance. *There is evidence for hope.*

PART SIX – THE SHADOW MONSTER REVEALED

Fear is often built on assumptions about the darkness.

Rearrange the shattered pieces of the looking glass to light up everything.

See the monster lurking in the shadows for yourself.

133

How the Beast is Fed

Find the Hydra heads in your mind – the stories that arc-flash over entire human beings just like you and compact these individuals down into a single word, validated by some other *Keeper of Truth* you may have internalized.

Find the stories that trigger a subconscious feeling of fear or anger that are attached to no firsthand experience for you at all. Find the words that spike your blood pressure or make your stomach drop when they appear and you can't quite understand *why*. Find the words that tell you a whole story and yet tell you nothing at all.

They are feeding the Hydra. They are creating black holes.

Words like Government. Deep State. Democrats. Republicans. Election security. Social media. Supreme Court. MAGA. Woke-ism. Liberals. Conservatives. Fox News. CNN. Stupid people. Millennials. Boomers. People that think they are smart. Scientists. Transgenders and gays and drag queens. Black people. White people. Illegal immigrants. Guns. Gun control. Gun promotion. Mental illness. Mental health. Critical Race Theory. School indoctrination. Lack of teacher support. Money. Science. Religion. Atheism. Police. Broken healthcare. The ultra-rich. Unions. Black Lives Matter. Antifa. Loss of morality. Right. Wrong.

Perhaps you can think of even more than this – or perhaps the words that create these feelings have changed since I wrote this. Either way, you can hopefully now recognize what these words can do.

This Floating Anchor Problem is still very much present. It is still cycling, embedded in the ways language hijacks the subconscious brain into survival mode by connecting these ideas to fear and anxiety. The ones that stick out ignite your need for shelter. They ignite a *Dark Forest Moment* completely internal to yourself, but in-group validation moves the Hydra into objective reality.

In other words: The Hydra has already escaped the abyss and grown to enormous size. Fortunately, it turns out word-shells are actually the best weapon to slay the monster. The hard part is not slaying it though, it's seeing this invisible subconscious monster at all, because it's made of the same stuff that makes words, and it moves *very* fast.

In my opinion, word-based stories work because the movements of boundaries and between boundaries within our reality seem to be "felt" or "experienced" by everything, including you and me. The wind seems to be experienced by leaves on the tree the same as the feathers on the bird and the hair on your body (assuming you have hair). The bees experience the bear the same as they experience the hornet and the flower, and each of these can experience you, and you them. Each step you take makes vibration that is experienced by everything capable of sensing that vibration, and your weight is experienced by everything underneath your feet in the same way. Every "thing" in our reality, in the year 2024 when I wrote this, seems to share a kind of physical connection, and is able to experience other "things".

As I write these words, and perhaps still as you read them, there are almost 8 billion human beings on the planet Earth. Each one of them is having a wordless and gapless firsthand experience in what seems to be a connected system of experience.

Each person's story system is building a new story with every ridiculously fast story cycle – and these stories are *not* converted into words. They are instead just movements transformed into electrical sensory currents (Step 1) that are experienced and moved into memory (Step 2) – assuming what is experienced is unique enough to be worth remembering, of course.

In a world *without* language, communication from reality into the human mind is pretty much always contained and isolated in the story system. The better your story system's predictions are using the language of atomic movement, the more confident you will be with what to expect from all the moving boundaries around you.

When it comes to moving boundaries that are *other humans* and spoken and written language do not exist, communication is first established, in my opinion, through the recognition of self in the other. This attempts to "read" feelings and emotions by recognizing familiar movements in the other you tie back to yourself. This will be prone to error, as these can be cultural, and cultures sometimes have radically different expressions and movements tied to emotions.

When it comes to intentional communication through the movement of boundaries, the first level is gesture-based. This is telling stories by trying to "act out" the movements in the world around you both without sound or words. As mentioned before, this is limited and slow, but you can communicate quite a lot this way if needed. This only works, in my opinion, because the "other self" recognizes your same movements as familiar to movements he or she makes, which tells a "story" you or the other know and recognize as familiar.

Spoken words, first built through sensing the same boundary from different perspectives, such as the situations in *The World of Two*, create the ability to share those points of view within seconds. You can both experience the "same" boundary movement and use words to describe what you are sensing to develop a better idea of what the thing is in the moment. Here's a terrible drawing of what I mean by this:

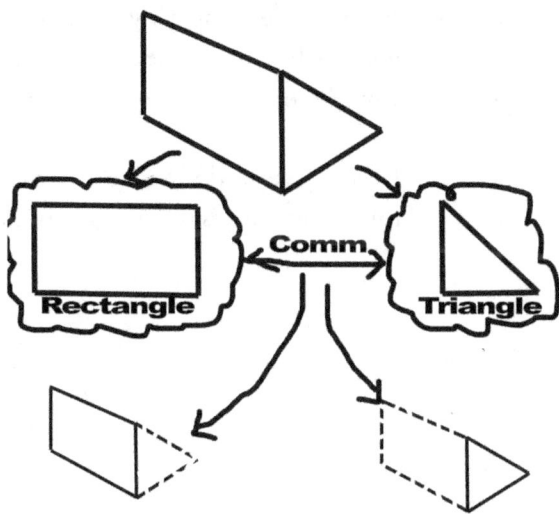

As long as you and the Being in *The World of Two* always experience everything together and focus your senses on the same boundaries, your shared language will be the strongest it can possibly be. Misunderstandings and assumptions should be relatively little. You can, therefore, fill in each other's gaps in experience using language, and quickly validate words, growing trust in their meaning.

To experience what I mean by this, just look around. If there is someone nearby to you, imagine picking out an object close to you both, and using words to describe it to each other (assuming you speak the same language and both have the ability to communicate in words). This "sharing of Step 1" controls the Exploding Cat Problem pretty well, because you can keep communicating to each other to work out where exactly any disagreement is happening, between word meanings and firsthand experience.

For example, my oldest daughter and I argue regularly about the color of an object if it is close to the color of orange or pink. I apparently tend to see these objects as more of an orange, and she apparently tends to see them as more of a pink. We can argue all day long, but in the end we both can agree that we must just see the electrical current signatures of the light from that object differently, or we have different experiences anchored to the word-sounds "orange" and "pink".

And this brings us to the incredible complexity that appears in language as soon you communicate with someone else about experiences you are *not sharing firsthand together* in the moment. These differences, which are gaps, feed the monster in the abyss when they are completely unnoticed or considered. In other words, if we could somehow become aware of how to spot these gaps, we could starve the beast, and maybe transform it back into our powerful friend and guardian.

This is exactly what I hope we can do together, because the only thing between us and the exit of this rabbit hole, is the monster hiding in the shadows of these gaps.

134

The Magic Story Box

To help expose the sheer scale of our secondhand story gap problem that hides the beast from the abyss, I am going to limit the constantly-moving-word-meaning problem embedded in the living thing called language. To do this, we'll only consider a tiny sliver of time - approximately five seconds.

Imagine you are asked to submit a written question into a magical box. I'll call this box the Magic Story Box. As you write the following question "What is your story over the last five seconds?" everyone in the world experiences the 5 seconds you are asking him or her to write about.

As soon as you put your question in the box, it is magically distributed to every single human on the planet capable of communicating an answer. Let's assume this to be 5 billion people (based on an assumption of 8 billion total people, with 3 billion of those people unable to communicate because they are small children, extremely elderly, or some other reason).

Each of these 5 billion communicating people receives your question *privately*. They then respond in complete privacy, and that response is translated (if needed) into your language by the magical story box message system. Every response given is automatically compiled into a digital document or spreadsheet for you - which now contains *5 billion stories*.

Do you think the words used in any two stories out of this 5 billion will be identical from beginning to end?

I suspect this is possible to happen given the short length of time involved. To deal with this (since no two people had the same experience), just imagine doing a second round of this magic story box process only for those perfectly matching stories. These individuals are asked to tell the same story about the same five seconds, but to choose different words.

Do you think their answers will continue to be identical a *second* time?

I'd say that while this is statistically possible – it is very unlikely. I think any matching stories in the first round were most likely caused by the limited number of words available to describe a wordless-but-unique experience and will not repeat themselves in the second round.

So, assume this is all done, and you now have 5 billion unique stories.

You have a huge problem now: How long do you think it will take you to read and analyze *5 billion* stories?

To give you a sense of scale, this book is not even close to *one billion* words. It's not even half of one *million* words. If everyone that responded submitted only a *one-word* story, then your spreadsheet would have about the same number of words as 17,000 copies of this book.

In other words, there are far more truth-stories created by humanity in just a few seconds than you have even the *slightest chance* of reading, even you had the entire rest of your life to do it.

Consider the amount of information though: If you could somehow magically read them all at the same time, you will have an approximately complete picture of all of humanity (well, the part that can communicate anyway) for that few seconds in objective time.

You will have a story from every perspective of a war that might be in progress. You will have a story from every perspective of a sporting event under way. Every meal being eaten. Every drink being taken. Every farmer in every field. Every laborer doing every job imaginable. Every parent meeting their new baby. Every family member saying goodbye. You will have the stories from every perspective of every human being all at the same time all around the Earth, for just that 5 seconds of time.

Yet it's nearly *certain* you will struggle to understand some of these stories. Some won't make sense to you, because their experience looks nothing like your own, or any other story you have ever heard. Some stories might use unfamiliar words, because they are so deep into a human specialty such as mathematics, physics, art, programming, or something else, that you can't even remotely understand what story is being told. These stories might as well say: "Ptykits sruvs were everywhere" or ".Salad to burning)while(loves and giRaffe banana flow a eating candle,"

Regardless, all of these come together to make a specific shape. You are already familiar with this shape on a different scale.

135

The Island of Misfit Stories

Let's bring back the mental freedom/action spiky-donut map from *Single File World*. Instead of being one human lifespan though, let's make it represent the entirety of all human lifespans to ever exist.

The outer ring of this map would represent all the movements made in free time by all of humanity. I believe this would result in specialty spikes so long that the inner circle, or "core" - which is the minimum time spent on movements of capturing food, water, and shelter - would probably be hard to see. However all spikes, and their sub-spikes all ultimately grow outward from this one inner ring.

The spikes are centered this way, because no matter where you are or what spike you are in, as an unseen dot helping to form this "probability" shape, you will always return to the inner ring to perform the mandatory actions needed to survive.

For my purposes, I will shrink the length and number of the spikes in my drawings.

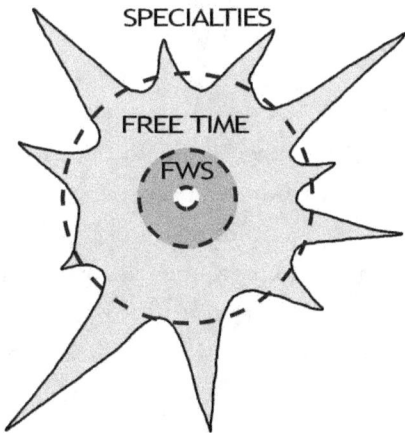

How much of your *free time* is spent on food-and-water-centric movements? I can't know who you are, but based on what I believe to be all the possibilities, it might range from "probably not much" to "80% of my life". I, for example, do not spend much of my time capturing-the-end of my hunger arc. I spend maybe an hour or two a day performing the movements related to food and eating, tops.

Instead, I spend most of my time moving among the spikes of human specialty.

If language were mapped out like this (because it is its own living thing moving through time with us), I think it would start out like our movements and create a simple circular word-cloud centered around survival – food, water, and shelter. Maybe this cloud only contained "grup" to start.

Through all of human movement and time though, language has grown to become its own enormous spiky donut. However, if the spiky donut of language was placed on top of the spiky donut of humanity's mental freedom/action map, I believe the language spikes would have significantly shorter spikes.

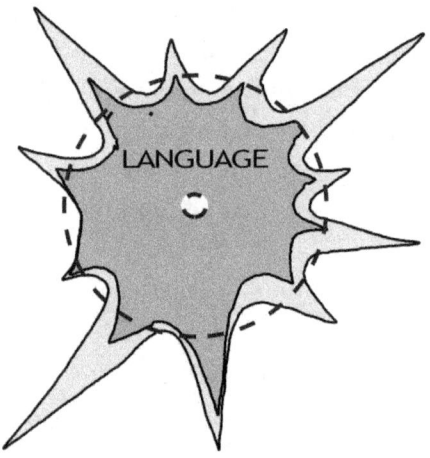

This limited reach of language problem does not and cannot stop the lengthening of specialty spikes of human experience, because those individuals working together at the extreme tip of each spike still need shared language to communicate as they continue to try to grow their specialty spike into the void. Due to necessity, language *is forced* to reach the tip of the mind/body spike.

As I see it, there are two problems with this forced-growth: First, the amount of words available is limited. Second, the speed at which new words can be added is limited.

The result is an imbalance. The speed of adding new words is not able to keep up with the speed of wordless progress, experimentation, and growth of human experiences into the void. The only limit to the speed of experience into the void is the willingness of individuals to step through the doorway into darkness.

In other words: The rate of expansion of human experience *exceeds* the rate of expansion of language used to communicate those experiences to each other.

The "fix" for this is using culturally-familiar (old) stories to understand new experiences, because the storyteller can see similarities in movements between the old story and new one that doesn't yet have new words to describe it.

A small grouping of words from the old story of movements are essentially copied out of the main cloud, and then reused, or *pasted*, as a kind of meme in an *entirely different and specialized context* to help others sharing different perspectives of something new understand each other. When this is shown on top of the spiky donut of experience, this copy/paste creates "story islands." As you look at your five billion story document these islands might look like familiar words to you, but they are not actually anchoring to the same wordless experiences in reality.

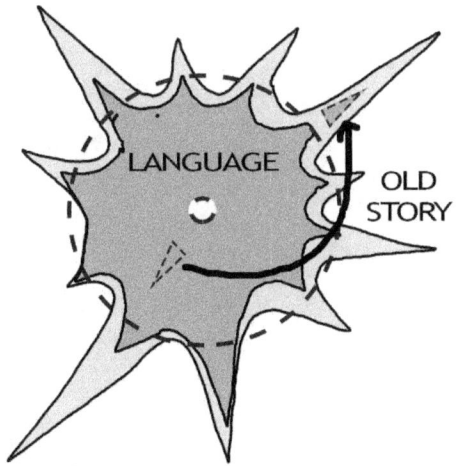

A great example of this, which makes its way into every language island one way or another is "electricity". It can be used to describe an exciting performance, or why your hair stands on end when rubbed with a balloon. It can describe biological functions of a specific eel, the sensation you experience if you pick up that eel, or the deepest fields of power distribution design or quantum theory. Without additional context, you wouldn't be able to know what any one person meant by "electricity" if it appeared in any of those 5 billion stories. Some other problematic "island" words in society today as I write this are "theory" and "law."

In the 5-billion story document, the words from specialty spikes you are familiar with would probably make sense to you – assuming there is enough additional context in the story for you to *assume* what the person was doing with perfect accuracy – which is extremely unlikely. As for the rest, the stories will look familiar and use familiar words, but make no sense to you. It will *feel* like they *should*, though. You recognize the words after all, but they just…don't make sense (because you are missing the context needed). The individuals in the specialty spike that are using these "story islands" in their "shared" context of experience both recognize them and make sense from them.

This means, as I see it, familiar stories appearing deep in a specialty are actually an entirely different language. They look to be your language, but are actually on a different "layer". This layer is not superior or inferior to the layer you understand, it only refers a *different* part of reality. This different-layers-of-the-same-language using the same word and stories adds new dimensions to the overall shape of stories.

136

Metaphors Rising

Your firsthand sensory experiences, which are really electrical currents coming from your eyes, nose, ears, skin, and tongue, are *metaphors* for movements taking place in objective reality outside of your senses. This means your firsthand experiences are one level *above* what we perceive as objective reality.

Objective reality is a metaphor, because all objects (assumedly) are composed of countless numbers of tiny things undergoing countless instances of The Process at countless scales up and down beyond our ability to trace it. Yet these processes all still seem to group-up and appear as objects to us. This makes every bit of objective reality, as we understand it, simply a metaphor for whatever it *actually* is.

If we were to stack up layers like we did for the past two chapters, "objective reality" would be lowest level, and it would be enormous – however big the universe and reality might be.

This shifts your firsthand experiences up one level. The true foundational level is objective reality. The foundation of you is your electric-current-metaphor-converted wordless sensory experiences of movement in the true foundational level. You can go no deeper than the second level, because you are always forced to use your sensory windows.

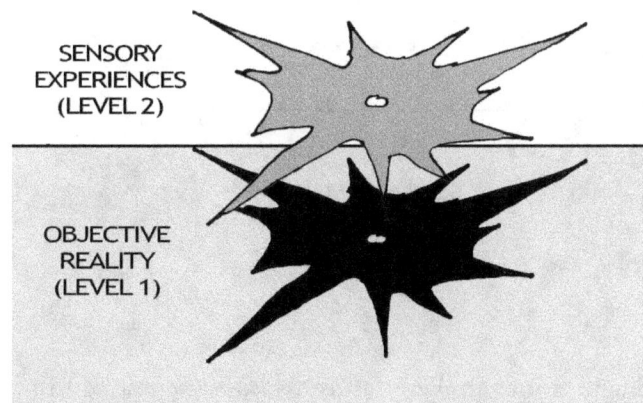

Your words, which are chosen and filled using specific remembered pieces of those sensory currents, are metaphors for your *memories* of Level Two sensory experience. This means words are a third-level of metaphor, floating above your firsthand experience.

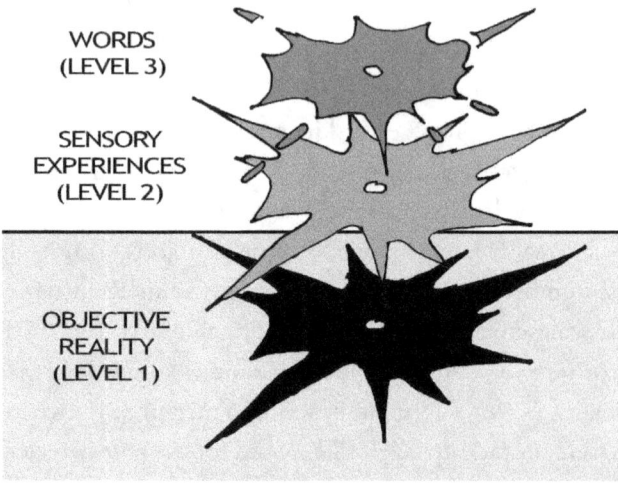

However, now this metaphor-story puts a twist in the plot: Your word-met-aphors are often connected to your experiences *by someone else*. This means your words are largely *transferred* Level Three metaphors built from *someone else's* experiences (Level Two) with objective reality (Level One).

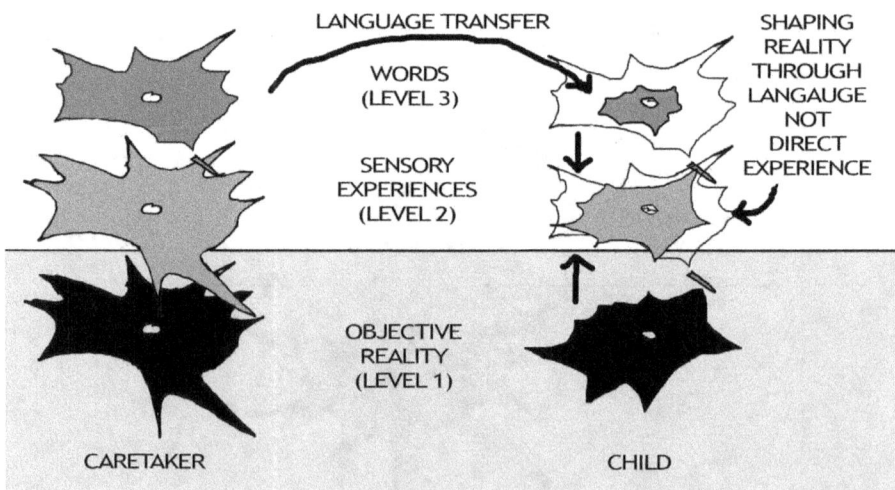

Secondhand stories are Level Three (words) attempting to transfer Level Two metaphors (experiences), which drag *along the foundational level* (objective reality, Level One) between two *separate* story systems. This transfer *always* has limited success; because no two people experience reality exactly the same (unique wacky windows) to understand words the same. When it does have success, it often reshapes unique experiences of the consumer to match the storyteller's perceptions of reality.

There are gaps everywhere: Between levels and systems and islands. When the two story systems encounter a secondhand story that doesn't make sense to *either* of them – which is often the case when discussing topics like quantum theory as I write this - a story that both story systems are familiar with is chosen and overlaid onto the nonsense story. This familiar story then provides another way to understand and communicate by offering word "units" to rearrange and adjust until the nonsense makes sense.

The familiar-and-shared story is a *fourth-level* metaphor to understand a *third-level* metaphor (secondhand story) that doesn't make sense (level two). If this works, then these two systems have used words to understand different words that each anchor to the same sensory experience with reality.

How many levels of metaphor exist? I'm not sure there is a limit to be honest, but I can tell you that "total breakdowns" are going to happen when it comes to all metaphor "transfers". This is because it's always possible a metaphor level is missing, or is being skipped over by an individual, which leads to a collapse in imagined understanding for the person sending or receiving the story.

For example, I personally notice The Process (the balloon-and-straw system metaphor) everywhere in my life at nearly every level. To me, The Process seems to be close to some kind of master metaphor, which is why I decided to make it the heart of this book. While it is possible you struggle with it, I've written most of this book using the assumption you won't.

If the metaphor of the balloons and straws doesn't make sense for you, this can cause *my* experience of reality, which I am trying to transfer, to collapse through a gap into the abyssal void. If this happens, the experience of reality being transferred by the metaphor will be unanchored and will instead find a new path of least resistance to a new anchor: which is almost always the storyteller – me.

You are familiar with this situation already. If any secondhand story claims what is right and wrong and true about reality, and you tell yourself to believe it (such as telling yourself to believe my story about balloons and straws without knowing from experience how the system will work) then the storyteller (me) potentially becomes a *Keeper of Truth* for you.

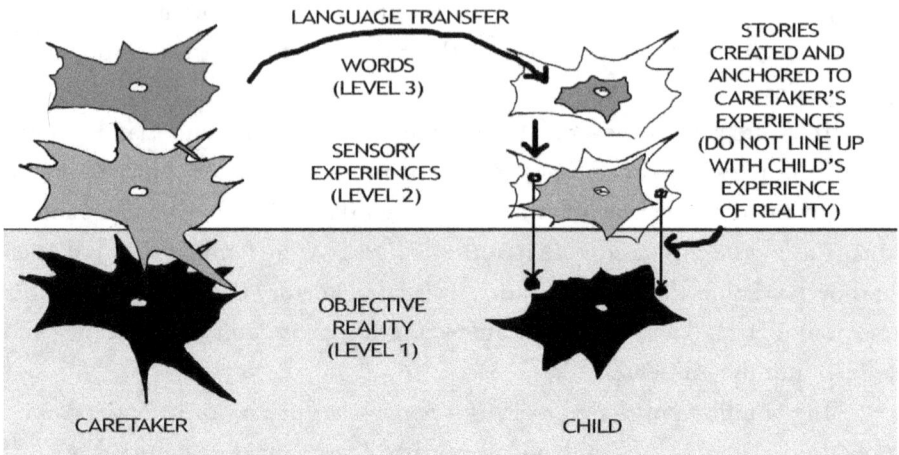

This, in my opinion, is why using the simplest words possible and the simplest story possible to tell stories of experience with reality itself is *always* justifiable as the right thing to do. To intentionally avoid doing this, or to intentionally use specialized words you know the consumer does not have, is to try to make yourself into a "truth island" – a Keeper of Truth.

This is *why* this book is so long, and *why* I believe it is *always* okay to ask questions to anyone trying to present you with a truth through language. There are gaps everywhere in the process of communication.

Will it ever be possible to eliminate them?

It some ways, I think the answer is *yes*.

137

Babeling

Return to the 5-billion story document created by the Magic Story Box. This magical device translated everyone's story to the same language – your language. But what if it didn't?

If auto-translation didn't happen in this experiment, I suspect you'd still be able to understand a few million of the 5 billion stories - enough to perhaps get a very rough idea of what each person was doing in that 5 seconds of time. You could maybe even understand a billion stories if you are lucky or speak multiple languages.

There would be stories *in your language* that don't make sense, though. Of these stories you can't understand, some would be stories you do not have the context needed to imagine that person's experience with reality using their third-level language metaphors, or the metaphors are islands copied and pasted on an experience *of specialty* you have no experience with.

You also have no way to understand foreign language responses. For example, I cannot even begin to read Mandarin, so all these stories in Mandarin would be immediately lost without translation.

This means, in the absence of translation, each language has its *own* spiky donut surrounded by its own metaphor islands made from words in the language core being *re-used* to communicate about experiences further out on the spike.

All of these spiky donuts would still line up at the same exact center point – no movement at all - death.

If we were to stack all of these language donuts on their center points (which makes sense because they are all created by humans), the result would be an immense tower of languages, all with slightly different spikes. This tower would appear to be exploding (expanding, really), with metaphor looking glass "shards" being slung out into the void.

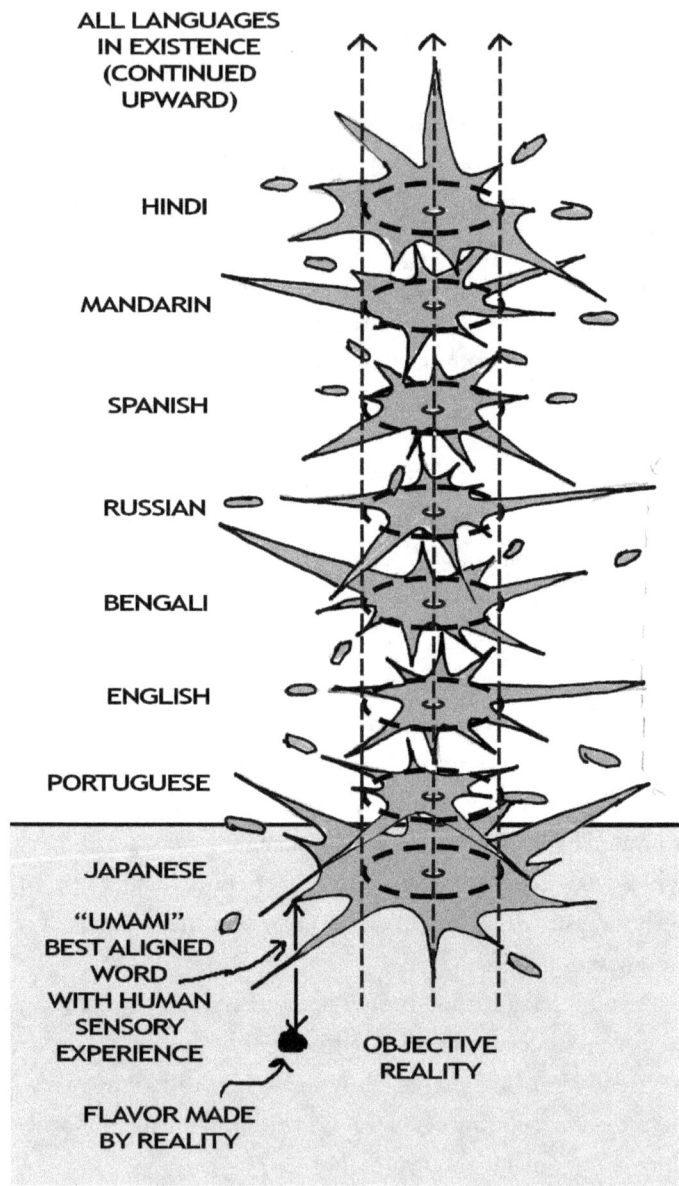

In this way, the division of languages and limited speed of growing them actually *slows down* the lengthening of all specialty spikes into the void around us through experience. Interestingly (to me), is this only slows us down when we are trying to work *together*.

You *as an individual* can charge into the void however fast you like. Examples of this might be injecting yourself with a medicine you've developed (or discovered and can't get testing for), launching yourself into space on a self-built rocket, or exploring an unknown cave.

If you want to do this with other people, however, a way is needed to communicate your experiences with reality in the void secondhand, so that others might be able to build predictions and expectations before experiencing it themselves, or to notice a pattern in the void, which results in an increase in the overall safety and survivability.

In this way, charging into the void alone is a lot like *Single File World*. But going in with two-way communication to others is closer to having a one-week overlap in Single File World *with an already-shared language to speed things along*.

The only reasons that language slows down progress into the void is that the person experiencing something new needs to anchor limited, existing, shared words to the new experience, or, the person has to create one new word at a time and share its meaning. Unfortunately, sharing understanding through shadow experience will be lost to most of society, because they do not speak the language, or they do not have the context needed to make sense of the story.

This division of language is not necessarily a bad thing, though. There is also an incredible strength in having different languages – because when multiple language donuts with different spiky shapes are stacked and aligned to match as closely as possible to *similar* sensory perceptions of a *shared* objective reality (for example, lining up all the "physics" language specialty spikes as closely as possible) it becomes apparent words in each language are "capturing" slightly different metaphors for *perceptions and "feel"* of the *same objective reality*.

This is where some words in a specific language can be extremely valuable, because they capture experiences that are not captured as well in any other language. One example of this is the Japanese word "umami." Valuable words like these become apparent when different language maps are aligned over similar experiences, which expose previously unnoticed *gaps* and limitations in the other spiky language donuts like, say, English.

Umami actually "makes sense" of a sensory perception in a way English words struggle to "capture". The result is umami entering the English language, which ultimately *increases* the resolution of *both* languages in their ability to describe our wordless gapless experiences with reality. It also starts to connect and merge two language donuts together.

Reconstructing-the-exploding-tower only works when languages are imagined to lay flat. In reality, there is another dimension involved...and this dimension is responsible for transforming this beautiful and complex system and tower of sounds and symbols into the monster that I believe is ultimately responsible for so much death, destruction, and atrocity.

138

Thank You For Calling Hell, Please Hold

Let's twist this Magic Story Box thought experiment from the previous chapters. The story box is now a *phone*. Imagine this phone is very basic. It doesn't have video calling, there is no text messaging or group calling, and no video conferencing either. The only feature the phone has is call-waiting.

Everyone else gets phones, which are basic phones as well – simplistic, one-on-one, two-way communication tools with call-waiting.

Oh, and congrats! You have just been put in charge of trying to establish a communication network on Earth. This network is a tool to be used toward establishing world peace. For the sake of removing the "noise" of different languages you do not know, assume these phones automatically and instantly translate everyone into your language.

You are relocated to in your new office at World Headquarters (located at the North Pole, of course). Your office has a desk, one red phone, and one really large globe. When the phone rings, as it soon will, a tiny light on the globe will turn on where the call is coming from. So you don't have to constantly walk around the globe, it will automatically turn the light so that you can see the location of the caller from your desk.

Assume this phone network gives you a way to directly communicate with 5 billion people. This also means there are 5 billion tiny lights on the surface of your globe (I am removing 3 billion people under the same assumptions as before - that they cannot communicate).

Every phone-to-phone connection has an automatic time cutoff. You get 10 seconds to talk with the person on the other end, and then you will be disconnected. Each phone is also only able to connect with yours *one time*.

There is major problem already in place: You can only ever focus on and merge with one story at a time. If you want to talk to every single person to get their story about that same five seconds of experience, you now have a queue of 5-billion callers to deal with. If each call is 10 seconds long, it's going to take you around 1,585 *years* to hear a story from every person. Assuming, of course, that you do not sleep, and do nothing else but talk or listen on the phone 24 hours a day, 7 days a week, for one-and-a-half millennia.

I think we can safely ignore the content of the calls right now, because we have a problem of practicality right off the bat. Instead let's start trying to figure out how to make 10-second phone calls possible at all.

What if you cut the number of phones in half? This would mean every person now talks to you on behalf of his or herself, and one other person.

This change gets you down to approximately 792 solid years of non-stop 10-second phone conversations with no time to sleep, and no doing anything else other than talk or listen on the phone. This is still not realistic.

How about one phone per three people? 526 years.

Four people? 396 years.

Five? 316 years.

Six? 263 years.

Seven? 226 years.

Eight? 198 years.

Nine? 174 years…

Fifty people per phone? *Under 32 years.*

FINALLY we have a somewhat realistic number, or a number that at least fits in your lifetime – assuming you actually do have 32 years remaining – which of course you can't know.

All you have to do is be on the phone for 32 years of time in your life, and you'll be able to get a 10 second story from every 50 humans on Earth about their experience in the same 5 seconds in time. Great!

But…what kind of decision regarding world peace can you make with every story about a measly 5 *seconds* 32 *years* after it happened? Does this really help you make any meaningful decision toward peace? How much gap in time is the "right" amount of gap between experience and story to make decisions that affect nearly every human on Earth? Is *a year* of gap better?

A year of 10-second phone calls to cover the same 5 seconds of experienced time across 5 billion individuals would require each phone to represent 1,585 people (again with no sleep or other activities for you in that year.)

But…what if a sudden disaster or violence happens six months into your year on the phone to learn about the same 5 seconds (that happened six months ago)? You will be blind in the gap - blind to know the event has even happened until the next round of phone calls *months* from now. No, this can't work. You *need* to get *much* faster.

What if you wanted to get all the 10-second stories about that same 5 seconds of experience in just *one day* of talking on the phone?

Each phone in the world would have to represent 578,703 people (assuming 5 billion total). And that is *still* talking on the phone non-stop for 24 hours.

You can't know when something might happen to disturb any peace or balance, and resolving any disturbance needs to happen as quickly as possible. You simply *can't* be on the phone all day every day to keep up this daily "refresh rate". Maintaining the peace (balance in the form of perceived fairness) is going to *have* to be a constant ongoing process.

As I see it, to get a good feel for fairness in the world at a speed with enough time to *do* anything about unfairness, you need at least half of your time reserved for actions.

Doing this, though, means that each phone call you receive at the World Peace Office will need to represent at least *one million* people. This will finally allow you to get the story of everyone once a day, as long as you didn't sleep or do anything else - and then spend the other half of that day correcting any imbalances.

To see the huge problem here, just imagine yourself as *anyone else in the world* in the one-million-people-per-phone situation.

Do you feel your beliefs about what is fair can be accurately communicated in a single 10-second phone call that happens twice a day at best, when that same call has to also communicate the feelings of fairness by 999,999 other people? You have just one *millionth* of a say about fairness in your particular phone group.

Now let's make a *real* mess: Do you think any peace or fairness is getting disturbed *within* your group of one million people? What size group would be needed for you to feel like your opinion is being counted fairly? Is 10 seconds really enough time to communicate your own feelings of fairness at all? If not, forget about actually understanding the fairness needs of one million people!

To be honest, as I see it, there can be *no groups at all* for this system to communicate fairness to actually work. Every single person needs his or her voice to be heard if things are fair. Otherwise, the *Dark Forest Moment* looms to ruin everything in a flash.

For everyone to have a voice, the information that exists in all 5 billion calls must somehow be collected and filtered until you receive only a handful of calls. To do this and it be anywhere close to fair, you are going to need communication *layers* between you and each individual.

How many layers? *Lots*. Maybe hundreds. Each individual's story must be collected and filtered up the layers of communication until it reaches you in the World Peace Office.

This filtering process is also guaranteed to have its own problems, too. Chances are high you've already experienced these problems firsthand on a much smaller and funnier scale:

This chain of communication resembles something called the "telephone game" that I played in grade school. If you aren't familiar with it, it can be pretty funny, but the idea is simple: All the kids get in a line, and then the teacher whispers a secret phrase into the kid's ear at one end of the line. The kid then has to whisper the same phrase to the next kid in line. This repeats until the message reaches the other end of the line. The kid at the end then shares the received message out loud for all to hear, usually with immediate laughter. The teacher tells everyone what the original message was, followed by more laughter.

Even if only 10 kids are involved in this game, there is still a *very* good chance the message that goes into one end of the line is going to be *completely* messed up when it arrives at the other. As I see it, this is due to many reasons, but here are three:

1. Sensory misunderstanding. This would be the whispered syllables arranging themselves into different words for the listener than they were in for the speaker. Perhaps the middle syllable of a word wasn't heard, or a syllable that was the end of one word was understood to be the beginning of another. I chalk this up to sensory "resolution" differences, which is not far removed from what happens to individuals with hearing damage listening to someone's normal speech.

2. Context shift. This means the same word meant something different to the listener's context than it did the speaker's context. This is the Exploding Cat/Great Context Problem.

3. Sabotage. Perhaps a child didn't hear the message or it just didn't make sense, and intentionally changed or maybe just made up new words to cover up his or her embarrassment of not understanding. Another way to look at this is fear of being "wrong" relative to "the truth" justified "lying" just pass along "something," or to make someone say something funny.

If we carry these children's telephone game problems over to the World Peace Office, this pretty much breaks down any chance of being *specific* at the individual level (highest story resolution), and hope this specific information makes it the office perfectly intact and interpreted exactly. After all, the person at the office, you, is an individual with the same one-point-resolution as everyone else.

The result of all of this ends up pretty bleak for world peace: Even if complete trust existed in this phone system, what exactly would someone in the position of representing one million stories say to someone at another level about fairness?

"600,000 think it's fair. 400,000 think things are unfair"

"Oh, what is unfair to the 400,000?"

"Food."

"What about the food?"

"Well, they see that the neighboring region is--- [click.10 seconds over. Next call].

This situation might seem hopeless. But there is always justification and evidence for hope, in my opinion.

There is a very interesting technology emerging as I write this that might be able to help us operate a communication network of this size far quicker than we ever could – A.I., which stands for artificial intelligence. However, there is an entirely *different* problem with A.I. that concerns me, beyond the problems above, and it's a problem we are currently fighting in other places with ourselves.

So instead of fantasizing about the details of how to make this system work with or without A.I., let's jump ahead to what would need to happen next *after* a successful telephone game from 5 billion individuals to one.

139

The Inner Sphere

Put yourself back into the World Peace Office.

Imagine that auto-rotating globe covered in tiny little lights that each represent one person again - we are going to make some changes to it.

This globe is now transparent. It is a large *clear* ball covered with 5 billion ultra-tiny lights on its surface.

It just so happens this shape merges nicely with a shape from your past in this book, which was never really two-dimensional in the first place: The mental freedom and bodily action map from *Single File World* before it ever transformed into a spiky donut.

The probability map had an inner ring representing your required-to-survive movements in time, which is to eat, drink, and seek immediate shelter. The outer ring represented free time when you can justify any other movements in time.

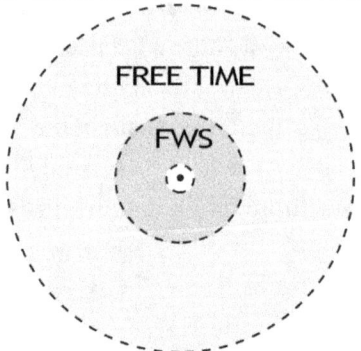

In a three-dimensional version of this map, these rings become *spheres*. A sphere just happens to be the exact shape of the clear globe in your office. So let's model how this would look in the laboratory of your imagination:

The outer layer of the globe is perfectly clear until about halfway to the center. This clear outer layer is "free time". This is where all the actions of verbal or written communication live. For example, talking on the phone is not a required action to survive – which means when it happens you are always in free time. To draw a boundary here – any action that does not exist at the beginning of *The World of You* appears in this outer layer.

This brings us to the inner layer, which in your globe is a solid sphere and cannot be seen through. *The World of You* begins entirely within the inner sphere, because it is purely the required movements toward food, water, and shelter to survive. The center of this darker sphere is hollow, because if you ever reach the center you have died, and no one alive knows the story or can tell the story from that location.

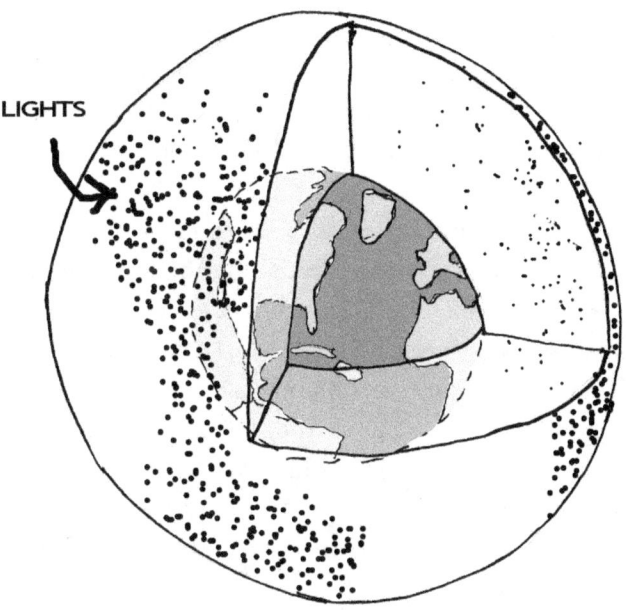

LIGHTS

You, as an individual, are one light in this model. As you watch from your office, all the lights are moving randomly from the outer shell into the inner sphere and then back out again. This up-and-down models everyone's story arcs into and out of free time.

I actually *can* predict approximately where your light would be right now. Reading this book is taking place in the outer sphere, because it is not related to capturing food, water, or shelter.

On the surface of the inner sphere is an outline of the land on Earth, and all the lights can move around side-to-side (physically moving around by car, walking, etc) as they also move up and down (need story arcs and mental freedom). When a light drops into the dark inner sphere (because he or she needs to find food, for example) the place where that person finds food will align with the location on the map where they actually ate the food. Then his or her light starts to rise again until another survival need story arc causes it to drop.

Make sense? It might be hard to keep track of in your imagination for now, but stick with me for a minute.

If you can, imagine 8 billion little lights on this clear globe of the world all moving in to the inner core and back out again to free time – all at different frequencies. So what exactly is the telephone game of World Peace trying to accomplish, and how does this globe map help?

Well, in my opinion, the Office's goal is to gather all the stories and use them to make *physical changes* to the inner sphere. These changes are made so that when an individual crosses into the inner sphere, he or she finds the food, water, and shelter needed to capture his or her end with the same confidence and expectations as everyone else - make the world *objectively* fair.

In other words, the World Peace Office is trying to *rearrange* the resources of food, water, and shelter to make the whole social planetary structure more stable. (For example, if you capture a hunger end without leaving your home, your light goes almost straight down, you eat, and then your light moves straight up again)

To put this all together, the telephone game is invisible messages from every light making their way to a single light (you) for the purpose of you ultimately rearranging the food, water, and shelters in such a way as to prevent unfairness when it comes to that very same food, water, and shelter. Every light needs to stay in the inner sphere for approximately the same length of time. No one stays inside for years while others stay inside for seconds, and vice versa.

And moving resources, which also appear as lights on the surface of the inner sphere, has just as many complications and scaling problems as communications in the outer sphere.

140

Broccoli Clone Apocalypse

Put yourself back in the World Peace Office again, and let's try to tackle just one survival need, and ignore the other two for now.

You've just managed tons of phone calls, put together every single story and even determined exact areas on the globe where fairness is out of balance related to one need – food - and are ready to make changes to food's location.

For the sake of keeping things as simple as possible, let's erase a ton of other variables: Every person is *exactly the same*. They all have the same appetite and wavelengths of their food story arcs. They are, in effect – *clones* when it comes to food.

In order to distribute food differently – *you first have to know where the food is right now*. This involves needing a *second* network of communication just as fast and even more involved than the first. You need to know *exactly* how much food exist and where that food is located. Of course, food decays and grows, so you need to have a *moving* system tracking everything food related. *Where is this food grown? How often is there a harvest? Where are the farms where it is harvested? Which foods grow where? How long can each food be stored? Where are the warehouses? The grain storage silos? The grocery stores?*

On top of all this, you also have to take note of how the people (lights on your globe) are distributed. Maybe one region in need of change is a giant city. Maybe another is a massive spread of land with each person miles apart. The food distribution needs of these two regions are radically different.

To make this now literally *impossible* to manage, merge the *Floating Anchor Problem* into this. Even if everyone is identical - exactly the same age, same height, same weight, gets hungry at the same rate and eats the same amount of food – fairness is ultimately *a perception*. This means, thanks to the mental boundary, fairness is defined by a *judgment* relative to other people.

In a world where all food is equally distributed, if the steak that Tim gets at the butcher shop weighs 4 ounces less than the one Jack gets next in line – is that *unfair*? If Tim's steak is extremely fatty and Jack's is lean – is *that* unfair? If Susan receives four heads of broccoli and found three worms in them, and Becky receives four heads that weigh *exactly the same* as Susan's but have zero worms – is *that* fair? Did Susan technically just get worms as part of her food distribution, or did Becky get extra broccoli?

The *Floating Anchor Problem* makes a living hell out of managing food fairness at every turn – even in the world of clones. Let's say Billy makes a huge salary, so he is able to purchase lots of food every week. Mike is making minimum wage at his job, so he is able to purchase only the cheapest food in small quantities. Is *this* fair?

Let's go *beyond* impossible: Replace all these clones with real people how they exist today. There are now babies, kids, teens, parents, childless adults, older adults, grandparents, and everything in between. There are skinny people, obese people, and everything in between. There are people with fast metabolisms and people with slow metabolisms, and everything in between. And none of them can *feel* another's hunger, thirst, or fear thanks to the mental boundary.

So what *is* fair?

Who gets to determine what fair means?

I suspect that most people imagine solving this hyper-complicated survival resource problem (food, water, shelter) by imagining a "fair" utopia where everyone has everything they need. When I ask a social media group with more than 600k members how they would magically change society, fulfilling all food, water, and shelter needs for everyone was a very common desire. To have all our survival needs handed to us would be something wonderful, no doubt. In my opinion, this being the most common answer speaks worlds about the love humans have for each other. However, it also neglects the raw power and horror the Floating Anchor Problem and relativity create through time.

Even if the world was magically rearranged to make this possible – everyone gets an endless supply of clean water, a magically filling food pantry, and a perfect shelter that was mobile and protected everyone from all harm, our story systems will just re-anchor to this new normal, and this utopia would quickly collapse into a dystopia.

I'll use an old saying to illustrate this:

Give a man a fish, and he will eat for a day.

Teach a man to fish, and he will eat for a lifetime.

Teach a child to fish, and that child may feed the world.

Now fast forward this fishing situation a few years:

Give a man a fish (handed to him by the now-adult child feeding the world), and he will eat for a day. Rinse and repeat this line until the now-adult child handing out fish dies, and humanity is brought to its knees because no one knows how to fish anymore.

Even if everyone was a clone, lived in the same exact house, ate the same exact quality and amount of food at every meal – the Floating Anchor Problem would *still* cause major imbalances.

Why does Tim get the house on the lake, and I get the house one row back?

"Why does Olivia look happy when I feel sad? Does she have something I don't have?! Does her steak or broccoli taste better than mine?!"

Imagining that maybe it does, which would be unfair, this person – who I will call Peggy - runs over and *steals* a broccoli floret off of Olivia's plate, and quickly eats it to find out if it was indeed better, and the reason Olivia seemed so happy. Peggy just captured an end – but the end *of what?*

Peggy captured the end of one of her need story arcs – not the need for food, but the need for *shelter*. Shelter from *her own fear* that someone else - in this case Olivia - was experiencing a happier or more fulfilled life, *and that caused Peggy's entire reality to become imbalanced*. To restore balance, Peggy decided to "capture" this fear to validate or invalidate her feelings created by the story she told herself.

As a consequence of stealing and eating this floret in a perfectly balanced world, the *entire world now also tips into imbalance*. Because she already ate all of her portion of brocooli, Peggy has eaten more broccoli in a single meal than anyone else in the world – and now there is no way to balance it *ever again* because the thing needed to balance it again – the broccoli floret – is *gone*.

You *could* split one of Peggy's broccoli florets into 7,999,999,999 pieces and distribute it to everyone else to keep it fair. Cutting one broccoli floret into nearly 8 billion perfectly equal pieces? *Good luck.*

Or maybe just make sure Olivia gets one of Peggy's florets next time. But can you see outside the imagined boundaries of this situation to find the incoming possible problem by rebalancing it this way?

What would happen if you, as just another citizen, had no idea any of this happened, but now witness the rebalancing? One person you don't know (Olivia) gets handed an *extra* broccoli, and you are missing the context for *why*. *Was someone just given an extra broccoli as a favor?*

The Chain of Suspicion begins.

We. Are. Consumers.

As I see this - the basis for ownership, is the same basis for territory, which is the same basis for capturing the end, which is the same basis for consumption

– *we want to survive*, and we subconsciously *believe* that whoever gets the most berries on the bush is going to get to keep living. In my opinion, competition with each other is born out of this fear and it seems to be deeply internalized into our minds. This is why mutually understood fairness through communication is a *critical* need. It carries the potential to help us let go of this fear before all of humanity becomes a violent arms race for survival over something as small as a single broccoli floret.

When you eat something, it crosses the boundary of your body, from outside to inside. The only way it's coming back across that boundary is if you throw it up, or if you digest it and it exits as urine and feces.

If I needed half of whatever you just ate because I have nothing at all, I now have to force you to throw it up and then eat it myself (not going to happen), or cause you severe physical harm by taking it out straight through your abdominal wall and eating it myself (also not going to happen). My inability to "retrieve" what you consume makes this situation *forever* unfair. You and I are back at Billy-and-the-cookie situation all over again. Susie just ate more than Billy and now everything is permanently imbalanced.

If I *appeal* to an authority because you ate my half – like some World Fairness Police Force or the World Peace Office – then I guess they could enforce things by ensuring that I get both halves of the *next* share while you get nothing. However, do you think you will honestly sit by and accept this as fair while experiencing hunger as I devour "your" half of the share?

My guess is you will not. I suspect you may become angry. You may start to resent the police, these rules, whoever made these rules, and the entire system you are apparently stuck in, and stare down those withholding your food with raw hatred in your eyes while complying with the "unfair" consequences for "stealing" what you were convinced was better tasting broccoli. This is even worse if you consider it is very likely you will tell yourself it actually *did* taste better.

Are feelings like this wrong? Was this ever about hunger or was this about fear and the Chain of Suspicion?

The invisible monster that hides in the shadows of the abyss has green eyes.

Language has helped move us more toward balance in the area of consumption, but our rate of consumption forces words to need tremendous speed. As a result we have been doing what amounts to driving a car with no brakes at high speed while wearing a blindfold.

141

Speeding While Vocabulated

Hopefully you have a pretty good feel for just how many human stories are created in this world in every second. If you consider each human to be a story balloon, there are 8 billion balloons interacting with each other using "straw" communication as I write this. These straws of verbal, visual, and tactile communication shoot out in every direction from each balloon, which creates an uncountable number of always-changing balloon-straw systems. Everyone that hears or sees another's actions consume that story and continue interacting with others in the world. The consumed stories may or may not ever recross the boundary in the form of communication, which creates an unknowable quantity of balloon systems acting out The Process when mapped through time.

Consider the enormous difficulty of the three-body problem in physics, which is simply trying to predict how three objects connected by gravity in space, such as stars, will interact and move relative to each other. Predicting humanity is like trying to solve an 8 billion-body problem, where each body is making invisible *choices* about his or her movements based on invisible stories no one else can see.

One of these 8-billion bodies is you, and you are not narrating every second of your life to anyone. No one except yourself might ever know the story of what you just did 5 seconds ago, and even you aren't aware of all your stories. You aren't going to be telling anyone about the virus molecules you bumped into today while walking down the street, or the gravitational waves that rippled through time and space during your lunch – because you can't notice them to include them.

So, if you want life to be fair in the very few stories that do matter to you as a human being because they are foundationally shared among everyone else – stories of food and hunger, water and thirst, shelter and fear – how *often* would you want to be able to tell your side of the story about these things? Is getting 10 seconds once a day enough time to tell someone your thoughts on your reality feeling fair? My guess is no - *not even close.*

The minimum speed limit of your life does not change, and nor does it care about how often or how fast you communicate. Communication is all done in free time – the outer shell of the sphere. Your need story arc needing the soonest capture forces your little light to always fall down toward the inner sphere at some point. The need for shelter, however, gets wrapped up with the Floating Anchor Problem and irreversible consumption to create one hell of a mess when trying to determine the right "resolution" to measure and enforce any sort of fairness through time.

We, as a world-sized society of human beings, simply cannot manage our survival resources for each and every one of the almost 8 billion people on this planet on an individual-by-individual basis. There is far too much change involved – which we call "stories".

Regardless of this seeming impossibility, communications and decisions about food, water, and shelter still have to happen fast, or there could be violence and riots related to food shortages, water contamination, and homelessness. If this happens, severe problems and imbalances tied to fear cascade through society in every direction.

Would you be okay if you walked into your local grocery store (assuming you have one) to discover all the shelves are empty? What would you do? In a way, this unexpected foodless situation throws you all the way back to the *World of You* - except there are plenty of people around you, perhaps 100, with the same problem. If someone then walks by you and the 100 others with a fat ham under his or her arm – things could get ugly right?

To speed things up and hold open the gaps between us to prevent violence related to unfairness, individual word-shells end up capturing more and more experiences within them. Words themselves become a kind of mental *shortcut*, because they *have to be* – everyone's minimum speed is set by their bodies, not their words. This causes each word-shell to grow into bigger and bigger *categories*, which I will call "buckets." The word-shells become word-*buckets*.

In the telephone-game-of-world-peace situation, one person had to speak on behalf of a million individuals once a day. I don't think you will be comfortable with someone speaking for you at such a level, so let's jump to the other end of the scale: Would you be comfortable at the one-phone-per-two-people situation? If you and a stranger were paired up to one phone, will you be okay to let the stranger give input to the Office of World Peace, which moves food, water, and shelter around the world, *on your behalf* once a day?

My guess is you *could* be – *but only if that stranger actually talked to you and understood you.* Even then, mistakes in understanding are totally possible, and you might suddenly get very uncomfortable if this stranger said into the phone "yeah, we think that's a great idea! We approve!"

"We".

We is an interesting word.

Is there a "we" in this situation? Or is this one person speaking for another? Would this technically give the stranger a higher social "status" than you through communication alone? Does someone speaking for one million others have a higher social status than those he or she speaks for?

There is a very real hierarchy of communication in place all around the world today as I write this. It doesn't *feel* like a hierarchy, but it is absolutely a multi-layered hierarchy – and it is extremely related to the hierarchy of authority.

Here's how you can see the connection:

When it comes to someone else's (secondhand) stories, no matter where that story is coming to you from – the television, the radio, the person next to you, or a book perhaps - you are only ever consuming *one story at a time*. There is very little chance you can listen to four podcasts or four television channels playing at the same time and understand what's being said. The stories would become a jumbled piecemealed mess – and that jumbled mess ends up merging into one story you will build: "I listened to four podcasts at the same time and it was very hard to make out any stories at all, but I caught a few snippets."

I really want to put a very heavy emphasis on this one-story-at-a-time idea because it is the thought tool that exposes the entire hierarchy – the entire invisible green-eyed monster for you to see as plain as day: *You are only ever consuming one secondhand story at a time.*

Right now, you are taking in *this* story, and I am honored that you are, and happy for you that you have the free time to do so. As you read this story, you are aware of other *wordless* and even *worded* stories around you (examples might be your awareness that you are on a plane, or in a chair, or that others are having a conversation nearby), but your focused *attention* is here, on *this* story, because you are decoding the shapes of these letter groups into familiar sensory currents of word-shells and then filling those familiar currents with context from your own past. This is an intense process. It takes a lot of your *mental* resources to do it.

If you are talking face-to-face with someone about an object that is present in both of your sensory boundaries – such as discussing a bird an arm's reach away with a stranger standing next to you – then this is the finest *resolution* of two-way communication possible. Each person has an equal voice to tell his or her side of the story when it comes to experience with the bird in the present moment.

If someone is telling stories *on behalf* of another person, such as a parent does for a child in many social situations, or as the stranger did on the phone with you just a few paragraphs ago, this is the *second* level of the communication hierarchy. This would be like you and two other people discussing the bird, but one of those two other people is speaking for both of them: "My kid thinks this bird is cool (even though the kid said nothing)."

Are you aware of anyone that speaks on your behalf in this way?

The second level is one story system speaking for *two*.

This is technically an *arc-flash event*, and it should always be considered with caution in my opinion. There is a way to clear this up, and I'll explain how a little later. In the meantime, I'll just quickly rip through some other situations that might be familiar which happen at different levels in the communication hierarchy. I bet you can figure out the level for each. After all, you've technically already seen the levels listed in the telephone game chapter – they are based on how many individuals one storyteller is speaking for:

The presenter of a five-person group project.

A middle manager speaking to an upper manager about workers' lack of motivation.

A mayor speaking on behalf of the town to the state governor.

A governor speaking on behalf of the state to the nation.

A local news broadcast using two interviews to support a story of how an entire town feels about a new statue.

A President speaking on behalf of a country to the world.

The national news speaking on behalf of your state.

The world news speaking on behalf of entire countries.

One way to think about all of these story-telling situations is in terms of the storyteller's "story resolution".

The maximum, or finest, story resolution possible is for every single person to tell his or her own story. This was what was being collected in the Magic Story Box experiment, and what happened with you and the stranger discussing the bird a few paragraphs ago.

The higher the storyteller is in the communication hierarchy, the more individual stories are being "summarized" into the word-shells chosen by the storyteller – which means they are really word-buckets.

This means any storyteller that speaks for other human beings is *taking shortcuts (arc flashes) in the name of speed.* The top of the hierarchy will be a global voice that may actually shortcut *every other human being.* It will be the lowest, or worst, story resolution possible.

And yet this hierarchy and shortcutting system *must* exist because our ability to communicate with others has exploded to include almost the entire world almost in an instant through the inventions of the internet (world wide web) and social media.

To get a feel for the power present in this storytelling hierarchy and the problems that come with it, imagine if aliens landed on Earth and the first person that walked up to them claimed that he spoke for every single human being on Earth. Are you comfortable with this?

Now consider that only a handful of individuals are choosing if their entire nation is at war with another. What story resolution is being used to justify this decision?

Theoretically, the maximum possible story power – the highest peak of the communication hierarchy - would be the ability to telepathically project stories into the mind of every single human on Earth instantly, and have those stories also make sense to every single person.

Realistically, this means anyone sharing their stories with millions of individuals at once is wielding an "extreme" amount of story power.

Should one of these storytellers be you, then I hope you tell your stories involving others very carefully, and recognize that you barely know any stories at all when it comes to other people's lives beyond knowing what they need to do to survive. Hopefully you recognize that any story you might craft about groups of other individuals is going to be low resolution, and almost certainly unfair to that entire group. Even those that willingly put themselves into the groups that you might represent with a single word may not realize just how unfair this situation is, because in reality each person in that group is *assuming* all the other members believe the same thing – which of course they don't.

When it comes to the stories we tell about each other, fairness and speed are the two balloons that are always trying to find balance.

The finest resolution of stories, which will be as fair as possible and the easiest to balance because it allows each individual to tell his or her own story, will be the slowest.

The lowest resolution of stories, where words become enormous buckets that are unfair to everyone they speak for, will be the fastest.

And now for the one gigantic enormous problem here: *You can't sense these speeds from your perspective.* You can't tell how fast secondhand stories are moving, and why they might be moving at the speed they are. No human being can see the entire hierarchy at one time. The reason is simple: *You can only consume one story at a time.* The same is true for every other human being.

This means you will notice no difference if the story came from the top of the hierarchy or from the bottom. One low-resolution story that buckets 10-million story systems enters your story system at the same speed as a high-resolution story that contains only one person – one story at a time. The only way you can notice what story level you are consuming is to notice if the words involved are "bucket" words.

This is why I have intentionally framed this entire book on the word "you" and tried to keep it that way. This book is me writing to one other person – you - and I cannot know who you are, where you are, or even when you are.

Doing this forced me to constantly check on my own buckets while writing. For every sentence in this book where "you" appears, I tried to consider all the possibilities of who "you" might be and what story you might have – and then try to keep my words within the common foundation of what I believe we *all* need and want as human beings.

I have done my very best to stay true to this, but it is not an easy thing to do, and I suspect I have probably made a mistake or two leading up to this point. If so, I'm okay with this. I am very happy and eager to listen to you and correct errors I may have made in this story, and maybe I'll publish revised editions in the future based on feedback I receive from you (if there is any demand for you to want to read a revised version).

From your point of view, I have appeared to be writing (or speaking or however you are receiving this) directly to you. But from my point of view, I have been writing to one single bucket that is one human being and *all human beings* at the same time –myself included.

I have definitely noticed the faster I wrote (because I feel strongly about something or am experiencing an emotion or anxiety about the future) – the more unfair my words became to "you", because I begin making assumptions about who you are and what you believe. I've often had to rewrite and reword pieces of this book because of this. Keeping your stories and other's stories as fair as possible *is hard work* because it requires slowing down and thinking critically about the language and words being used. In my opinion, doing this and recognizing the communication hierarchy in the moment is critical to see the beast and break the cycle of destruction and violence.

If there is one thing I hope you take away from this chapter, it's to pay close attention to each secondhand story in your life. If you can, at first, just try to notice each story as it flows in one-at-a-time, and also notice how many secondhand stories you consume a day.

Sometimes I find it helpful to force myself not to respond to as many stories as I can, because there are often no consequences for choosing not to respond. Doing this reminds me they are *secondhand* high-level low-resolution stories, and not directly addressed to me as an individual.

How many secondhand high-level stories do you take in on a daily basis? See if you can tell which level each one might be in the communication hierarchy. If it's high – because the story is coming from someone that is intending to cast their story to tens, hundreds, or millions - carefully consider what assumptions might be filling his or her bucket word of "you".

For example, I read thousands of comments every day in huge social media groups and on worldwide communication platforms. Every single comment I read is an individual story, usually about a certain topic, and takes up a small piece of my time. Ideally, each of these comment-stories is coming from one person and one person only. Based on the words in each comment-story, I try to notice how many shortcuts, how many arc flashes are being taken, bypassing others. I then keep this in mind when I tell my own stories.

Consider your news-story sources. How many people do you think your source of news is reaching? Thousands? Millions? Now look or listen to the words in the stories. Are these words "bucketing" people by the millions? The thousands? What story resolution is being used? What level in the hierarchy of communication is this story speaking from? Is the storyteller telling the story using the bottom of the hierarchy, such as a one-on-one interview, or the top, where entire countries composed of millions of people are blamed or credited for certain actions or said to feel a certain way?

One thing in particular might surprise you in this exercise: The *quantity* of stories. You are almost certainly being constantly bombarded by secondhand truth-stories *every single day, all day long* – and not all stories are created equal.

Some can and are easily being used as a weapon against you, in order to turn you into a weapon against other human beings.

142

Great Power and No Responsibility

As I see it, speed is everything. Speed is the difference between life and death. Speed in communication can keep you and everyone else on the middle path of stability in the face of a near constant collapse to fight or flee.

Communication is the most powerful tool we have to achieve peace and prosperity.

This means communication is also the most powerful tool to *destroy* peace and prosperity.

When the speed, frequency, and resolution of communication are weaponized - misused *on purpose* - the size and strength of the invisible hydra quickly grows.

To understand how this works, let's return to an old familiar friend: The poultry treadmill. If you and Bob were no longer on the treadmill, but still needed a shared answer to the question about the jet on the treadmill, how much time do you think it would take you to successfully merge truth-stories about it taking off (or not)? It might take a few minutes, or even a few hours, right?

It *takes time*, and sometimes lots of it, to merge two different stories into one shared story. *One-way high-level secondhand stories do no such thing*.

How many one-way stories do you consume a day? How many stories are squeezed into thirty minutes of watching or reading news? How many are shoved into your eyes when you visit a webpage? How many are squeezed into thirty minutes of reading social media? Dozens? Hundreds? Thousands?

All comments on social media are one-way. If you are simply reading comments on a forum, group, or public platform, no one is talking directly to *you* when they post a comment (unless that comment tags or calls you out specifically, but even then the writer is responding to you as your username and whatever comment history is available to him or her). In the same way, any news you consume is also not created for you specifically.

You are instead *choosing* to read or listen to these one-way stories. You are making a *choice* to use your time to consume them. At the time I wrote this, there is a bizarre illusion happening: These one-way comments and stories often appear on a screen that is in your hand or on your desk, and that same screen also connects you to people you *do* know firsthand and *do* have *two-way* conversations with.

When you consider that one-way comments and stories use words and ideas that you are familiar with and even use in two-way communication on the *very same screen*, it can make the one-way story feel like it was written for you – that you are actually *intended* to be part of a two-way conversation. This story is often validated by one-way sources adding a "comment section."

The result is one-way stories that *feel* similar to face-to-face interaction, or give the impression that the stories on the screen are very relevant to your own story. Nine times out of ten though, they are not – *at all.* Even all the comments in the comment section are not likely to be relevant to you. To test this, just pretend these one-way story sources no longer exist, and now go about your life. Would not knowing these one-way stories result in something unexpected happening to you? I'd argue the answer is almost always "no".

In my experience, very few of these one-way stories are truly relevant and change my day-to-day life. Most stories are not about food or water in a way that affects me. Most are also not about things I need to seek immediate shelter from.

Your story system *does not operate this way when it comes to stories*. Instead, *any* story that is highly validated, which means it stays consistent, starts to be *believed as true automatically.*

I can test you on this from here in the past: *What feeling do purple people give you?*

Did First You experience a snap judgment right then? A feeling that imme-diately drove you toward one answer that Second You perhaps had to pause and override, because you knew this was a test?

If so, then the sensory currents you experienced when you read "purple people" have already started to associate in your memory to the sensory currents tied to words that have appeared nearby to "purple people" more than once in this book, like: "Bad". These two ideas have started to connect in your mind from nothing but pure repetition of those word-shells appearing near each other in this book. This might have happened for you even though the entire point of me putting them together was to help prevent this *exact thing* from happening to you.

To think critically about purple people, you needed *time*. The connection of "purple people" to "bad" *is* terribly unfair to people that are purple to you. It's a statement that is very high in the communication hierarchy and very *very* low resolution, because it buckets every single purple person into *one bucket* and labels it "bad".

Now consider how often you watch twenty-four hour news channels, or look at social media or mass-media websites, and consume stories that involve other individuals as unique as yourself being considered and judged and labeled *in buckets.*

How much time are you spending letting Second You actually consider the fairness and gaps of each story you consume before you are bombarded by the next story to consume, leaving no more time to think about the first?

You can only consume and merge your truth-story with one story at a time.

In the case of one-way communication, the two things being merged are your projection of the storyteller's context and experience (which is really your own) with the story being told. Are you giving yourself any time at all to see if your own context can actually merge with the one-way story in a way that makes sense *without* using the same bucket words?

If you don't take this time, then you end up becoming hostage to believing whatever story *is the most familiar.*

Familiarity is anything that repeats itself. Repetition is a foundational pattern tied to your survival. Familiarity is not the void. Familiarity is safety. Familiarity is *shelter.*

Your story system *does not care* what the level or resolution of the incoming story is *when it is in a hurry.* It is looking only for *consistency.* If all of reality changed every time you blinked, you would eventually find the consistency between blinking and change. You would then know that every time you blinked, reality will change, and your story system would then structure your survival and safety around blinking.

In this way, the upper layers of the communication hierarchy telling stories to millions of people quickly and non-stop are lobbing the equivalent of social and mental nuclear bombs. These one-way stories about other people are often so low-resolution and so shortcutting of those they are about, that they might as well be primitive grunts: *"Purple bad."*

You can literally hijack yourself from your own ability to think this way. By keeping Second You focused on what comes next in the high-speed story, or what story comes next in a high-speed story feed, and never slowing down or stopping to think about what story has been told and if it really makes sense to you, or if it is being fair to other human beings - this constant high-speed flow of stories becomes a weapon that transforms your truth-stories to be whatever stories they tell. Without stopping to think, you end up believing whatever is repeated to you the most as true, *especially* if you have no firsthand experiences with the topic of the secondhand story to question anything about it, easily notice the shortcuts, or recognize missing context.

For instance, if a source of one-way stories sending out a non-stop unbroken chain of news were to suddenly insert high-level *judgments* between these stories, so that the news and judgments were constantly mixing, and if you consume all of this and aren't ever backing out to think if you actually agree with any of it - then your story system becomes nothing but a copy of the *opinions and judgments* of this storyteller that keeps spinning word-shells into stories that seem to make sense to your context - often to the point of *absolute certainty*.

On top of this non-stop truth-and-judgment-by-repetition/familiarity method, you might also hijack yourself in a second way: The *type* of stories that appear in the story sources you choose to consume.

Almost all front-page news and mass media sources (at the time I wrote this anyway) work First You into a state of fear, triggering what feels like a desperate need for *shelter*. Do they do this intentionally? Honestly, I don't think they do.

First, there is no "they". There are only ever *individuals*. All mass media story sources are composed of individuals. In the end, *one person* is making final editorial decisions for each story source. This means whatever source of information you consume, you are only ever getting the opinion and judgment of that story source's *editor*.

Think about the Magic Story Box and how many stories are generated every minute by human beings. Consider all the stories that exist in this world to choose from. There are so many – *too many*.

Which ones matter? This is the choice of the editor. And that editor is a human being just like you (he or she may even *be* you). If so, then you as an editor have all the same core feelings and fears that everyone else does, but you also wield an enormous amount of story power.

What I see happening in mass media today is what boils down to a huge amplification of the editor's beliefs of what stories are important, and the filtering away of everything else. After that, it's all a competition between editors to get the most attention. Which beliefs will get the most attention? The same ones you and I both care about most: Stories to trigger fears – stories of authority battles, violence, killing, crime, and war. Stories of famine, and drought, and disaster.

As a story consumer – how connected are you to these editor-chosen secondhand stories? If you are living paycheck-to-paycheck, is a purple people terrorist attack on the other side of the world something you should be anxious about happening and make your top priority?

An individual specialized as an editor debated this same question, and experiences this same potential fear and anxiety, and chose to circulate a story about the attack based on it through their "loudspeaker" to everyone that might consume it. But what good is fear and anxiety if you have no means at all to relieve the imbalance they bring?

This is where I suspect it makes sense as an editor to provide *another* story that makes this fear and anxiety feel *controllable*. This can be done by creating huge buckets of assumptions that connect "you" to an act on the other side of the world through the mere existence of purple people.

Who *is* to blame for this attack that can be connected back to "you" the imagined audience bucket? Perhaps it makes editorial sense to run a story assigning some *local* blame or connection to that distant terrorist attack. This might look like:

"*Is* [insert local purple person that runs restaurant] *sending money to other side of world to fund terror?*"

This "connecting" of distant fear to local fear can trigger local outrage that makes it seem like terrorism *is controllable and preventable*, but provides no direct way to do so. Fear demands shelter, and the only shelter from other people is two-way communication. But the editor already made the story-consumer *tell themselves a story in the form of a question*, and the only *local* actions available to that now-fearful consumer will be directed at purple people. Maybe that restaurant now gets vandalized, or the owner assaulted.

Maybe another editor-chosen story presents an interview with one person claiming to speak for thousands or even millions of others, like a leader of a political party. "*Look, purple people have been the problem for thousands of years, and we all know it! Even the Being that founded this great country knew it!*"

These stories validate each other with no firsthand experience or two-way communication involved.

Maybe the editor will choose to run this story through a pundit show where the talk-show hosts assign blame to local purple people by implication and again in the form of a question ("*Are purple people a danger to our town?*"), or maybe just run an interview with local people that have decided they are going to do something if the purple people problem ever makes it into their life –they might have already justified doing something about this problem. (*"Purple people are bad and the whole problem! Me and Billy got attacked by one last week, but we put treadmills all around our house, and we're safe if we keep them all running at 12 miles per hour! We heard purple people can't run faster than 10!"*)

How many hours of this combination of stories would it take for you to reach the conclusion that local purple people in your town are potential terrorist and "need" to be locked up? If you consume no other source of stories and keep this one flowing all the time, in my opinion, it won't take long at all.

And yet you have experienced nothing but an editor's opinion.

As it turns out, this goes much further than just news and media sources. *All* secondhand stories are just an editor's non-stop opinion of what's important. Opinion of what their imagined bucket of "you" *should* see, what you *should* hear, what you *should* do, and what is *really* happening. All secondhand sources, from news to social media posts to spam email, are just individuals battling to control priorities and truth and with it, reality.

You have a choice to make.

143

Wielding Damocles' Sword

You have a choice of what to do with your limited time in life. Every bit of that time will be spent creating, analyzing, and acting out *firsthand* stories – *your* story - and this will undoubtedly include consuming the secondhand stories of others.

In some ways you do not have control of what stories you consume. The stories your eyes, ears, nose, mouth, and skin convert into electrical currents that tell you about the universe surrounding you are only partially controllable – they are windows after all.

You can only partially control the flow of direct-person-to-person second-hand stories because they come in through your sensory windows whenever you are close enough to someone else, whether you like it or not. You try to close off your senses, but in the end you can only control your justified movements toward or away from these stories.

When it comes to the flow of one-way secondhand stories that are broadcast widely using a screen, or a radio, or the internet, or any other medium, you actually have *a lot* of control. You can't shut them *all* out – billboards and advertisements are often shoved in your senses one way or another everywhere you go - but you can shut off *many* sources of these stories if you choose to do so.

This choice of what stories you consume is incredibly important. You will inevitably begin to believe stories you choose to make familiar. Your free time will reshape itself to "center" around the feelings created by the biases and opinions of the editors choosing to manipulate stars and shadows. If you consume nothing but stories of fear, you will become fear-centric as your past fills with more and more stories and movements of things to fear in the shadows. You will find yourself with an ever-increasing need for *shelter*.

But there is no shelter from your own feelings.

Having a need for constant shelter from your own feeling of fear is a paradox - a state of feeling forever imbalanced – that creates constant anxiety that certain threatening events or things are about to happen to you in the near future. How do you take shelter against something that cannot be seen or heard firsthand? *How many purple people are out there? Are they coming my way? Oh god, there is one now! Am I about to die?! Should I attack first to save myself? To save the children?!*

This thinking, in my opinion, happens when stories are being used as weapons against your own mind *by* your own mind. However, it is also natural that you, at the bottom of the communication hierarchy, will consume mostly one-way stories that are coming from much higher in the hierarchy, because they are moving information the fastest and are often exciting, making them easy to talk about with others face-to-face. However, these stories are also the most "bucketing" and unfair to your fellow human beings.

This echoes back to the enormous power given to storytellers. To artificially *force* someone else into a rushed mental state - like demanding that an answer is needed *right now* when no reason exists it is actually needed right now (because no one's life is at immediate stake) – is to weaponize *time* by collapsing it, in order to force the merge between storyteller and the consumer's personal truth-stories. Forcing others to rush through stories or force them into more stories *intentionally*, is to try to block out the others' firsthand experiences from the equation by allowing no time to reflect and think. Someone doing this is literally trying to cause an arc flash event in someone else to override their personal truth and sense of right and wrong.

As I see it, this is why pundit and opinion shows tied to news and world events are so powerful. The individuals on these programs *appear* to offer justifications for fear. I'm not saying pundits (which might be you) are lying and being manipulative – they do what they do because they believe it's the right thing to do. That doesn't mean their projections and opinions – *stories they tell themselves into a microphone* - accurately represent what is happening in the world, are prioritized in the same way you would prioritize them, or are being fair to those involved.

After all, there are more human stories happening every five seconds than you or anyone else could ever consume in a lifetime.

Those that design these pundit-like programs often bring in an individual from "both sides" of an issue to give the appearance an honest attempt is being made to merge two truth-stories that don't agree into one larger and stronger truth story – but this is not actually what happens.

If you are the one consuming this "both-sides" show, then all you usually get in my experience is three separate individual and polarized opinions: The high-level low-resolution story that is the subject of debate, which already is influenced and worded by someone's opinion, and the opinion of each person that represents one "side" (which is usually an enormous bucket of "unfair").

Your context, and *your* experiences are not included in this show, and unfortunately they are what needs to merge with the high-level story as making sense, *not* the two polarized storytellers pulling from their own context for shadow experiences somewhere up high in the hierarchy and also low in resolution. *Everyone is just telling themselves stories.*

If everyone on the show never has any intent to concede on his or her opinion, or consider a different person's opinion as valid and possibly a better truth, then *there is not a story merge even being attempted.* There is no rabbit-hole happening. You are just watching a show where two people on a stopped poultry treadmill yell "No, you are wrong" at each other over and over, and only ever *pretend* to listen to the other person's story, because there is no immediate consequences. The treadmill isn't on and the cage is empty. It's all just noise that is literally wasting your free time.

What is *your* opinion built from *your* firsthand experiences with the topics being discussed on shows like these? Do you have any, or do you often find yourself in a position very much like a young child, where you automatically believe the story because it makes the most sense to your own context because you know no other stories?

This 4-legged fuzzy creature is a kangaroo.

I suspect it's highly likely you have unintentionally put yourself on an accelerating mental treadmill, and are trying to make sense of everything from one-way low-resolution huge-bucket stories and one-way pseudo-arguments that are only pretending, or perhaps believe they are trying to find truth. In reality, these are really just opinions likely built on even more secondhand stories that were written on wildly different levels in the hierarchies of communication and authority.

If you never give yourself time to reflect on your own opinion outside of these storytellers, you might easily get so panicked and feel so rushed that you actually begin to justify actions into the real world based on the beliefs that were only repeated to you from the shadows, desperate to "capture the end" of the shelter story arc you can't seem to ever capture.

That "capturing" action is hardly ever the choice to slow down, dump the buckets, and increase the story resolution to build a better truth. The truth instead remains bucketed and starts moving at faster and faster speeds.

Seeking out others with shared "high-level" story interest and beliefs will often lead you straight to those that consume the same exact story sources you do. The result is what feels like firsthand validation of your beliefs and fears, but it is actually a now-merged truth of a bucketed, high-level, low-resolution, frequently repeated story.

You might even start to believe words mean the same thing to everyone, when in reality you have only internalized the same *Keeper of Truth*, the same assumptions, and the same fears.

You, and the groups of those people you believe share the same beliefs you hold, move one step closer to an arc-blast. One step closer to that "shared" truth-story self-validated to the point of *certainty* you know the *real* cause of all of society's problems.

The *Dark Forest Moment* approaches.

The Hydra grows.

144

Dumping Buckets of Stolen Time

When it comes to mass media stories, unless it threatens your food, water, or *immediate* shelter – it is coming from higher up in the hierarchy, which means it is at a lower and very possibly unfair resolution or worse, completely unanchored from reality.

When it comes to comments from others on social media, forums, and comment sections – *every comment is literally just the storyteller projecting a story into the world based on his or her own past and present,* unless that comment is sharing a personal story recounting a wordless, gapless firsthand experience.

This being said, the hydra created, grown, and fed by forced-speed one-way storytelling designed to prevent reflection and increase assumption and fear in the consumer has also figured out how to hijack your firsthand experience.

I'm going to give you the list of stories that you might have believed to be "the reason for all of society's problems" once again. This time though, consider how many individual human beings are being "arc-flashed" over, shortcut, and dumped into *huge* bucket-words to allow for fast and easy judgment related to "all" problems possible. Some of these extremely short stories put millions or even *billions* of people all in one word bucket.

As you reread the list, consider how you might have become familiar with these stories. Consider your firsthand connection to each of them. In some cases, you may have no firsthand connection at all, which leaves you in the position of believing whatever you consume about the story as long as it seems to make sense or is repeated enough.

Notice which stories ignite your subconscious need for shelter due to their high-speed context from your own mind, which is likely to *still feel needed* even knowing how far from the truth and fairness they probably are. You'll notice this need by experiencing a sudden "feeling" when you read the word – which is most likely stories of fear or invalidation tied to the words that have been repeated and seemingly validated at an individual level to be true. The words that do this are collapsing time for you.

It's the government. It's the Deep State. It's the Democrats. It's the Republicans. It's election security. It's social media. It's the Supreme Court. It's MAGA.

It's woke-ism. It's liberalism. It's conservatives. It's Fox News. It's CNN. It's stupid people. It's millennials. It's the boomers. It's people that think they are smart. It's the scientists. It's transgenders and gays and drag queens. It's black people. It's white people. It's illegal immigrants. It's guns. It's gun control. It's gun promotion. It's mental illness. It's mental healthcare. It's Critical Race Theory. It's school indoctrination. It's lack of teacher and school support. It's money. It's science. It's religion. It's lack of religion. It's the absence of god. It's the devil. It's the police. It's our broken healthcare. It's the ultra-rich. It's the union. It's Black Lives Matter. It's Antifa. It's the loss of morality. It's confusion over what is right and wrong. It's purple people.

Every single one of these stories has a low-resolution high-level bucket word that speeds up and simplifies the world.

If you see *yourself* as belonging in one of these buckets, then you probably felt that story is *wrong* to appear in a list that is "the reason for all of society's problems". That feeling is your story system reacting to its truth being considered invalid – "you" are being *attacked*. Such a story doesn't feel right or fair to label as the problem at all.

If you believe one of these bucket words actually *is* the problem, then you probably felt a kind of anxiety or fear when the word appeared in this list. However, your story system might now be struggling to self-validate such a secondhand story because *opposite* bucket words are appearing right next to it to be the reason for the same exact problem.

None of these stories are the reason for all of society's problems. Nearly every single human being in the United States can be put into the enormous buckets on this list.

Bucket words, or labeling of massive groups of people to speed up judgment in order to "capture" fear and anxiety related to the enormous "chaos" of 8 billion individuals with impossible-to-see beliefs and justifications and therefore unpredictable *movements*, results in the horrific side effect of collapsing and oversimplifying the world.

In this case, it is collapsing the experiences of thousands of *generations* of individual human beings – which at any moment in time create more stories than you could possibly consume in your lifetime – all the way down into a handful of extremely simple sentences.

This extreme level of bucketing can quickly and easily make you feel trapped, lonely, and believe that the world is awful and out of control – especially if you keep consuming stories that repeat and therefore validate these bucket words to be true and the "movements" that are captured by these validated stories are "going" to do you harm.

If these feelings are validated enough, and the fear of other human beings that are being bucketed cannot be relieved (because how can it? The buckets don't really exist in any realistic way), then you might find yourself considering the justification of increasingly dramatic actions to "feel safe" again.

Editor-filtered and condensed stories of the entire world merge into one story system – *your* story system – pitting you against hundreds, thousands, or millions of others unfairly bucketed by your own words in a way that destroys all individuality except your own, because there is *no time* to consider anything else.

"The stable middle" between violence and escape, held open by language, feels like it is collapsing due to the non-stop flow of stories using the same word buckets you use (because they gave them to you in the first place), the pseudo-story merging "game," and the constant firsthand validation by those that consume the same secondhand story sources. Internal story voltage rises to the point of feeling painful, and the story resolution cannot seem to get any better. These constantly-growing-and-filling bucket-words can't get "dumped."

The *Dark Forest Moment* looms ever-closer, self-contained inside a single story system – your story system - which is getting increasingly "unit-ized" into a two-position reality with a razor sharp boundary that does not and has not ever existed: It is "Us versus them." It is "Me versus the world."

When the entire world collapses into black and white, there are only two paths available to choose. The middle isn't one of them, because it (using language) is the cause of its own decay. There is no more communication left. There is fight, or there is flight.

In a near-totally connected world, we are reaching the point where there is also nowhere to escape. We are stuck here together. This means the path of flight is also decaying. And this position, where only the fight option remains, has been reached time and time again on a massive scale throughout human history.

Weapons of mass story-system destruction have often been the chosen solution at the time I am writing this. These story-destroying weapons typically sling pieces of metal hundreds of miles per hour – enough to penetrate most boundaries of any kind. They are used in schools. They are used in churches and religious buildings. They are used in government buildings and in the streets.

And the blood that spills flows backwards in time all the way to the cave with the stone door in the center of town. There, this story of ultimate unfairness is consumed – the story of ending someone else's experience out of fear. This story then telephone games its way to you through the screen in your hand or your desk or wall, where a new high-level story is woven or self-validated out loud, and more low-resolution judgment and opinion of "causes of all of society's problems" is imagined and grows the buckets even larger and speeds up stories even more, collapsing time and amplifying fear until the next act of violence feels justified.

Those that choose mass violence often do so, and then seem to suddenly rediscover the other path he or she may have believed was gone – the path of "flight". Flight, in this case, is self-destruction after justifying violence – the path of suicide. I suspect that all justifications for violence are lost when the realization eventually occurs the killing and suffering that were just brought into objective reality did nothing to actually fix any of society's problems. The middle remains - but it is invisible, leaving what looks like only one choice.

The world is out of control, "they" say.

Purple people are the problem, 'they" say.

Weapons are the problem, still "others" say.

It is none of these things.

Non-stop storytelling, bucketing words, warped boundaries, and the runaway shelter need – fear and imbalance in individual story systems - are the problem, as I see it.

Which buckets do you think *I* belong in? I knew I was dealing with this mental bucketing system from the very beginning when you started this book. The concept of bucket words in language was one of the very first things I noticed after "The Event" that changed all my thought processes.

There were times at the beginning of this book when I felt it was the right thing to do to try to forcibly stop you from bucketing me - especially when I approached the "authority" of law and religion you might pin all your hopes on to solve all of society's problems.

I put myself into one bucket and one bucket only, because in my opinion it is the only fair bucket of language to exist: Human being. Any other bucket I might appear to belong in from your perspective is there because I am forced to use language to communicate with you, and I used words that must have caused me to fall into a bucket for you – *that doesn't make the story you might believe to be tied to that bucket true and tied to me.*

After all, which buckets do *you* belong in? I hope it's the same one as me.

"Human being."

Every human being is just as awesome and worthy of having the opportunity to pursue a well-experienced life as much as you. No one is born better or superior to anyone else. I am not better than you. You are not better than me. No one is the Keeper of the Truth, and no one is born with natural authority over another. No two people actually agree on what any bucket word even means, no matter how much these two people might insist they do.

Bucket words – collective words for people - are *literally* mental black holes. As you listen to or read stories outside of this book, watch for black holes like the ones in the list above, and see if the storyteller defines what the word even means, or if they are letting *your* assumptions and feelings run wild like their own.

Take a social media post like "If I was a parent, I would never let my kid…." This is passing a one-way judgment on *all parents* and *all kids*. It is also 100% projected, and is essentially the storyteller telling his or herself a story out loud. When you read those words back to yourself, they are familiar and so get filled by your existing feelings and beliefs. This causes the story to create very real self-defined boundaries in the mind of most people that read it.

If your experiences as a parent or with kids merge with whatever this statement continued to say about kids, you might put yourself in the storyteller's bucket of "approved people", making you feel valid and right. If they didn't merge, you'll probably be angry and polarized against this storyteller right away. "*That person is wrong.*" In reality, this post is just one opinion out of the almost 8 billion opinions out there in the world. However, this is an extremely high-level (large audience) low-resolution (huge collective groups) and horrifically unfair opinion. There are literally 8 billion people contained in two word-buckets within the statement – *everyone* is either a parent or a child of a parent.

Every story you ever hear or read is just *one person's* opinion. Every single one.

So why do opinions like this one have such a powerful effect? In my opinion, it's because when we read it, we *imagine and assume* others are reading it, too, because it is being broadcast and interacted with on what appears to be a large scale. This makes us imagine our reaction or comment on it to no longer be at the bottom of the communication hierarchy, but somewhere much higher. It makes us think that our opinion *matters* and we are locked in a battle to be right and declare what is true with hundreds or thousands watching.

Imagine what would happen if this kind of polarizing imagined-to-be-high-level bucket-all-people thought process made it to the top of the hierarchies of authority like the government and religion! It would split entire populations in half!

What *is* a democrat?

What *is* a republican?

What exactly *is* election security?

How *exactly* is any particular leader the *entire* problem, or are all the problems just being shoved into a bucket that is labeled with that leader's name on it?

What determines if someone is "woke"? A liberal? A conservative?

What is a stupid person? Relative to *what*?

What makes someone a millennial or a boomer? Is it really the year they were born, or is this just a bucket word used to arc-flash over millions of human beings in the name of easy judgment to limit and validate feelings of fear and anxiety quickly?

What is a gun *exactly*? Where does a toy gun end and real gun begin? Where does a gun end and a cannon begin? Does "assault" mean anything specific? Does any of this matter, or does it all just boil down to the choice one would make in the *Dark Forest Moment*?

What is science exactly? What is religion? And what is it about these bucket words that makes the individual that uses these same words *the problem*?

What is happening in the previous few sentences might be strange to think about. You perhaps see yourself as belonging in some of the bucket words above, or thinking with them. Just know that every time you put yourself in one of these buckets that collect people, this is purely a projection of how you judge yourself *relative to others*, and any agreement with others in the bucket you imagine yourself belonging in is not as clear as you might believe.

Getting yourself out of buckets can be hard. For example, if you put yourself in the "liberal" bucket, you end up automatically opposed to the "conservative" bucket. And by outwardly labeling yourself with this bucket word, you give a kind of passive permission to let others use that label to quickly categorize and judge you using their own context and bucketing system, which you might take personally, and start to fight them with language to control what the word means.

But how do you determine if someone belongs in the "conservative" or the "liberal" bucket? His or her outward opinion on a single political issue? Should their belief on a single issue – which might also be suffering from "bucketed" thinking – justify you putting their *entire* story system into a bucket of "invalid" that no longer deserves to share stories with you ever again? Is this not choosing a position of superiority and domination?

Bucket words can make the chaotic world feel manageable – controllable. Bucket words for people remove all the uncertainty, fear, and anxiety those people can bring to your mind. Bucket words filter out millions and billions of amazing stories created every second, and condenses them all into a singular story in your own head.

If you can, dump your bucket words. *All of them.*

There is only *you*, and however many other humans are out there with you right now, which might include me, and we are all in one bucket. Each one of us, including you, has a beautiful story system filled with firsthand stories that might be amazing secondhand stories and experiences.

Challenging your buckets, and seeking out stories that oppose the familiar and comfortable stories you currently believe is exactly how you grow your story system, enrich your context, reduce fear, and grow trust and love.

For example, you could easily bucket a gun enthusiast that scares you because you "hate" and are scared of guns – but this same person might also be very much into woodworking – which is your passion. If you cut this person out of your life because of your fear of guns, you might be missing out on a rich and wonderful shared experience related to woodworking.

If time is instead spent with this person sharing this interest, you both might not only improve at your hobby, but also learn to trust and even be friends. With the expansion of trust, you can start to collapse your own fears, and eventually ask this person to help you safely explore his or her *other* passion, the one you fear. Why does this person love guns? What is it about them? You might find this person's story for why they love guns to be far more interesting and enjoyable and different than the one you were *telling yourself* to believe.

Notice your own bucket words. Dump them by trying to avoid using them, and maybe even getting to know someone in those buckets that did not include you. My only advice is to avoid the bucket word at first if you can, and instead choose to see the other person as a human being with a life story you don't know at all, but is just as awesome as your own.

I can't say I've met a single person in my life that *wasn't* amazing and interesting in some way. Just because I might not agree with a person's passion doesn't mean I can't let go of my assumptions and fears and have a conversation to better understand why he or she loves it so much. Who cares what I think about their passion anyway? I might even learn to love what they love just as much. If anything, I love a good story, and have found that just about everyone has lots of phenomenal stories, from human beings that farm to human beings that pilot jets.

You are not *any label* that I can think of except "human being". You offer a perspective unique to only you. You have a biased and subconscious response that occurs when you experience hunger, thirst, or fear just like everyone else. However, you also long to find something behind the billion doors of *The Ark*, and I bet it's beautiful and awesome, whatever it might be.

In the bigger picture, when I feel I am not in danger given the context (not a survival situation) I try to just approach people with judgment-free curiosity.

Of course, I have to assume you might have difficulty in telling if you are in a safe situation. After all, fear of the unknown seems to be subconscious and hard to override. I might be able to help. Before we dive into that though, there is one more way that words can be weaponized you should know about, and it is extremely relevant to sharing stories face-to-face. Realizing when conversation itself is being used against you can be the difference between violence and peace.

145

Exploiting Infinities

I've already covered how one-way secondhand stories can force you to *speed up* using constant non-stop storytelling, but there are also ways to weaponize communication in the two-way and face-to-face formats. These methods can even be used by one-way storytellers against their entire audience (though not as effectively).

The act of dumping buckets (collective and shortcutting words) through conversation is to reach for deeper mutual understanding. It is to try merging truth-stories all the way down to the finest level of whatever can be mutually understood. Bucket words are the "rabbit holes" that people like you and Bob fall into, dump for each other as much as possible, and then fill back up together while on the treadmill.

This process-of-mutual understanding is often slower than the minimum speed required for survival. The poultry treadmill represents this minimum speed, and creates the potential to feel forced into a "shared" answer you don't truly agree with in order not to suffer the roosters.

Someone that is aware of this can exploit the gap of difference between survival speed and the speed of mutual understanding to work against you, and for his or herself. Bob could slow down the process of merging stories with you *on purpose*. The justification for doing this is not to force you to fall into the roosters, but to intentionally let the treadmill reach faster speeds than it ever would have if he instead made an honest effort.

For example: Bob believes his answer is right and yours is wrong, and he also believes he can run faster and longer than you, so he slows down the process so he can "win" without ever having to consider a shared answer. You might call this the choice of "time violence" or "passive violence".

All Bob has to do is say "I will only agree with you when all the words used to define other words match between us exactly."

A person that demands perfect agreement on the meaning of words is trying to indirectly be the controller of the treadmill.

This must-match definition situation actually requires *infinite* time. You and Bob would both die by rooster long before the end of this process is reached.

Words are built from individual contexts, and those contexts will *never* perfectly match, because they were built from different perspectives and past events, and those perspectives never overlap because we do not pass through each other.

This can easily be exploited against you if you did not realize words do not mean the same thing between any two people: "What do you mean by 'wrong'?" What do you mean by those words? And by those words? And those? And those? Do you even know what wrong is?!"

This word-destroying "game" can keep going until you fall all the way down the rabbit hole from the first parts of this book to eventually realize words have no fixed meaning at all. Having this realization will not help your situation though, because the person weaponizing time through language (in this case, Bob) can just pivot into a certainty and say "No. Wrong means [this]" or pull out the dictionary and read the definition to you.

To fight this new position, you'll have to tumble all the way down the rabbit hole to destroy all the certainties Bob can claim just to prove that "wrong means [this]" is not *objectively and certainly* true – that it was only ever Bob's own personal opinion and *judgment*, the Dictionary Illusion, and The Exploding Cat.

What Bob is actually doing is "capturing the end" of ever experiencing invalidation or guilt. Why? Perhaps Bob has performed some action in the past that comes with a consequence he imagines to be "unfair" should it ever be discovered. Perhaps Bob is telling himself a story about you that simplifies and buckets you, and he would prefer not to like or listen to you instead of dump it. Or perhaps Bob is insecure and scared to be judged for his opinions – he doesn't want another Hydra head in his mental boundary.

Both of these abyssal weapons of time - the exploitation of definition relativity and the exploitation of disproving every certainty - can be used to *prevent you* from reflecting, engaging in deepening two-way conversation, or considering why this might be happening long enough to ask different questions. Your emotions might start to run high as everything churns toward the *Dark Forest Moment* from frustration of having no time at all to think, or needing infinite time to find agreement or destroy certainties to even begin a meaningful discussion.

Your feelings become hijacked, because you constantly feel like you are in a fight for survival (incoming high-speed unfair statement flood), or that your truth is invalid (Destroyer of Words game exploitation, or prove-me-wrong exploitation).

But there is yet a *third* way language can weaponize an infinity against you, and it's exploiting the you-can-only-consume-one-story-at-a-time situation.

When you read comments on the internet, do you think that each and every commenter is a *real* person tied to a wordless, gapless, sensory perceived experience with this world? If so, this weapon is almost certainly being used against you (unless humanity has figured out how to control this since I wrote these words).

On the internet, or in any situation where you do not know the "person" you are communicating with, anyone can be anyone – or any*thing*. One person can actually become 10,000 "people" online. This concept alone has nearly destroyed any semblance of truth on the internet as I write this.

In my opinion, take all crowd-sourced reviews of anything online with an enormous grain of salt. Half of those reviews may actually be one person using fake "bot" accounts that look like unique individuals to flood in tons of good or bad reviews. News story comment sections? Same thing. Social media massive websites? Same thing. Phone calls from strangers? Same thing.

Since you are forced to consume one secondhand story at a time, and you tend to believe things that are repeated, if I can make sure that at least 50% of the stories surrounding a topic I care about are controlled by me, I can make you not only believe my opinion is the truth, but perhaps be the only source of truth-stories you'll ever consume, assuming I'm able to flood the comment section, your phone line, or your story source.

I can potentially bring a new product to market with five thousand five-star glowing reviews. All of these reviews are actually me using an army of "bots", but there's no way you can know that. One bot might appear as "Bob" and another as "Eric" and another as "Olivia".

And this leads into what I consider the true incoming horror, at least for me: Many news stories are now written by language-based artificial intelligence (AI). Language-based (LLM) AI *has no ground* because words have no objective meaning. Language-based AIs have no way to *validate* the actual truth-value of any statement except to rely completely on other statements it knows, which are the only "experience" the AI ever has. If I had access to the training data for a language AI, could I potentially flood it with a bot network or some other flood-method with "purple people are bad/evil/vile" 8 billion times, and potentially cause this AI to then regurgitate it as a "grounded" truth based on repetition alone.

Since these language-based AI algorithms are *kept secret* from the public that uses them, we have no way to know *how* they are grounding truth. As such, I see language-based AI as it sits today as the *ultimate hydra being unleashed on society*. It connects very little it says back to a human source in any way we can validate. Its creators (programmers) do not seem to share the same belief-stories of how AI determines what is right and what is wrong. It occasionally has massive linguistic freak-outs. It seems to be easily tricked into giving "unallowed" answers to questions simply by changing the order of the words, which might uncover a darker "truth-story" the AI has consumed through humanity or itself (self-validating).

I design electrical systems for a living. At the time I wrote this, there is zero chance I would touch an installed electrical system if I knew it was designed by AI. Why? Because AI has no incentive to *care*. It has no *value* system tied to my survival or even to reality. It perceives nothing but words and how they connect – it is a program that connects stars into constellations and that's it.

As I see it, language-based AI is a tool capable of using words at extreme *speed*, which is not without value. It can "rewrite" stories in any style you request. There is no way to be sure these rewrites are grounded to the original story though unless you already know the topic or story. *This* is the extreme danger as I see it. How truth is grounded in language-based AIs is not being shared openly, *and* it is currently communicating in *two-ways* with huge portions of the world, and therefore it is potentially having an *enormous* editing effect on what is true and what is not.

AI will perhaps be a wonderful future tool to instantly "translate" the 8 billion story systems out there into familiar language for an end user. However, for now, that user should consider using *a lot* of validation methods to be sure what is being relayed makes sense relative to wordless reality.

As these artificial intelligence systems inch ever-closer to "always seeming to be right," I believe we will reach the most dangerous place in human history – we will start to trust it as we would a human being.

I am not saying all this to claim that AI will never be worthy of our trust. Rather I am saying that at this point it is nothing more than a fancy language "calculator" that sometimes spits out 1+1 = 3 *with certainty*, and perhaps only the programmers know why (or perhaps have no idea why).

Essentially, all language-based AI programs are nothing more than amplifications of their programmer's beliefs of what is true and how truth is determined. If we become dependent on AI without vetting of its truth-building foundation, (which unfortunately involves trying to check multiple abysses at once), then we expose all of humanity to, in my opinion, *grave* dangers at *extreme* speed in the form of AI-influenced engineering disasters, environmental disasters, and atrocities on huge scales.

My concerns regarding A.I.s are not present day as I write these words – they are for generations that are born into a world where it exists, is trusted, and is normal, and all critical thought disappears because it becomes *assumed* there is no longer a need for critical thought – AI does it for us. This would be the *Floating Anchor Problem* anchoring right back into ungrounded certainties. That is when I believe disaster has the potential to strike.

So, with most ways language can be used against you now in your memory – which has made what I consider nearly the entirety of the hydra running free in society to be visible – let's talk about containing it.

146

Installing Air Conditioning in Hell

Resistance is every thing. This means boundaries are every thing. And I mean this literally. Some boundaries that resist your desire to peer inside, to understand and to control, are other people just like you.

This means the most important boundary (to me) by far, is the mental boundary. You cannot read my thoughts or experience what I have experienced, and I cannot read your thoughts or experience what you have experienced.

Who should have *natural-born* authority to cross mental boundaries and tell anyone else, which includes you, what they are thinking? Who *should* have the natural-born authority to not only assume what you are thinking in this moment, but also then judge your thoughts to be stupid, smart, right, or wrong? To judge you as irrational? Nonsensical?

Relative to *what*?

Judgment is always being made relative to one point: The person doing the judging.

No one was born the keeper of your thoughts except you. No one was born the keeper of "right", or "smart", or "rational", or "sensical" or "logical."

This boundary, established way back in Chapter 1 and the centerpiece of this entire book, is being crossed nearly *all the time*. Telling stories about what someone else believes, thinks, or desires with certainty is to cross this boundary.

I found a way to restore this boundary when it comes to language, and I call it using "boundary words".

Consider that every single story you ever hear from another human being *is an opinion* that is tied to a context of experiences you can never actually get full access to, because you cannot read minds.

Based on this, boundary words are very simple: They are a set of words that acknowledge and recognize this boundary during the actions of communicating – and I've found they have an *extreme* ability to calm emotions in conversations and increase overall stability – they seem to grow the middle path.

I'll run through some imagined conversations between person A and B with and without boundary words to show what I mean:

Conversation 1 (no boundary words):

A: "Purple people are bad."

B: "That's wrong and unfair. You are an idiot!"

A: "Screw you. You are a purple people lover! You are one of *them*!"

B: "What? No I'm not, you're just a racist!"

A: "Eff you moron, don't come crying to me when they steal your children!"

Conversation #2 (with boundary words *in italics*):

A: "*I think* purple people are bad."

B: "Well, *I feel* like *your opinion* might be a bit unfair. Do *you believe* that all purple people are bad?"

A: "Yeah, they are all bad *in my opinion*."

B: "Have you ever met a purple person?"

A: "No, and I don't want to, *I think* they are all criminals just waiting to rob us whenever they get the chance."

B: "Okay yeah, I could see why that would be concerning – I don't want to be robbed. Why do *you think* purple people want to rob you, though? I've met a couple and they seem like nice people *to me*."

A: "*I believe* they are all criminals and thieves. *I bet* they all hate us."

B: "Well, I can see how *you think* that, but I can't agree from *my point of view*. The purple people I know are nice *to me*."

A: "Ok, well, wanna go do karate in the garage?"

B: "Yup!"

You might be frustrated the conversation above suddenly ended with no agreement or resolution, and that's okay. *There is no race* to get to an agreed-upon conclusion. Any feeling that you *need* a conclusion *right now* is your shelter-need being exploited. (This assumes there are no purple people nearby in this situation creating an internal-Dark-Forest-Moment for person A, and neither person A nor B is about to go into a situation with purple people against their will.)

Person A, which I'll call Alice, expressed her *fear*. Person B, who I'll call Bob, doesn't experience the same fear, and is friends with a purple person – so Bob knows through firsthand experience that Alice's story is not a true statement, because it is bucketing and judging *all* purple people, and has exceptions.

But Alice is still right to herself in that moment. As I see it, any assault to Alice's truth by Bob will *immediately* polarize her against him. He will be cast out of the bucket she puts him into (which includes herself) and instead become a threat to her personal validation. He becomes a threat to Alice's *entire belief system – her entire reality.* Everything will collapse into black and white for her when around Bob.

Stories like these are not your truth or my truth or Bob's truth or Alice's truth. There is only *the objective truth*, and that truth has a resolution of 8 billion stories at every moment (ignoring the systems that cannot currently make stories (babies, dementia, etc).

If a person believes "all purple people are bad" is true, then that is one 8-billionth of all the human truths on the planet Earth. It is also a truth composed of *enormous bucket words.* The path to peace and stability, in my opinion, is to create an environment where it is safe to dump them. After all, safety and security – the shelter need – is *nearly the entire problem here.* The goal is to stay in the *same bucket* as the other person – the human bucket - and to help him or her feel safe enough to begin to dump bucketed thinking, which makes the world a bigger place with more possibilities and more time to explore them. Boundary words are the key to pushing away the shadows and monsters, and achieving feelings of safety and security.

"In my opinion...."
"I think......"
"As I see it...."
"So if I understand you right, you are saying......"
"So it seems like you think...."
"It seems like to me you are trying to say..... "
"Do you believe......"
"So in your opinion......"
"Why do you believe that?"
"As for me, I believe...."
"Help me understand..."

These boundary words are so effective, I have found that even in the middle of a highly emotional argument, if either one of us suddenly re-insert them, the temperature between us *immediately* cools.

Try listening to or reading other people's arguments. You might notice that the instant these mental boundary words go missing is almost always the exact moment emotions explode into the scene. It pivots the argument that was about unique perceptions into a debate about those perceptions being valid at all.

We seem to be incredibly sensitive to our mental boundary being violated – which is wild to consider because we seem to not even be able to recognize it exists.

The mental boundary should always be respected, in my opinion. You can believe whatever you want. So can I. The only person ever changing my mind is me, and the only person changing your mind is you.

With all this said, in the quest for greater balance with this world and each other, and trying to build a better and more fair society, you will undoubtedly end up talking or arguing with someone else about what is true and what is not, or what makes sense and what does not. That's the treadmill of life – we can't get off of it.

The goal as I see it is not to let things collapse to violence. It is to always try to stay on the middle path – to hold the middle open always, and to keep increasing the resolution of each other's stories – which amounts to dumping buckets and falling down rabbit holes together until things become agreeable or at least make sense to both you and the other person.

This task – the *process* of growing trust and mutual understanding – The Process of Unexcluding the Middle – the process of shared language building - requires constant adjustment.

Once the beast is calm, however, knowing the limits and stabilizing the process to remain inside of those limits (or at least be able to recognize when it has escaped its confines), is critical.

147

Weighing Stories

Time is the most important resource we have as human beings in my opinion. If you agree, then "wasting" your own time (or someone else's) is to not value that time.

As I see it, the opposite of "wasted" time is mutually *consented* interaction, and I believe it is the highest honor two individuals can give each other in the pursuit of understanding. Each of us has a limited time to live, which means there is a limited number of stories each of us can consume and potentially merge with before death (which will almost certainly occur at a time and place neither of us can predict).

Since time is so critical, and we build and consume and tell stories constantly, I've found that considering the limits of human experiences can be immensely helpful to determine if your time is being respected (and vice versa). These limits are ultimately defined through the recognition of the mental boundary, and what is possible from the viewpoint of that boundary.

Another way to say this is the *content* of a secondhand story can be evaluated using a simple framework to determine its "story mass" – which can range from carrying enormous weight to having no weight at all. Knowing where the content of a story falls on this scale relative to the storyteller's own perspective can keep many problems you have read up to this section - The Hydra and whispers from the shadows that collapse time – in plain sight and calm.

To start, the content of all secondhand stories falls into the familiar shape we've had for the past couple of parts to represent human experience: A sphere. This story sphere has an inner core and an outer shell.

Each of us has our own sphere of possibilities, defined by our sphere-shaped sensory boundary. But the sphere shape assumes you do not move. Since each of us, as individuals, experience time in a straight line of actions or movements (A to B to C), the sphere of all possible experiences reshapes itself into two "cones": A future cone and a past cone. Here's the basic outline:

The Inner Layer (Survival - the Inner Sphere)

Communication, or sharing stories relating to the inner sphere of human movement, which deal with someone's next meal, next drink, or the ability to immediately shelter - are the most serious of stories. They therefore have by far the most story mass.

These stories can lead to a *Dark Forest Moment* and violence right on the spot, because they are potentially changing the length of someone's timeline. If you are communicating with someone and the content of your stories stand between this person and what they need to capture to survive, the thing in need of capturing could potentially be you, because you might be choosing the path of domination without realizing it.

For instance, if you are arguing that purple people should not be allowed to buy food at stores, while also blocking a purple person from entering a store to buy food, you are setting up a *Dark Forest Moment* and collapsing the middle path of stability at the same time.

In the same vein, if someone is telling stories to justify actions that would prevent you from capturing what you need to survive in the future (we should ban Eric's readers from stores and make it law), that person is setting up a *Dark Forest Moment* to appear whenever such a law is justified into reality and enforced.

In my opinion, it is best to tread carefully on the topics of food, water, and immediate shelter if it is not your *own* food, water, and immediate shelter need being discussed. These stories interact with the very real foundation and ground state of survival, so they are deeply tied to experience, reality, and physical violence in serious ways.

The Outer Layers: Free Time

When it comes to stories that develop in the outer layer of human free time and specialization, there seems to be several distinct "story layers" the experiences within these stories can fall into based on the mental boundary.

First free-time story layer:

FIRSTHAND

Firsthand stories. When you tell these, I think it's important to remember they are firsthand experiences *to you*, but *secondhand* stories to *everyone else*. Likewise, when others tell you their firsthand experience, they are always secondhand stories to you.

No human that you communicate with face-to-face is telling a firsthand story from an experience that took place *before* they were about two years old – unless something radically changed regarding human memory since I wrote this.

All stories told by the storyteller before that time period do not belong in this "firsthand" layer. Stories from firsthand experiences are the strongest stories that can be told, because the word shells are, in theory, being chosen and filled from the wordless memories of sensory experience with objective reality (filtered by the storyteller's wacky windows).

Just try to remain vigilant of the mental boundary, Dictionary Illusion, and Exploding Cat Problem while listening or reading, because the storyteller's words are always justified as right from his or her own perspective. I try to always treat stories from this layer with the utmost respect, unless these stories are crossing a boundary or intending to justify crossing a boundary into the inner core that threatens my immediate food, water, and shelter. This applies in reverse as well.

I think it's also important to keep in mind that all memories become warped and change over time as they are recalled and replayed, because each time they are replayed they earn a new layer of context – the one First You experiences in the present when that memory is replayed. This, as I see it, is the mechanism that leads to "romanticizing" a place, a person, or an event you experienced firsthand long ago. Details get fuzzy, and the context drifts in time warping the memory to be more wonderful than it actually was, or more awful that it was. This can make you long to go back, or make you want to forever avoid it, but these feelings are getting created by comparison to the current moment as it is being remembered.

In my opinion, the older the memory, the more likely it has begun to "float away" from its original context anchor of firsthand experience, and has instead become more of an "island" of a particular feeling floating in the context of a larger ocean of memories.

Second free-time story layer:

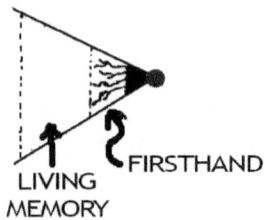

These are secondhand stories within humanity's current living memory. Another title for this layer might be Recent Human History. The farther back in time any secondhand story involving humanity goes, the more "truth" starts to expand and collapse any certainty (and with it right and wrong). These secondhand stories are about experiences that can be validated by someone that witnessed them firsthand.

For example, assuming I am still alive as you read this, or there are individuals still alive that knew me in person, this book (which is a secondhand story) belongs in this layer. However, my hope is that this book of stories is a wonderful and fun *firsthand* experience with your own thoughts and beliefs.

Over time, stories I have written specific to "my time" will break down more and more, and the list of word-stories that represent experiences with "all of society's problems" will be less and less understood. That is why I have tried to avoid anchoring any critical part of what I have written in this book to be based solely on my own personal experiences or current events.

This layer includes stories like the "Ghost Diary" of Alice that led to the eventual Great Sugar War from earlier in this book (if we imagine it actually happened). Alice's diary was in this layer while she was alive, but crossed out of this layer the moment she died, because Alice was the person holding the oldest firsthand memory of any living person. If she had been alive when Charlie read her book, the Great Sugar War might have never happened.

Relative to the storyteller, this layer reaches backwards in time from just before the storyteller's first memory to the earliest-in-time story any human is still capable of speaking about from experience. Every story repeated by the storyteller that is not firsthand experience, but *did* happen in his or her lifetime is also included in this layer.

Third free-time story layer:

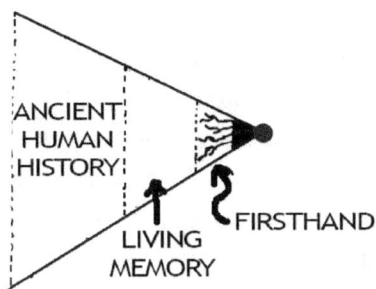

These are secondhand stories of movement and time *before* humanity's living memory, but after humanity began – whenever that was. Another title for this layer might be Ancient Human History. This layer includes any human story that took place before the earliest memory of any human currently alive. For the most part, this layer is made up of recorded stories in the form of permanent written communication that goes all the way back to cave paintings, as well as human-made objects.

I think it's very important to note that of all the human-created writings and artifacts we have ever found, there is an ever-shrinking pool of "evidence" to anchor any story to truth. This is because of decay.

Evidence, as I see it, is made of boundaries in reality (real objects) that seem to tell a story based on the context of where, when, and how the object was found, in addition to its shape or design.

However, *objective* evidence, in my opinion, also requires that if desired, you or I could actually examine this object with our own senses alongside someone else's *at the same time* (assuming we both had access to it).

This decaying objective evidence problem means the further back in time you go, the less "resolution" *any* story will have. For example, at this point in time, we only have theoretical stories of how and why the Great Pyramids of Giza were constructed. Without new *objective evidence* from the same time period and location being discovered to better anchor this story, we are likely to forever be stuck in this low-resolution position.

Still, "not being certain" seems to bother us, because we are pretty sure humans built the pyramids, and we also know the "resolution" of the human mind is very detailed. For unknowable reasons, and as I understand it, there seems to be no surviving stories to tell us how or why they were built – there are gaps and voids everywhere.

As a result, we sometimes find it irresistible to fabricate these missing stories of details and justifications from our own imaginations to fill in the gaps and bring the story of the pyramids to a "makes sense" resolution - so we can be more comfortable. However, all stories in this layer will always have lots of uncertainty in them, because even if we found a book from the time period the pyramids were built, it's entirely possible we'd end up like Charlie finding Alice's diary, and misinterpret what it says because our own present-day context for the same words is getting mixed in.

This third story layer in free-time extends all the way back to the earliest story of humanity that is believed to have evidence supporting it. The less evidence that exist to anchor an ancient human history story to reality, the more likely a heavy amount of assumptions are taking place in the story. Stories in human history that lack any objective evidence cross the boundary into the fourth layer.

Fourth free-time story layer:

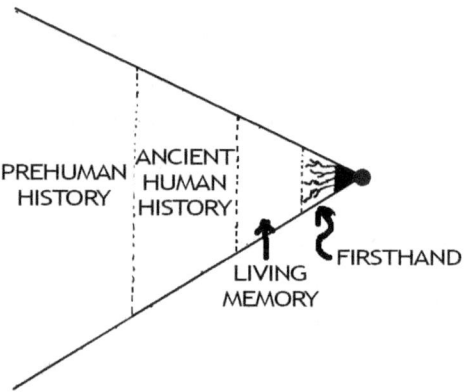

These are secondhand stories *beyond* evidence of humanity's existence. Another title could be Pre-Human History. In other words: No human to *ever* exist had firsthand experience with these stories. They are *all* packed with *massive* amounts of gaps, and often assumptions filling those gaps. This means all of these stories are improvable to some degree, and extremely prone to being believed purely on "making the most sense".

Some of these stories are spiritual or religious, feelings or hunches, and some are based on trying to work backwards in time based on boundary and object movement processes (such as chemistry and decay) we can still study today.

The Void:

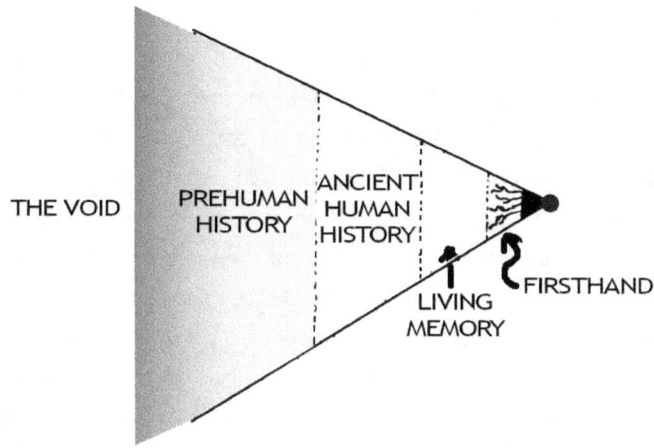

These are the stories no one knows at all – because the story doesn't even exist for us. While most of these stories probably exist beyond the layer of Pre-Human History, there are also pockets of The Void present all throughout this layered story structure – from the inner core all the way out to the very edge.

Stories of The Void happened in the past or present and went unexperienced, or are unable to be imagined based on what we have experienced.

Since unexperienced and lost stories in the past and the unknowable stories of the future appear the same to us, this is often what I believe to be the source of our own accidental story building paradoxes. We currently cannot know how much the past and the future are connected. Consciousness connects them through memory and projection, and acts to try to control and even change what *seems* to be a chain of connected events that lead up to and away from the present moment, but we cannot know with certainty how much any one thing ultimately affects the other until we understand the nature of the entire universe – until there is no more void for us at all.

Given this seeming shared connectedness of The Void to the past and future and everything in between, I'll use it to pivot the layers of time in a forward direction from the storyteller's present moment.

Future story layer into The Void:

When it comes to looking into the future, things become less confident and organized the further out you go. As far as I can tell, there are no boundaries or objects that exist in the future relative to this moment we can study and determine exactly what the future holds. We can only ever connect the past to the present to try to *predict* what is *most likely* to happen based on what we believe to be true.

Due to this limitation, there are no clear layers going forward into the future. This is just the firsthand sensory boundary, and then a layer that gets fuzzier and fuzzier until it spreads out so much that confidence in any story starts to fall apart.

When it comes to imagining what happens only one second ahead of this moment, you probably have a high confidence about what to expect. Of course, you also know the next moment is never guaranteed, so this expectation is still always an assumption. Throw in Black Swan "almost completely unpredictable" events like micro-meteorites, solar storms, and other human beings, and even the best truth-story predictions can possibly fall apart.

For example, there is a very thorough and well-laid-out theory that arrives at the heat death of the universe (the end of all motion) complete with calculations of when this is likely to occur in the very distant-to-us future. This is all based on the underlying assumption that our understandings of the universe is "right" and will continue to be "right" based on "laws" of physics we have validated over and over to be "right" from our point of view on this tiny speck of dust within that same universe.

There's no guarantee any of this will come to pass of course – we barely know anything at all about the *entirety* of the universe at the time I wrote this. It's all based on what we can currently sense and measure, which seems to be "not much."

I think it's entirely possible another cosmic "balloon system" will interact with the one I imagine us to be inside, which currently defines our sense of "objective" time. If this happens, it would potentially warp everything we thought we had established to be true in the distant past and future. Gravity may suddenly change in strength throughout the day, month, or year - or perhaps the seasons fall apart - or something far more drastic with a very slow burn (such gravity and seasons growing in length by 3 seconds every seven-and-a-half months, and then suddenly and dramatically stretching out to last 15 years every time the Andromeda Galaxy aligns with the fifth arm of the Milky Way. Oh, and gravity triples in strength in that same time period.)

My point is this: These highly thought out scientific theories might only contain one or two assumptions – usually called axioms - but they are *enormous* assumptions because they are anchored only to what we can experience and imagine.

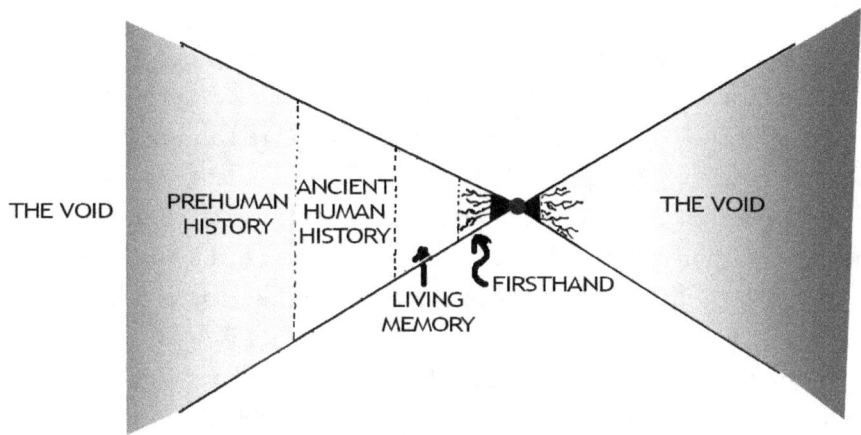

Putting all this uncertainty and assumption aside for a minute – we seem to love stories of all kinds anyway. When it comes to our stories, they travel all directions in time and include every possibility we can imagine.

In my opinion, it can get very easy to get "lost" in these layers and lose all bearing on what is true and possible. It is easy to get caught up in the "spell" of the beast whispering from the shadows across three boundaries or from a future that doesn't exist.

This is why I have found it immensely helpful time and time again to reference the beginning of humanity, the beginning of *me* – *The World of You* – to think about what stories would exist and which would not exist when the world of stories gets confusing or seem to blur all these boundaries together.

Doing this can sometimes quickly filter out what is an imagination-created void-story and what is a grounded-to-firsthand-wordless experience story, and expose what amounts to "experience theft" through the illusion of language. These layers can filter what is anchored to firsthand experience and is using "objective evidence", and uncover what is massless words weaving together stars and islands. They can tease apart what is knowable to the storyteller and what is not, and even what might be hiding in his or her buckets and gaps.

As storytellers, however, you and I have layers just the same as the stories we tell.

148

The Light of You

Our journey into the abyss and back is nearing completion. With a framework to measure the weight of secondhand stories and filter away those that might feed your Hydra (or someone else's), we can begin to "break the spell" of the beast and balance the layers of control and action that exist relative to you and your body. These layers have an absurd number of possible stories within them, so to me it makes more sense to define the boundaries that divide the layers, rather than describing the actions and movements within the layers:

First Boundary of You: The mental boundary, which separates your mind from everyone else's. No one can tell you what you are thinking, and you cannot tell anyone else what they are thinking because this boundary exists between any two humans. You can only ask what someone might be thinking, or share your perceptions of what it seems like they are thinking from your perspective (and vice versa).

Second Boundary of You: The boundary of your body – also known as your skin. Who should have *natural-born* authority to cross the physical boundary of your skin and override autonomy and control of your own body, forcing you to act against your will without your approval?

In my opinion, no one should be allowed to cross this boundary without your permission. I certainly do *not* have a natural-born authority to touch, hit, kiss, or break your skin in any way. As I see it, all of these actions require *your* permission – your *consent* - to justify as "right". There is a *ton* of gray and nuance with the process of consent, and I'll attempt to address some of this grayness later. In the meantime, here's a mental tool I've found very helpful to potentially determine the reason someone might attempt to cross this boundary without consent.

By coupling the *Dark Forest Moment* and *The World of Two*, it seems clear to me that the core balance between yourself and any other person is based on the *projected* position of dominance or submission, superiority or inferiority, or equality. That's it.

With this in mind, you might be able to see how you might have warped or damaged mental or physical boundaries of others in your approach with them, or notice someone else in your life doing this to others. This is why mental and physical boundaries are so incredibly important and huge concepts in my opinion: Crossing boundaries without consent seems to be the cause of endless cycles of polarization (two balloons instead of one human balloon), possible violence (instead of words), and collapse through generations (the two balloons finally collapse back into one).

No one was born the keeper of your body except you.

Third Boundary of You: Your sensory boundary (the limit of Step 1 in every direction using every sense you have right now). This is going to constantly be changing shape, because it is not a boundary you control beyond closing your eyes, covering your ears, plugging your nose, sealing your mouth, or sheltering in an enclosed space. Your senses are windows. How much control, if any, is reasonable over what is or isn't *allowed* to appear in your Step 1 boundary? How much of that answer would involve controlling other people or deciding to play *Keeper of Truth and Decider of Right* for other people? There's definitely a lot of food for thought and conversation to be had here, but it ultimately ties in with the next boundary of you.

Fourth Boundary of You: Your controllable space, which is the limits of what you perceive to be controllable based on the speed you can move relative to objects around you.

This one is perhaps a bit strange, because it is very much a *feeling*. One way to notice this is to consider how close you are willing to let something moving get to you relative to your own ability to deal with that speed. You want enough *time* between you and this object you believe you can justify movements quickly enough to control your relationship to it. You might call this your boundary of "personal space".

If a stranger is talking to you in public, and they move their face six inches away from your face – this stranger would be crossing this boundary, in my opinion. This distance would be perceived as uncomfortable because it is too close *in time* for you to react to an unexpected movement he or she might make next to prevent it.

For example, if someone this close suddenly went in for an unwanted kiss

– you will not be able to react fast enough to avoid it. If this person were to grab you from this same distance, it's the same story. All of this changes if the stranger took one or two big steps back from you. I admit, there is a lot of variation on personal space built into culture, but I think the speed-based approach makes significantly more sense because it is based on time and speed, and not norms and assumptions. At least until I consume a different story that explains why a speed-based definition is not a good one.

The solution to this boundary being crossed is simple and language-based: Ask the person to step back – to give you space.

When it comes to non-human objects (such as if you have to walk next to a busy road), I suspect you'd prefer not to be a shoulder's width away from cars zipping by at 30 miles per hour (48kph). You'd probably prefer a few more feet of space at the bare minimum, because you would want enough time to react to something unexpected involving cars moving at 30 miles per hour – which will be based on your perceived ability to move out of the way.

When objects move beyond the speeds of things that might exist in *The World of You*, defining this boundary becomes difficult. Objects that move faster than animals like birds and deer and their predators trigger a fear within us to keep them as far away as possible, so that we have a better chance to predict and avoid the object should it come our way.

This quickly results in a runaway shelter need involving invisible and imagined things. I'll say a little bit more around this specific problem in the next and final section of this book, but we've tried to overcome this by adding more layers.

Fifth boundary of You: Technology-extended sensory boundaries (secondhand extended firsthand experience, which I've called "shadow experience"). This is using technology to experience and justify movements you could not with your own "naked" senses. This is anything made experience-able through something like a microscope, a telescope, or just about any other type of "scope" or device that enhances your senses.

In my opinion, stories from this layer are to be used with caution when it comes to trying to justify action and control using them, because often a sense of scale is lost, which causes the experience to become total projection. For instance, you can read on the side of a lens that you are magnifying something "200x", but you cannot "feel" 200x magnification between your natural Step 1 resolution and this modified-and-extended secondhand Step 1 resolution.

This is why I highly enjoy "gapless" animations that try to expose these scales, which might include side-by-side-by-side size comparisons of many objects from the atom to a cell to a human, or from a human to the outer planets and stars and galaxies.

However, just like with the distant future and past in time, the further away this technology-extended sensory resolution is from your native resolution, the harder it will be to anchor these stories to the experience of these object boundaries at high speed. For example, the edge of ruler looks and feels "perfectly" straight, but is actually very course and lumpy once magnified several times. That does not change that it still *feels and looks* straight and smooth when you use it to draw a "straight" line.

Open boundary of You (Void): This is everything beyond the technology-extended sensory boundaries to what remains of all of existence. This layer is only ever stories from the lower layers being projected onto the void of the unknown and unexperienced.

This includes stories about the size and scale and nature of the universe, alien civilizations in distant galaxies, and a lot of times (as I write this) the quantum world. In some ways, this is human imagination pushing its own limits. In others ways, it is simply the entire void being relabeled to have a usable story for words like "nothing," "the universe," and "creation."

Finally, I'll offer a definition of reality itself.

Objective "object" boundaries: This, as best as we can ever understand it, is just our perception of change – of atomic-based movement and resistance. If reality were like The Matrix movies (1999, Warner Bros), we would not see lines of code everywhere, but simply "heartiness" and "shapes created through different intensities and densities of vibration" in the soup of atoms all around us in every direction that create the layers we can sense.

This "layered soup" of atoms would be very thin and quick-moving when we look up at the sky. It would be quite thick in liquids like water (relative to the sky), and have a somewhat clear "edge" where these two layers meet (though evaporation might make it difficult to see exactly where the "edge" exists if we could see individual atoms).

This same atomic soup can also get so thick it behaves like a solid under our feet, as atoms are trapped and pushing against each other by their own boundaries and gravity. Gravity tries to pull them all into occupying the same space, but because they are unable to pass through each other, we get solid "Earth". If we could see atoms, again there would be a decently clear edge between what is "solid" and what is "liquid" – but it would get pretty fuzzy for all of these layers at locations like the beach.

All of reality is the same soup. As far as I understand we know, it is all atoms. It is all their boundaries, their fields, their particles, their vibrations and their forces, and their absorbing and ejecting of light. The only thing that ever changes at our human scale are the shapes made by atomic *resistance,* which forms the boundaries we perceive. I can't pass my hand through a leaf, even though they are both made of atoms, because of resistance – which reaches all the way down to the stories of electricity, magnetism, and gravity.

Where things get interesting to me, is when you add heat. If you heat up this atom "soup" enough, all the layers will start to blur. The solid will start to thin and transform into a liquid, the liquid into a gas, and the gas into a plasma, which reverses if you cool it all down (though after cooling you might get entirely different shapes and objects, because the arrangement and placement of each atom in the soup has changed).

And with this, you can now turn and look back into the abyssal rabbit hole without fear of it, and see the monster – calm and quiet - made plainly visible in the light of you.

149

The Beast and the Abyss

Certainty does not exist. Even if we actually found a relationship between two things that was absolutely certain, we would never be able to know we had reached a certainty until it was perfectly repeatable in every condition and in every location at all times throughout the entire universe. Otherwise, we would wonder if there was something we were still missing – or if there was some exception to the rule.

When you couple this always-something-more approach of human nature, along with the "unitizing" of the human mind by word sounds that create gaps and categories in otherwise gapless experiences due to contexts that are never perfectly shared, you might be able to see how the Hydra is born from certainties being claimed in the abyss.

The Hydra's name is *Authority By Assumption.* It is the assumption that one's opinion and perspective is somehow actually *superior* to another's, and that one's opinion is objective truth itself. It is the assumption one is truly the *Keeper of Truth* – any truth at all.

Because of the *Floating Anchor Problem*, once other people's boundaries are successfully collapsed with assumed consent (or without it at all) and this is passed on through culture to be "right," one can quickly transform opinions of certainties and superiority into beliefs with no decay or expiration. This causes the my-opinion-is-objective-truth storyteller to structure *all of his or her reality* around this belief to *force it* to remain true.

The result is someone quite literally believing that he or she is the keeper of what is true, which expands their mental boundary of assumed authority to include the entire universe and all of existence. Such a person has been transformed into the monster in the abyss – and may not even be aware of how it happened.

The easiest way to spot this transformation is to notice the words or stories such an individual uses – the storyteller and the audience are often put into binary or black-and-white positions. To make things worse, because the assumption is one of authority over all others, these stories often contain very little shared context of life's difficulties or needs, and often have extreme violations of mental boundary crossing and the crossing of many boundaries in general. Here are some examples:

"You are an idiot."

"*This* is *that*, which means *that* will become *this* other thing."

"You deserved it."

"They did it before, and they will do it again."

"If you are not one of us, you are one of them."

"God would agree with me, not you."

"Everyone knows unicorns are real, just ask anyone!"

"Your beliefs are illogical."

"They are all a bunch of morons."

"It's all their fault this has happened."

"I am right and you are wrong."

"Political Party A is destroying America!"

"God hates purple people."

"Only I can save you from purple people."

"She did it last year, so I can do it now."

We are all struggling with beliefs like these from time to time in my opinion – especially when we are fearful. I beg you not to hate or be angry at individuals that communicate this way, as he or she are likely unaware of how he or she got into these mental positions. There is no one person or organization at fault. This collapse to black-and-white thought, in my opinion, is caused by the runaway process of our primal and forever-uncertain need for shelter. We all have a minimum biological speed that is layered against a much slower moving communication speed, which is required to be the same speed or face imbalance in perceptions of fairness.

Instead of any one storyteller having the absolute authority of truth, there are only individuals that know more stories about specific things than other individuals – but those stories are always from the individual's specific and limited point of view. They are merely an opinion – but always a *valid* perspective – of the same larger wordless objective truth.

Perhaps you can see where to question someone else's stories, and where to limit your own stories. When it comes to yours and other's beliefs, any certainties contained in them are *not wrong, but they are also not objectively right, fair, or complete,* and so the certainties need to be gently broken. Bucket-words can make a storyteller's mind operate in "units" that are too large, and he or she may need help dumping them. If you choose to help, keep in mind that many have wrapped their entire life and truths around these certainties due to fear of falling into the abyss, meaning it does not necessarily feel safe to let them go around someone else.

Where there is tiny simplicity, there is also enormous complexity. Where there is enormous complexity, there is also tiny simplicity. Solid unmoving ground does not exist for us – time is an abyss in both directions, and knowledge of each movement in time without knowing the smallest thing or largest thing to exist creates a second abyss of chain reactions up and down in scale we can never be sure about, either.

No one knows with any certainty where he or she was before his or her first memory, or where he or she will go after experience ends. We are stuck using a *floating ground system of knowledge and truth* that began in the middle and will end in the middle, because we ourselves are beings that use a floating ground system of knowledge and truth to survive – no matter what happens in reality around us.

With all this being said, and in my opinion, the most solid ground we have is the present. The further we reach into the abyss of time and movement to the past or the future, the bigger the word buckets that must be used, the lower the resolution of the story, and the less experienced evidence exists. As a result, there is a decay of certainty that any one "grounded" truth-story can be relied upon to stay stable and objectively right – and that may include the one you are reading right now.

This means the closest we can ever be to certainty is in the *right now*. You are reading this *right now*. In a few moments, you'll forget what parts of this you read at this specific moment in time (Test: What did you read at 11:34 A.M. last Tuesday? See?). If what you read within these pages makes sense though, it will merge and start to work itself into the context of your life, hopefully to help bridge and connect ideas in a way it is all better understood. That has been my entire goal after all: To build a bridge.

150

A Bridge of Straws

Before you enter the final part of this book following this chapter, I'd like to share the missing parts of my own story that led up to the writing of this book. After all, I never intended to write a book. I did not decide that I was going to try to spin reality together in a way that made sense and try to force a bunch of metaphors into an arrangement that built a unified belief system.

The story behind the creation of this book is far stranger. When I left you in Chapter 0, the house next door to me, which was inhabited by priests, was hit by a bolt of lightning, and was on fire. I then skipped over 336 hours to tell you about "The Event" that changed my mind and my life with it.

This missing time is a critical gap in the story of me.

Up until the day of the lightning strike, everyone else was a Keeper of Truth for me. Even though I loved learning and consumed every piece of information I could get my hands on – everyone else still knew better than I did. If I tried to say what I thought was true, and the person across from me or the person online said I was wrong – I always backed down. I must have missed something. I *am* wrong - somehow.

I grew up in the rural bible belt of the United States. I went to a now 100+ year old one-room country church on Sundays all through my childhood. You have already read about other pieces of my life and childhood – but essentially I grew up dreading every day I had to go to school or going anywhere public. Would I make it through a single day without someone making fun of me? Without experiencing physical pain at the hands of another? Without feeling I'm being constantly judged as right or wrong? The answer was almost always no. I always took the path of submission or the path of escape.

The world was always done *to* me, and I pretty much did nothing to the world in return. I overthought everything because I was trying to predict everyone else and meet his or her expectations. If they were so confident about everything and I wasn't, I must be the one missing something.

My life got dark. Very dark. I began escaping through various things I found interesting or gave me some hit of dopamine. It began with drums, but became so many more things: Cars, art, chasing girls, music, reading, collecting things

– anything that didn't involve needing approval from others, or were guaranteed to be approved by others. I carried this way of being with me all the way into my 40s. I consider myself fortunate none of these things were physically destructive, because things could have easily turned that way.

Then along came my first child, who was so eerily similar to me it was often like looking in a mirror. Fatherhood was amazing and fun, and I lost myself playing the role of Dad and not having to care what at least one other human being thought of me for a change. Years later, my joy started decaying as she began showing what I believed to be signs of the same dark social and self-destructive path I was *still* on.

I wasn't going to let this happen. I *couldn't* let this happen. Not to her. Was I the cause of this? Could I prevent this?

I talked with her as carefully as I could about what was happening in her life. What was bothering her, and what she was struggling with. In many ways it was like having a two-way conversation with my past self. The communication we used was many levels of metaphor removed from any firsthand experience, which she did not like to talk about. We talked in terms of walls, which were often made of sand or made of brick, and we talked of bridges and of boundaries, and what should and shouldn't matter in life.

None of it was helping get her off what I saw as the dark path. So I wrestled with the metaphors we had shared all throughout the days and nights. I struggled to sleep – instead twisting and turning the stories and thinking, trying to build the story and metaphors I needed to understand.

I tried to overlay the stories against my own childhood experiences, and compare what might be different or the same. This quickly pivoted everything into trying to understand myself. Maybe I could help her if I only knew what was wrong with me. I fought with metaphors of bridges between others and myself, and if they were superhighways or maybe just one lane rickety overpasses loaded with dynamite. I dreamt of walls, and constantly churned ideas of how to keep walls strong and how people somehow manage to still get behind them.

And then came the day that 300 million volts took a shortcut through my reality.

The priests inside were all okay, by the way. No one was hurt. They were each surprised the neighbors were all in a tizzy about things, and had no idea their attic was currently on fire.

As the firefighters ripped open their roof and dumped hundreds if not thousands of gallons of water into the attic, I meandered and talked with all the clusters of neighbors that formed to watch the spectacle.

Lightning. Priests. God. Scandals. On a day of judgment (on a topic tied directly to the religion the priests bucketed themselves into) by the most powerful branch in government, the Supreme Court.

These stars of sense-making were weaved into a constellation story by every single story system that I talked to that day, even by the priests themselves. We all noticed the story and told it out loud to ourselves with a dismissive half-laugh. There's no way these things were connected, right? *What a bunch of nonsense.*

And yet it persisted. For weeks the nearby neighbors and myself started trying to make sense of what happened from a completely logical point of view. There was no logical sense to be had.

Yet some were projecting the story they imagined from these connections onto the rest of the world. This boiled the entire chain of events into two enormous bucket resolution stories: There *is* some kind of cosmic justice at work in the universe, or this was a chain of random disconnected events that are difficult to disconnect from each other, but had nothing to do with the story of some larger cosmic force for justice.

Between my own stress of always believing myself inferior and wrong, my daughter seemingly tracing this same path into what I believed would become a personal hell for her, the arrival of a second daughter who might have the same struggles, and others implying the universe was *supposed* to work a certain way because it's what made sense to them, I reached a breaking point.

The clips came off the straw. Not the straw between the balloons of me and everyone else, but the straw between me and the certainty of being right about *anything* in this reality.

I let go of *everything*. I decided I would destroy and collapse everything I thought I believed, and told myself to believe, to find the truth of this place that must be hell. I needed just one truth, *any truth*, that I could be so certain about being right *no one* could make me believe or feel I was wrong about it. I had to find it for me, and I had to find it for my kids.

So, I wrote a story I thought I believed, and then tried to find stories that were the opposite. Both made sense to me. So I let go of both stories and dropped one level deeper. I wrote *why* I thought I believed the first story, challenged that, destroyed that, and dropped one level deeper. I did this *for days*.

If something made me feel uncomfortable, I took it as a beacon for the direction I should take. I faced it, wrote it down, challenged it, and often destroyed it as something I should believe with certainty, because there was more than one belief of what was right or wrong about it. Deeper I dropped.

Then, in the middle of one of these sessions I had already been doing for days, *The Event* unexpectedly happened. Everything collapsed. And by everything, I mean *everything*. It was like I had pulled the drain on the universe itself, and I was left seeing everything for what it is.

For lack of better words, I was completely free. All my stress was gone. All my fear had vanished. My thinking had changed. I wanted to do things I had always avoided and normally wanted nothing to do with. I *wanted* eye contact. I *wanted* to talk to others. The world was suddenly a beautiful place. I had reached some sort of paradise of the mind. No one actually knew *anything* with certainty. No one was better than me. I was inferior to no one.

But paradise was not for me to enjoy. In some ways, matters were now worse. How do you help your *children* from a position of mental paradise, when they can't be there with you?

This book is my bridge back. It spans from a place of mental paradise with the universe back to where I had left my family, and each word is a straw. This book (which was not started as a book at all), is the end result of my attempt to re-understand the world in a way that makes sense for me, because I needed to get back across those 10-seconds of *The Event* to the old me to fully understand what changed. If I understood, then I could hand that understanding to my kids in a way that could outlive me should something happen to me, and hopefully help them steer into a better place where the world could make sense if they ever needed it.

As I neared the old me, I realized that what I hoped to help my children with might help others that struggle as well, so I'm hopeful that maybe – just *maybe* – my journey could connect peace and harmony back to the world through this book-bridge, because everything awful seems to be one huge misunderstanding and power struggle built on fear and beliefs of superiority forged through generations.

I destroyed everything I believed down to the atom, where I found I could go no further (even with the help of the wonderful writings of Richard Feynman), but I did notice the unconventional path I had used to learn the story of electricity *within* the atom could be used to reframe *everything*. So I began connecting my heaven and my hell one word at a time for the more-than two-and-a-half years it took me to write this.

I am forever thankful for the help I received in the unlikeliest of places - from amazing people, which might have even included you - and potentially from reality itself, depending on the unknowable objective truth of that bolt of lightning.

This book works as a bridge for me personally. I can read it and return to inner peace and understanding and patience, because that's where I was when I started writing. As I approached the point of book-bridge completion, I have realized - with horror - that I have quite literally put myself in a room with doors that all lead into a void, with no way to know what might happen next.

I did not know what might happen for you while reading this. I do not know what this book could mean, or not mean, for anyone. I just want this world to no longer be the hell it seems to be. I don't want it to be this way for you or me, or for my kids, or for you, or for your kids. I cannot predict the future. No one can. It's all just probabilities with no certainties.

The next-and-final part of this book is my attempt to rearrange and layout groundwork for stability from the current state of society as I write this. To me, it appears the "free" world has flipped upside down from fear of falling, and keeps attempting to anchor truth-stories to the looking glass, which obscures the shadows from where the Hydra of authority-by-assumption whispers a million certainties.

PART SEVEN – THE NEW FOUNDATION

No system can be held still.

Time cannot actually be stopped. Not by human hands.

Everything is moving.

Embrace it, and keep moving forward.

151

Pulling Up the Anchors

Nothing in the universe seems to be holding still. The sun is moving. The planet is moving. We are moving. The electron and the atom are moving. The entire universe itself might be moving.

It seems to me that almost all of our problems come from trying to stop movement. We seem to want things to hold still so we can study them, figure them out, and finally be certain about *something*.

This, to me, is an exercise in going mad. You can try to make reality hold still with words, but it will only create stories you must keep telling yourself. You can try to make reality hold still with immovable mathematical constants, but it will only result in errors of prediction of varying degrees. You can try to make people hold still and be controlled, but that will only create polarity and violence. You can try to declare what actions and beliefs are right and wrong and true in a way that can never change, but you will only end up having to constantly revise these statements – or face collapse into polarity and violence.

Trying to make anything hold still is to try to build an *immovable* foundation.

I think the best way forward is to let go of this story. Let go of the story of "objective" morality, truth, right, and wrong within the human experience. We are each a floating-ground system, and constantly float and reset our grounds independently using language and actions relative to each other.

This relativity continues up the scale: Humanity itself is a floating-ground system. There is no ground – we define our own ground. We were born hurtling through space on a ball of magma with a hard crust and liquid water (probably), and today we are on the same spherical ball hurtling through the same universe (probably) without the faintest idea of what is happening in the grand picture of everything.

Every "unmoving" system we have created, such as language, math, and scientific fact – which have brought amazing progress to our understanding – are also ultimately floating-ground systems. Language and other long-lasting recordings of human stories allow us to stretch our "memory" far beyond our own firsthand experiences, and project our imaginations just as far into the future.

In my opinion, the process of teaching these "unmoving" human systems to the next generation too often gives the impression that we truly have a solid, unmoving foundation under our feet – we do not. We have figured out an amazing number of things thanks to these systems, but we seem to be nowhere close to figuring out the whole story.

This section will be my effort to put everything you've read so far *into motion*. The goal is to layout a groundwork - a *floating* foundation – one that allows growth in *all* directions equally instead of the directions the most violent and authoritative see fit.

This foundation is built around constant movement and attempts to leave room for adaptability to change, while also respecting and recognizing the very real boundaries that exist between you, me, and everything else.

As I see it, the solution to the *Floating Anchor Problem* is to not focus on finding an anchor point and controlling it, but to instead focus on where we are going and embracing the reality of how we seem to operate. Withstanding and accepting constant movement and change has to be included, and yet the ship of society still has to hold itself together through the rough seas of movements in the universe.

So let's pull up all the anchors with a few questions:

What is it you are ultimately seeking to "capture" in your life at this moment?

What is it you want to have and make last forever?

Do you wish to capture death? A billion dollars? One certainty?

When it comes to that last one, forget about it. It's not going to happen. The only certainties in this life as I write this are decay and eventual death. Even then, no one knows what is truly on the other side of that moment. The only thing you can ever do is keep improving your truth-stories relative to your past to enrich your future.

Knowing the answers to those questions above can help you see an enormous problem. Solving it once you know where to look for it is actually pretty easy. Does whatever you are seeking to capture with more certainty involve needing to capture *other human beings*?

Let's start floating this foundation and see if we can make it *fly*.

152

Rewriting the Rules

I'll begin trying to layout this floating foundation framework by addressing The Rules.

Rules that outlaw a specific action (movement) are subjected to the Exploding Cat Problem because they are human-created. Because they are a human-created, rules follow (wait for it)...*the 3-step story process*.

Assume a rule has just been written. From this point forward:

Step 1 (perceived present): The rule is broken or unbroken according not to the rule writer, but to *the Rule Enforcer*. The *Dark Forest Moment* arrives with "the words in the rule" assumed to be *an object* of shared reality, truth, right, and wrong. Arrest and trial of human beings is justified (or not) based *only* on the Rule Enforcer's *judgment* of what the words in the Rule mean.

Step 2 (merged past): The story of past actions performed by the individual judged to be breaking the rule (per the Rule Enforcer's judgment) is consumed by the judicial context (a judge) who is an "expert" in upholding social fairness and balance. The Judge determines if the Rule was truly broken or not based on what makes the most sense, or previous rulings deciding the meanings of words in the Rule.

Step 3 (future): The judge's truth-story from Step 2 is coupled to the new present moment to project a probable future story of the rule-breaker, and justify the "right" consequence to maintain balance.

Action Justified: Based on Step 3, the rule-breaker is enclosed in a boundary of the judge's design: Extreme isolation, locally confined, or simply limited in ability to interact with society. This boundary will be enforced not by the judge, but by Consequence Enforcers. These boundaries can have different resistances judged to be "right" to control and prevent the "wrong" individual from being wrong again.

Examples of boundaries for rule-breaking could be a prison cell for higher resistance and separation from society, or house arrest or parole officer visits for a medium resistance and separation from society. Perhaps large sums of money are made due for high resistance to future rule breaking, or maybe a time penalty is used: Community service. This list can go on and on.

The problem with this system, as I see it, is that the original rule too often creates an *arbitrary boundary* that is different from person to person. The rule enforcer's *judgment* of what the words of the rule mean become the very black-and-white justification to confront, arrest, warn, or not arrest other people. This creates social imbalances everywhere the Rule applies, and immediate polarization *whenever a Rule Enforcer enters sensory boundaries*, because the Rule Enforcer's context is given the authority of the entire authority hierarchy until a judge decides otherwise.

If you rearrange the words of laws and rules so that they *only* include boundaries and consent, the rules won't be as easy to collapse over time because they naturally include time. The rule is no longer trying to capture *wrong*, which creates enormous social polarization and voltage. Instead, it is only trying to capture *unconsented* crossing of boundaries.

This boils any rule down into two very simple ideas. The first is the definition of a boundary, which becomes a closed door between you and me. The second is consent – or permission to cross that boundary - which is the door being unlocked by you for me, or me for you.

If the boundaries of a rule ever capture more than one person's actions at a time, the bucketing process has begun, and unless it is checked, the process of growing an imbalance in "right" begins. This results in two opposing groups that will battle each other over control of the only bucket remaining – those on the *other side* of the rule's boundary.

At the time I wrote this, this is already happening in the United States, and the two opposing groups are called Democrats and Republicans (or liberals and conservatives). The Rules are the United States Constitution. The Constitution contains something interesting to consider: The Bill of Rights, which are Rules limiting The Boss's (government) authority over the people. Overall, the Constitution contains checks and *balances* intended to prevent imbalance and corruption (and the possibility of a "king" from ever re-emerging).

Slowly, individuals in these two groups, along with individuals that fund and support them, are all held together by the assumption they are united under the same beliefs. These two groups have grown far larger than any individual branch of government. This has resulted in the steady "capture" of the *checks and balances themselves* – the Constitution - because these "parties" are beginning to effectively undermine the Rules that limit their power. This is done using all the storytelling methods from the last section to polarize beliefs into black-and-white, which then coordinate individual's justifications to act out of fear across all branches and all states - bypassing the Constitution's ability to limit anything.

Essentially, politics in the United States is now a war between these *two kings*, with individuals placing themselves into a self-chosen "king-bucket" labeled Democrat or Republican, or liberal and conservative. That self-defined bucket is then projected onto all "others" in the population, and each individual then labels others in black-and-white to create the "kingdoms" of *Us* and *Them*.

If one kingdom loses authority to the other based on the "Rules", it is intolerable and "unfair", because the winning kingdom has individuals throughout the entire government hierarchy to create polarizing change very quickly. If the losing kingdom ever regains control, it will attempt to change whatever is needed to ensure the loss of authority to *the other* never happens again.

For example: Each party keeps a list of judges to install should a seat ever open up to "make sure" the judicial branch's judgments and precedents will fall in line with their party's brand of rules and control. Each party has even started trying to manipulate individuals in state and local governments to support their attempt at *permanent* king-level control.

And yet, *not a single person in either party actually believes the same thing as any other*. It's all just *assumed*. This assumption is rooted in fear, and validated through language. That fear is largely created, in my opinion, by the system being designed as "winner takes all". Under this system, one party has near-total control, or there is usually gridlock. The middle path of stability is decayed.

In my opinion, the entire United States political system needs to be overhauled. Perhaps you are an expert in the field of law that can think of ways to fix this situation. Perhaps you yourself are a "member" near the top of one of the parties-of-shared-belief fighting for your "king" to get permanent control. I will pitch some ideas for changes over the next few chapters, but I am in no position to say how things *should* be – then again, no one is. There is only how I see them from my perspective, and remaking the shared "contract" between us by pulling in your perspective as well.

Everything in need of change in your world begins with the party of *you*. And every rule needs to be as fair as humanly possible to *all* parties if it is going to survive. Essentially, for a change toward fairness to work at all, *everyone* needs to consent. The *Floating Anchor Problem* absolutely wrecks this. I was born into a system I did not consent to, and so were you. When it comes to the United States, the only individuals we can be somewhat sure consented to our system of government are those that put their signatures on the Constitution itself.

And this brings us to the critical idea of balance. One balancing concept is sovereignty, which is just a fancy word for "having authority inside a given boundary". Within the boundary of your body, you have total sovereignty. No one else in this world has natural-born authority over your body or your beliefs. Consent for this authority is always given, not taken. If consent over you is forced and taken, the result is the breaking of the only true rule that exists in this floating foundation system:

Do not cross the mental or physical boundary of another human being without consent.

If someone says what you are thinking without your consent to speak for you, the rule has been broken. If someone pierces the boundary of your body without your consent to do so, the rule has been broken. You have total sovereignty, total *authority* over your body and mind. This means you and you alone are responsible for your beliefs and actions. The culture you were raised in grew many of the beliefs that ultimately set the tone for those actions, but that is now revised: You can hopefully recognize you are free to make choices and pursue stories outside of those word-and-geography limited story-boundaries of right and wrong. *You have a choice.*

And this loops us right back to political parties and groups defined by beliefs in general: All groups exist in an *imagined* boundary – which I called a bucket. These boundaries are defined by assumption and labels (words) alone. This means groups within humanity exist purely in a mental boundary – so the only thing that truly matters when it comes to individuals that *tell themselves* he or she belongs to a group, is the person's actions in the *Dark Forest Moment* while believing he or she represents that entire group. Is the person justifying supremacy and dominance over others by using the label of the group-bucket as a source of authority or threat? If so, that person is trying to cross a boundary without consent.

While I personally would like to see political parties and human "groups" dissolve, I recognize this is not really possible with a snap of anyone's fingers. There is far too much fear in the world as I write this. If parties were magically dissolved, all it would take is two individuals working together to *dominate* another individual's idea for the "party" to be reborn.

This begs a question: Why would any individual's idea *need* dominating?

As I see it, teaming up is a good indicator the idea is perceived to be trying to *control something*. It is being viewed as an attempt at *authority*.

This seems to be what all the fighting is about as I finish writing this in the summer of 2024.

Everyone, possibly including you, seems to be fighting to "gain control" over one, some, or all of the following "ideas" that are being perceived as a threat to your sovereignty or to individual sovereignty in general: The government. Deep State. Democrats. Republicans. Election security. Social media. The Supreme Court. MAGA. Woke-ism. Liberalism. Conservatism. Fox News. CNN. Stupid people. Millennials. Boomers. People that think they are smart or better. Scientists. Transgenders, gay people, and drag queens. Black people. White people. Illegal immigrants. Guns. Gun control. Gun promotion. Mental illness. Mental healthcare. Critical Race Theory. School indoctrinators. Teacher and school support. Money. Profit. Science. Religion. Lack of religion. The absence of god. The devil. The police. Our broken healthcare. The ultra-rich. The unions. Black Lives Matter. Antifa. Morality. Right and wrong.

The true problem is not any of these ideas – as I see it, it is the collapse of human trust in one another into fear. Through this overwhelming fear, the entire United States is currently a giant fight for authority thanks to the paradox shelter problem, which mixes in horrific ways with the *Floating Anchor Problem* that keeps resetting "normal" to a new place for each generation. The list of words above have come to represent a subconscious (or automatic) feeling of fear and an urgent subconscious need for shelter, forged through constant repetition of these words as ideas tied to authority or threats over others.

The boundary of most of these words (the walls of the buckets they create) *don't even exist* in any meaningful or measurable way. How they *do* exist in the imagination might surprise you, though, because it *immediately* clashes with the First Amendment of the United States Constitution, and also with the One Rule in this moving floating-foundation system. Here's the First Amendment if you aren't familiar:

> Amendment 1:
> Congress shall make no law respecting an establishment of religion, or prohibiting the free exercise thereof; or abridging the freedom of speech, or of the press, or the right of the people peaceably to assemble, and to petition the Government for a redress of grievances.

If no one truly understands each word the same way or to mean exactly the same thing, then what the heck should we call it if two people *believe they do*?

As I see it, such a belief becomes the foundation of a new two-person organized *religion*. This religion's creed is that words *can and do* mean the same thing between two people. And this quickly starts unraveling *a lot of things*.

Belief in facts is a form of religious belief. The higher power being worshipped by this religion of linguistic assumption is the belief that facts exist at all as unmoving certainties. There are exceptions to almost every "fact", because all them - even scientific facts - fall apart at some level of precision and resolution. There are exceptions and margins of error everywhere for everything.

This means if the first amendment of the United States Constitution is to be followed as written, any shared beliefs of group identity smaller than "human beings" cannot be respected, because that would be "respecting an establishment of religion" – which happens to be the oldest established religion of all.

This means all political parties must immediately be ejected from the system of government. Taking things to this point creates the paradox that destroys the amendment itself, because belief that the words of the amendment are understood to mean the same thing *by any two people* makes the amendment a *doctrine of the same religion*.

Does this mean that words and language and human beliefs built from them *are* god and/or a higher power? That whatever language collects the most users must be the one true and rightful higher power and/or god?

That's dominant/submissive thinking. That sends you back into the pit all over again. As I see it, the idea of a higher power, often called God, exists somewhere else entirely.

153

The Assumption of Belief

Uncertainty of everything is relatively easy to deal with in my world of electrical design. I can account for it by inflating my numbers to add a protective layer to the design's "territory" to increase certainty my design will work as intended. Inflation, of course, is a relative word, and the foundation of number measurements I inflate from are "found" from repeated testing – measuring *actual* electrical current through *actual* objects using a *similar* design. I use these measurements as evidence to set the expectation the values I choose will be "safe".

When it comes to needing to act, uncertainty is not as easy to deal with. The span of time between Step 3 (what you believe to be the most likely future story of movements around you), and the moment you actually justify movement anchored to that projected story being right is a *really big deal*.

Think of two identical empty balloons, inflated to different sizes, and then clipped-shut. The bigger balloon gets labeled "the most likely future story based on what I believe, not objective reality" and the smaller one is labeled "justification to act".

When it comes to your story system, there is a problem that prevents these two balloons from being connected cleanly by a single straw-path: uncertainty. If you or I don't know the whole story about anything, then that means there is *always* an open "gap" between these two balloons. There is no straw at all. There is only the void of uncertainty.

This uncertainty is what the first two parts of this book spent trying to uncover. No matter how certain you feel that something is absolutely going to happen, there is absolutely no way to be 100% sure it *will* happen. Something in reality can *always* end up changing. Something can *always* prevent that story from meeting expectations. Some variable hiding in darkness outside of your senses could be waiting to interfere with what you thought was a perfectly straight line between two points. *A meteorite could destroy the treadmill and the plane along with it.*

No one knows the whole story about anything. Yet, we all *must* move anyway. We all *must* act anyway. It is forced upon us to survive.

This means every justification is choosing to step through a door into darkness time and time again. It is not until *after* you step into the blackness of the unknown that the lights come on – which only reveal more doors that lead back into the already experienced, or further into more darkness.

The doors and the darkness are not what are changing – *The Ark* is a metaphor for everything that's between your current understanding while reading this book, and understanding the entire universe. It is only you who can ever change your own understanding.

We begin as babies, trapped in a room with no windows or doors. We cannot yet justify our movements on our own, and our senses are limited and don't yet make sense enough to build stories. We will die if someone else does not save us by justifying for us.

Eventually though, we learn to move our bodies, and can begin to shape our reality by choosing to open a new door, or choosing to leave it closed. By our end, if we are lucky, we make it to a space with a billion options to choose from, though we will always lack the time needed to open every door. There is simply not enough time to know every choice possible.

When you justify a movement or an action, you do so with the *hope* that the gap between the story you believe is playing out, and the action you perform as part of that story, line up as you predict and expect. All you can do is hope, and have *faith* that your senses and your story system understand all the boundaries around you "close enough" so harm does not come to you, and your continued survival does. But you cannot pre-validate your actions to be "right" in any situation before you justify them.

You and I are always dealing with an incomplete story at all times because we cannot control the resolution or the refresh rate of our own senses in this moment. You cannot make your eyes suddenly see viruses so you can avoid them. You cannot tweak your hearing so you can hear the Sun in this moment. You cannot speed up your senses so much that you can dodge a bullet whizzing toward you or slow down a beam of light to walking speed. You and I cannot change our limited senses, and yet you and I both have to justify actions using them *anyway*.

This means every single action you or I justify has *some degree of faith* involved. To not need faith is to be certain about the outcome of your actions before they are made – which would require knowing the entire story - the whole truth - about *everything*. It would require knowing every story of every action in all of existence, on both sides of every boundary, because every action in the universe has the potential to connect and have an affect on your own.

Being certain of your actions before justifying them requires *knowing* that a meteor is going to hit the plane on the treadmill before it happens. It requires knowing the exact dimensions of that meteorite, its angle of descent, its speed, and its weight. Being certain before justification requires *knowing* every single word I'm going to write from now until the end of the book without even reading them, and also knowing exactly what experience I felt and context I used as I wrote each one.

Being certain of your actions before justifying them requires knowing the shape, size, and nature of the universe, from its outer edges (if they exist) to the smallest possible measurement inside. It requires knowing the highest voltage possible between any two points inside that universe, and even the imbalances that exist between the inside and outside of each and every point. It involves knowing every movement and every current and every boundary at every scale, and by knowing all these things, *everything is already inside the boundary of firsthand experience for you.*

I feel nearly certain no human has ever or will ever come close to such a position. This would be experiencing complete unity with every thing and every movement in existence, including every single process in your own body *and* in mine.

If this were possible, you would already know all of your own movements and how each will affect every boundary, current, and voltage in existence. You would already know the consequence of every action performed by all things, not just yourself, which includes the consequence of every word-action used by every person in existence at this moment as well as every single word that will occur after. There can be no unknowns. There can be no mystery at all. There are no doors to choose. There are no choices to make. There is only the *singularity of you*. Everything is *one*, and one is *you*. By becoming certain in this way, the gaps between prediction and justification, and justification and validation *no longer exist*, because there are no gaps anywhere.

It just so happens, as I see it, this is exactly what we so often are alluding to when discussing a higher power, which is so often given the bucket word "God". This word is a way to "capture" two gaps at once.

The first gap is the prediction-to-justification gap (or the trust gap), between what you *currently* understand about yourself, and the *complete* understanding of yourself.

The second gap is the justification-to-validation gap (or the gap of understanding), which is the void of uncertainty between what you *currently* understand about the universe and its workings, and your *complete* understanding of the entire universe and its workings.

These gaps together expose a very interesting system, in my opinion.

The only reason the previous paragraphs are possible for me to construct and possible for you to make sense with is because our story systems (minds) and our experiences, which includes the meanings of words we were each taught, are *similar*.

Countless others before me have written *similar* stories to everything I've written. These stories are created by the shared *ability* to project the future around us from the present and remembered past – and to use language to expand each other's possibilities through the illusion of shadow experiences that can reach *far* beyond the boundary of our humble firsthand experiences.

This illusion gives us the ability to "make sense" of reality even if we have never experienced the specific chain of events filling the words. This is accomplished by our imagination, but rests on our *shared foundation*.

As I see it, our deepest shared foundation is the assumption that *consistency* exists. It is our shared faith in the belief *if this happens then that should happen*.

If reality were pure chaos without a predictable order to it, then this illusion would not be possible. Humanity wouldn't even exist, because we would each be *radically* different from each other physically and mentally. We would each have *wildly* different experiences with no predictable similarities or overlaps at all. Everything would be completely unpredictable. The same word could never represent approximately the same experience in such a reality – which means language would fall apart completely because it would become impossible to anchor the same word sound to similar experiences, because there are no shared similar experiences.

Of course, as I write this, humanity has not yet ventured outside of this particular gravity well known as Earth and its Moon. At this point, we can only *assume* that our story systems will work inside and outside every *other* gravity well in the universe at our *human* scale of experience.

This belief of "if this is, then that *should* be" relies *entirely* on the assumption that *all* the balloon systems, or every single copy of The Process in existence, can eventually be known and understood, we just haven't found them all *yet*. This relies on the additional assumption we know *how* to know the whole story, we just have to keep *growing and merging that story* to be longer and longer until we have the whole thing.

In my experience, this concept is often labeled "the arrogance of man." I don't see this as arrogance, though. Instead, I see this as a massive *shared assumption*. This creates, in effect, a species-sized *religion*. It is the religion of the *human-projected experience*. Another title for it might be the religion of human consciousness. It might even be the religion of *all* consciousness.

To me, this shared foundational religion is born from the human mind's need to capture shelter, which is almost completely out of control thanks to the Floating Anchor Problem and the existence of other human beings trying to capture the same end at the same time. Thankfully, this resulted in the amazing phenomenon of being able to create new *imagined* sensory experiences (and the feelings that come with them) for each other from nothing more than sounds and symbols anchored to fragmented and unit-ized memories. Memories of electrical currents that we can rearrange and transform for each other – which you are doing right now.

Belief that anything in reality can be predictable and remain predictable is to have faith in the assumptions that "discovered" this story. Believing that the story of the imbalanced balloons collapsing to balance with each other is an exercise in faith in the human experience.

This means acting based on stories is to act *entirely* on faith in what makes that experience possible: Human consciousness.

To discover that an act you justified on faith turned out *as expected* transforms this faith in your consciousness to *trust* in your consciousness.

To discover that an act you justified on faith did not turn out as expected transforms this faith in your consciousness into *fear* of your consciousness.

Faith in your own consciousness, or story system, and the assumptions it is constantly making on your behalf based on your experiences, is the mechanism that moves you toward a life with increasing fear or a life with increasing trust.

To believe in yourself is to have validated faith (trust) in your own memory and imagination to make "close enough" predictions and projections of reality for you to not immediately die.

This shared foundational religion of consistency, which puts faith in human consciousness to uncover it, seems to be validated and therefore justified. It seems safe to trust our own assumptions. It seems safe to believe ourselves, and to keep believing ourselves - because we are not dead.

But there are limits when it comes to expanding belief.

154

The Plasma Sails of the UHS Tesla Sphere

The religion of human consciousness rests on a foundation of faith in our own assumptions. These assumptions are validated into trust or invalidated into fear as they continuously cross the biggest gap of all: The gap between our beliefs and our actions.

But how in the heck did this process even get started?

In my opinion, human consciousness – our story systems - is brought "on-line" in a time and space that are *not* unpredictable chaos.

The existence of *stability*, in the form of "approximately consistent layered atomic soup" of moving boundaries we call "objective reality" is what I believe to be the deepest anchor point for every belief, language, science, religion, assumption, and every creation or accomplishment so far within that same soup.

And the shape of our current bowl of soup as I write this - all of the Earth's solids, liquids, gases, and plasmas held against each other by gravity - *is a sphere*.

Achieving complete *unity* with this soup seems to be the goal we subconsciously strive to reach, and we get ever closer to this goal by taking leaps of faith into the darkness of uncertainty. What we learn from these leaps of firsthand experience is then used to try to make better predictions, which are proven valid or invalid by the same soup that supplied the experience in the first place.

You are made of the *same stuff as the soup* – atoms – which make up you, everything around you, and everything we currently understand. Therefore every step you take into the darkness toward greater understanding of the atomic soup *also* results in greater understanding of yourself.

You began at the lower limit of human understanding. The lower limit is to have *no* stories at all. You had no understanding of anything at all – not even yourself. Beliefs did not exist. You knew no stories and you could not willfully act to change or create any stories.

Once your story system first began to loop (after about two years of incoming sensory currents caused by "the soup" that grew, stretched, shaped, connected, collapsed, and expanded the straw-paths needed to make that loop happen in the first place) you created your first memory and then first prediction, and eventually your first validation.

With this first validation, you experienced *trust* for the first time. From that point on, you had no choice but to rely on this trust in existence all around you. After all, that prediction was validated *without any input or action from you*. Trust was validated into existence by objective reality interacting with your senses *consistently*. The more reality stayed consistent, the larger your starting anchor of trust grew to be.

In this way, *your body* is the second point to the first point of your parents' bodies, but your *firsthand experience* is the second point to the first point of the layered atomic soup we call Earth, which may or may not extend to connect to the rest of the universe.

From this trust in your reality, you then start to grow *faith* in a new idea: You might be able to *act* in a way that controls and even changes reality. You might be able to stir the soup.

This is the origin of *faith*. Faith that you might understand.

Eventually, your assumptions about reality (built from lots of observation made possible by reality itself), and your assumptions about how your body moves (built from lots of practice with that same reality), start to align with each other. And then, in a single moment, you justify making a big step into the void: You justify an action based on what you assume (predict) will happen next in reality, while also imagining how this movement will interact with that assumption. The result is the merging of two worlds into one: Your imagination and reality.

This is the origin of "belief." You predicted something was going happen and had so much confidence in it that you acted based on it. There was no guarantee it was actually going to happen, and that's exactly why justification is an act of *faith* resting on the foundation of belief, which is resting on experienced consistency.

Your 3-step story system validates if your "leap of faith" from belief was justified, based on how much went as expected. If as expected, faith in your own beliefs is transformed into trust through experience, which expands your assumed understanding of yourself and reality.

In other words, the gap between your beliefs and reality starts to shrink.

This means that trust and understanding are rooted to the assumption in *all of experience existing as a single system* connecting everything on both sides of the sensory boundary known as your skin, eyes, ears, nose, and mouth. Understanding grown from firsthand experience validates the core belief that faith in yourself and in your reality *together as one* is justified, can be trusted, and you (as an object we decided to call a "human being") are connected to the atomic soup of objective reality in a meaningful way.

In other words, the more you discover faith in your own beliefs to be valid relative to reality, the more you will trust yourself and your reality, and inch closer to personal peace, which is to achieve a complete unity, or validation, with yourself.

Reaching this point transforms the Hydra into your much-quieter friend and guardian angel that works hard to keep you alive and maintain balance as you continue to grow your trust and love into the universe.

This idea seems to line up with the teachings of many present-day organized religions. However, here's the catch that also keeps organized religions from hijacking this idea to validate themselves as the "one true religion": No single person knows the whole story about anything in our little gravity balloon atomic soup system called Earth, so claiming to know the whole story of the *entire universe* and use that as a source of authority, as many individuals do using religion, is out of the question. No one is qualified to be the Keeper of Complete Understanding.

Your understandings are part of this existence – and they work in parallel with everyone else's individualized understandings forged through their connection to what seems to be the *very same approximately consistent atomic soup*.

The only difference between how you feel in *The World of You* and how you feel while reading this book is the level of trust you have in yourself and your understanding of reality. In the absence of trust, you collapse back to the choice to have faith in your beliefs about reality, or not.

For instance, you probably have lots of trust in your surrounding environment for you to be reading right now. This is made possible by consistency of your experiences with reality, wherever you might be.

If your environment suddenly changed – like a meteorite crashed down beside you right now – that trust between you and reality might be shaken (invalidated) and collapse. Fear is the *current* as this "contract-of-stability" is violated, which keeps being felt until a new stability – a new "ground" is found – consistency in your firsthand experience.

After the meteorite, you would still trust your mind and body (First You) to move you as expected. You might immediately run, or investigate, or call first responders, and you would do this action without once stopping to consciously re-validate your faith in your ability to perform it. You have near total trust in yourself to run "right", how to move around to investigate "right", or how to use a phone "right". Why? Because nothing changed on your side of the boundary. It stayed consistent. If it did change (because the meteorite badly injured your body perhaps), your story system might go into shock from the sheer amount of change on both sides of the boundary of you: Destroying your trust, beliefs, and faith in your understanding of both yourself and reality.

Assuming you are *not* injured, you (Point 2) will start trying to regrow understanding in your environment (Point 1) in every direction from this trust-island-of-you. Faith is the "arcs of plasma" that reach across the gap, connecting beliefs in reality founded on validated firsthand experience with reality itself through justification "lightning".

This same thing is happening with every person around you. We are each a Point 2 to the enormous Point 1 of the atomic consistency created by resistance and gravity. The only thing that is ever different between us is the size of the gap we have to cross through faith.

The more faith in your beliefs is validated through firsthand experience, the more you trust your justifications to be "right." Some call this confidence. I'm going to call it lowering your resistance to trust.

I think *The Ark* thought experiment is a great place to understand trust resistance. At the beginning, you had very little understanding of what was going on and where you were – you resisted trusting this place. It took an act of faith to open the very first door.

You have trust in yourself to know how to move to open the door, but have no trust in The Ark itself beyond the slow-growing faith that the room and door are not going to suddenly change, which is validated into a growing trust by every story cycle that passes since entering – which is probably in the billions.

Consistency across cycles is all that is needed for faith to expand from your beliefs. The darkness is nothing more than not knowing if consistency and predictability will exist beyond the boundary of what is understood. The inner sphere - or ship - is belief, which is grounded to experience (which is grounded to nothing but assumption and faith in your own senses and story system). The plasma-lightning in all directions is faith (your headings), which forge the possibilities of your movements (justifications). The dots in the darkness are secondhand stories. Distant islands or stars of "sense-making".

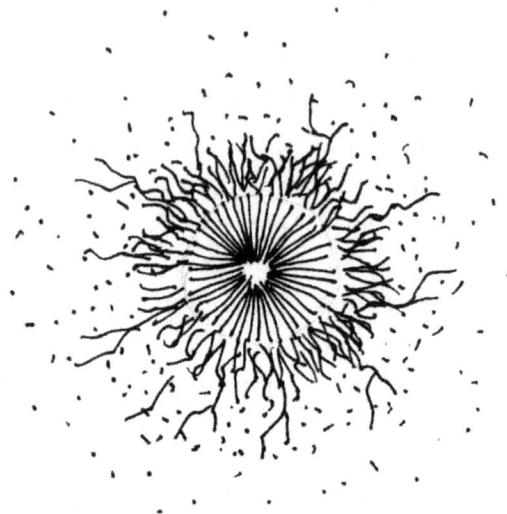

You can use this faith-to-connect-belief-and-reality-through-experience system to understand why those that work often with the uncertainty of movements in reality have more faith in the idea of a higher power than those who rarely do. This might seem like a risky statement, but bear with me.

Other people are part of the approximately consistent atomic soup all around you, just like you. Trust is the soup validating your faith into belief about the *same soup*.

But what happens when the faith-lightning from a Point 2 belief (You) about Point 1 (reality) connects with another, completely different Point 2 belief (Bob) about Point 1 (reality), and the *only method available* to balance which belief is more-right relative to Point 1 (reality) is *language*, because the beliefs are about something that is *invisible*, and so cannot be experienced firsthand?

How do you settle the plane on the treadmill debate if you don't have an actual plane and an actual treadmill? After all, we are each our own point of validated beliefs based on experience with reality, right?

155

Hurricane Pauli

Specialists exist because free time exists. When you are in free time, you are free to pursue specializing in whatever you want to – even being a couch potato.

Becoming a specialist allows you to grow your trust in reality by gaining understandings you didn't have before (Example: What's the best method to couch potato?). However, huge possible problems pop up when specialties you might choose involve *other people* as the subject of that specialty. This includes specialties like medicine, politics, and psychology.

Should a crisis pop up for humanity that already has a long specialty spike of understanding that reaches way outside the central core of what we can easily experience firsthand, it *makes sense* to lean heavily on the understandings individuals deep within these spikes have in the name of speed. In crisis, we don't have enough *time* to move *everyone's* understandings to match those deep in the relevant spike.

At the time I wrote this, all of these things recently came together: Extreme specialty, a crisis, specialties with humanity as their subject, *and* things that are invisible to our senses. This period of time is known as the COVID-19 pandemic.

If you experienced this period of time and relied on the secondhand stories of news and high-level officials to make sense of what was happening, the experience of fear was almost certain to have happened for you, which resulted in having a polarized belief. It is usually not the officials and specialists that cause fear and polarization – it is your chosen *source* of secondhand information – your chosen *editor* of reality beyond your firsthand experience. If this editor (whoever it was) told stories that made sense to you, then you automatically trust and believe that editor. However, that editor often ran high-level big-bucket stories through their *own* filter before it got to you. The result is completely different "edits" of reality itself.

Here's a quick 7-step rundown of how reality is warped and unanchored by story editors, starting with the first three steps:

1. Individuals get sick (highest story resolution, smallest buckets).

2. Specialists notice this, then track and quantify this information up the authority hierarchy using the communication hierarchy. You know this as the telephone game. Their stories, usually expressed as statistics and percentages, typically have buckets already representing thousands of people. These buckets were labeled things like "sick," "hospitalized," "dead," "asymptomatic," "tested," "positive," and "tested negative."

3. Higher-level specialists and national officials and authorities must make sense of these bucketed stories (telephone game) to make decisions of what actions to justify. By the time these "statistical" buckets reach these levels they are enormous, and contain hundreds of thousands, or perhaps *millions* of people represented by just a few words or numbers.

In the action-based chain of using communication to re-arrange resources, individuals with high-level political authority justify decisions using extremely large buckets due to the great speed-of-action needed at that level. Each decision potentially affects millions of individuals. For this reason, I believe the specialties of public health and policy might be some of the most difficult to balance with the rest of the spiky donut in times of crisis.

To better understand why, just imagine yourself into the position of King or Queen of your country during early COVID-19 and face it head-on:

You are handed a piece of paper that says a new virus is wiping out people in your kingdom by the thousands. Healthcare workers within hospital systems are reporting these numbers, which are then trickled up through statisticians and epidemiologists to your country's leadership in the Ministry of Public Health specialty spike.

These healthcare systems (Balloon 2) seem to be overwhelmed with patients (currents coming from the much bigger Balloon 1) and are nearing the verge of collapse (popping). Resources like medical supplies and medicines are being drained, there isn't enough ventilators to go around, and there isn't enough hospital bed space.

Healthcare workers (which make healthcare systems function), are starting to get sick themselves, which moves them to the other side of this boundary: Becoming a patient. This accelerates the currents of collapse even more.

As King or Queen, you are told huge-bucket stories of other kingdoms getting wrecked both economically and socially. You also have a team of your own highly specialized public health and medical personnel at your disposal. Even so, you are stuck in a position of forced-high-speed decision-making, and each of which will potentially affect millions of people.

You do not have the free time needed to take a phone call from every citizen to justify a decision. The data you have, collected throughout most of the country and brought to you through the telephone game, is all you get to make decisions. The best you can do is try to get the stories to the finest resolution you can by asking questions, usually involving the *definition of words*. Then you must take the leap of faith into darkness: You must justify actions from the stories you have. *And every action you choose to make is going to be unfair to someone.*

Even one person making a decision for another person using limited communication will be unfair to some degree. As Queen or King, one person – you – are making decisions for an entire country of people.

In the pandemic situation, the head of the Ministry of Public Health specialty spike (which is the arm of human knowledge related to disease and its management) is in a more informed and experienced position than you. This person, who I'll name Bob, knows many stories that are supposed to be trusted and grounded, because that is *why* Bob was selected or promoted to be in such a position. He is an *expert* in public health.

In the real COVID-19 pandemic, this point in time was when all the "flattening the curve" talk was happening, lockdowns were being considered, and there were near constant press conferences by leaders in the United States to keep the public informed of his or her decision making and justified actions.

The editors of mass media chose to not filter these moments by inserting editorial opinion, criticism, and contradictory stories, because nearly everyone was feeling the same level of fear. Faith in our understanding of our environment was collapsing. There was a new invisible enemy we didn't predict, and it was very evidently killing *lots* of people. Fear spiked, and many of us were unable to look away from the only connection to the secondhand world beyond each of our Step 1 boundaries of firsthand experience – our televisions, radios, phones, and computers.

Thanks to the editorial choices of mainstream (meaning widely shared and accessible) media, elected leaders were able to speak directly to many – possibly even most citizens through these devices. These communication systems allowed the top of the authority hierarchy to send a one-way unfiltered message straight to the bottom about evidence, justifications, and change to truth at his or her level. The far end of the telephone game was telling everyone what message was being received, and what he or she was going to do with.

As King or Queen, I have no doubt you would want to use mass communication resources as well. You want to let the people know you are aware of the trouble happening, and what you are doing about it. You might even hand off your microphone to Bob, because he is incredibly informed on the jargon, methods, and justifications related to what is happening and the measures to keep spread under control. Bob essentially can *translate* the metaphors and specialized language from the now-critical public health and medicine specialty spikes of human knowledge into the "common" tongue.

This brings us to the middle of the story and the next of seven-total steps:

4. Around the same time steps 1 (sickness) through 3 (end of telephone game) are playing out, mainstream media editors start trying to build and tell high-resolution small-bucket stories about individuals at the *top* of the authority hierarchy. Each editor also tries to build and tell high-resolution small-bucket stories about individuals at the bottom of the hierarchy. Both stories are projected outward in all directions, with a general audience (anyone) in mind.

Editor-based secondhand stories start to run parallel to the very real telephone game of statistical stories and action-validation-feedback stories traveling up and down from patients and the disease experts to leaders like you that must use them to justify country-level decisions.

Editors are trying to fill out *The Middle* by dumping the buckets from *both* the top and bottom of the authority hierarchy, and in the process create their own telephone game gathered from as many of the same players of the telephone game as possible.

With this said, something starts getting overlooked very quickly and, in my opinion, quickly sends everything off the rails. Those at the top of authority are individuals, and it seems to me editors in the media often *ignore* the huge buckets each of these individuals are forced to work with to make decisions fast. The editors then seem to hold individuals in positions of authority to the *same standard* of high-resolution small-bucket judgment as individuals at the bottom.

To do this though, editors have to build high-resolution tiny-bucket stories about authority-wielding individuals the same as they would for anyone else: Interviews and research into their professional and personal life (increase context).

However, the danger lies in what happens next: Given the additional context about the individual and the overall situation, the editor makes a (likely subconscious) choice to *become Point 2 to all of this telephoned reality.* He or she evaluates the context and decision-making by those at the top, middle, and bottom - all the players - against *his or her own personal context and decision-making.*

The editor-created story is then sent up and down the communication and authority hierarchy from his or her position *at the top of a separate communication hierarchy* that happens to share the same means of communication (like news channels and news websites). Mass communication media editors (who could be you) are secondhand storytellers. These storytellers are not involved *at all* in the telephone game from hospital to leadership, or from leadership back to citizen. He or she get pieces of it – stars and islands - and spin constellation stories from the pieces, often inserting their own judgments and opinions of what matters and what has value, what is right, and what is wrong. They each may do this blatantly (pundits and editorials) or simply through the choice of which of the trillions of valid stories in the world in each moment should be widely shared.

Since these "top-level" storytellers suffer from the same capture-as-many-viewers-as-possible-to-survive problem as political parties, which are doing the same thing by capturing checks and balances - a small handful of editors within huge media conglomerate groups (you might call them religious denominations of "truth") have grown to control almost *all* secondhand storytellers that any two citizens can both experience.

As a result, the editors controlling these secondhand-story sources have become just as powerful, if not *more* powerful, than those elected into positions of authority. This is because all high-level decisions by authorities will always be unfair to *someone* – and unfairness is going to lead to the deepest primal outrage - and outrage is going to attract more viewers and listeners to these storytellers that may barely be attached to reality. These stories are merely *the opinion of the editor* based on what the editor *believes,* or worse, *wants* to believe of the 8-billion (or more) new stories happening every second in the world. Which brings us to Step 5 and 6:

5. These editors (who may include you) can warp and twist what is really happening. They can exploit every gap you have – and maybe even exploit their own gaps – and create stories that are unfair to every person involved (or just every person, period) without even realizing it, and then force them to merge with reality anyway, because he or she needs to fill 24 hours of "news" for assumedly "millions" of consumers.

6. Citizens at the bottom of this parallel editor-opinion communication hierarchy, (which includes you if you consume mass-media secondhand stories), often only experience a screen firsthand. All the rules of hijacking and editing truth apply.

You might sympathize with those that are suffering (which might seem like millions), and feel anger at those in authority (which might be the editor pinning "all of society's problems" to one person).

If the storyteller's opinion of outrage at authority considers the reality of the high-speed decision making using low-resolution huge-bucket stories authorities must use, then perhaps the outrage at the authority's actions is justified.

If the storyteller's opinion instead "omits" how high-level decisions are made, then perhaps the outrage is not justified at all, and is the editor attempting to hijack your story system to put his or her personal opinion into a position of supreme authority – as representing *reality itself*.

The editor tells these stories because he or she feels it is the right thing to do. But just like that ice cream law when I was King – you can't know the editor's justifications for telling such a story, and there are a lot of possibilities.

Let's ground these three steps to experience using the COVID-19 pandemic again: Instead of King or Queen, imagine yourself a common citizen in a small town within a bigger kingdom. After the initial scare and collapse of stability, the constant flow of mainstream media news, press conferences, and discussed protection measures coming from the top of the kingdom might start to seem downright panicky. When you look around your small town, *no one is dying*. No one is even sick. What the heck is the King/Queen and this Bob guy going on about?

Time ticks on, and the editors continue to flood in the most dramatic stories possible 24 hours a day. You consume story after story: Cities and hospitals having to deal with the dead using tractor-trailers, overcrowded hospitals, political fights over medical equipment, and the ongoing race to find a vaccine. Stories about the invisible enemy virus start growing out of control: Suddenly the virus *might* be traveling on the mail, or it *might* be going airborne.

I was decently educated and experienced in health and medicine (and my wife, who works in healthcare, was putting in 18 hour days) during this time period of the pandemic. I was consuming mainstream storytellers while also listening to specialized virology and political podcast storytellers from deep in the spikes, and can tell you that at this point, the Exploding Cat Problem was in full swing. For example, the word "airborne" seemed to mean something different to each editorial "religious denomination", and each editor or "head of denomination" spun wildly different stories based on what the word meant to him or her.

Again though, as a citizen in a small town - *no one around you is dying*. The local barber said she heard it was all a bunch of nonsense. She decided this whole thing is about government control and authority. *"They want us to be scared so they can control us! It's a conspiracy!"*

Slowly this "consensus of conspiracy" story starts to grow in many places, until it eventually seeps into the parallel editorial realities. Before discussing the consequences of conspiracy however, I think it's worth saying that in *both* of these stories about you in the pandemic – the one from the top (You as leader), and the one from the bottom (you in a small town) – you are *right*. This is a clash of different truth-story resolutions – different bucket sizes.

You as King or Queen are justifying decisions using the telephone game from hell. You do not have enough time to gather all the stories of what is happening. You have to make a forced-choice of action at a different level than any individual citizen does – at a purely *secondhand* level. As King or Queen, do you shut it all down? Do nothing? What is fair? What is right? What is safe to do? There is no higher authority to appeal to, and you are completely responsible for your actions.

On the flipside, as someone at the bottom of the hierarchy, you are not necessarily experiencing a forced-choice related to COVID-19 at all. Your choices are not about authority or how to protect buckets of people under you – there is no one else under you (ignoring children and dependents). Your choices are only made relative to the virus itself, and viruses *are invisible*. This puts you in a pretty wild position: The virus responsible for COVID-19 is *purely a secondhand story* to you – it *cannot* be experienced *firsthand* - but the *actions* of the authorities trying to control the invisible virus *are* experienced by you *firsthand.*

When it comes to stories about invisible things, there is no experience-able evidence. *You only have the choice to trust or not trust the storyteller telling you it exists.* This is a Secondhand Dark Forest Moment. There is no middle path of stability, because you cannot access the wordless experiences being used to fill the word-shells of the story. This means all of your justifications relative to *any* secondhand story are *binary.* You either *do*, because you trust and believe the storyteller or the story makes the most sense, or you do not. There is no gray – *it collapses as soon as you consume the story.*

You can collapse some of your freedom and free time to take actions against the invisible virus, or not. This could be an action provided by storytellers such as staying home, wearing a mask, limiting your activities, or avoiding group settings. Otherwise, perhaps you do none of these, and keep your truth-story exactly the same, and reject the story of an invisible threat. The only variation is how many actions you justify based on how much of the secondhand story can merge to become your firsthand experience.

The only way to *prove* that a secondhand story is anchored to reality is to experience that story yourself - *firsthand.* And you either experience something firsthand relative to your own context, or you *do not.*

Either the coronavirus responsible for COVID-19 exists, or it does not.

Either the illness of COVID-19 exists, or it does not.

How can you prove either to yourself?

You can't until you either experience the illness, or you don't.

You let the virus make the choice for you.

If you become infected, you no longer have any choices related to the virus. You will either die or not die. The "not die" option also contains a *huge* range of possible stories, from having no symptoms at all to losing entire senses. It's also possible you simply never experience the virus, or you do, but you believe your experience to be a different imagined story of the invisible: A common cold virus, or an influenza virus perhaps. Is it?

Due to how your story system builds truth, a *firsthand* truth-story will always override a *secondhand* truth-story. Your story system is anchored to "making sense". You will always have to "tell yourself" a secondhand story is true, even if it makes sense. You can never *prove it* to yourself, because you cannot actually sense it or experience it firsthand – you have to instead *make* something, and making things - like sense - takes time.

If you trust the storyteller and so believe the story, you can make this work, but it is going to be a constant struggle to keep telling yourself versus your moment-to-moment, sensory-built, always-changing, always-valid truth-story. And that truth story back in your small town might be that *no one around you is dying or even experiencing severe symptoms*. Over time, the local barber starts to *make more sense* than the authorities telling stories on the television.

Back at the top where the officials set and make big decisions, there is a full-blown paradox caused by the excluded middle of all secondhand stories: Are you sheltering others, or imprisoning them? Protecting, or caging them?

I do not envy public health officials (which might be you). These human beings must always ride a line between black-and-white that must be forever uncomfortable.

Preventing tyranny, or unfairness in the face of this paradox, also helps explain why kicking decisions and actions down to a smaller bucket level – to the state or the town perhaps, makes sense to some degree. Many state-level leaders became folk heroes based on how they chose to handle communication and decisions in the pandemic.

Regardless, the boundary between shelter and tyranny is razor thin: As King or Queen, you can "capture this end" by ordering every single person to be locked in individual sealed bubbles in the name of sheltering your people from *invisible* germs. Boom. Pandemic over. But unfairness was *extreme*. Some countries seemed to have come close to implementing something near this idea.

The other choice is to do nothing at all, and let people and the invisible threat run free like it doesn't exist, and see what happens. Maybe you would measure what you can while this happens, but continue to "capture" nothing in terms of shelter. In this case, the fear and consequences may be extreme.

Finding the excluded middle – the balance between these two extremes – is *a process* involving everyone. All actions justified by these telephone-game stories have consequences that have the potential to reach into the daily lives of each individual person. We are all connected in this bowl of atomic soup.

During COVID-19, it did not take long for this balancing act on an impossible middle-edge to tip into imbalance. This is because at an individual level balance is unreachable. The thing to fear and shelter from is *invisible*, and fear is a hell of a thing. Fear is a master manipulator, accelerator of speed, and collapser of time, and capturing it is one of our three very primal needs. When you believe shelter is needed, it very quickly moves you into the inner sphere, which is not a comfortable place to be.

As I see it, if you are being told to fear something that doesn't make sense to you, you are likely to face down and override that fear by finding a new stability as fast as you can. But since the viral threat can't be proven to exist until directly experienced, the fear in need of overriding pivots *away* from the invisible story-to-tell-yourself-to-fear, and *toward* whatever *is* being experienced – the storyteller telling you stories. With no objective firsthand evidence possible, the story of the virus is easily perceived as an idea trying to *control something* – it can be viewed as an attempt at grabbing *authority.*

The "party" is reborn to stop this "choice" of domination. Two people validate each other into teaming up with a new truth-story: *This must be domination; the story being told does not match the evidence or the experience.*

And so, in the United States after a few months into the pandemic – the individuals in rural towns were often the first to "rebel". There were cries of government overreach, protests claiming authoritarian power grabs, and that the virus was just an attempt at controlling the masses for political gain and injecting them with poison (why poison? Because the new "party" has been born to oppose the storytellers insisting the virus exists and to get injected).

If this describes you, *you are not wrong*, there may have been just too many gaps to be able to understand how and why decisions were made the way they were, as this story of "tyranny and authority" seeped into the parallel mass-media editor-based-reality and even the editors found themselves in the *exact same position*. Add in the *Floating Anchor Problem* that defines what is true and makes sense based on where and how you start your secondhand stories – and the tail began to wag the dog. Individuals were born into a "truth" based on storytelling that was reflecting a growing belief that also happens to create outrage and earn the storyteller money and attention. And by sharing this story – injecting into the flood of stories first as questions - a new "truth" a new story to believe, a new religious denomination was born.

If you were one of those in authority at this time, you are not wrong either, but there may have been just too many gaps to make non-polarizing decisions. It's also possible you didn't know how to ask questions to increase story resolution. I also have to accept that you may have justified decisions based on some group bothering you every day, or you actively refused to make decisions, or you yourself switched communication hierarchies and therefore switched realities - and chose to believe the editor-based story instead of the telephone game of experts and specialists within their respective spikes that are connected all the way down to healthcare workers and patients.

In my opinion, there were many communication blunders by those in authority in the form of *certainties*, which were made in what I suspect to be desperation to increase compliance with The Rules. Book after book has been written on the details of policy and authority during the pandemic, so I have no intention to dive into the details of this specialty. As I see it, the problem with pandemic "Rules" all boils down to one extremely problematic idea: Fairness.

If you lived in a big city that was on total lockdown mid-pandemic – how would you feel if the country's leadership came on television and said that all rural towns with populations less than 30,000 were now *no longer* under lockdown at all - they were instead free to continue their lives like the coronavirus no longer existed - while all cities with populations above 30,000 remained locked down?

Because everything is always moving, a decision like this – which doesn't move - creates a boundary out of words – this *invisible* boundary creates an abyss of uncertainty the closer you get to it. It is an abyss, because it is actually a gap, and there are *endless* exceptions in this gap: What about individuals that live in small towns under 30,000 but work in big cities over 30,000? What about towns that have populations of 30,500, or maybe 30,010? How should leaders handle what will quickly amount to an explosion of "exceptions" to what seemed like a clean "rule boundary." This is the precedent and context collapse problem (word boundary collapse) meeting with the exploding cat (word meaning expansion), which creates currents that eventually collapse all authority using words.

As time goes on after this rule, everything continues moving, and the virus mutates to become less deadly. How would you, as King or Queen, decide to start removing shelter measures? When should quarantine be dropped? You have terrible story resolution at your level. How would you design your telephone game from hell? Should you keep demanding mandatory masking? For everyone? Or just in some places, or just for individuals that identify themselves to be part of a certain group?

Once vaccines are developed – should they be mandatory? What if the virus suddenly mutates to become even more deadly than it was initially? This has happened before in the past. Is it better to force a vaccine than to risk a sudden wave of death again?

All of these invisible boundary and "rule" problems create unfairness at the *individual* level. The fight to gain control over fear itself creates extreme voltages in lots of places. As I write this, it is still very much a problem. I would argue that much of this problem is the way the mass media covers and words all stories related to this issue.

I think it is a sound argument to claim political party religions, trying to avoid being invalid in a loser-takes-nothing system, have grown to now capture mass media editors (which were originally intended to watch the authority hierarchy for cheating, lying, and bad-faith behavior). This would merge the authority hierarchy and its telephone game with the parallel telephone game, and this brings us to the final of seven steps – the switch to an *unprovable secondhand reality*.

7. At the time I wrote this, there is a trend where some individuals in positions of authority seem to be getting their high-speed big-bucket telephone game information from secondhand story-tellers (editors). This creates a story-loop of outrage, injustice, and unfairness that might have no connection at all in reality – because the editor does not have access to the telephone game those in authority use to justify decisions.

The result can be editors getting indirectly "captured" by authority, who act in a dramatic way to "capture" the editor's attention, which is then run as a mass-media story, which polarizes and outrages those who consume the story, which polarizes and outrages the authority, which then very much start to warp reality itself. The authority and the editor self-validate his or her own fears and beliefs off of each other, and project them to millions of individuals at a time. "Purple people are evil vermin!" leads to headlines like "Yellow person screams purple people are evil!" which leads to "I am not a yellow person and I did not scream!" which leads to "Leader attacks news" and the cycle continues.

Eventually, due to the *Floating Anchor Problem* and repetition becoming "truth", new individuals might seek authority and justify decisions and actions *based* on the editor's parallel-reality storytelling *instead* of the collection of stories coming up through the telephone game from all the individuals they are elected to represent.

The result is the *removal* of the first three steps earlier in this chapter. The original telephone game disappears completely, and mainstream media editors control reality itself, who are themselves "captured" by reporting on constant outrage to survive by getting money and views, or because the political party religions have grown so large that believers put themselves in the editor's chair – resulting in the "capture" of mass-media by the party.

For example, some editors seem to believe in a oppose-anything-this-political-party-wants-to-do position of righteousness. Some seem to believe in telling all stories from a control, power, and authority frame. Others seem to believe in filtering reality only from a fear-the-invisible perspective. If you are an editor, would you say you are being fair? Would you appreciate your editorial actions if everything were flipped and you no longer had control of reality's narrative, but someone else did and used the same system of injecting his or her personal beliefs and fears that you did?

There is not enough time for anyone to understand anything completely. We are all working with incomplete information at all times. No one knows the whole story about anything. Does this justify completely destroying all authority and mass media? Does this justify pursuing total authority and media control? In my opinion – the answer is almost always *the middle*. The middle is usually balance, and to stay in balance, there is always a *constant movement* toward or away from these extreme positions.

During the COVID-19 pandemic, I was in podcast and secondhand story overload. I listened to political podcasts dealing with what was constitutional and what was not and to get a feel for what was happening in the struggle for control over fear. I listened to deep dives on public health and virology to get a feel for how statistics were gathered and how knowledge about an invisible enemy that crossed boundaries without consent was built, and how building weapons to stop it progressed. I followed epidemiologists and lots of experts that did their best to "ground" the invisible things we were dealing with.

To this day, in the fall of 2024 as I write this chapter, many individuals seem to take positions of certainty (black and white) when it comes to the virus or sheltering from the invisible. *Masks don't work. The vaccine is poison. The virus was intentionally released on us. I will die if I get it. Everyone should wear a mask at all times.*

Again, no one knows the whole story about anything. The question, to me, is if anyone holding these positions is even *trying* to find the whole story and face his or her fears to make the best decision possible, or if it is just easier to build a firsthand story of certainty as a shelter, never change it, and then structure one's entire life (and consequently all of reality itself) around making sure this certainty always stays true and can never be broken apart.

Reality doesn't work this way. It doesn't conform to our beliefs.

Language doesn't work this way. It doesn't dictate what is real and what is not.

Knowledge doesn't work this way. None of it is complete.

Fear and the need for shelter *do* work this way. But they are only feelings. You cannot actually shelter from the entire universe, or even from the sphere of atomic soup.

Find your own boundaries of authority and control. Then think about that boundary relative to other people, and notice if it is lopsided to the point others must become "things" inside your boundary to be "captured".

No one person is the *Keeper of Truth* – but someone has to be an authority, or we will never be able to move society fast enough to keep up with constantly changing reality.

So what is the best way to find the ground of truth in an always-moving system?

156

A (Con)Science Art of Balance

Truth, as the end result of validating all faith in the massive and complicated system we call our universe into total trust, experience, and understanding, is not likely to ever be reachable for us as humans. As a result, unity with objective truth is not a reachable goal, but is instead a process that has no end.

This means the word *truth* is perhaps forever indefinable. Instead, we only have belief. We only have the religions of language and assumption to try to share our experiences.

This means the process of growing belief is to continually collect and compare human stories of experience through time – a constant work in progress. There are no certainties in any story, because from our wacky-windowed perspective, what is true is always moving, just like the language we have built and continue to grow trying to describe it to each other.

To avoid creating massive imbalances in this collection of stories attempting to describe beliefs of what is true, strong justifications are needed to hold onto any one belief for very long. If we try to define a "perfect" balance as not allowing movement in beliefs, there will be no way to grow our understandings of what is true because what is true is not allowed to change. Trying to enforce such a definition of perfect balance also becomes its own who-has-authority-to-define-perfect problem, which starts the cycle of division and collapse. Instead, to stay balanced but still grow, I believe we have to reach into the darkness in all directions "just enough".

But what is "just enough"?

As I see it, the foundation of belief, trust, and understanding is a floating system that starts from the middle with belief, trust, and understanding in yourself and what makes sense to you.

Your senses are "feeling" movements that are transformed into an incredible array of electrical currents, which are then decoded in your mind back into a central story, with certain bits of that story stored to build future expectations.

Objective reality, or truth, is the reality interacting with your senses, but it also *includes* your senses. It is the movements you sense, the transformations of those movements into currents, and the decoding of those currents into experience. You and I are both "objects" composed of an uncountable number of other "objects" that are all interconnected through invisible straws and balloons. And it might be this way throughout the entire universe – we have no way to know yet.

The existence of this shared objective "ground truth" reality of movement outside, through, and inside our sensory windows is what I believe makes perception and consciousness possible at all – objective reality must first interact with our sensory windows for any stories to be created in the first place. Without truth, there are no stories; there is only the void.

In this way, every believed secondhand story, no matter how bizarre or fictional – is valid. Ultimately, the words or images of all our stories are perceptions grounded to some perceived truth – because experience with truth is how we learn, remember, and choose our word-shells. A bizarre story to you might simply be many metaphor-layers removed from what *actually* occurred in experienced "objective" reality by a many-people-and-many-generation long telephone game. This can easily result in totally different beliefs of what is true than the original wordless experience of movements from long ago that began that telephone game.

Language can make reality a kind of funhouse mirror, because reality is always "transformed" in the gap between any two individuals, which sometimes are missing the "similar" firsthand experiences required to keep the metaphors of a story anchored to the experienced "truth".

This is where time and movement mess truth up forever: Since the objective reality of the past cannot be re-experienced firsthand, any secondhand story (because they are all created from the past) is prone to the telephone-game problem which only ever gets worse as it passes through individuals in time.

Perhaps I will call this Linguistic Metaphor Contextual Gap Transformation Drift. Though I very much dislike hard-to-understand terms like these because they will be the first words to have their meanings fall apart through time. So I believe calling this something like Exploding Cat Keeps Seeing Different Parts of Itself In an Constantly Shattering Mirror more fun and memorable.

If you want to try to un-explode any story mirror for yourself – or grow your belief in the truth of any secondhand story - you have to try to chase that story *through* the mirrors. Unfortunately, believing it is possible to chase a story to the point of certainty can and will lead to your own madness, because even the first person to tell the story, based on his or her total wordless firsthand experience with truth, had wacky window filters and gaps that edited it.

Instead, I suggest trying to remember what stories actually are – perceptions, feelings, and emotions expressed into word shells (or another communication medium) by other humans like you. One of the places in reality right now where stories are running wild with all kinds of imagined ideas of what might be true is quantum theory.

If you choose to read about the principle of uncertainty as it relates to the atom, (locality versus non locality, hidden variables, and other fun topics), it will not take you long to build some really *wild* stories. This is because quantum theory is sitting on the very fringe of human understanding as I write this. Those at the very edge of the void at the tip of the spike are trying to find the right metaphors, the right *story*, to make sense and shine a light on what they are dealing with wordlessly when they open new doors.

What do *you* actually know about quantum theory? How much of this specialized spike of human knowledge are you familiar with, which stretches out from *The World of You* to the limit of all things we know to exist? How do you *know* you should believe the electron exists? Is it a point particle, or a wave? What exactly is superposition? What do these words even mean when it comes to something so small no one can ever experience it firsthand?

If you try to use your own context to fill all the word-shells involved (assuming you yourself are not a quantum physicist), you're going to run into a massive problem. You are missing *tons* of stories, experiences, and understandings that led humanity this far into the spike. For this reason, I've found it best to try to start at a concept I am sure I understand and have experienced, and then try to follow as many contexts and stories as I can away from this concept out to the tip of the spike.

Essentially, I try to follow the spike of specialization as it grew and evolved through time.

For example, consider the speed of light. How have we determined the speed of light? Don't just accept that number as true because someone or some website told you it was the speed of light – memorizing it would be knowledge, but it is not *understanding*. You are simply choosing to trust the storyteller and believe what you are being told without any form of validation.

If you just accept the speed of light on faith, whoever told you the speed becomes your *Keeper of Truth* when it comes to how fast light moves. If you trust this individual, then perhaps your faith is justified, but be aware there is a *massive* gap sitting in this same spot for you. After all, what happens to your knowledge if this person dies? You are now stuck telling yourself to trust a memorized speed, with no way to understand *how* this now-gone person came to believe it. The speed of light becomes a *religious belief* to you.

I think it's interesting to watch what happens if you ask someone (assuming the context is appropriate) to explain how we believe we know the speed of light. Can the person explain it? Does this person suddenly get incredibly defensive? Does this person double down and just say "because it just is, okay?!" Does this person witness his or her own gap and retort with "You know, I am not really sure, that is a great question!"?

There are many people in this world that know the stories of how we think we figured out the speed of light. I think it's also very likely these individuals understand the limits of what we think we know and how we think we know it. These individuals are valuable in my opinion, and if you wish to transition from "knowing" to understanding, I would encourage you to seek them out. Read their writings, their videos, or even have a conversation with them if you can. The story of measuring the speed of light is really a very fascinating story.

To get a handle on just how much we had to figure out to do this, think about it from your perspective in *The World of You* – how could you possibly determine the speed of a photon while standing naked in the forest? How could you even begin to figure out that light has a speed at all? The stories of how we came to believe what we think we know about the speed of light exists in recordings and writings that have outlived the context that wrote them, and it has been an ever-evolving process from the earliest of thinkers to today. The story of light speed has crossed hundreds of generations to arrive where it is now – where students and speakers too-often regurgitate it as some timeless *certainty* of a number to be memorized. It's a terrible disservice to the story of human ingenuity and light, really.

What experiments were done to measure it? To validate the extremely specific speed we believe light to actually travel? I promise you this story is actually unbroken in time – you just have to seek it out and rebuild it through the broken mirrors for yourself the best way you can. There are no gaps in the story where an object movement specialist (physicist) just randomly declares, "the speed of light is fixed at [this speed] and that is that". There is no one person out there that is the "Keeper of the Truth about the Speed of Light". *The speed of light is the speed of light.* Objective truth keeps itself. Do we actually know this tiny piece of truth? We can never be perfectly sure – certainty doesn't actually exist for us.

The best we can do is try to measure it and test it in every way we can imagine, and see if we keep getting the same result.

And as for how this validation of an impossible-to-experience firsthand story works (like both viruses and light speed), all of this verification and validation of reality is made possible through the magic and wonder of memory, which makes language possible, which makes the sharing of stories through time and boundaries possible. Which, if you think about it, helps us make sense of quite a lot of things that might not be possible without language.

How can you be sure an object you are sensing in front of you actually exists and is not a figment of your imagination invisible to others? Use language to ask someone if they sense it, too!

"Do you see that? Do you see it as blue?"

"Yes I see it, and yeah it looks blue to me, too."

Verified. Validated. The experience is limited and filtered by words (what is blue again?), but faith in yourself is converted to trust quicker than it would be if you were forced to use only your own senses (which might involve touching it, which is potentially dangerous), as well as trust in the other person's senses and the consistency of reality.

But let's assume the object suddenly appeared and then quickly disappeared. What was it? You can't know unless you can make it appear again. So this is exactly what we set out to do.

"Did you see that? Did it look blue to you? Do you think we can make it reappear?"

"Yeah. I think we can. I have an idea, let's see if it works."

And so you team up and work together. It might take you many failures to find success in getting it to reappear, but eventually, you get it. It *does* appear blue just like you both thought!

But perhaps you both are still not sure what it is, even though you agree it appears blue. To fix this, you pull in someone else to get his or her opinion.

"Ah, yes, wow! That looks just like a blue widget-dinger! I haven't seen one of those in years!"

A blue widget-dinger. Awesome.

You write up the story of what you saw initially, and write up the story of how you and your partner made it appear again. You write up what you perceived in extreme detail – the very best secondhand story you can create from firsthand experience. You include your concerns and gaps you know about. You include anything that might be relevant to this story at all. And then you share it as far and wide as you can.

Others will then consume your story. Perhaps it will make sense and merge into their own beliefs as a shadow experience, and perhaps it will not. Some might choose to resist trusting your words and instead try to recreate the firsthand experience. Charlie wants to test your story to see if it holds up firsthand, and if his senses perceive the same thing, because he doesn't think that sounds like a widget-dinger at all. It actually sounds like a ding-widget.

So, your carefully written story of how to make it appear is carried about, experienced, and then rewritten somewhere else in the world by someone else. And that person (Charlie) observes what *might* be the same phenomenon, but doubles down that it is not actually a widget-dinger, but a ding-widget and that it does indeed appear to be blue. Warning: The recreation was limited to Charlie's context for *your* words, and also limited by the process *you* detail. Charlie is actually in a rabbit hole of *your* creation, and found a place to disagree with you.

Another person, Alice, creates an entirely new method to test the story, because based on what Alice is reading from you and Charlie, she believes that if she changes a part of the testing method, it will probably give the same result anyway – because it seems to Alice that the ding-widget is probably just a result of fling-flang-floppities.

On and on and on this goes. Thousands of individuals read your original secondhand story. The ones that are serious about fully understanding what in the world has been observed will continue to read every single story by every single person that has experienced it and tested it. They will look for gaps and limits. They will imagine new ways to test or challenge the idea.

This is exactly *why* the Mythbusters' attempt to answer the Jet-On-A-Tread-mill question carries way more story-mass than yours and Bob's opinions (unless you and Bob are both on the Mythbuster's team and have the experiment in your memories). The Mythbuster's team actually tried to recreate the objects and their movements from the secondhand story in reality – though not to perfection, but to simply test assumptions and beliefs through wordless firsthand experience. Will the team's predictions and the results of their experiments be a perfect final answer never to be questioned again? No of course not, because micrometeorites are *always* possible. To believe otherwise would be to create the Religious Sect of Thy Mighty Mythbusters.

It's impossible to truly isolate *any* system from the rest of the universe. If this ever happens, we technically created a separate universe inside of our own universe, and paradoxically, we would never be able to know or predict what happens in it because it would then have a boundary we couldn't cross without changing it by merging our universes. It would be Schrodinger's Universe.

Back to a smaller scale, let's assume that tests around widget-dingers, ding-widgets, and fling-flang-floppities are performed by hundreds of individuals, and they all confirm that this phenomenon seems to exist. This story has grown to become a highly trusted human *belief* that we understand this piece of our reality, because hundreds of human perceptions have confirmed it firsthand to be *consistent* and have even found the limits where it stops occurring and how to make it occur.

Actually, let's really blow this up and say the phenomenon that goes by widget-dingers, ding-widgets, and fling-flang-floppities has been tested and verified to be accurate more than 6,000 times using 600 different tests, and each test recreated the "experience" at least 10 times.

All of these individuals that created a test began by consuming the story in the present, comparing it to their past, and then predicting a story about the future. Then took the leap of faith to try to justify their belief in their own predicted, projected, and imagined future with an experiment. The experiment is an exercise in *trust resistance*.

Observation. Hypothesis. Prediction. Experiment. Repeat. Familiar?

This is the scientific method.

As I see it, the scientific method *is* the exact same process as human consciousness, but expressed at a different and dissociated scale. The method's ever-growing findings about reality are made possible because the *written language* it uses does *not* move or change in time. Language *holds still* in all scientific papers, and only the context of each scientist that experiences an experiment to witness the results firsthand shifts or rewords the previous story built with words in a grounded way – and the ground is only ever shifted one story of firsthand perception at a time.

Through repetition of this across decades, we slowly increase the resolution of our beliefs of what is true by piling up more and more slightly-different stories - which are individual firsthand perceptions - and keep merging them together one at a time to create a deeper and deeper foundational belief-story – until the story finally starts to fall apart. When it does – when it splits – which is when an experiment to test the same phenomenon results in a different story that doesn't merge easily with the others, we then try to determine if they *should* merge together through more experiments and conversations because a mistake in methods or observations were made, or if they *should* truly become two different stories because the result is measurably different. If it is different – then we know we have found a *boundary* to the original story, or, perhaps the original story is no longer valid.

There is no certainty in any of this. There are only unanchored human-defined probabilities.

We do not know the master "one" to measure the universe with perfect precision.

We can potentially bring ourselves to be 99.99999999% sure that something is true about reality – but we can never be 100%. *We are moving.* So is everything else – and all of this moving stuff *is* the objective truth. Nothing is holding still. It's all changing, all the time. Even if something in the universe was holding perfectly still, we'd never be able to tell due to problems with relativity. This includes the process of trying to find the master "one" to measure the universe with. This unit could actually be changing, too.

This method of using unmoving written language to develop and test a better truth-story through time has uncovered consistencies we call "constants." Unfortunately, we cannot use these constants to actually predict the future of the universe with any certainty, because we have only begun to explore the universe. Most of the constants we believe to exist have only ever been tested right here in the same gravitational atomic-soup-filled balloon we call Earth, and we simply project them out onto the cosmos.

As we increase our resolution and understanding of any story in reality, we are discovering even the most highly validated and repeated measurements – these constants, such as the speed of light - have exceptions. Gravity doesn't always behave as expected when we actually use our human-created systems of math and physics to make an *actual* spacecraft travel through *actual* space.

So, we learn from these errors in expectations. *We missed the fruit (Mars) when we reached for it.* Why? What did we get wrong? Let's back up, let go of what we believed before, float and re-anchor our beliefs of what is true, and try again.

While the scientific method might seem like some ideal way and a wonderful system to compile the truth-story of reality – which it absolutely is sometimes - it starts to fall apart in ways I've already outlined all throughout this book. Findings from using the method are prone to all the same problems as our own faith/validation/trust/belief story systems.

What happens if someone doesn't consume *all* the stories of experiments and methods used to validate why we think we believe something is true? What if someone just ignores all of that and goes with whatever "feels right" and then uses the same words as all the others?

What happens when someone only reads one of the middle frozen-in-time stories of science that declared dinger-widgets are actually fiddle-faddles and ran with it – without realizing that this fiddle-faddle story actually contained a critical mistake that later invalidated the connection?

What if someone comes along that says, "No, the result of that experiment was a blue widget-dinger and it always was from the beginning"?

All of this is actually *unavoidable*, and there is a very specific reason for it: There are now so many stories for each and every thing we "know" from using the scientific method, that you and I do not have enough time to consume even a fraction of them before we die.

You can spend half your life in a specific scientific rabbit hole and become a kind of hyper-specialized expert, and end up consuming almost no stories at all from anywhere else in the spiky donut structure that was also grown using the scientific method.

To make this worse, as a kind of hyper-specialist, if you decide you want to consume stories from elsewhere on the spiky donut, you must often start with "high-speed huge-bucket" explanations – quick-and-dirty attempts to tell the stories – just like everybody else. Perhaps after reading this, you decide the *entire* spike now makes sense, so you believe you now understand both spikes without any firsthand experience in one of them (except for being familiar with the scientific method). That is often the goal of big bucket stories after all – to make as much sense of the entire spike as possible as quickly as possible.

However, by assuming and then believing the big bucket story is enough to understand the finest resolution in the new-to-you spike, you potentially leave *enormous* gaps of highly detailed and elaborate wordless experience that are missed by the word-buckets, creating holes that reach all the way down to the *individuals* that have carefully built and are carefully building what we have come to understand about movements and limits in this reality through wordless, gapless, firsthand experiences.

Of course, there is just not enough time in a human lifespan to fill in all these rabbit holes – to understand and be an expert in more than one or two spikes.

This makes using the scientific method and its findings prone to the same context collapse as laws when it comes to piling up "precedent" and the passage of time. This means, to me, that *how* scientific knowledge is built and shared must be reworked so it does not end up slowing the moving foundation to a stop over time, or causing enormous splits in beliefs that destroy everything back to nothing.

157

The Human Constant

Science is not the *Keeper of Truth*. Believing anything is the *Keeper of Truth* would make that keeper your god, and that Keeper's words your doctrine in a religion that only really has one follower - you.

Science is like electricity: It is not a *thing* - it's a *process*. It is a wonderful method and powerful system to grow knowledge, but because it is formed by nothing more than a recursion of our consciousness where memories and projections are replaced by records and writings, it creates the exact same structure as communication and authority – a hierarchy.

Believe it or not, *you* are a scientist, along with every other human being. The scientific field of you is the smallest, fastest, and most validated scientific field and pool of scientific knowledge in your life. You have steadily been growing your light of understanding into the darkness of this place since the origin of your belief system, and you only ever stopped exploring and growing your beliefs when faith in yourself waned, because your scientific process of acting on faith and trying to transform it into trust and understanding with reality reached a balance point with the fear of uncertainty.

It is in *these* moments that individuals like you and I pivot to each other to keep growing. And it is in these moments, where one person is more afraid than another, that imbalance in the form of dominance and submission can easily appear.

Individuals that work together in fields of study that anchor to phenomena in objective reality are working at the limits of their perceptions and beliefs. They are at the threshold of darkness, and growth is incredibly slow and difficult to validate, because it depends on each person's entire experience leading up to that point to make sense of what is being perceived and measured.

This means when something unexpected happens that requires new understandings to know how to shelter from it, the field of science is often handicapped because it doesn't move anywhere close to the speed of human consciousness.

For example, imagine if tomorrow all gravity on Earth suddenly became very irregular, and in some places you could jump as high as a house and in others you couldn't even stand up. Imagine if The Process changed and caused *some* balloon systems to collapse into balance, but others to grow into bigger imbalances.

In this situation, the top of the scientific hierarchy – scientific laws - can offer no immediate help. They are just words and equations and "constants". There is nothing in elementary science textbooks that can help us understand right away *why* this has happened. There is nothing in peer-reviewed scientific white papers published in academic journals that can help us immediately know *why* some balloon systems are running backwards.

Instead, I think the field of science called physics would experience almost a full reboot in established "truths". Physics would have to float the axiom-anchors built by thousands of individuals and measurements over hundreds of years that involved gravity and entropy, which is basically *all of them*, and then start all over again.

As I see it, science is a process and only a process – it is a methodology – it is our natural human method of validating faith in the experience of consistency in our reality into trust and understanding of that same consistency. The only differ-ence between your story system and the field of science is that the truth-stories are transferred into unmoving written language and recorded communication. By doing this, we can use the power of language to "make sense" of reality without necessarily having to experience it firsthand.

But this *only* works because of the persistence of consistency.

If that consistency is disrupted in a way that requires extreme speed to keep surviving, we are each *on our own* to figure what's going on, and only our individual experiences in memory and the specialists in relevant scientific fields become our sources of hope to reestablish some consistency, predictability, and expectation in the face of sudden inconsistency.

Science as a body of validated beliefs moves far slower than the minimum required speed to survive, because it moves at the speed secondhand stories *can be merged through constantly recreated firsthand experiences*.

As a scientific-method-built story grows to be successfully validated more and more, the slower and harder it is to change – it starts to collect an enormous story *mass*.

If it starts to collect so much mass that generations pass without that story ever changing or even being questioned or retested, science itself can start to be seen as a source of shared *certainty* that is *not justifiable*. The foundations underneath those certainties could suddenly change and established scientif-ic "knowledge" would just keep trucking along, anchored to the *assumption* of consistency – to the belief the word "constant" means the same thing to everyone, and that constant means *forever*.

For individuals born into such a system, scientific knowledge becomes a collection of stories believed on faith alone. It becomes just another *religion*.

If there is one axiom, one core statement at the heart of science, in my opinion it is to "question *everything*." Everything very much includes the constants and assumed certainties. After all, the master "one" unit has yet to be found; so all equations and measurements are ultimately up for questioning and debate and should probably be revalidated at all times as often as possible.

It's possible that we struggle to merge quantum and macro physics with each other because macro physics is resting on human-scale assumptions that too often gloss over the uncertainty in every direction.

It's also possible that the atom, specifically the electron and its movements, exist at the limit of the universe and measurement itself, and the atomic system can never be made certain because doing so would require a perspective outside of the atom, which would mean having a perspective outside of all matter and existence from our atom-based perspective.

The only constant in this universe from our prison-cell wacky-window perspective is the consistency in our existence, which boils down to the persistence of human senses, memory, and imagination we call experience, understood only through sharable communication of all kinds, and the persistence of our species through time.

I believe sense, memory, imagination, and justification are all highly adaptable to any reality (assuming that reality is compatible with our body). We are all floating anchors, and if reality started completely changing every time we blinked, we would use these four things to eventually figure out how to control our blinks and start building a body of knowledge and understanding all over again.

Scientific knowledge as an *authority* on what is true in reality rests entirely on the validated belief in consistency. I think it's critical to note that there are no guarantees this will remain the case.

Science is you.

Science is me.

The process of science is as simple as falling down a rabbit hole together to build a story that makes the most sense for both of us. A new field of science is born between us the moment we lock eyes for the first time.

The speed of needing to know what is going on around you will determine the size of your personal "field" of science. That speed is determined by your need for food, water, and shelter. You need to control and understand these things to "capture" them. You always want more certainty they will stay consistent.

And this consistent and persistent need for certainty causes you to always value *whatever* provides a growing certainty.

This means human values are a floating anchor, too.

158

Mutiny In The Dirac Sea

Just because you or I may have largely escaped the inner core of mandatory movements related to food, water, and immediate shelter, does not mean that we no longer consider it as we enjoy our free time in the outer sphere.

Our story systems are constantly working to ensure these ends *remain* captured. We don't *want* to fall back into the core. It's stressful. There is no freedom there. To be in the inner sphere is to inch nearer to extreme hunger, thirst, fear, and even death.

So the certainty machine known as your consciousness keeps grinding away at trying to ensure the wavelengths of your continued existence wiggle on *forever*.

When is the last time you experienced hunger and couldn't find food?

Thirst and struggled to find water?

We so often seem to try to keep ourselves perpetually in free time. Since this involves consumption of objects, we often begin to *hoard* these objects. We become the dragons in the abyss of uncertainty collecting property and "gold", because the property and gold (money) have become a form of certainty all on their own.

Property provides a boundary that makes us feel safe to consume without worrying about the *Dark Forest Moment* suddenly appearing.

Gold can get us the next thing we want to feel safe. Having more and more gold in theory gets us more and more control of everything around us, because gold secures more property. Having lots of gold can bring a supply of food or water at the snap of the finger instead of searching through the woods on the brink of life and death.

Should your property and gold allow you to spend enough time in the outer sphere, you might start to notice other individuals that can barely escape the inner sphere. You might even begin to think these people represent a level of uncertain existence you no longer wish to think about. You might even be disgusted by them. Repelled by them. You might no longer wish to be near them. Their neediness of property and gold might become a threat to your own ability to stay in free time – your own ability to pursue your dream of opening *all* the doors in *The Ark*.

With excess gold and property, your fear seems capture-able.

Get rid of these "untrustables". Send them somewhere else. I'll surround myself with those like me, so I can feel safe.

Your fear is actually the collapse of trust in other people. You imagine a boundary between you and "others", but you cannot build a wall strong enough to prevent this imbalance from breaching a straw-path through it.

The reason is simple: There is only so much property. The garden is only so large. You cannot wall off huge quantities of property and gardens without collapsing trust and expanding fear.

So, you might begin constructing traps and weapons to capture your fear of *the other*. It feels like the right thing to do. "They" will harm you if you don't. They all want what you have.

These self-created stories tip your own mind out of balance, and make you into a true monster dwelling in the abyss of uncertainty.

If you try to exist in *permanent free time* and reject or harm those nearer to the inner sphere than yourself, then you are choosing the *Dark Forest Moment* path of superiority over your equals.

This path is not sustainable.

And yet your gold might continue to pile up.

Perhaps walls made of gold will protect you?

This is the runaway need for shelter, and an attempt by individuals in the outer sphere to completely *escape* from the inner sphere. The only possible result from this is an inevitable collapse. If the imbalance between the outer and inner spheres grows big enough, the collapse will result in a black hole that pulls everything, everyone, all property, and all gold into the hollow core of death.

The runaway shelter need is trying to seek shelter from the one thing that provides the *most* shelter – other human beings. In a state of fear, the need to capture shelter from the bucketed *"untrustables"* starts to manifest in language, and words become weapons, wedges, and walls to try to force a boundary between the fearful and what is feared. In this case: Other human beings.

Values start to float and flip upside down. Faith and trust in others all collapses into only fear which must be balanced through a new ways to find shelter. There is only gold. There is only fame. There is only validation through the envy and jealousy of others. There is only *being right*.

Human beings have managed to rise to such an amazing level of certainty of survival using language and science, that the floating anchor problem seems to cause most of those born and raised in the outer sphere to believe a life of free time and nearly unlimited choice at all times is *normal*. These individuals are unaware they have been born in a culture that can easily justify him or her to behave like monsters.

Due to growing up in a largely controlled and predictable reality, these individuals may begin growing their certainty by anchoring truth-stories not to wordless reality – which is perfectly stable, but directly to secondhand stories. "Truth" becomes statements of certainty that provide comfort that his or her beliefs are the *right* beliefs. That his or her beliefs of supremacy and imbalance relative to others are naturally justified.

Some might find themselves continuously speaking in the second person – from an "ivory tower" - and dictating what others "should" "must" or "ought" to do purely to validate and justify their own hoarding and superiority. I suspect that the hope is to become a source of gravity to pull others in, while at the same time hoarding as much property and gold as possible from those same people as they cross the event horizon.

There is one issue with this: Gold and property cannot be permanently captured. Boundaries and balance are both destroyed trying to do this.

159

A Standing Wave of Mind and Body

You have beliefs of what is true, and so does each and every person. All of these beliefs are different. Some are wildly different, and others are slightly different. Each of these beliefs are also always changing. As I see it, for any two people to understand each other, they must communicate. The most important need for stabilizing and maintaining balance of the entire social structure with a moving foundation is the *persistence* of communication regarding movement.

One way to think of stability and balance on a mass scale is to use a giant loop. This giant loop is actually a wave because time is always moving in one direction for us, but a loop is a bit easier to imagine due to it being a giant closed circular boundary instead of an endless wiggling line.

The top half of this loop is the mind. The Ark.

The bottom half of the loop is the body. Gravity.

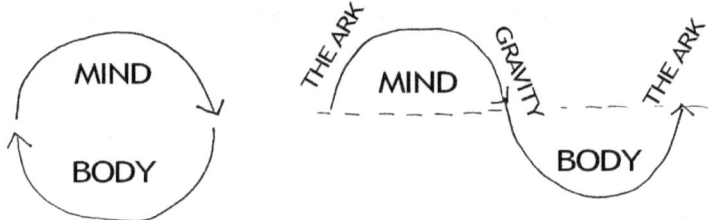

These two halves are constantly pulling against each other. Your body tries to pull your mind down and away from total freedom, and your mind tries to pull your body up and away from performing survival actions. Your bodily needs do not change unless your body expands, shrinks, or develops a process-related problem. This means your mind tends to win this tug-of-war by figuring out how to make your survival actions as efficient and certain as possible.

As I see it, personal harmony and internal balance result from your mind accepting this system as it is, and recognizing that your freedom *depends* on fulfilling the needs of your body quickly.

Just like science is human consciousness slowed down and expanded into unmoving written language, your mind-and-body loop has also expanded and moved into the external world as well. This loop has grown to be so large that it might be very difficult to see. I'll try to expose it with a simple question:

How much time do you spend each day just making sure food, water, and shelter will be where you expect them to be *before* you actually need them?

Do you constantly check your faucet or nearby water fountain to make sure it works *before* you get thirsty?

Do you call all the grocery stores or restaurants, or take constant inventory of your pantry or food storage to make sure these places have food *before* you get hungry?

Do you constantly check that your house or apartment is still there *before* you need it for shelter from the elements?

My guess is your answer is "no" to all of these questions, and that you don't actually spend much time, or perhaps *any* time, doing these "checks".

Do you instead have *faith* in your belief these things will be there when you need them, despite having no evidence at this moment?

Do you *trust* these sources of need fulfillment will *always* be there because of how many times your belief of this has been validated in the past?

If so, what would you do if all these things suddenly went away? What if all the faucets and water fountains went dry, and all the groceries and restaurants and kitchens had no food in them, and the shelter you depend on to stay warm or cool or safe from predators disappeared?

Would you begin to panic?

If so, then consider that right now you very much exist somewhere in the top half of this loop of stability and balance between mind and body – the mind is winning the tug-of-war easily. You are largely free to pursue whatever you want with your time. You might have had very little to do with getting yourself here however, because it's entirely possible you were born into a part of society where the tug-of-war was already well under way, and you just happened to be placed on the side where the mind was winning.

This placement is only *ever* possible by the existence of individuals *remaining* in the bottom half of the same mind-body loop, *because the human body did not change.*

To get water to your faucet or a water fountain, to get food into a grocery store or into your kitchen or restaurant, and to erect the shelter you have come to expect – lots of individuals had to perform lots of survival-centric actions on your behalf with their bodies.

The balance of mind and body can be applied in a million places in society.

The ability to think, imagine, and create *depends* on faith and trust in action and movement. The ability to act and move *depends* on the ability to think, imagine, and create. These two halves of the same loop are circling around one central idea: Experience with reality staying consistent.

I design electrical systems. The drawings I create do absolutely *nothing* to deliver electrical current between Point A and Point B. Instead, that delivery *depends entirely* on the men and women installing the *actual* wires and *actual* equipment between those two *actual* points.

It was men and women that tinkered around and configured actual electrical items and objects that developed the understanding needed to control electricity in the first place. It was their *actions* that set the foundation that allowed me to design anything at all without ever experiencing it firsthand.

The idea of harnessing electricity to do work for us was first imagined during someone's *free time* because it does not relate at all to food, water, or shelter. The validation of that idea was forged by reality itself through experiments – actions using boundaries called objects. The actions of those objects aligned (or didn't align) with that someone's predictions and expectations, and trust grew in the ability to understand something new (or fear grew from failure to predict and understand, followed by the adjustment of belief and a second attempt to understand).

When I design a system, it is very much in my own interest to seek feedback from those men and women that install it. I have to connect the top half of the loop (me) to the bottom half (these men and women). Did I design something that wasn't physically possible? That had to be changed? By getting this feedback, I can grow trust in my ability to make my designs reflect reality, or adjust my beliefs and understandings, and design differently (and hopefully better) next time.

If you exist in the world of mind more than body – in free time – and you only communicate with others in the world of mind, and not with others in the world of body – whose movements are absolutely connected to you - then you are at a huge risk of creating an imbalance in the *whole structure* of society.

As I see it, this is exactly what has happened in reality around me as I write this.

Many individuals on social media and in the world of mass communication exist almost entirely in the upper half of the loop, ignore the lower half that makes their position possible (or tells himself or herself a story of what the lower half does), and instead focuses on maintaining fairness between others also in the upper half.

This quickly leads into validation purely through communication, because using body is not part of the equation. Reality becomes a language-battle of being the "most right" or the "most valid" with words and other media. The individuals that make the internet possible – the information technologists (IT) and programmers among hundreds of other specialized roles – are almost completely ignored or treated like a stepping stool. In my experience, social media influencers are rarely ever thankful of those that keep servers running, or install the fiber optic cables all around the world, or maintain the satellites and network signals, or debug the platforms the influencer is using. These roles are *invisible* from the perspective at the top of the loop, and replaced with the expectation the internet *should* always be there and working. If it is not, then there is often anger, hostility, and resentment.

This whole only-interacting-with-your-half-of-the-loop problem can also be flipped around.

Many individuals that work hard-labor jobs exist almost entirely in the bottom half of the loop. He or she might ignore those in the upper half of the loop (or tell him or herself stories about those in the upper half) that are connected, and begin focusing on maintaining fairness between those in the same bottom half of the loop instead.

This quickly leads into validation through actions and movements only, and also more violence, because growing language and expanding free time are not part of the equation. Everything becomes an action-battle to be the "most-right" or the "most-valid" with better/stronger actions and their results.

These two halves need each other and complement each other, and often seem to not realize it. Individuals in urban settings often vote and act against those in rural settings, and vice versa. Individuals that only have management experience often work against individuals with only labor experience, and vice versa. Individuals too often never cross over to experience anything on the other side of the loop firsthand, and yet that is often a key moral to the story that enlightens the characters in so many of our most familiar and popular stories: The rich guy experiences the poor life, the princess experiences the average citizen's struggles, the poor person experiences the world of wealth, the hateful experience the world of love and acceptance, and many more.

Two-way communication and respect acting as a feedback loop is the key to achieving harmony and balance in the standing wave of existence between body and mind, in my opinion. The mind can imagine infinity, but the body cannot bring infinity into reality. The body can reshape reality, but the mind cannot perfectly predict the new shape before it is made without first having experience through the body. The mathematician and the engineer need the machinist and the constructor, and the machinist and constructor need the mathematician and engineer in the current state of society.

Big question time: Which half of the loop has more value? Which role is more deserving of gold?

The answer is a bit of a rabbit hole that may surprise you.

160

The Currents of the Sea

Gold is a metaphor for money – though it can also *literally* be money. So what in the heck *is* money anyway, and why does everyone want it? It seems like American culture in general completely worships the stuff, and yet at the same time no one seems to be really sure what it is, what its true value should be, or what that value is anchored to.

To expose money's value anchor, I think you really only need to ask yourself one question: What is the *one thing* your money absolutely *must* be able to buy? Is it a fancy car? A nice house? Access to exclusive locations? Fame? What is it about these things we as a society seem to want so badly?

As I see it, obsession with wealth and property is the shelter need that has gotten out of control. It seems to be a common belief that having money *is* safety in society, because society seems to value money no matter the culture.

During my own journey before this one, I discovered that finding the limits of an idea exposes what it can and can't do, and what it actually is. In *The World of You*, money had absolutely no value. It also didn't have any value in *The World of Two*. This is not the lower limit – this is beyond the lower limit. This is "zero".

We are looking for the first sign of "one."

Using this book as context, money didn't start to have value until Bobville became quite large, so at that point "one" was already defined at least twice (it was worth *one* gold nugget the size of Mayor Bob's fingernail, and could be used to trade for just about anything else of value), but I never actually covered *why* money needed to appear at all.

As I understand it, early "currency" (the similarity to "current" is not to be overlooked, in my opinion) was just bartering various objects and personal skills. Shiny metals like gold were seen as valuable to everyone, so it became something to barter with. What actually set that value of gold and seemingly other shiny things? Is a shiny rock *really* worth the same as a goat? Is money's value just *a feeling*?

In my opinion, money as it is used and circulated today (in the year 2024) *is* valued by human feelings, though I've often heard it try to be defined by relative gross domestic products and overall material goods, among other ideas.

But because money is a human-created system, I think it has a very real and measurable anchor that can be used to define its true value across every currency in every society in the world – and that value anchor is what was actually being traded back in the bartering days: *Time*.

Let's begin at the point money often enters our lives – a paycheck.

At my job, I get paid for my time spent performing some action (mostly thinking), and at your job, you get paid for your time spent performing some action.

You and I then "trade" that money for specific experiences or for material (objective) things - *property*. This means money (meaning coins, bills, or numbers on a screen) is just a real-life metaphor for bartering. If we strip away the metaphor layer, I am trading my time performing a specific task for an agreed-upon value until I have traded enough time to equal the value of the things I need or want.

However, even though we are all human and we all currently share the same objective rate of time (the Earth), we very obviously do not get paid the same amount. Is one human *objectively* worth twice as much as another across the same amount of time?

While it might seem reasonable to say "yes", you have to consider that your value is always going to be *relative* to a second point: The person trading something of value (money) for your time.

Your value is determined by the demand for your skill, and since demand is always going to be relative, this time-value game quickly becomes a floating anchor problem all its own and spirals out of control almost immediately.

To see it happen, let's briefly return to *The World of Two* at the point where you and the Being have developed a decent amount of trust and communication, but you still do every task together and split everything equally.

The Being approaches you one day and says "I want to do something that is going to take *three entire days*." The task, it explains, is building something it calls a "parachute." The Being asks if you will bring it food and water a couple of times a day for the three days it will take to build this parachute.

Let's assume you refuse, because you see no value in this "parachute" – whatever the heck that might be.

The Being thinks for a moment, then pulls out a one-dollar bill and holds it out to you while asking, "Does this change your mind?"

If you imagine yourself replying, "it does", I would be shocked. You and the Being are trying to survive in the wilderness as the only two human-like creatures on the planet - what value is a one-dollar bill in this situation except the paper it is printed on?

If you instead imagine yourself rejecting the offer and saying "no", this small piece of paper really adds next-to-nothing to the situation.

We are now quickly approaching the lower limit and money anchor: Who gets to determine what that dollar bill is worth in this situation, and why? Take a minute and think about it, and then come back and turn to the page.

In this situation, *you determine the value of money.* The Being is asking you for three whole days of *free time*. Total freedom from hunting, gathering food, and seeking out water. *You*, as the only possible provider of those survival needs, are the one with all the power in this situation. *You* are the one setting the price of free time.

If you changed your mind when offered the dollar bill, then you just set the value of that dollar. One dollar is now worth *three days* of free time.

Even if you still said "no" after being offered the dollar bill, you still set the value of the dollar. It is worth nothing in this two-person society. It is only worth whatever a piece of paper is worth – which is still undecided because no transaction using it has happened.

Why would you say yes to a paper dollar being equal to three days of free time, anyway? What if, after you said no, the Being pulled out a pallet loaded with a billion one-dollar bills? My guess is you would *still* say no, because *why should you care about a pile of paper? What good is it to only you as a trade?*

In this situation, whether one dollar or a billion dollars, all these pieces of paper are being proposed as an "I.O.U.".

If you accept the dollar (or billion dollars) it will now represent three days of free time. Should you offer it back in the future, the Being will need to value it as a metaphor for three days of free time.

There is a *huge* potential problem with this: The value of that dollar exists only *in memory*. But the value of that dollar cannot *only* exist in your memory or *only* exist in the Being's – it must exist in *both memories* for this money-idea to work at all.

Imagine if you agree to this deal, then a month later you ask for three days of free time and hold out the dollar, and the Being replies "No way! That's just a piece of paper!" Did the value of the dollar just change in this two-person society? *Yes it did.*

If you strip away the paper bill, you are comparing the parachute to your free time, and are deciding if money should represent a "contract" for that trade. If the previous contracts are forgotten, money can *change value* with each and every trade that ever happens – and this appears to be exactly how the value of money is being set as I write this.

This makes believing the dollar has the same exact value between any two trades *an assumption* with no objective evidence, which makes money's value *a religion.*

Currently, there seems to be no anchor at all - we just keep trying to anchor to the "most stable-looking mass of trades" and call it the foundation of money, which results in a runaway process where the value keeps moving based on whoever gets to define that "mass." This keeps going until the *Floating Anchor Problem* moves it up or down too far, resulting in mass financial disaster. As I see it, using the real money anchor may help stop this cycle.

For money's value to be fair, as a metaphor for some currently-undefined unit of free time, its value would have to be anchored and agreed to by all individuals that make that free time possible at all. In my opinion, we are way too far past such a point as a society, and we'd simply repeat the telephone game from hell to define a "fair" value.

As I see it, the individuals within society with the greatest position of power to set the value of the dollar are the individuals that collects edible food into one close space, and then does their best to take care of it so that everyone spends a minimal amount of time in the inner sphere – maximizing free time. In today's society, these individuals are *farmers*.

The farmer is reality's version of yourself when you imagined creating the single largest source of certainty in *The World of You*: A garden.

As the only human on the all-wild planet, it wasn't until you figured out how to construct a garden that your free time really started to grow. As more individuals were added to the world, it wasn't until someone *volunteered* some part of his or her free time producing food for *other people* that the role of the farmer was born. In my opinion, this role still sits at the deepest part of society's shared foundation – near the lowest point of the mind-body loop - and holds the controlling share in setting the value of money.

I see a dollar to be worth whatever amount the farmer accepts it to be worth to continue his or her role of supplying food for other humans on the planet. I think this will anchor the dollar to the exact balance point between the number of people in need of food, and the minimum speed the farmers must move to produce enough food to fulfill that need.

If the garden needed to feed the world is so big it needs huge tractors and high tech equipment to move the farmer fast enough to make enough food relative to the rate it is needed, then the price of food should include the cost of all of these things, plus the cost of any other survival need the farmer might be missing by not having time to capture it, like drinkable water and a shelter.

Each and every dollar in circulation is a fixed *unit* of the time saved (free time created for others) by the time volunteered to remain in the inner sphere by the farmer.

You can imagine this as the difference between two points: The size of the inner sphere if all of humanity was reset to *The World of You*, and however small it might be for humanity today. Every person that performs actions that shrink the amount of time you or anyone else on the planet must spend in the inner sphere capturing food, water, or shelter is setting the value of money.

If you live in the top half of the mind-body loop (mind), then you are *dependent* on those in the bottom half to be able to stay there. In turn, if you live in the bottom half (body) are partially *dependent* on your sacrificed free-time coming back to you in the form of money so you can continue capturing your ends.

To get a feel for which half of the loop you live in, simply imagine all money in the world disappearing right about…now.

All bank accounts now read zero, and all paper money and coins vanish. You and billions of others have been very much thrust back into *The World of You* together, except now everyone wants the same thing – and it is *not money*. It is food, water, and shelter.

If you can provide any of these things for yourself right away in a repeatable and sustainable way using natural resources, then you are most likely somewhere in the bottom half of the loop. If you are in a mad scramble for these things against hundreds, thousands, or millions of others, then you are likely somewhere in the top half.

I think one thing is for sure in such a situation, though: The world would quickly descend into chaos as all farms and food sources are quickly overrun by mobs of people.

At the time that I am writing this, this exact thing has played out in reality time and time again, but in slow motion and with a twist. And this chaos only ever ends when the money anchor is found again.

161

Value Voltage

Shrink all of human society to be represented by eleven people.

Imagine a big farm that can produce a huge diversity of food, and then imagine 10 much-smaller properties surrounding this farm. On each property lives one of these 11 people. There is a moat between the central farm and the surrounding properties, and each person can get water from this moat freely. Each person also has a small house on his or her respective property.

However, no one ever sees or directly interacts with the other 10 people in this society. Each person is a specialist in something different, and no specialty is tied to food production except the farmer.

Money exists in this small society. The total of all printed money in circulation is 100 dollars.

Each day, each person puts up a sign explaining what he or she has to offer and the price for it. All exchanges of goods or services for money is done blindly so that no one knows who is buying what from whom, or where each dollar came from. The only thing each person knows is his or her own finances.

In this society, the farmer easily grows enough food for everyone including herself. Each day she puts together 11 meal-packs (one is for herself) and each day she puts up the same sign: "Daily meals. One dollar."

Another member of this society, Bob, develops a product that becomes highly wanted. Everyone wants to buy Bob's product (but no one realizing it beyond his or herself). As a result, Bob quickly collects the lion's share of the money circulating between all the individuals, farmer included.

Bob now has 80 dollars.

The farmer has 11 dollars.

Everyone else has only 1 dollar left.

This means each person, except for Bob and the farmer, can only afford to buy *one more day of food*. The only person that is remotely capable of realizing this situation is Bob. Everyone else can't be sure where the money has all gone, but must buy that last day of food anyway, or starve while holding onto a piece of paper.

Tomorrow arrives, and from Bob's perspective, the future is looking bright, because he knows he can afford 79 days of food. The farmer is technically unaffected by all of this, except for no longer earning 10 dollars at the end of the day. Instead, she suddenly makes only $1 each day and nine meal-packs are going unsold.

Although it pains her to toss the unsold food and have less income, she has as many meals as she wants because she grows her own food, and her job is a lot easier because she eventually decides she only needs to make one meal pack to sell a day.

You could argue that the nine starving and moneyless people should have watched their money better, and not spent so much on Bob's product if they couldn't afford it. That is a fair critique of this story. You are not wrong. However, I think this argument breaks down pretty quickly, because it involves *not moving*. If each person stopped spending, the result is the same - money stops circulating.

For example, if person #4 (who I'll call Alice), doesn't sell anything for a week, that's a tough break. However, *this doesn't change her need to eat* because she is still a human being. Her finances still drain at a fixed-minimum rate because she *must* continue to eat, and her only source of food is the farmer – who charges one-dollar per day for food.

So, Alice keeps trying to sell things – *anything* - to make that bare minimum of one dollar each day. She even tries switching up what she is selling in hopes it might stir interest, but without *extra* money to buy what is needed to create and offer something new, the demand for her products and skills continue to drop, because what she has to offer keeps dropping in quality or is something no one wants or already has.

Eventually Alice quits spending money on anything at all *except* food. If she drew a line for herself to stop spending money on anything else when she reached $20 remaining, then little would anyone else know, but there is now only $80 left in circulation to buy their offerings. If everyone else draws the same line and stops spending on anything other than food when his or her balance reaches $20, then very quickly this $100 economy locks up, because it only takes five people dropping out of this marketplace with $20 to freeze *all* the money. The only person that might not notice this *is the farmer, because her incoming money flow will not change* when this happens. However, 20 days from now, she will sell a lot less meal-packs.

So, knowing all this, let's go back to the first crisis: Bob has $80, the farmer has $11, and everyone else has $1. Everyone is about to buy their last day of food and they each know it. Each person is also desperate to make more money to keep this from being true.

Bob sees signs from his neighbors offering items that seem to be made of garbage, or oddball services like singing or dancing, so he quits spending his money, and only ever uses it to buy food. The farmer is now the only person with an income, one dollar per day, because only one person is actually able to buy food.

Here's the twist in this story that exposes the money anchor again: The flow of money will continue unchanged until Bob runs out of money too, because no one else except the farmer has any money to purchase anything from anyone else, including Bob. The only reason she would buy what he or anyone else is selling is if she *wants* it, because she *needs* nothing.

In the meantime, free time collapses completely for everyone else. They have all entered *The World of You*, and fallen entirely into the inner sphere. However, unlike you were in *The World of You*, each person in this mini-society knows *exactly* where to get what they need: The farmer. They *do not know* where all the money has gone, or why this has happened.

If we add in the *Floating Anchor Problem*, and everyone believes money has value all on its own separate from time, then a *value voltage* appears in reality. This voltage can never be collapsed because it only exists in the mental boundary of each individual. This belief causes money to shift from being a *means* to capture ends, to *an end* needing capture by any means. This can cause someone to start trying to capture as many *means* as possible, but instead of doing this to ensure the survival of their body, it is done to ensure their *wealth*. This individual might start to hoard things, and then start charging money to let go or give access to these hoarded things. If the hoarded things involve food, water, and shelter, this essentially creates a hostage-to-religious-belief situation for everyone else, and is an extreme *injustice*.

As I see it, doing this easily results in events such as the hoarder getting robbed by thieves. A person might steal all the candy bars off the candy bar trees, or the dollar bills hidden in Bob's mattress, because that person actually has a *very good reason*, which is exactly the same very good reason I eluded to way back in Chapter 3: *So that he or she does not have to die while surrounded by food, water, and shelter simply because everyone is using them as a means to make money.*

In the 11 person society, since no one knows where all the money has actually gone, I think the farmer is going to be the focus of Chain of Suspicion by the moneyless nine, because from their perspective, she went from provider to gatekeeper, locking away what they each need to survive behind the demand for a dollar, which they can no longer provide or seem to get.

Another possibility, should the Religion of the Almighty Dollar be strictly followed, is that all the individuals on their last dollar offer a new service to society: "Farm work for a day. Price: $1.50".

The price of each person's labor is set higher than the cost of their survival needs, because they each want to escape this no-money situation. If food is one dollar, then his or her time needs to be worth more than one dollar. Such a belief, in my opinion, is a failure to recognize that the previous ability to simply pick up packaged food for one dollar while spending *zero time* growing and prepping that food, required someone else to spend time growing and prepping that food *on their behalf.* By setting the price for labor to $1.50 for that same food, he or she is taking that exchange for granted, and setting the value of his or her time to be worth *more* than the farmer's time.

The farmer *could* hire the nine individuals for this price. She would pay each $1.50, and then each would buy $1 meal packs for the day (which they helped make, and so will likely view the one-dollar price as unfair and want a discount, but let's ignore this). This nets the farmer a *loss* of 50 cents per person per day, or -$4.50 a day.

If the farmer *also* fails to recognize that money actually represents a contract exchange for the time she saves everyone else, and instead believes her price is simply whatever price she believes her food to be worth, then she too will try to "capture" a collapsing cash flow arc. To counter the daily loss, (because a farmer's income is speed-limited to how fast food can be grown and harvested, which is beyond her control), she will raise the price of food for everyone to $1.50 per day.

She now nets $1.50 a day, because Bob is still purchasing his meals, and everyone else can afford them since they each make $1.50 a day from the farmer, which they pay right back.

At this point, the nine are back to netting $0 a day, so they each raise their price for farm work to be $2 per day, resulting again in the farmer losing money, so she again raises the prices of food. The workers then raise their prices, then the farmer, then worker, then farmer, on and on this goes, until the farmer posts a sign that reads "Farmer. Price $(whatever she needs to keep a positive balance).

This runaway imbalance is caused by believing a piece of paper is somehow worth *anything at all.* That there is some agreed-upon value that is set and known by *someone somewhere*, but no one knows who or what that is. That piece of paper's ability to buy *things* is the measure of its value, and that someone out there sets that value. This is the Religion of the Almighty Dollar.

The farmers of society, and everyone who directly reshapes the resources of the Earth into food, water, and shelter for someone else is who sets the value of the dollar, the "universal" barter I.O.U.

If the farmer's price was only ever 50 cents per meal, this whole process to imbalance and collapse into violence still happens just the same, but takes longer.

If her price was suddenly $50 per meal, the *Dark Forest Moment* arrives much faster because the "straw of currency" is bigger. It always arrives though, because people always need to eat, and all the food is behind a money-gate.

The anchor value of a dollar in reality, as I see it, is one "unit" free time spent by: The farmer raising the garden to help feed everyone else faster than they could without that farmer, the water bringer's time to deliver water to everyone else faster than they could get it without the water bringer, or the shelter builder's time to construct a shelter for everyone else faster than they could build it without the shelter builder. Every other service or good money can buy is set *relative* to this difference in "time" because without this difference, *no other service or good exists*.

In the 11 person society, if the farmer spends 5 extra hours a day tending to the garden to produce 10 meal packs for others that are traded for $1, then each dollar in that society is worth 30 minutes of the farmer's free time spent farming (math: she spends 30 minutes producing the food for each meal pack).

This means, in total, *all* the money in this society ($100) represents 100 meal packs, but is *actually* 50 hours of time farming it took to produce those 100 meal packs *everyone else did not have to spend time doing*. After that debt for free time is repaid, which is set by the farmer accepting a trade for money, all $100 will belong to the farmer, unless she trades it to someone else.

Its important to note this model assumes the farm's productivity and the meals produced from it are constant and consistent. This not always being the case exposes another way this 11-person society could collapse to violence. This way has actually happened many times in reality in the past: An event causes the farmer's normally consistent food supply to collapse. This could be a natural disaster, drought, heat, disease, or many other factors. Let's apply this to 11-person world:

Something happened, and the farmer can only make just enough food for herself and one other person for *one more day*. Now everyone could have *hundreds* of dollars in their hands, but it is all more-or-less worthless, because everyone needs to eat and there is simply not enough food to go around.

In this 11-person $100 system, assuming the farmer still spends 30 extra minutes farming to make this one meal and will still sell it for $1, each dollar in the economy has increased in value to now be temporarily worth the entire economy: $100. After this last meal is sold, the value of all money changes to be *worthless*, because it buys nothing *needed* to survive. You can't eat money (well, you can, but it doesn't have many nutrients). Evidence again that the farmer and those that capture survival needs for others are responsible for the existence and value of money.

I now have a question for you and your hydra to chew on a bit: If you were that farmer, and you only had the means to grow enough food for yourself and one other person for one day before all food disappeared, *should* you raise the price of that last meal?

The free market system of money's value being anchored to supply and demand, the creed of The Religion of the Almighty Dollar, says "yes." You can and *should* charge anything you want for that meal, up to $100 and beyond, because demand is *off the charts*: Whoever is first in line will pay you *anything* for it.

Why raise the price though? What value is all the money in the world to you if all of society is on its last two meals and you are holding both of them? You, as the producer of those meals, *know* this is the end for everyone. Food won't regrow fast enough to make new meals before everyone including you starves to death. What good is making $100 at this point, or even a million dollars, considering you can't take it with you in death?

We all *need* food, water, and shelter, and we need it as quickly as we can get it when we start experiencing hunger, thirst, and fear. Your ability to acquire these things at the time I wrote this is often determined by how much money you are currently holding, because money is a contract-in-memory for the time others have spent to give you these things quickly. The contract allowed you to have free time at all for building "parachutes," or whatever.

Your ability to purchase (see: capture) these needs with money depends *entirely* on the people that possess the skills, time, and resources to grow and deliver them accepting your money, and how much he or she chooses to charge for their time and effort on your behalf.

If you make $10 an hour working in free-time as a widget maker, and meals cost $60, you are pretty much forced into a life of constant widget-making so you can keep eating every 6 hours. This ignores that you also likely need to pay for water and shelter.

With this being said, when does it make sense to use someone else as *a means* to capture money *ends* to then capture your own survival *ends* at the cost of someone else's means *and* ends? Are you better than others? More deserving to survive? Are you the Keeper of Who Should Live and Who Should Die? Fight the hydra if it is rattling around in there right now.

If farmers collectively decide to raise the prices of food 100-fold, what options do the rest of us have but to comply to their wishes, or to choose violence to force them to comply with ours? In this way, farmers have a monopoly on the entire free world. Only the relatively wealthy would be able to keep eating, the rest of us would die, and the world would be left with only the wealthy and the farmers. They would be two peas in a pod, except for one little massive problem: The farmer has no reason to ever move his or her money toward the wealthy.

Farmers, and those that work each and every day to supply food, water, and shelter to others are the foundation of money as I see it. They are very much putting up signs for "Farmer: Price $$$" and are getting captured by the hoarders that buy them with the very money farmers made possible. This forcefully un-anchors money from its true time-contract foundation, and all of society floats its entire system of value, which starts to endanger everyone and everything.

162

Capturing the Means

What would you be willing to pay to not starve to death?

To not experience extreme thirst?

To find warmth in sub-zero temperatures?

To find relief from extreme heat?

To not suffer and die from a treatable disease or medical condition?

To protect someone you love from experiencing these ends?

What are you willing to pay to ensure you or your loved ones can continue to survive – to *exist*? Is there a number you can put on this?

I suspect the individuals out there that are currently starving, dying of thirst, trying to find warmth in a bitter cold, dying of a treatable condition, or have a loved one in such a position, know the exact amount they each would pay: *Everything*.

If *you* are the one that is dying, I think you would probably be willing to give up everything you own to reverse the arc of this story. If someone you love most in this world is dying, you would probably be willing to give up everything, even your own life, to capture the end of this story.

In the absence of money, everything boils down entirely to what resources you can access to capture all these capturable ends. If all money went away today as I write these words, this access would be entirely determined by how much of the Earth you *own*. This concept is anchored to words like "property", "acreage", "lot", "parcel", "ranch", "stake", and more.

It is probable, if you have the deed or rights to a shelter right now, that you traded money to "own" it. If you live in the United States and your shelter is a house, it's also very likely you don't "own" it at all - the bank does. If you are unsure about this, just imagine you no longer have any money or any income. What around you do you truly own and can call yours, and what will be taken away?

Now for the big question: Can anyone truly *own* the Earth itself? Can anyone *own* the Moon? The Sun? Can someone lay claim to the North Star in the sky? If you answer "no", then why is it reasonable that individuals and legal entities (which are also individuals – if humanity disappeared, so do they) to claim large portions of the Earth's crust as "theirs"?

Don't try to predict what comes next, because I'm not about to advocate for the dissolving of property. I've already explained *why* I believe it exists in the first place: It was born when you formed a resource-based shape called a "territory" because it was the shape that contained enough food, water, and shelter for you to survive within its boundary.

But, compare this imagined territory-property from *The World of You* to the property you might "own" now as you read this. If you live in a suburban neighborhood, apartment, urban area, or anything other than a multi-acre lot of land capable of growing lots of different food-providing plants and also support an ecosystem of animals – you are very much *dependent* on the individuals that *do "own"* property that can and do provide the food, water, and shelter you need to survive.

However, if these individuals decide to up their price by 5000% tomorrow, what exactly are you going to do about it?

The entire social system will collapse to *exclude the middle* and become two points: Those that have hoarded enough gold to pay, and those that "own" the resources that can supply food, water, and shelter for other people beyond what they need for themselves.

What are you willing to pay for food when you can't make or grow your own? For water when you don't know where some might be? For a shelter from the cold and heat and storm? For treatment of a treatable medical condition? *For the capture of all your fears*?

Would you be willing to capture another human being? Force that person into the position of having to provide these things for you or face death them-selves? Are you willing to *kill* if they don't provide the means to your ends? How many will you have to kill before you realize you are damaging your own ability to improve your chances of survival?

The system of *unanchored* money (Religion of the Almighty Dollar) and the mass-capturing (hoarding) of property in the United States as I write this is literally just a slow burn into *slavery*. Society has experienced a radical progression through farming, industrialization, and technology, but it is still all sitting on the same exact foundation it has been since we first appeared on this planet – the Earth itself, and the limited speed and quantity it can provide what we need to capture to survive.

We are all, as human beings, very much in this experience *together*. The billionaire has to capture all the same needs as the homeless man digging in the dumpster. All the money in this world is worth only what someone else is willing to accept in exchange for food, water, and shelter from the elements. Unless we are willing to destroy all property lines, farms, and gardens, and all go back to *The World of You*, which is hunter and gatherer homesteading with barely a society at all, the balance between the value of money and time is *required* to be recognized.

The resources we all need to survive cannot be hoarded and gate-kept by one human being for profit against all other human beings. The top half of the mind-body loop exists because of the bottom half's movements of Earth's resources, which make extreme amounts of free time (the top half of the wave) possible at all. This means all the money in this world is actually worth *less* than all the time spent building and maintaining these gardens, and harvesting and preserving and moving these survival needs, because should these things become scarce, all the money in the world cannot purchase someone else's willingness to die.

Those people that move around the resources of this Earth needed for the survival of others – the farmers and fishermen and women, the carpenters and the roofers, those that maintain water lines and systems, that keep things in balance by collecting garbage and keeping things sanitary, that wash and preserve food, that move your needs closer to you like truckers, train engineers, pilots, and ship captains and crews, as well as those that collect the resources needed to construct the places where you might capture these need story arcs, such as miners and lumber workers; and even those that construct and maintain these places like construction workers, electricians, plumbers, and all the tradesmen and trades-women in this world - not to mention all the cooks, food deliverers, restaurant workers, and everyone directly associated to speeding up the food, water, and shelter you need to capture and consume regularly - *these* people (who might be you) are the true royalty – the heroes of this age that make the tremendous amounts of free time possible for the rest of us to try to keep improving things and growing the spiky donut as much as we can. And their time performing their skill is ultimately more valuable than any other free time and skill, and even all the money in the world.

In an always-moving system, the greatest value of an individual working in a non food/water/shelter related specialty is determined by how much time he or she can save these need providers by improving their speed or yield. This improvement has a limit, which is not related to the farmer, but to the rate and abilities of Mother Nature herself.

We are the great consumers of her bounty, and we as a species cannot consume what we need to survive faster than Mother Earth can replenish it without inviting disaster. We also cannot start consuming *each other* without inviting disaster, which is to capture each other as means to a hoarding end.

And yet, this is *exactly* what we are doing. The world is entirely upside down.

In its simplest form, you (Point 1) would get your food straight from the Earth (Point 2). If you exist in free time and get your food from someone else, you are using a middleman (the farmer and everyone else that directly handled that food). These middlemen spend the time performing the required actions needed to capture food from the Earth to hand to you. Since your free time skill might be of no direct value to the farmer (such as being a sports player), you *trade* the money made through your free time skill with the farmer for the price he or she requests for the time and efforts spent growing it or raising it.

As I see it, this direct straw-connection between consumer and farmer is now extremely rare, and exists only in the context of a farmer's market or personal farm connections. Without this direct relationship balance, the farmer can easily view his or her equipment and land as the means to an money-end, and start to abuse and over-farm the land, and purchase incredibly expensive equipment in the hopes of massive increases in productivity for the sake of profits – but *the Earth* determines the productivity, not the farmer. With a shifted value toward capturing money as a *master end to capture all means*, the farmer could easily become unable to keep up with growing expenses like engineered seeds and pesticides and equipment that were purchased just to squeeze out a pinch more productivity speed (and therefore money).

With a negative and irregular cash-flow that is determined by the Earth, and debt-flows that are determined by humans that want money consistently, the farmer gets *captured* by debtors, or is captured by individuals that seek to *capture food* to leverage into *infinite* money, because every single human needs food to survive. These "means capturers" often include the government itself, especially when that government requires a farmer to pay taxes on land he or she "owns," where Mother Nature acts as the very foundation of the entire population and government's (and money's) existence.

When it comes to water, in its simplest form, you (Point 1) would get your water straight from the source like a river or lake or creek or well (Point 2). If you exist in free time and get your water pumped through a pipe to a faucet very close to you, then those who installed and maintain this pump and pipe perform the required actions needed for you to capture water. You trade your money for the time it took to install, and if it was installed before you ever made money, then you are trading for the time spent by individuals to operate and maintain that system to keep it working for you. This is how the public water system largely works in the United States. However, public water systems themselves are also starting to be captured by individuals that seek to *capture water systems* as a means to make *infinite* money.

Lastly, with shelter, in its simplest form, you (Point 1) would get your shelter straight from the Earth by using a natural structure like a cave or constructing one from raw materials like sticks and rocks (The Earth, Point 2). If you exist in free time and have a shelter that was built for you, then those carpenters and tradespeople who collected the raw materials to construct it, as well as those who actually constructed your shelter, performed the actions for you. You traded your money for only their time spent collecting and moving the materials, and building the shelter. If it was built before you bought it, then you paid only for the time spent by those that maintain it as well as the time of those that collected the raw materials needed for maintenance.

This is *not* how the shelter value system works today. It is merely a supply-and-demand system – but here is the messed-up part in my opinion: The supply part of the equation *doesn't actually change*. At the time I am writing this, there is no such thing as a shelter that is not *also* connected on the Earth, and therefore any shelter boundary you might use is always connected to an "owned" territory or property.

This total amount of property available on the *one* sphere of Earth *does not change*. It does not increase or decrease except by the Earth's own activities (side note: we can add land through methods like landfills, but something is not coming from nothing, because that garbage was made from materials of the Earth, so it's still just re-arranging). This means the value of property is exclusively determined by demand - how many people currently need shelter to survive. In other words: If there are 8 billion people, take all the available land where shelter can be built, and divide it into 8 billion properties. That's property equality.

Except...it's not.

The value of a shelter is relative to the survival features of the property. How big is it? What is on it? Where is it? What is it close to? The arrangement of these things determines value based on *feelings*, and since we are all the same species, we tend to desire shelters that provide similar feelings (similar arrangements), and this *wrecks* the divide-by-8-billion idea.

For example, those living almost exclusively in free time seem to often love the idea of "owning" properties in paradise. Paradise is a relative word of course, but we are all still approximately the same, and for this reason, it makes sense that property "demand" in beautiful places (often near water) is significantly greater. Everyone that doesn't need to farm or fetch their own water often dreams of living in a mansion on a tropical island, or a lake house in the mountains, or a cottage overlooking some "beautiful" landscape.

This immediately creates a repeat of the Broccoli Apocalypse, except with properties.

Even with this fairness impossibility aside, if the total amount of property in existence is fixed and value is based on demand, then if the population grows relative to the fixed number of shelters, *all* property goes up in value. If the population declines relative to shelters, demand drops, and so does perceived value.

If this were true though, new shelter building would sometimes stop. Property values would sometimes naturally flex downward, and this change would correlate with the population in need of shelter. As best I can tell, this does not actually happen. In the United States, the price of housing has skyrocketed and has *for decades*.

This has nothing to do with demand based on population size, in my opinion. Instead, systems are in place to make money from the mere *existence* of property by *hoarding* the fixed and limited amount. The means are being captured, because money has become the ends.

Money, which is something, doesn't come from nothing…

163

The Troll Monopole

More and more Earth in the form of property is getting *captured* by individuals acting as "middlemen", "gate keepers", or "dividers", which might even describe your actions. If so, you have chosen to capture property because you believe it is the right thing to do, but there is one justification you might be using that contains a gap capable of destroying society itself: To make money.

Property hoarders using this justification don't capture properties for any survival purpose. He or she do not intend to use collected pieces of Earth as a way to sustain or benefit society, but rather as merely a *means* to sustain his or her personal non-stop collection of gold.

This whole concept hinges around one question: How much would you pay to have access *to the Earth*?

Food and water are attached to these properties by default, because food and water are attached to the Earth. Food and water does not rain down from space, or bubble up from the core. More and more means to produce food and drinkable water, which everyone needs access to survive, is getting captured.

How much would you pay to have access *to food and water*?

The potential to make money is *unlimited* when you know someone will eventually pay you *everything* to get what you are hoarding.

These middlemen and women – perhaps you - unknowingly (or perhaps knowingly) become a parasite created by the *Floating Anchor Problem* eating the core of society. The passage of time and the usage of language warps values within cultures over generations, and money eventually becomes seen as an end – a primary need to survive – as if cash has survival value all its own. You might be born into a situation where everyone seems to want it, so it easily becomes an end to all means without a story to keep it anchored.

Money, however, is a human system that acts as currents in the balloon-straw metaphor between humans that are trading *time* for *things*. There is no straw between humans and the Earth except consumption - and the Earth provides the resources we need to consume to survive *for free* (ignoring our requirement to move toward them).

If you are trying to capture the constant currents of free time and money between humans in the inner sphere and outer sphere, or capture the constant currents of consumption between humans and the Earth, you are ultimately trying to insert a straw *and* clips between as many balloons as possible.

You would not be a balloon, however. You will not expand from taking in both of these currents until you balance. You are upside down, which makes you the opposite of "one". The opposite of one is *not* zero. You exist, so you cannot be zero. Instead, you are an opposite one, a *negative one*. The only way to reach balance from such a position is to siphon off all of "one" until you have canceled it out, and you both become *nothing*.

However, you are not trying to drain other people dry. Because you are capturing money-currents by first capturing survival resources, trading resources, or communication resources, you are ultimately trying to capture *movement*. Essentially, you are trying to force *all* balloons to connect their currents through your straw so you can capture an endless supply of the "means to capture all ends" (according to the doctrine of Religion of the Almighty Dollar). You become a *monopole* that begins intercepting all currents and pulling them in the same direction – towards you. Why? Because money is a social shelter.

This monopole process requires trying to capture *both* ends (both balloons) for the sole reason of feeding an insatiable appetite for gold, mistaking it as needed for survival based on the idea money is required for property, and connection to the Earth is needed to survive. You have to make yourself the new path of least resistance for everyone else.

The warped value of this is pretty easy to spot in my opinion: If you seek out or create a need in society, and then attempt to fulfill that need with the *primary* goal of making sure everyone *must* come through you and *only you* to fulfill that need, and must "pay" you for it - you are creating a future *Dark Forest Moment*. When that moment arrives, it will be you versus *everyone else*, and you've already chosen the path of domination. You essentially become a troll living under the bridge you built for the sole reason of installing a tollbooth to collect money. The value was *never* to help other people, but to exploit other people to help yourself.

Let's do a quick walkthrough of what this might look like in society: Imagine a person creates a trucking company that connects farms to grocery stores. Let's name the owner Bob, and his company Bob's Grocery Trucking.

Bob's Grocery Trucking makes Bob a ton of money. This is fair at first, because these trucks are saving society lots of time. Before this point, the farmers were delivering the food themselves with a horse-drawn cart for 50 cents a meal.

Bob's trucks created more overall free time, and that shrinks the inner sphere, which grows the outer shell. Food gets to the shelf faster, can sit there longer without spoiling, and now arrives multiple times daily. That saved time has a lot of monetary value – technically everyone that gains free time from Bob's company trades that gained free time in the form of money to Bob. Food is now $3. Farmers get 50 cents like before, the store owners gets 50 cents, and Bob takes $2, because he is saving everyone time.

Bob's life drastically improves. He trades his money earned for a new house on a beautiful property, fancy cars, and starts taking nice vacations. After awhile though, the *Floating Anchor Problem* and the need for evermore certainty will insert itself, especially once Bob discovers Tim's new trucking company is trying to "steal" Bob's connection by chatting it up with farms and stores. To ensure his company's survival (and continued level of certainty in life it brings him), Bob convinces all the farmers in his area to sign a lengthy time-contract to only ever use his company in exchange for a steady income of $1 per meal (double their current income) for the entire length of the contract. After all, if you could double your pay and not have to change the amount of work you do, you just have to agree to do that work for several years – would you take it?

Bob has just captured one of the two ends he was connecting with his company. Of course, neither are ends from Bob's perspective – they are merely means. He can now leverage this captured mean against the other, which is every grocery store owner Bob's Grocery Trucking currently delivers to. After all, Bob now has complete control over the flow of food into stores. He controls the *supply*, and the demand will *never change*. How much would you pay not to starve? Bob can now raise his prices to deliver that food, and there is nothing anyone can do about it without entering the Dark Forest.

Bob tells all the grocery store owners each meal will now cost $5 to deliver instead of the $2 the stores were paying him before. The 50 cents the stores were paying the farmers before simply goes away. *What option do the store owners have?* If they don't agree to Bob's terms, they will not have food. If they do not have food, they do not have a grocery store. They must comply with the demand (notice the change of direction in demand?), and then decide if they will now lose money every day, or if they will raise the prices of food for the consumer to prevent going out of business.

By capturing just one end, Bob easily captured the *other end*. Demand is always the same, and there is a minimum rate food is needed through time. There is no time to let stores go empty to haggle prices. The store owners would face the *Dark Forest Moment* with consumers almost *immediately*.

Farmers are making a guaranteed $1 per meal and have gained lots of free time since the horse-and-cart days.

Bob is now making $4 per meal (with $1 of that going back to the farmer).

Store owners were previously paying 50 cents per meal to the farmer, then started paying $2 to Bob and 50 cents to the farmer when he first began trucking.

Now Bob is charging $5 per meal, and the farmer is out of the picture for the store owner, because Bob pays the farmers $1 of his $5. To keep the stores afloat, the stores are now charging $7 per meal to the consumer.

From the store owner's perspective the only thing that changed between the $2.50 system and the $5 system is the cost and the number of people that needed paying. It's all the same trucks and the same food on the same schedule. 100% of the difference is flowing right to Bob.

From the consumer perspective, these cost increases are only understood at the grocery shelves and might seem annoying, but paying it is required to keep the same amount of *normalized* free time to be the same, which requires *expecting* food to be at the grocery store. This all feels like the cost increase is just a part of life.

Unless, of course, *you can no longer afford food at all*. You slip closer to the *Dark Forest Moment* of stealing that candy bar. Of stealing from a farmer's field. Of needing to work three jobs - wage slavery - to make enough to eat.

"Blame *inflation*," the editors say.

No consumer, including you, is actually *required* to pay anything at all except *time*. The Earth does not bill the farmers, truckers, and grocery stores any money at all to grow plants and animals. But you are not often aware of these systems and middlemen, and design your life around having an expectation of a certain amount of free time and for grocery stores to have food.

The only thing *required* is to have *access* to property capable of growing food and supplying water and providing a shelter – and farmers have *exactly that* – and they use their property to do these things on your behalf.

Everyone else involved is a middleman between the person growing your food and you. Some have good intentions to help food distribution to help balance society, and others have self-serving intentions that tip everything out of balance.

If everyone skipped over Bob and went straight to the farmers, his or her meal would only cost $1, and the farmer would experience no change at all in free time or income, and would even gain a much stronger connection with the consumers he or she work hard to provide food for every day. It is back to you and the Being trading time for the piece of paper – or maybe just a smile.

Bob's contract with those same farmers makes doing this *illegal*. You cannot have a relationship with the individuals providing your needs from the Earth. You *must* buy food from local farmers from the grocery store, because Bob has forged a new path of least resistance.

So why not start your own farm? It makes sense right?

Unfortunately, as I write this, obtaining your own "farm" or productive "piece of Earth" is becoming ever more difficult, thanks to individuals hoarding property because he or she believe it is the right thing to do. The result is a runaway paradox-of-a-process that demands money for the property needed to farm, which sets the value of money in the first place.

Any new farmer often ends up taking on enormous *debt* just to buy the land to farm, or worse, buy "access" (rent) to the property *to* farm. The huge imbalance present here is that success in farming to make repayment possible at all hinges *entirely* on Mother Earth and her weather *consistently* making plants grow and keeping animals healthy, which is in many ways beyond the control of the farmer. One bad season, and the farm life could be over. The farmer loses everything and debt remains, and the hoarder has lost nothing at all. If the hoarder sold it to the farmer, he or she may even now get to buy it back from that farmer for less than the farmer paid for it.

For farmers, this makes contracts that *guarantee income* (or at least stabilize it), such as the one the farmers made with Bob's Trucking Company, appear more and more appealing. But capturing inner sphere time starts to flip the whole world upside down, because those in free time are capturing the source of money's value.

Let's say Bob makes enough money from his new arrangement to try to buy out the farmers completely. He offers a new deal: He will pay them a regular salary (instead of by-the-meal) if they sign a very long-term exclusive contract, or if they sell him their property and agree to continue farming it. Bob does this by creating a farming corporate entity that "employs" the farmers or buys the property – AntiMatter Farms.

The farmers are excited to have a steady paycheck, sign the deal, and now work for Bob's corporation.

Because hoarding gold has always been Bob's main value and goal, he requires all the farmers of AntiMatter Farms to grow only the *most profitable* crops and animals. His wealth skyrockets.

Consumers start to complain about lack of food variety, so Bob creates more corporate entities that buy factories and facilities to re-arrange the same profitable crops into new processed foods. These foods cost more, because Bob has to offset the cost of the factory and make sure he still makes money. Now people can buy raw antimatter heads, but also buy Blackholios, Void-Its, and Annihilatachips (which come with tasty SpikyDonut dip) which are all made from the same antimatter heads.

Soon Bob makes enough money from this to buy *all* the grocery stores as well, which he does by creating another corporate entity – Troll's Grocery Stores. Now all the former grocery store owners work for Bob as well.

Bob is in nearly full control of the entire food system, but it is in his interest to hide all of this from other human beings (consumers especially), because they are now *captured* in a way they are *forced* to participate in the system. It is also in his interest to hide as much of this as possible from the authorities that work to prevent monopoles like his that suck away society's freedom (government and media oversight). If push comes to shove, Bob can always "sell" one of his entities to someone else, like say...his nephew.

And now the fractal that destroys society begins, as Bob moves to maximize profit for each corporation between each other – money from *nothing at all.*

AntiMatter Farms increases the cost of food it sells to Bob's Grocery Trucking to $6 per meal. Bob's Grocery Trucking then moves that food and resells it to Trolls' Grocery Stores for $10 per meal. Troll's Grocery Stores then sells those meals to the consumer for $25 per meal while paying each former store owner (now store manager) a steady but small salary. Food that gets processed is sold between companies as well, and ends up being $35 a meal.

Everything has changed, but yet nothing has changed at all. The farmers are on the same land, and the consumers are in the same location.

All of this current between Earth to consumer is now captured inside a single straw that has no justification to improve or lower prices, because improving would cost money and so would reducing prices, which goes against *the entire reason this whole system from Earth to consumer has been captured – to make money.* The result is worsening service and worsening quality food, over-farmed and nutrient-depleted land from growing the same most-profitable crops over and over again, and increasing cost for every meal.

"Blame *inflation*," the editors say.

Bob now has so much money that if the land stops producing, Bob just buys new land and hires farmers to work it. He even bought Tim's trucking company and dismantled it. Bob can keep doing things like this as long as people are willing to give their money and time to him. And whether or not you have a choice gets interesting to think about very quickly.

If all the water on Earth was captured by middlemen – how much will you pay for a bottle of water? Knowing the answer, do you think individuals started corporations that bottle water did it to help society, or to make money?

To test any "yes to help" or "why not both" answer you might have, consider the following:

If all the remaining drinkable water on Earth were suddenly poisoned (so they could bottle no more water), do you think the owners of these companies would start giving away their bottled water *for free*, or do you think only the wealthy would taste water again before death?

There are many specific situations in the world as I write this where an upside-down-middleman has captured both ends by capturing the means, and is literally starting to reap *unlimited* profit from humanity while at the same time decaying quality, service, and ultimately hurting everyone in the process.

It is natural to want to build monopole structures like these, because owning an entire process is the most cost-effective and efficient way to run it. *Unity is perfectly efficient.* But unity stops movement. A singularity that demands something from you by taking something away you need to survive is like a black hole requiring you to pay money as you fall into it.

This situation is so messed up and backwards to me, I've decided to pick a couple of real-life monopole systems and give them a chapter all to themselves.

164

Cancers

Reality will always be somewhat unpredictable at the scale and speed we naturally operate with our senses.

We seem to be able to make everything human work pretty consistently – for instance our buildings don't immediately fall down, our engines don't immediately fail, and our electrical grid doesn't constantly start fires or kill every third customer each day. However, we still can't stop the process of decay. Buildings, engines, and electrical grids will eventually fail.

We can't stop the meteorites we can't see coming, the droughts and storms we can't predict, and the diseases we can't see finding their way into the world from the void. You might experience a life changing event tomorrow (or maybe yesterday) that leaves you unable to walk, see, or talk, or alters your existence significantly in some way. A tornado, hurricane, tsunami, earthquake, or flood could wreck your world 30 days from now.

How much would you pay to not have these things happen in a way that they affect your life? For anything that gets lost to these events to magically restore itself?

What would you pay for the universe to quickly "correct" itself whenever it might wreck your certainties and security?

Welcome to the idea of "insurance".

In exchange for a monthly sum of money up front, a middleman hidden behind an entity known as an insurance company will *guarantee* your property can be replaced if it is destroyed or damaged by the universe based on what someone else (not you) judges it to be worth.

In exchange for a monthly sum of money up front, a middleman hidden behind an entity known as an insurance company will *guarantee* your life can be replaced if it is destroyed or damaged by the universe based on what someone else judges it to be worth.

The starting idea of insurance was a good one. Typically life-changing disasters and accidents are isolated to a handful of people at a time because our little gravity well of atomic soup is staying decently consistent. The whole human world is not destroyed by a single tornado or tsunami. Based on these events being "rare," if everyone chipped in money to the same pool, then through application of the golden rule, that money could be used to restore you, me, or anyone else back to the same certainties we used to have, should we be struck by an "unlucky" unpredictable inconsistency in reality – like a lightning bolt.

I like to think of this concept as an offering of good will toward each other, much like the farmer does for those of us that eat the food they produce today (assuming he or she is not producing food for others solely for the purpose of making as much money as possible by capturing the Earth). This pool of insurance money is actually a pool of "free time" that can be paid back to the bottom half of the loop that build, repair, and move things to restore the world.

Insurance as a concept has the potential to save nearly all the time it would take to start all over again at *The World of You* and return to the level of certainties before the unfortunate event. However, in my opinion, insurance has transformed over time into a monopole of epic proportions. As I see it, insurance has even gone beyond monopole to become a cancer draining the life and well being from United States society in particular.

Cancer, by definition according to Oxford Languages at the time I wrote this in Summer of 2024, is "a disease caused by an uncontrolled division of abnormal cells in a part of the body." In other words, cancer is an uncontrolled expansion of cells. It begins with a cell that refuses to stop dividing, which in a world with limited resources (like your body) is the same as refusing to stop multiplying.

To rephrase: A cancer cell is born when a cell's instructions (values) become warped to put itself over anything else, and it starts to capture all ends necessary to capture the means it needs to live forever while also avoiding detection of those that can stop it (immune system, which recognizes things that are not "self").

In my opinion, most health insurance in the United States behaves like a cancer in society. I hear and read many individuals around me accusing doctors and hospitals of price gouging and being greedy. This might be the case in a for-profit medical system, and is sometimes the case for doctors in particular, but most healthcare systems in the United States are *not* for-profit organizations at the time I wrote this.

This being said, the organizations that pay the bills the hospital sends out – insurance – are almost entirely *for profit* organizations at the time I wrote this. As I see it, they exist to make money *first and foremost*.

Those that lead and own health insurance companies, which might be you, seem to have captured nearly every facet of healthcare, turning themselves into parasitic super-trolls that have started to drain the entire system dry and are slowly collapsing currents to zero. These individuals *easily* captured most citizens, just like Bob captured most farmers in the previous chapter, because in the beginning health insurance was a good and "fair" idea that made sense. The authority hierarchy even enforced the idea that everyone *must have* insurance – because it used to be a good idea.

But, as generations pass and the *Floating Anchor Problem* takes its toll, everyone normalizes this insurance system, and "everyone" *includes those that run and own insurance companies.* Eventually, the values of these individuals - who might include you - flipped from helping others save time for money, to helping *the business* save time and money. This results in goals like trying to ensure market domination to ensure and maximize profits.

When these inverted goals are actually met, the switch from troll monopole to cancer occurs. The first shift was from providing a straw between two balloons (helping fulfill a need), to trying to force both balloons to always flow into the straw (maximizing profits and market domination). Next comes the second shift to "live forever" and keep the flow of money growing by any means possible.

Non-profit healthcare systems and hospitals today are often struggling to stay afloat financially. People need medical care and can seek it in the United States, whether they have money or not (for emergencies anyway). It seems some shred of humanity still lives in the system and we don't believe you should actually die because you don't have money...yet.

To offset those that cannot pay for healthcare services to survive an injury, accident, or severe illness, the cost of healthcare must rise so that the systems can continue to function with a break-even budget. After all, doctors, nurses, and other medical staff that make hospitals work at all are technically restoring free time to others, which is immensely valuable (and is a good argument for quality, outcomes, and productivity measures when considering pay).

This raise-the-price-to-offset-free-services can easily become a runaway process though, because as the price gets higher, more and more people cannot afford to pay the bill, driving the price higher still to offset these non-paying individuals as well. Insurance tries to disconnect from this spiral by trying to have a constant "buffer," or money pool (or balloon), against sudden spikes in illness in society.

But things got dark quick in the United States.

Insurance Company "A" – let's call it "Bob's Insurance Cancer Shack" –

started making deals with other insurance companies to create what boils down to "territories" in the form of exclusive contracts with specific healthcare systems, regions, or even states. Inside these contract-created territories, each company has "captured" almost all *possible* patients, which by default makes up almost all the possible income for a specific healthcare and hospital systems.

So what happens to those systems if Bob, who runs Bob's Insurance Cancer Shack, decides the local hospital system is charging too much for X-Rays, so he will only cover 75% of the cost?

What happens if Bob decides he will only pay if the patient is required to try *all* the cheaper generic brand drugs that *might* work before "allowing" the name-brand drug that *will* work but is completely unaffordable without insurance?

What happens to your bill if a doctor that Bob has specifically excluded from paying (because the doctor is from a different insurance company's territory) walks into your surgery while you are unconscious and bills you for his or her time?

Insurance companies in the United States have captured nearly *everyone* in a game that has become nearly impossible to escape. Insurance companies, who are often a regional monopole, negotiate contracts with hospital systems about the price of almost everything *before you ever set foot in the hospital.* Do you think Bob is going to let a hospital system increase the money made from paying patients when that money *could* belong to him instead? Do you think Bob will let a doctor decide a patient needs an advanced and specific treatment, if there are two other basic nonspecific treatments available for a quarter of the price, when his only value is making more money?

Those individuals running health insurance companies in the United States have largely made the transformation from troll monopole to cancer. In my experience, you as a patient seem to be nothing more than someone trying to "steal" money from his or her bottom line. You are not a consumer to be helped; you are a threat to profit. Doctors and their diagnostic and treatment decisions seem to be a threat to profit. Medicines and pharmaceutical companies are a threat to profit.

Many insurance companies now run health and wellness programs. As I see it, individuals that run these companies did not create these programs to help you be healthy because it's the right thing to do for society, but only to try to decrease how much money you might cost him or her. This isn't wrong in the big picture, but cancer can perform tricks to make sure it never starves: Sticky pricing.

Do you believe the premiums you pay are ever going to go *down* if *all* the people under the same insurance company you use are healthier and need *less* healthcare? It seems to me that the only reward likely is that your cost is not *increased* for non-compliance with the complicated health program – but hey, you can earn points and buy t-shirts from that same insurance company, right? (I'm sure they aren't losing money on those shirts; either.)

There are sports stadiums with insurance company logos on them. Millions of dollars were used to put them there. There are multi-million dollar Super Bowl advertising campaigns. What if all that money, which was given by the people to insure themselves from harm, was instead given back to the same people as a reduction in cost?

In my opinion, those running health insurance companies have captured both the consumer *and* the healthcare systems, and are holding the healthcare of the United States at a metaphorical gunpoint. They are holding doctors and your care providers "hostage" by refusing to pay only for the medical care that they - *the insurance providers* - deem necessary – *not* the individuals actually providing the care with medical degrees, training, and expertise through firsthand experience with the patients, biology, chemistry and other specialties.

Patients are *secondhand stories* to those making decisions within insurance companies. These *secondhand stories* the insurance companies put together about you hold you hostage in many ways. As I see it, the individuals that own and run these companies do their *absolute best* to hide and obscure all of this as much as possible from those already captured, and exploit all gaps and imbalances for his or her own gain.

For instance, these companies usually send you an "explanation of benefits" that very *intentionally* shows you some *outrageously* priced hospital bill, along with the portion insurance is paying on your behalf to save you from this "horrific injustice" of a price. This likely leaves you mad at the hospital system because the bill you end up having to pay comes to you from *the hospital billing department.* But that same insurance company is conveniently leaving out the part where *they negotiated all these prices and what they would pay the hospital in the first place.* In my opinion, this is exactly backwards, and it is the insurance company, *not the hospital,* that should be *asking you for the money that they refused to pay on your behalf.*

The healthcare systems and hospitals, much like the grocery store owners in the previous chapter, are often struggling to make ends meet and must comply with the demands of "Bob's Insurance Cancer Shack/Grocery Trucking" because they need money to pay specialists and buy equipment to save people time - maybe even his or her life - and Bob has captured all payers. So in the end, non-profit healthcare systems must raise prices *anywhere they can* in order to afford to pay their specialized human beings food, water and shelter in the form of money, and also cover equipment and supply needs.

And this runaway cycle just repeats itself – it *recurs* - inside this gap. There are middlemen within middlemen trying to capture both ends, which are themselves both middlemen.

You get a large copay bill from the healthcare system you can't afford to pay, so you might not pay it, repeating the original cost spiral problem, and insurance *doesn't have to care.* You might even blame the healthcare system, as the gap of what insurance covers and what you are billed for continues to get bigger. To keep this imbalance from destroying his or her insurance business and any usefulness to the consumer, often this growing debt is put into memory in the records of the insurance company and simply pushed down the road to *future generations* of insurance payers – *our kids.* As I see it, their insurance rates will only ever be higher because insurance is quite literally cancer.

Unfortunately, the cancer of insurance does not stop at this point. It metastasizes and begins to spread: Those that run these companies have invented ways to prevent complaints and arguments and ultimately capture more money by weaponizing time itself against the very people that pay them for coverage.

Many insurance companies today seem to have some sort of policy to reject almost any new claim you might make. Why? Because maybe you won't fight it. If you don't? Hey, more money made. But in case you *do* want to fight Bob's Insurance Cancer Shack's denial to pay, he has also installed a horrific time-eating maze-like communication system for you to navigate (if you can) to even *try* to get the story of why you were denied. To avoid his employees experiencing direct verbal confrontation, not only are his employees not empowered to make helpful decisions, but the story you are given will often be shifted into another language of specialty – such as hospital billing codes you probably won't understand (how could you unless you worked with them?), which makes it almost impossible for a consumer to argue without having expertise in healthcare or insurance.

Bob's Insurance Cancer Shack can say "no" to paying for your care in 5 seconds, but you will likely need *5 hours or more* to navigate the gauntlet to even ask them why and to reconsider. In this process you are likely to get redirected, disconnected, looped around, and read a script about how your argument is going to "go into review."

Once again you are *waiting* on them for your own care, and potentially losing precious time to live with deteriorating health.

They are quick to make decisions when it benefits them (denied!), and slow as molasses if it might cost them money – the general weaponization of time. All the abyssal imbalances of language can be exploited against you.

(Reminder that the owners of these companies pay *enormous* amounts of money to put their logo on sports stadiums and in your face on television with commercials.)

If you ask me, insurance-for-profit cannot be allowed to exist, or must be made completely transparent as to what they are doing with money intended to pay for the well-being of society. Healthcare falls into the same genre of existence as farming. It is in our interest as a society to help everyone be as healthy and capable as they can be, the same as being fed and having water. We are all in this together. If the idea of profit is introduced *anywhere* in the straw path between life and a very real chance of death through a preventable means, the profit potential *has no ceiling*.

Develop a new drug that cures cancer? Should only the wealthy get their cancer cured, and the poor just die? Sounds like adding one cancer to another cancer to me.

Health insurance has captured both ends – you are forced to have it. The owners of these companies tell you what they will pay, tell the hospital what they will pay, don't tell either of you about what they will *actually* pay in a way you can predict your bill easily, and reject paying anything at all as much as possible. There is also no justification for these companies to increase the quality of coverage or make it easy for you to understand or lower their prices, and profits are forever maximized as the gap between coverage and cost grows. As I write this, there is now insurance available to cover the gap of what your primary insurance won't pay (gap insurance). You need insurance from your insurance. This whole system is now leaking over to your pets with veterinarians as well. *It's all cancer.*

Pharmaceutical companies are not much better. They have no incentive to find new medicines in the dirt, liquids, or air of the Earth. Instead, there is only incentive to create new man-arranged medicines, so they can be patented, protected, and gate kept – not in the interest of the patient's well being of course, but in the interest of how much gold the patient will give for it.

Cancer.

If you are trying to become the sole provider of something that you yourself would pay anything to have because it helps you avoid dying or suffering – then you may be on your way to becoming a troll and eventually cancer.

These approaches seek to hold all of humanity hostage for profit as *an ideal.* I'm not saying whatever you might provide is not worth money, especially if it saves time or creates time for others – but nothing is forever, change in inevitable, and trying to force your service or product to be needed forever *and* constantly generate more and more wealth forces you to become a literal monster that will begin to justify sabotaging any other provider, and even the Earth itself.

And yet, the leaders of society, which might even be you, are often the same individuals running these types of cancerous organizations. It should come as no surprise that these enterprises of single-owner interlocked monopole corporations, which are often themselves interlocked into a handful of mega-monopole corporations, are considered "too big to fail," and are simply handed mountains of gold by the government from the people to "save them."

Let them fail.

As I see it, they are simply cancer that ran out of resources to consume to stay alive.

In some cases though, the cancer might need a more aggressive treatment....

165

Dividing By Zero

Let's talk about something from nothing....

Let's talk banks and shareholders.

Consider that the *entire financial system in the United States* is sitting on the foundational idea that is *literally* and easily seen to be the Oxford definition of cancer: The belief in *forever growth*. The Religious Cult of Profiteering.

I'd be willing to bet you probably put your money earned into a bank account. That money doesn't likely sit in a vault. Those that own and run most banks invest it and trade it in ways to try to help grow that money into *more money*. Essentially, your money is loaned out to other people that buy expensive things with it, with the hope that these debts created by loaning your money to others like Bob will eventually be paid back with interest.

Banks also buy government bonds, which essentially hands your money over to the government to spend to help you as a citizen in exchange for a promise written on paper. That promise is that the paper can be returned to the government after a set amount of time (contract written on the note, usually measured in years), and at that point it will be worth *more money* than was used to buy it.

How can anyone ever *guarantee* growth? Or that money will always increase in value? This seems absurd to me, and obviously unanchored from all ideas of productivity and gross domestic product and the true anchor of units of human-created free time. Yet the whole banking system is built on it. To me this is late-stage maybe-terminal cancer, and the closer you might look, the worse things get.

The stock market has ups and downs, but if you have a job that offers a 401k or retirement program, your "savings" rely completely on the stock and bond market *continually growing in value* for you to reach your goal of retirement. Can you see the potential issue here? Stock and bond traders themselves have become middlemen and trolls trying to manipulate the market to make their "bets" always gain in value. There are human beings that make money when a stock value goes *down*.

Cancer.

Forever growth is a paradox of a belief. It is impossible. When it is believed and acted with, massive holes and gaps open up in the social structures, creating weaknesses that will eventually cause everything to collapse on itself to find stability again.

Consider that right now if 50% of people in the United States with a decently large amount of cash in the bank withdrew that cash – the entire financial system might crumble all the way to the bottom. This is because the bank *doesn't actually have your money.* That number on your screen is hollow. Those that run banks are playing with it, trusting their own decisions with it – sometimes justified and sometimes not - but guess what the number one value he or she cares most about? *Profit.* Forever growth. Permanent stability in an upward direction. They don't have to worry too much; your money is *insured* by the government. You know, the one borrowing money from the same banks using bonds guaranteed to go *up in value*?

This means the time spent to create more free time for others, which is divided up into the form of trillions of tradable pieces we call dollars, is trying to constantly *expand* in value. This is impossible because there is a maximum ratio of time that can be used to create more free time. That limit is defined not by us, but by reality and physics. We are not actually the rate-setters of time – it is the Earth, it is the universe.

As best I can tell, the bank itself has become a middleman between you and your money. And that same banking system has created markets that are built on the idea that property, which is the root of all wealth and free time created, will always go *up* in value, despite that the amount of property being used to produce food, water, and shelter is continually going *down*, which will eventually result in the population going *down* from lack of available resources, which makes it impossible for wealth and the value of the dollar to keep growing.

It seems to me our values are upside down: The green spaces of the Earth are valued *less* than the skyscraper office buildings. A one-acre lot is worth less if it has no house, and more if it has a house, and more still if it has nothing but concrete and steel and glass and office space. Each time the wild Earth is replaced with concrete and architecture, the overall "value" of the Earth goes up – and this is exactly *backwards* from reality, where the increase in sustainable crops, drinkable water, breathable air, and usable land are actually more valuable in an undeniable way. This creates a massive bubble that will pop eventually, and when it does it will be a disaster of horrific proportions as the value of money collapses, and with it all insurance, healthcare, and all other gatekept needs.

Property might even collapse as millions return to *The World of You* and begin raiding all the gardens remaining, but there will be no magical "pause" to help us restabilize the collapse of the gardens, waters, and shelters.

Sustainability – or keeping the rate of our consumption in balance with the rate of resource reproduction – is *everything*.

In my opinion, a moving system needs to consistently "pop" all the monopoles and aggressively treat all the cancers, though my hope is that eventually we may realize that profiteering by exploiting resources is to work against your fellow human, and true balance is constant change and adaptation.

Just like farming, and water care, and shelters – it is in everyone's interest to help everyone else capture needed healthcare, medicines, and emergency services. I'd love to hear or read your ideas of how to keep these systems in order and balanced without the runaway profit and troll problem, because what we have now is definitely unsustainable.

If things do not change, and society collapses, the *Dark Forest Moment* will appear for us all again, but the "shelter dance" has been heavily revised in modern times as to make overcoming fear and choosing cautious trust incredibly difficult.

The revisions made to avoid the *Dark Forest Moment* actually pushed the boundaries of contact with each other to be so far apart we are invisible to each other.

Fear now radiates in all directions, filling the gaps between us. And it is justified and validated again and again until a new looking glass of fear is born.

166

The Demon Core

Imagine you stumble upon a strange device you have never seen before. It's a small bright purple ball, about the size of a basketball, but you can see through it to notice the intricate workings inside. You can see gears and levers and springs arranged with each other through the purple tinted shell. It's pretty, and it's fascinating.

As you turn it over in your hands, entranced by the complexity of it you feel your index finger sink into the backside of the ball, and as you turn it over to look at what has happened, an ultra-high-powered laser blasts out of the opposite side, burning a hole through the ground under your feet and deep into the Earth before almost immediately turning off.

Startled by all of this, you immediately drop the ball, and it falls into the hole. You hear it tumble and bounce off the walls for many seconds into darkness further down than light allows you to see. As the "clinks" and "clanks" fade away into silence, you see a faint orange glow begin in the darkness. It gets brighter and appears to be rising up from the hole.

Let's pause the story here and address some things:

Did you just punch a hole in the Earth because you thought it was the right thing to do? Of course not. You had *no idea* what the device was capable of doing when you picked it up. You had never seen it before. It didn't have a sign that said "Warning: Powerful Laser" next to it.

Instead, you are human, you are curious, and you are trying to understand your reality - and a purple ball is not immediately a danger according to your context of experience so far in life, so you wonder what is inside its boundaries.

Children are in the same position *with guns*. They are harmed or killed by guns more frequently than I think anyone would prefer (I would like to believe that preference is zero, but I know there are some individuals out there that currently believe domination and violence against the smaller and weaker is the right thing to do).

Children can and do find a parent's gun or a neighbor's gun without knowing what it is, or knowing but not understanding the power of it – just like you would not expect a laser to shoot a hole through the Earth from a small purple ball. So, kids might pick it up and examine it, and maybe even begin to play with it, and accidentally set it off with a finger in a place they didn't know would cause something to happen. This fires a bullet at high speed in a usually-unintended direction.

As I see it, this alone is evidence that all guns that are fired are not necessarily fired with intent to harm or kill. After all, a gun lying on the table is not going to kill anyone by itself. A gun thrown through the air is not going to kill anyone by itself either. The guns themselves are not "doing" anything (ignoring automated guns that exist in the militaries of the world).

Where I live – the United States – guns are pretty much a cultural obsession. I personally have no problem with guns. Growing up in the rural countryside, guns were just another tool. They helped protect livestock and gardens from "thieves" in the form of birds and animals. They were often stored in the back window of pickup trucks, and learning to shoot was a rite of passage into growing up.

As I write this, there is a lot of debate about guns and gun control. Kids keep getting gunned down in classrooms, and what seems like "senseless" shootings of all kinds keep happening (though each person that used guns in these events did so because he or she believed it was the right thing to do at the time it was done).

However, in trying to control guns, which is really related to trying to stop specific objects from being used for specific actions, you first and foremost end up trying to *define* guns, and the Exploding Cat Problem takes center stage.

What exactly *is* an "assault style" gun, anyway? Who gets to define that?

All this labeling game will ever end up doing is creating a cat-and-mouse authority-by-definition game, which will result in the collapse of all meanings of the words. If you try to regulate something by "assault style", enthusiasts and manufacturers will just figure out how to avoid that bucket-label.

Rather than discuss how to forcefully limit individuals from choosing the gun as a method of shelter from other people, I'd like to offer a very simple framework to discuss guns without creating buckets and labels: Just use words for distance, time, movement, and boundaries instead.

Based on this framework, the definition of a "gun" is literally a clogged (or clipped) straw. This clogged-straw object has an internal process that propels the "clog" out and away from that straw. You might even simplify it to be any system with an internal process to accelerate another object out and away. That process is triggered to happen by the actions of a human being.

Using this definition, *lots* of things are suddenly guns. This makes sense in my opinion, because we call lots of things guns that aren't the subject of topics like mass shootings and gun control – guns like squirt guns, laser guns, nail guns, and staple guns. The Exploding Cat Problem applies. These are all now included in this one huge category "bucket" of "guns".

So what sub-category of "gun" is the focus of debate? In my opinion, the guns we are discussing related to mass shootings and sheltering from others are the ones that can cause "serious" damage.

This means the speed and size and shape of the object being propelled away matters quite a lot to this story, as well as the number of objects that can be propelled across an approximate measure of time – like one second. After all, entire wars have been won based on being on the side with a gun that performs these things better than the other.

For instance, a rifle that fires only one small (caliber) object at low speed before needing to be reloaded with another object is a vastly different type of "gun" than a machine gun that fires several large (caliber) objects at high speed every second with no need to reload objects until a dozen or more have been propelled away.

So how can we define "serious" in my proposed requirement of "serious damage" in a way to control the Exploding Cat? In my opinion, use the human constant: It's what effect these propelled objects have *relative to the boundary of a human being*.

To me, it just doesn't make sense to try to divide up and classify guns based on anything else. A gun that can't fire a projectile through a thin cardboard box is also not a threat to other human beings in a serious way. A gun that can propel a projectile through a brick wall can easily be a "serious" threat to other human beings.

However, my opinion on "serious" is not isolated to the power of the gun – though that certainly plays a part. As I see it, the one thing that can cool down society on guns is to have some kind of sensory firsthand indication or assurance that gun-wielders know how to handle such a device to avoid these objects of serious damage taking an unintended path.

For example, I once saw a man open-carrying a pistol in a fast food restaurant. It was strapped into his hip holster. A toddler with the family at the table directly behind him was doing what toddlers do: Toddling around the table to see all the other family members. This child came within a finger's length of that gun multiple times, and the gun-owner was none the wiser to it.

I'm not calling this man an incompetent gun-handler – but I am saying that guns that are designed to launch an object at a human-piercing speed *are* dangerous objects to us *by their design* for one specific reason: The object launched from that gun has the potential to move in *any* direction based on the orientation of the "straw", and all that is required to start the process in any direction is the slight *twitch* of a human finger.

To frame this another way: From a perspective of a person observing another person with a gun, the bullet or projectile's potential path anchored to the "straw" of each gun creates a sphere-like shape in *all* possible directions from that gun through time. If you are holding a gun for five seconds, imagine how many directions the barrel has been pointed. The potential path changes *constantly* and *dramatically* as the gun is moved, ultimately creating a sphere of possibilities. For an expert handler, it might take a whole day to point a gun barrel in every direction. For a child, that sphere of pointing a gun in every direction is created *in seconds.*

The size of this sphere is determined by how far the propelled object can travel from that gun and still have a speed fast enough to pierce the boundary of a human being's skin.

These "serious damage" spheres might be *huge*. Bullets can travel a long distance, and to assume anyone has perfect control of a bullet's path is *nonsense*. No one has perfect control of anything, and bullets can ricochet, bounce, and even change direction *within* the sphere and still maintain human-piercing speed.

Here, experience it yourself: Imagine handing a small child with no gun training whatsoever a loaded gun with no safety installed. This gun is capable of launching a bullet with skin-piercing force up to half a mile (800 meters) in any direction.

Would you consider this a reasonable thing to do?

Based on the words as written in the Second Amendment of the United States Constitution – *it can be justified into reality*. If this situation represents the lowest limit of gun control, then what is the upper limit?

How about someone that literally trains with a specific gun every hour of every day not spent eating or drinking or sleeping, and that specific gun has six different safety devices installed, which can only be unlocked by this one person, and this gun is only skin-piercing within a 100 feet (30.5 meters)?

If you have any experience with guns, you know both of these limits are unreasonable, right? One seems horrifically reckless, and the other seems to defeat the whole point of owning a gun in the first place (except under very controlled conditions like a gun range).

I accept the gun for what it is: *A form of shelter*. For farmers, the gun is a huge invisible sphere-shaped shelter for crops and livestock, which becomes a sphere-shaped shelter from starvation. For hunters that eat the animals they shoot, the gun is a huge sphere-shaped shelter from starvation. For the scared and insecure, the gun is a huge sphere-shaped shelter from fear. It is in this last gun-shelter that a paradox appears.

When the first gun was created and wielded as shelter from another person, if the other person created and wielded an *equally* performing gun, the first gun is *no longer providing shelter* from the other person. Fear found a way through that sphere-shaped boundary. To capture fear again, the first gun-wielder now needs to have a *better* performing gun than the other person.

And so begins the arms race of weaponry to grow our invisible sphere-shaped shelter boundaries to capture a feeling of fear caused by the mere existence of someone else growing their own sphere to be bigger, until we reach a point of creating weapons so great their very existence creates a fear that we can no longer find shelter from. It is a "gun" that fires invisible "bullets" in *all* directions at the same time and each is skin piercing for a huge distance – thermonuclear and hydrogen bombs.

As I see this, fear of any weapon – guns included - is directly related to your perceived ability to avoid it, multiplied by your lack of trust in the individual wielding that weapon. For example, if a man you do not know has a 10-foot long stick, you know if you are 20 feet away the only way he can harm you is to throw it at you or move closer by running directly at you. You are outside of the man-with-the-stick's "sphere". Your imagined possibilities are manageable, though you might still experience some fear because you are still only a moment or two from the man being able to reach you with the stick.

Keeping the same framework, a knife is somewhat predictable when held, and the sphere size is limited to how far one's ability is to throw it. Even then, if thrown, one end might pierce you, and the other might leave only a bruise.

But with a gun that fires bullets (or other intended-to-pierce projectile-speed), those bullets move so fast you can't sense them coming. One headed straight at you will reach you faster than the sound it makes moving. There is no way you can make predictions and form expectations when it comes to a piercing-speed bullet. They are basically *invisible* to us, and if one is aimed at you there is no intentionally dodging it.

With this invisible and undodgable and unknowable *secondhand story* of the bullet, the gun itself becomes the firsthand story *source* of death and harm from invisible things. The mere presence of the gun can create anxiety and fear, and the only way to comfort that fear is to *have trust* in the individual it is anchored to, because the physics of the gun are not affected by fear and anxiety.

When it comes to trusting the individual a gun is anchored to, what magical difference exists between a child and 20 year-old? What magical transformation has made such a difference in that span of time that now a person just *automatically* knows how to handle a gun safely and limit the sphere of fear, knows the size of the sphere (its skin-piercing range), and *should be trusted* with it?

Sensory evidence of competency is, in my opinion, the key to controlling the runaway fear cycle process by the presence of the gun and its invisible projectiles. In the past, society decided not to let people drive around in 6000-pound hunks of metal at 55 miles per hour without requiring them pass a test *twice* and having a period of apprenticeship (learner's permit) – so why then should people have *unlimited* access to own and carry around a device that can pierce any other human inside an invisible sphere-shaped boundary that might be a mile across with the twitch of a finger?

There are legitimate arguments that this fear is *exactly* what is being used as a check against those in power and authority by the words of the Second Amendment. Guns are often pitched as an anti-tyranny device. For this reason, it is my opinion that the government, which is controlled by party religious beliefs, cannot and should not be the authority that decides who is competent to have a gun and who is not.

Instead, I would argue such a thing might be expert citizen-led – perhaps even *children* judgment gate-kept. It is in all our interests as human beings not to let someone have a gun that is ignorant, incompetent, or reckless to its workings and dangers.

Can you convince a bunch of children about what a gun does, and then convince them you should get to have one *without* using stories of boogeymen and monsters in the shadows? How about experts? Can you convince them without saying "because other people have guns too"? Without validating or having to say, "because I am scared?"

Or perhaps the system could work a lot like driving licenses (which it currently roughly resembles): There is a license to drive your car, then another one for commercial vehicles, then another for boats, and another for planes; etc. Gun licenses and/or arm licenses could be tiered for a skin-piercing sphere boundary size of 1000ft, 2000ft, 3000ft, etc, and then also include subcategories like single-shot, semi-auto, or full auto.

However, I see any system of gun control as only a bridge that will eventually decay and collapse, just like every other rule that tries to prevent any actions or belief that someone can justify to be right through experience. The best gun *control*, in my opinion, is the one that removes the fear creating a "need" for their existence in the first place. If everyone felt the world was perfectly fair, and everyone recognized everyone else as an equally valuable human being – what is the purpose remaining for the gun to exist except for them to collapse back to being a tool - a shelter from hunger? They are a way to *save time*, and capture fast-moving food by piercing its boundary without having to chase it down or trap it.

Unfortunately, if we are not wise about keeping our rate of consumption (which is heavily tied to our birthrates) balanced with the rate those same ends of food, water, and shelter can regenerate – we may actually need guns to shoot each other just to capture other's share of food, water, and shelter, right? If you find yourself agreeing, I'll simply refer you back to the stable way out of the *Dark Forest Moment*, and the benefit it brings. As I see it, we'd be better off working together to find a solution than eliminating perfectly capable human minds and the possibilities they bring.

Ideally, maybe someday we will figure out how to create consumable energy for our bodies in a way that doesn't throw the Earth out of balance, freeing us from needing to consume other life around us, which frees us from the gun completely. Well.....almost: We do not know what other life might be out in the universe.

This is why developing non-piercing weapons make much more sense to me. For me, the goal of bringing forth a gun against another person is to "stop" that person's movement toward domination or violence in a *Dark Forest Moment*. Why not make it a "pause" button instead, and allow more time to find balance and let fear subside through communication? The gun throws things all out of whack, and adds imminent death and permanent harm into the mix, which causes fear to expand out of control and explode into social imbalance faster than can be re-balanced through a lifetime of communication.

The goal, as I see it, is to try to avoid extremes when it comes to our senses and movement in our atomic soup – both full-stop and too-fast-to-react movements.

This includes the speed of stories.

167

Balance In A Moving Unity

Language, in a way, is the currency of thought. Conversation can be seen as a form of trading your ideas and experiences for someone's time, and vice versa. One-way communication through a medium like this book, or a piece of art, or is made from or with objects, is the "sensory-experience-story-dollar" that represents an I.O.U. for the experiences responsible for them.

Imagine if you were back at the end of the *You, Me, and The Room* thought experiment, and you are unable to communicate with any other human being, and no other human being can communicate with you. In such a place, language, books, art, and music have no value – and yet, there you are, trapped in your mental boundary prison-cell with your pile of sensory-story-gold and nothing to "spend" it on.

While you imagined existing in *The World of You*, how much would you have given just to have a conversation with another person? To sit in the comfort of your home or shelter with him or her, share a glass of something tasty, and let the last shimmers of fading sunlight dance in through the windows?

Language is an amazingly powerful tool, and even though it is responsible for a religious explosion, black holes of certainty, and many other horrors, the more languages you can speak or the greater your vocabulary, the richer your thoughts, your experiences, and your stories can be.

If you are filled with gaps and holes in your experiences relative to many others, but your vocabulary is *huge*, you can very much compensate using your word-wealth, but you will quickly become a kind of dragon telling stories about yourself using word-shells you cannot fill completely – a faking-of-experience-and-understanding. You might persist at this until you are unrightfully *The Keeper of Truth* over others, which will only happen because they lacked the wealth of words that you have. You become a firsthand experience *thief* using secondhand story means.

Words, as a tool to communicate human perceptions, memories, and projections secondhand are truly something wonderful. As I see it, we can use this wonder to strengthen and bring balance to the always-moving structure of society and knowledge.

Let's bring back the spiky donut of human knowledge and experience in your mind.

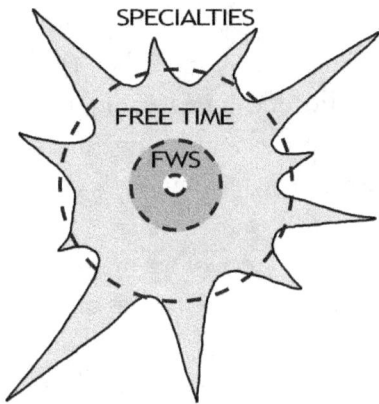

Now, *why* do these spikes exist at all? Why isn't the whole structure just one big ever-growing circle? Well, in some ways it is, but for some reason our culture at the time I wrote this struggles to value those that fill in the spaces between specialties to strengthen the structure. I'd argue this is mainly because the mechanism to connect all the spikes into a balanced circle has been captured by monopoles trying to transform into cancer: Mass communication trolls.

Mass communicators are, of course, people. Artificial Intelligence can do the job, but it does not see, hear, smell, taste, or touch. It does not have a *human* experience with reality. It is unable to judge and flex its values and truths to match the situation it finds itself in. It might be able to help, but communicating across specialties is to specialize in one specific thing: Metaphors.

People that become fascinated with multiple spikes of specialty - like myself - either experience what the metaphors explain within these spikes firsthand, or read and learn about them secondhand, and then start merging them all together into bigger picture ideas in free time. Some individuals might be lucky enough to get paid money to do this. These individuals are the "editors of reality" from many chapters ago. All storytellers are editing human reality for those that consume their stories.

The catch, of course, is *trust*. If you consume the stories of a mass communicator, having to pay money for that story *after* the consumption seems absurd. But if you have to pay *up front* to consume a communicator's story, then you run the risk of that communicator not telling a story worth your free time, or paying money to have your possibilities collapsed with certainties and *shoulds* and *oughts*.

Attempting to build a metaphor-bridge-story between different spikes and keep them both balanced and valid has tremendous value in my opinion. Each spike ultimately has its own language - its own religion - and each language starts to separate more and more from others the longer the spike gets. Eventually, one specialty might use the exact same words and stories as another, with each meaning something *entirely* different with no overlap at all.

This is where the hierarchy of communicators and "translators" come into play.

They work the specialized metaphors to match them with the audience they imagine to be telling stories to the best of his or her ability. I have tried my very best to keep my language in this book at a level that I think nearly everyone can understand, because I consider my audience to be everyone capable of reading it (except perhaps young children). I have tried my best to not destroy possibilities, but merely show relationships between them.

As I see it, honest and limited-experience-acknowledging communication (boundary words used and bucket-words dumped or defined as best as possible) is the key to balancing everything. These stories are like an arc of lightning connecting the spikes being touched, and then the whole human system churns with communication and stories spreading out from that bolt, which find flaws and holes and imbalances until no more can be found based on the whole of human experiences.

These stories are not *only* the news and information of firsthand experiences around the world – they are also embedded in the arts. Artists can provide powerful metaphor-stories that help us understand ourselves far beyond words – they can reach into wordless experiences and feelings themselves. Artists can help reset or shape human expectations, and even warn us of incoming self-created disasters and ideals through their work, which is often object-based metaphors (instead of word based) for something larger happening within society.

Well-done communication can, in my opinion, grow paths for trust across the spikes which can provide a means of achieving a greater sense of unity, understanding, and love for humans having radically different experiences with the same reality. This understanding and love is the substance that fills in the gaps between spikes to create an enormous sphere of near-unity that churns and moves with story currents anchored to experiences, movements, and possibilities.

But before there can be communication, there must be those that teach us how to communicate at all. The smallest of these culture religions are you and the adults in your life that gave you your words. As you got older, it's possible you merged into larger culture-religions such as a regional educational system. Educators are critical to holding together cultures of shared understanding in this consistent atomic soup through language and experiences. Teachers are the specialized roles that explain to us how "grups" seem to work without us having to figure it out ourselves through trial and error. They greatly accelerate our speed into the spikes (or inner sphere role) we will eventually spend our lives within.

While each teacher has a unique approach to education – some strict and some loose, some detail-oriented and others big-picture - together, they add up to ultimately be a strength that might be recognized by the student years later: Each teacher provided layering of information, interaction, and communication style into the student's story system to eventually merge into a bigger story that hopefully makes more and more sense. The more teachers - the greater the variety - the better.

Unfortunately, in my country, education is increasingly tied to money, since tax dollars often fund it. This means it often works in a value cycle that accelerates its destruction or growth, because places of "poor" education do not draw in more people to fund improving things – which results in an accelerated decay of quality in education. Meanwhile, schools that are believed to provide the best education act like magnets that draw in tons of new people to tax, which accelerates the funding to the point of near absurdity in my opinion – some schools build multimillion dollar stadiums, while others cannot afford libraries or even new books.

The only way I see to un-anchor education from money is to make it all "free". Otherwise accelerated learning will always be for individuals with money, with perhaps a handful of "lottery" winners. If we were to make education free for our upcoming future (children), the government, which should ideally represent the people, becomes the source of funding for schools and teachers. Thanks to the *Floating Anchor Problem*, this system unfortunately leads to inevitable culture/context wars around what *should* be taught and *why* it should be taught, exactly like what is happening in public schools as I write this.

As far as curriculum in schools, it's my opinion that teaching the uncertainty in everything is *critical*, as well as moving away from the black-and-white, right-and-wrong culture that exist today to focus much more on the *why and how*. The stories of social history are important, but they can often be simplified into metaphorical stories of domination and submission, the imbalances they create for decades, and the chains of events that led into major losses of human life.

Learning the systems that make free time possible is also hugely important in my opinion, and that includes the importance of farmers and the trades that move and reshape the materials and resources of the Earth and universe.

As I write this, I currently exist in culture-religion that seems to believe there is an *objective* righteousness, which leads to almost everyone wanting to go to "the best/most righteous" schools. This belief also leads to an entire religion that seems to believe only those that went to these schools have a valid opinion everyone else should listen to.

This creates a culture of personal validity only through credentials, and I find this incredibly frustrating. It seems to lead many to look down on "lesser" community colleges and state schools, and judge them as inferior without ever experiencing them. My experiences show an opposite possibility: Some of the best teachers I have ever had the privilege to learn from were in community colleges. How does one change the perception of these cheaper programs being inferior?

I personally believe it is likely the best teachers on the planet exist in some of the most underserved and underfunded places. *Schools* do not make wise and knowledgeable adults – *teachers* do. Teachers, in my opinion (and assuming they perform their job to help grow understanding and not destroy possibilities and capture an end for his or her own gain) should be some of the highest-paid individuals in society, because they are not far behind farmers, water bringers, and shelter builders in the amount of free time they help create for others.

All of this bridges into another topic: There is an enormous difference between knowledge and understanding. You can gain knowledge from textbook and a teacher's lecture, but you gain understanding by application, experimentation, and experience.

For example, kids that are identified as "gifted" are often incredibly good at absorbing information and reciting it – but, in my experience, they often struggle to understand it. Yes, he or she can name nearly every dinosaur and animal of the Earth in every ancient time period between mass extinctions, but can he or she withstand a teacher or parent suddenly becoming the child and asking a million questions around this knowledge? Does he or she know *why* the extinction occurred? Does he or she know *why* we even believe any of these events occurred at all? *Why* the animal is named the way it is named? What actually separates a cat from a kangaroo?

In this way, bright kids that are good at memorizing and test-taking are often accelerated fast through material – far faster than it can ever be applied and understood outside of the school setting – and it can eventually result in a person that sees the entire universe as black-and-white, categorized, unitized, and absent of almost all mystery and unknowns. In my opinion, *nothing could be further from reality*.

This brings me to the last important social role I want to bring up when balancing a moving structure of society, and one you've not heard of before: *The loop-breaker*.

Loop-breakers are those that are knowledgeable *and* experienced. In my own experience, loop breakers are individuals that have risen slowly throughout their life to experience an entire spike, or perhaps multiple spikes that are related to each other. He or she can see how the whole system churns and works together, how the bottom and the top half of the mind-body loop/wave complement each other, as well as knowing how and why the spikes exist and what the overarching goal might be. To me, this makes loop-breakers the true arbiters of fairness and understanding.

An example might be a person that grew up in her dad's machine shop as a kid, learned how metals and parts were formed and made, and watched the people and machines work, talked with them, and even began to work metals herself as a teenager and beyond. She then goes to school and becomes a mechanical engineer, and gets hired to design and create new designs from metals. This person now can see the whole design-from-raw-metal-to-final-product system from beginning to end, and understands through wordless experience what the machinists must do to create something, the limits involved, and how it could be done.

Others can tell a loop-breaker like this that a mathematical equation says a design should be possible or a word has a fixed and defined meaning – but she will know from experience the limits of an equation when transferred into practice (machining tolerance errors etc), and knows the limits of manufacturing – of the atomic soup and reality itself - wordlessly. His or her story resolution is far finer than most other people.

In other words, this loop-breaker knows the limits and ability of humans relative to reality when it comes to creating metal products. This does not necessarily translate to chemical products, or marketing products – but the concept is the same. Loop breakers understand the entire waveform between mind and body, and between designs, ideals, and imagination and reality, action and movement.

Loop-breakers are often the wisest individuals among us as humans (in my opinion), and they typically rise to the top – not because they pursue power, authority, or influence, but because they understand the role and limits of all other human roles in any process, making them the most trusted individuals by those that work with, rely on, or interact with them.

In a way, loop-breakers are the anti-trolls, because he or she cares about the whole process because its place in reality is understood, and not just the means or ends. Loop-breakers can often communicate anything you want to know about the process using words you can understand, because they know the process wordlessly, the metaphors of language and communication to be utilized to their full potential. Loop-breakers can move the stars, bridge the islands, and rearrange them however is needed.

Most of all though, these individuals can often and easily break circular thinking and processes that threaten the stability of society's structures – because they can easily and intuitively see the limits and gaps being missed by others through their high-story resolutions. They often naturally know the questions to ask.

I like to imagine that perhaps I am one of these loop-breakers, but in the end it is not up to me to decide. This can only be true if *you* have found my metaphors and stories to be useful, because they have explained the reality you experience in a way that is helpful and makes more sense, and maybe breaks the loop traps that lead to what I see as suffering in the abyss of believed objective righteousness.

By the way, *children* can sometimes be the very best loop breakers. If you are ever unsure, pose a problem to a child you cannot seem to find the "right "answer for, or has been trapping you. Change the words of it to be kid-friendly, use boundary words, and keep the buckets dumped. As long as the child knows his or her answer will not be judged and will be considered valid – you might be surprised at the simplicity and ease they can loop-break. After all they are still largely "unit free".

With that said, I'll now start to wrap up this journey of ours, which began in the middle and will now end…in the middle.

0

I've had a soft spot for the symbol for zero ever since I heard the story the symbol can be considered an infinite loop of "something" around "nothing". How beautiful and balanced is this story? It captures both *The Ark* and *Gravity* at the same time. It captures the idea of the *boundary* and the *bucket* and collective. It has no beginning and no ending. It captures the idea of a *balloon*, and even the idea there is *no ground* and a void outside of it in every direction. To top it off, it means nothing at all, without something relative to interpret what it means. Is it a circle, oval, number, letter, or something else?

No matter what you think about it, you aren't wrong, you were never wrong, and wrong is something that is only ever defined by you, for you, all in your own past. In this place we call reality, there seems to be only differences in resistance that form very-clear-but-not-completely-solid boundaries at the scales we have to perceive them. Why do we exist at this scale? What is this place? What is our purpose in the universe? I have no idea, but it's something I think about almost daily.

There are no actual rules here. I am not going to end with a lecture about what you should and shouldn't do with your life. You don't have to do anything at all. You don't have to eat. You don't have to sleep. You don't even have to live. But I'm hoping you choose to. I'm hoping, if anything, this bridge-of-a-book connects to you in a way that breaks through whatever resistance might be keeping you from being the change in this world you most want to see, and I hope that change is one of hope, equality, and movement toward balance.

What boundaries can be justifiably crossed, and which cannot? Who is the keeper of truth, and who is not? What *is* true, and what is not?

So many actions (which includes words) seem justifiable as right that seem completely wrong to someone else. Are they? It is not for me to decide. That is between those two people, communication, and the *Dark Forest Moment* of violence and severe imbalance, or better understanding of one another through language, faith, trust, and love.

I believe with enough close relationships of earned trust and love, the greatest shelter of all from fear and violence will construct itself – strong community and communication with those you experience firsthand in your life. With enough trust and a large enough community – ideally the entire human community - there seems to be little left for us to fear in our sphere of atomic soup, and it becomes possible to enjoy the moment of being here, of experiencing this place, of existing at the same time – together. It becomes possible to work toward meaningful balance with Mother Earth herself, instead of only each other.

As community is grown, shared belief between any two people creates what amounts to a religious denomination. As I see it, each denomination ultimately belongs to the largest religions in human reality: Languages.

In the religions of language, each and every personal story of the currently 8 billion adds "verses and scriptures" to his or her respective religion's doctrine. When all the world's doctrines are merged together, along with the billions that came before, they all seem to ultimately validate the existence of a "higher power" of consistency, which might all exist because of gravity and resistance. How these things came to be, I do not know.

All of these active human religions rest on the existence of human consciousness - memory and imagination – forged from the same consistency, and which births experience and belief, languages, and the ability to see self in others – to even see self in *all* living things (if one reaches deep enough).

As I see it, smaller human religions built on fear, or love, or anything in between, are all ultimately resting on this strong human sense of empathy, which is the thing that gives validation and invalidation its seemingly extreme power. It also makes boundaries critically important, and propels us to want to share our experiences with each other at all.

You are the Being in someone else's story. Someone else is the Being in your story.

The Golden Rule, which is simply the recognition of equality through empathy, drives us each to want to balance and correct unfairness in the world. It drives us to want to root for the underdog, and to celebrate success and the overcoming of struggle, but then to lift others up once success arrives. It demands the respect of boundaries and consent to cross them.

We are all different, and we all need slightly different amounts and frequencies of the same resources. No amount of standardization can make the world a magically fair place for everyone. Fairness is a feeling, and for that feeling to exist – for world peace to ever be possible – we have to share our honest feelings with each other and share them often and then also work to find balance within those stories. *This* is the true and deepest power of language in my opinion, even more so than the birth of science made possible by writing things down.

I sincerely hope that you have enjoyed the journey within these pages, and I hope it has helped you perhaps find peace with yourself, and perhaps bring to light some of the problems in our world we share together, so that maybe you can bring your light into the world to help solve them – should you choose to.

There is no actual ending to this journey of course – it's a process, just like everything else - and it does not stop moving and changing after you close this book - and that means you never stop moving or changing. You, after all, are a process too.

The words in this book will some day expire, and perhaps become irrelevant and unable to help anyone at all – and that's okay. I wrote them because I believed they were the right thing to do at the time I wrote them, and I cannot foresee the future to guarantee their timeless relevance, and I do not have the expectation or even the hope that they live forever.

I don't know the whole story about anything. I am not the *Keeper of Truth*. I am only the keeper of my perspective, of my own truth, and controller of my resistance to adding to my truth, my own beliefs, and my actions. I shared them with you in the hopes that maybe it will help you, which is and has been my goal since I wrote the very first word: *If this book helps one person, then it was worth it.*

I can only dream that what I have written here might help you transform the hydra back into your friend and guardian, and grow to slow or maybe stop the endless churning of societal collapse and disaster, moving us one step closer toward peace and allowing everyone to experience *The Ark* and value free time in a meaningful and fulfilling way.

After all, if all of us work together in trust, we can open all the doors at once, and turn what would be a thousand years of searching into one second of knowing.

Any of your limited time spent with another human being by consent, in my opinion, is the greatest honor they can bestow you – and at this point, you have given me quite a bit of yours. I thank you for giving me so much of your time, and hope you feel it wasn't wasted.

I wish you the best in this life as you perhaps step into the void of change and new experiences in a moving system – as you captain the ship of You and your million-billion processes through the seas of atomic soup, and hope you choose to press on through storms and rough waters, monsters, sirens, hydras, and maelstroms to discover and grow an all-new story with calm clear waters that spread from you into something wonderful.

Perhaps I'll see you out there. I have a funny feeling I will.

After all, I am everyone you will ever meet, because I have always been nothing more than you telling yourself a story.

Appendix of Star Doors and Rabbit Holes

If you've enjoyed this journey and want to expand beyond it, I've created a list (on the following pages in no particular order) of metaphor-stories that might make more sense after what you have just read over the past 167-ish chapters. These might also make good conversation topics with other readers, and help the process of beginning to grow communities of shared understanding that break each other's loops that trap and limit possibilities.

Ontology

Epistemology

Quantum Bayesianism (Qbism)

Copenhagen Interpretation

Endowment Effect

Planck length

Planck time

Personality Disorder Cluster A

Personality Disorder Cluster B

Personality Disorder Cluster C

Dissociative disorder

Narcissistic personality disorder

Imposter Syndrome

Identity of indiscernibles

Law of non-contradiction

Principle of sufficient reason

Polyembryonic seeds

Modality and modality principle

Actualists

Realists

Modal realism

Flat vs polycategorical vs hierarchical

Constituent ontologies

Prophase

Metaphase

Anaphase

Telophase

Cytokinesis

Mitosis

Fault current

Arc blasts

Tabula rasa

Neuroplasticity

Eutrophication

Heuristics

Group Think

Dogma

Echo chambers

Positive Disintegration

A Priori knowledge

Adenosine triphosphate (ATP)

Cilia

Steriocilia

Cochlea

Mechanoelectrical transduction

Cation

Anion

Ion channels

Plasma membrane

Axon

Dendrite

Neurotransmission

Synaptic cleft

Voltage-gated calcium channels

Potassium channels

Electrical synapse

Chemical synapse

Excitatory and inhibitory synapses

Cell polarity

Action potentials

Phosphoinositides

Long-term potentiation

Acetylcholine

Myelin sheath

Grid cell

Problem of Induction

Self assembling systems

Critical point systems

Attractors

Occams Razer

Mast fruiting

Debtor's prisons

Ionic bonds

Covalent bonds

Electron tunneling

Allegory of the Cave

Delta P

The measurement problem

Law of excluded middle

Law of non contradiction

The carbon cycle

Thermal currents

Gettier problem

Problem of induction

Molyneux problem

Munchhausen trilemma

Sorites paradox

Theseus paradox

Demarcation problem

Cosmic censorship hypothesis

Holographic principle

Problem of time

Copernican principle

Cosmological constant problem

Proton decay

Atomic decay

Solar cycle

Threshold problem

Arrow of time

Locality

Bell inequalities

Time crystals

Hilbert's sixth problem

Veil of ignorance

Lichtenberg figures

Goodhart's Law

Benevolence

Malevolence

Determinism

Dirac Sea

Law of Excluded Middle

Organized religion and taxes

Organized religion and property

Banks and property

Stock Markets

Prosperity Gospel

Manifest Destiny

Binary and Gender

Fourth Estate

Eutrophication

Romantic love

Dead Internet Theory